STRUCTURAL STEEL DESIGN

LRFD Method

SECOND EDITION

STRUCTURAL STEEL DESIGN

L R F D M e t h o d

S E C O N D E D I T I O N

Jack C. McCormac
Clemson University

Prentice Hall   PRENTICE HALL, Upper Saddle River, NJ 07458

Sponsoring Editor: John Lenchek
Project Editor: Cathy Wacaser
Design Administrator: Jess Schaal
Cover Design: Lesiak/Crampton Design Inc: Cynthia Crampton
Cover Photo: *Background image:* Letraset Phototone™
                Inset image: The Chicago Photographic Company: Russell Phillips
Production Administrator: Randee Wire
Compositor: Interactive Composition Corporation
Text Printer and Binder: R R. Donnelley & Sons Company
Cover Printer: R.R. Donnelley & Sons Company

Structural Steel Design: LRFD Method, Second Edition

© 1995 **Prentice-Hall, Inc.**
Upper Saddle River, NJ 07458

Library of Congress Cataloging-in-Publication Data

McCormac, Jack C.
    Structural steel design: LRFD method / Jack C. McCormac.—2nd
ed.
      p.   cm.
    Includes index.
    ISBN 0-06-501627-0
    1. Building, Iron and steel.   2. Load factor design.   3. Steel,
Structural.   I. Title.
TA684.M25   1994
624.1′821—dc20   94-29972
                                                     CIP

6 7 8 9 10-DOC- 01 00

Contents

Preface

The author's major objective in preparing this new edition was to update the text to conform to the 1993 LRFD Specification of the American Institute of Steel Construction (AISC). The design profession is gradually moving to the LRFD method from the allowable stress design (ASD) procedure. It would seem that this trend will continue and accelerate in the near future because LRFD provides a more realistic design procedure and its use commonly results in more economical structures.

Many changes were made in the new LRFD Specification from the first, or 1986 Specification. Of particular note are various changes in design stress values: information for the use of heavy rolled shapes, revised requirements for built-up sections, updated provisions for slender web girders, and so on.

Perhaps the most obvious changes made by the AISC are those pertaining to the LRFD Manual, which is now published in two volumes. Volume II is wholly concerned with connections. Many improvements have been made in these manuals. Perhaps the most pleasing one to the author is the improvement in the procedure for designing column base plates.

A large number of the homework problems in this book have been revised and their number substantially increased. A computer disk is now enclosed. It can be used to solve a good many of the rather common and mathematically tedious problems faced in structural steel design.

The author wishes to thank the following persons who reviewed this edition:

Lewis B. Burgett, *American Institute of Steel Construction*
Hany J. Farran, *California State Polytechnic University-Pomona*
Hanz Peter Huttelmaier, *Colorado School of Mines*

B. K. Rao, *Idaho State University*
Wallace Sanders, Jr., *Iowa State University*
Subhash C. Goel, *University of Michigan*
Terry J. Wipf, *Iowa State University*

The author extends a special thank you to James K. Nelson for the software program included with this text.

Finally, the author thanks the reviewers and users of the first edition for their suggestions and criticisms. He is always very grateful to anyone who takes the time to contact him concerning any part of the book.

Jack C. McCormac

Chapter 1

Introduction to Structural Steel Design

1-1 ADVANTAGES OF STEEL AS A STRUCTURAL MATERIAL

A person traveling in the United States might quite understandably decide that steel was the perfect structural material. He or she would see an endless number of steel bridges, buildings, towers, and other structures—comprising, in fact, a list too lengthy to enumerate. After seeing these numerous steel structures this traveler might be quite surprised to learn that steel was not economically made in the United States until late in the nineteenth century and the first wide-flange beams were not rolled until 1908.

The assumption of the perfection of this metal, perhaps the most versatile of structural materials, would appear to be even more reasonable when its great strength, light weight, ease of fabrication, and many other desirable properties are considered. These and other advantages of structural steel are discussed in detail in the following paragraphs.

High Strength

The high strength of steel per unit of weight means that structure weights will be small. This fact is of great importance for long-span bridges, tall buildings, and structures having poor foundation conditions.

Uniformity

The properties of steel do not change appreciably with time, as do those of a reinforced-concrete structure.

1

Erection of steel joists. (Courtesy of Vulcraft.)

Elasticity

Steel behaves closer to design assumptions than most materials because it follows Hooke's law up to fairly high stresses. The moments of inertia of a steel structure can be definitely calculated, while the values obtained for a reinforced-concrete structure are rather indefinite.

Permanence

Steel frames that are properly maintained will last indefinitely. Research on some of the newer steels indicates that under certain conditions no painting maintenance whatsoever will be required.

Ductility

The property of a material by which it can withstand extensive deformation without failure under high tensile stresses is said to be its *ductility*. When a *mild* or *low-carbon* structural steel member is being tested in tension, a considerable reduction in cross section and a large amount of elongation will occur at the point of failure before the actual fracture occurs. A material that does not have this property is generally unacceptable and is probably hard and brittle and might break if subjected to a sudden shock.

In structural members under normal loads, high stress concentrations develop at various points. The ductile nature of the usual structural steels enables them to yield locally at those points, thus preventing premature failures. A further advantage of ductile structures is that when overloaded their large deflections give visible evidence of impending failure (sometimes jokingly referred to as "running time").

Fracture Toughness

Structural steels are tough; that is, they have both strength and ductility. A steel member loaded until it has large deformations will still be able to withstand large forces. This is a very important characteristic because it means that steel members can be subjected to large deformations during fabrication and erection without fracture—thus allowing them to be bent, hammered, sheared, and have holes punched in them without visible damage. The ability of a material to absorb energy in large amounts is called *toughness*.

Additions to Existing Structures

Steel structures are quite well suited to having additions made to them. New bays or even entire new wings can be added to existing steel frame buildings, and steel bridges may often be widened.

Miscellaneous

Several other important advantages of structural steel are: (a) ability to be fastened together by several simple connection devices including welds and bolts, (b) adaptation to prefabrication, (c) speed of erection, (d) ability to be rolled into a wide variety of sizes and shapes as described in Section 1-4, (e) fatigue strength, (f) possible reuse after a structure is disassembled, and (g) scrap value, even though not reusable in its existing form. Steel is the ultimate recyclable material.

1-2 DISADVANTAGES OF STEEL AS A STRUCTURAL MATERIAL

In general, steel has the following disadvantages.

Maintenance Costs

Most steels are susceptible to corrosion when freely exposed to air and water and must therefore be periodically painted. The use of weathering steels, in suitable design applications, tends to eliminate this cost.

Fireproofing Costs

Although structural members are incombustible, their strength is tremendously reduced at temperatures commonly reached in fires when the other materials in a building burn. Many disastrous fires have occurred in empty buildings where the only fuel for the fires was the buildings themselves. Furthermore, steel is an excellent heat conductor, nonfireproofed steel members may transmit enough heat from a burning section or compartment of a building to ignite materials with which they are in contact in adjoining sections of the building. As a result of these facts the steel frame of a building may have to be protected by materials with certain insulating characteristics, or the building may have to include a sprinkler system if it is to meet the building code requirements of the locality in question.

Susceptibility to Buckling

The longer and more slender the compression members, the greater the danger of buckling. As previously indicated, steel has a high strength per unit of weight, but when used for steel columns is not very economical sometimes because considerable material has to be used merely to stiffen the columns against buckling.

Fatigue

Another undesirable property of steel is that its strength may be reduced if it is subjected to a large number of stress reversals or even to a large number of variations of tensile stress. (We have fatigue problems only when tension is involved.) The present practice is to reduce the estimated strengths of such members if it is anticipated that they will have more than a prescribed number of cycles of stress variation.

Brittle Fracture

Under certain conditions steel may lose its ductility, and brittle fracture may occur at places of stress concentration. Fatigue-type loadings and very low temperatures aggravate the situation.

1-3 EARLY USES OF IRON AND STEEL

Although the first metal used by human beings was probably some type of copper alloy such as bronze (made with copper and tin and perhaps some other additives), the most important metal developments throughout history have occurred in the manufacture and use of iron and its famous alloy named steel.

Today, iron and steel comprise almost 95 percent of all the tonnage of metal produced in the world.[1]

Despite diligent efforts for many decades, archaeologists have been unable to discover when iron was first used. They did find an iron dagger and an iron bracelet in the Great Pyramid in Egypt, which they claim had been there undisturbed for at least 5000 years. The use of iron has had a great influence on the course of civilization since the earliest times, and may very well continue to do so in the centuries ahead. Since the beginning of the Iron Age in about 1000 B.C., the progress of civilization in peace and war has been heavily dependent on what people have been able to make with iron. On many occasions its use has decidedly affected the outcome of military engagements. For instance, in 490 B.C. in Greece at the Battle of Marathon, the greatly outnumbered Athenians killed 6400 Persians and lost only 192 of their own men. Each of the victors wore 57 pounds of iron armor in the battle. (This was the battle from which the runner Pheidippides ran the approximately 25 miles to Athens and died while shouting news of the victory.) This victory supposedly saved Greek civilization for many years.

According to the classic theory concerning the first production of iron in the world, there was once a great forest fire on Mount Ida in Ancient Troy (now Turkey) near the Aegean Sea. The land surface supposedly had a rich content of iron and the heat of the fire is said to have produced a rather crude form of iron which could be hammered into various shapes. Many historians believe, however, that human beings first learned to use iron that fell to the earth in the form of meteorites. Frequently the iron in meteorites is combined with nickel to produce a harder metal. Perhaps early human beings were able to hammer and chip this material into crude tools and weapons.

Steel is defined as a combination of iron with a small amount of carbon, usually less than 1 percent. It also contains small percentages of some other elements. Although some steel has been made for at least 2000 or 3000 years, there was really no economical production method available until the middle of the nineteenth century.

The first steel was surely obtained when the other elements necessary for producing it were accidentally present when iron was heated. As the years went by, steel probably was made by heating iron in contact with charcoal. The surface of the iron absorbed some carbon from the charcoal which was then hammered into the hot iron. Repeating this process several times resulted in a case-hardened exterior of steel. In this way the famous swords of Toledo and Damascus were produced.

The first large volume process for producing steel was named after Sir Henry Bessemer of England. He received an English patent of his process in

---

[1] American Iron and Steel Institute, *The Making of Steel* (Washington, D.C., not dated), p. 6.

1855, but his efforts to obtain a U.S. patent for the process in 1856 were unsuccessful as it was proved that William Kelly of Eddyville, Kentucky, had made steel by the same process seven years before Bessemer applied for his English patent. Although Kelly was given the patent, the name Bessemer was used for the process.[2]

Kelly and Bessemer learned that a blast of air through molten iron burned out most of the impurities in the metal. Unfortunately, at the same time the blow eliminated some desirable elements such as carbon and manganese. It was later learned that these needed elements could be restored by adding spiegeleisen, which is an alloy of iron, carbon, and manganese. It was further learned that the addition of limestone in the converter resulted in the removal of the phosphorus and most of the sulfur.

Before the Bessemer process was developed, steel was an expensive alloy used primarily for making knives, forks, spoons, and certain types of cutting tools. The Bessemer process reduced production costs by at least 4/5 ths and allowed for the first time the production of large quantities of steel.

The Bessemer converter was commonly used in the United States until after the turn of the century, but since that time it has been replaced with better methods such as the open-hearth process and the basic oxygen process.

As a result of the Bessemer process, structural carbon steel could be produced in quantity by 1870, and by 1890 steel had become the principal structural metal used in the United States.

Today approximately 80 percent of the structural steel produced in the United States is made by melting scrap steel (primarily old car bodies) in electric furnaces. The molten steel is poured into molds which have approximately the final shapes of the members. The resulting sections are run through a series of rollers to squeeze them to their final shapes. The resulting members have better surfaces and fewer internal or residual stresses than newly made steel.

The term *cast iron* is used for the very low carbon content materials, while the very high carbon content materials are referred to as *wrought iron*. Steels fall in between cast iron and wrought iron and have carbon contents in the range of 0.15 percent to 1.7 percent, as is described in Section 1-7 of this chapter.

The first use of metal for a sizable structure occurred in England in Shropshire (about 140 miles northwest of London) in 1779, when cast iron was used for the construction of the 100-ft Coalbrookdale Arch Bridge over the River Severn. It is said that this bridge (which still stands) was a turning point in engineering history because it changed the course of the Industrial Revolution by introducing iron as a structural material. This iron was supposedly four times as strong as stone and thirty times as strong as wood.[3]

---

[2] American Iron and Steel Institute, *Steel 76* (Washington, D.C., 1976), pp. 5–11.

[3] M. H. Sawyer, "World's First Iron Bridge," *Civil Engineering* (New York: ASCE, December 1979), pp. 46–49.

A number of other cast-iron bridges were constructed in the following decades, but soon after 1840 the more malleable wrought iron began to replace cast iron. The development of the Bessemer process and subsequent advances such as the open-hearth process permitted the manufacture of steel at competitive prices which encouraged the beginning of the almost unbelievable developments of the last 100 years with structural steel.

1-4 STEEL SECTIONS

The first structural shapes made in the United States were angle irons rolled in 1819. Steel I beams were first rolled in the United States in 1884 and the first skeleton frame structure (the Home Insurance Company Building of Chicago) was erected that same year. Credit for inventing the "skyscraper" is usually given to engineer William LeBaron Jenny, who planned this building apparently during a bricklayers' strike. Prior to this time, tall buildings in the United States were constructed with several-foot-thick load-bearing brick walls.

For the exterior walls of this 10-story building Jenny used cast-iron columns encased in brick. The beams for the lower six floors were made from wrought iron, while structural steel beams were used for the upper floors. The first building completely framed with structural steel was the second Rand-McNally building, completed in Chicago in 1890.

An important feature of the 985-ft wrought-iron Eiffel tower constructed in 1889 was the use of mechanically operated passenger elevators. The availability of these machines along with Jenny's skeleton frame idea led to the construction of thousands of high-rise buildings throughout the world in the next 100 years.

During these early years the various mills rolled their own individual shapes and published catalogs which provided the dimensions, weight, and other properties of these shapes. In 1896 the Association of American Steel Manufacturers (now the American Iron and Steel Institute, AISI) made the first efforts to standardize shapes. Today nearly all structural shapes are standardized, though their exact dimensions may vary just a little from mill to mill.[4]

Structural steel can be economically rolled into a wide variety of shapes and sizes without appreciably changing its physical properties. Usually the most desirable members are those which have large moments of inertia in proportion to their areas. The **I, T,** and [shapes, so commonly used, fall into this class.

Steel sections are usually designated by the shapes of their cross sections. As examples, there are angles, tees, zees, and plates. It is necessary, however, to make a definite distinction between American standard beams (called S beams) and wide-flange beams (called W beams) as they are both I-Shaped. The inner surface of the flange of a W section is either parallel to the outer surface or nearly

---

[4] W. McGuire, *Steel Structures* (Englewood Cliffs, N.J.: Prentice-Hall, 1968), pp. 19–21.

Erection of tower for the Holy Name Church, Edensburg, Pa. (Courtesy of the Lincoln Electric Company.)

so, with a maximum slope of 1 to 20 on the inner surface, depending on the manufacturer.

The S beams, which were the first beam sections rolled in America, have a slope on their inside flange surfaces of 1 to 6. It might be noted that the constant,

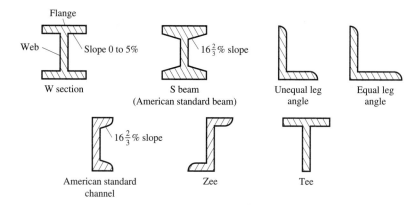

Figure 1-1 Rolled-steel shapes.

or nearly constant, thickness of W beam flanges as compared to the tapered S beam flanges may facilitate connections. Wide-flange beams comprise nearly 50 percent of the tonnage of structural steel shapes rolled today. The W and S sections are shown in Fig. 1-1 together with several other familiar steel sections. The uses of these various shapes will be discussed in detail in the chapters to follow.

Constant reference is made throughout this book to the 2nd edition of the *Manual of Steel Construction Load & Resistance Factor Design* published by the American Institute of Steel Construction (AISC). This manual, which provides detailed information for structural steel shapes, is referred to hereafter as the LRFD Manual or simply the Manual or the Handbook. Reference is made here to the 2nd edition of the handbook, which is based on the December 1, 1993, "Load and Resistance Factor Design Specification for Structural Steel Buildings." The Manual consists of Volume I (*Structural Members, Specifications & Codes*) and Volume II (Connections). The two volumes are divided into 13 parts which are numbered consecutively. In this text the author refers to part numbers rather than referring to Volume I or II.

Structural shapes are abbreviated by a certain system described in the handbook for use in drawings, specifications, and designs. This system is standardized so that all steel mills can use the same identification for purposes of ordering, billing, etc. In addition, so much work is handled today with computers and other automated equipment that it is necessary to have a letter and number system which can be printed out with a standard keyboard (as opposed to the old system where certain symbols were used for angles, channels, etc.). Examples of this abbreviation system are as follows:

1. A W27 × 114 is a W section approximately 27 in. deep weighing 114 1b/ft.
2. An S12 × 35 is an S section 12 in. deep weighing 35 1b/ft.
3. An HP12 × 74 is a bearing pile section which is approximately 12 in. deep weighing 74 1b/ft. Bearing piles are made with the regular W rolls

but with thicker webs to provide better resistance to the impact of pile driving. The width and depth of these sections are approximately equal, and the flanges and webs have equal or almost equal thickness.

4. An M8 × 6.5 is a miscellaneous section 8 in. deep weighing 6.5 1b/ft. It is one of a group of doubly-symmetrical H-shaped members which cannot by dimensions be classified as a W, S, or HP section.

5. A C10 × 30 is a channel 10 in. deep weighing 30 1b/ft.

6. An MC18 × 58 is a miscellaneous channel 18 in. deep weighing 58 1b/ft. which cannot be classified as a C shape by dimensions.

7. An L6 × 6 × $\frac{1}{2}$ is an equal leg angle, each leg being 6 in. long and $\frac{1}{2}$ in. thick.

8. A WT18 × 140 is a tee obtained by splitting a W36 × 280. This type of section is known as a structural tee.

9. Rectangular steel sections are classified as wide *plates* or narrow *bars*. In general, sections wider than 8 in. are said to be plates, while the narrower ones are called bars. Detailed information on these sections is provided in Part 1 of the LRFD Manual. A plate is usually designated by its thickness times its width times its length as PL$\frac{1}{2}$ × 6 × 1 ft 4 in.

The student should refer to the LRFD Manual for information concerning other rolled shapes, such as pipe and structural tubing and other items. Additional sections will be mentioned here as it becomes necessary.

In Part 1 of the LRFD Manual the dimensions and properties of W, S, C, and other shapes are given. The dimensions of the members are given in decimals (for the use of designers) and in fractions to the nearest sixteenth of an inch (for the use of craftsmen and steel detailers or drafters). Also provided for the use of designers are such items as moments of inertia, section moduli, radii of gyration, and other cross-sectional properties discussed later in this text.

There are variations present in any manufacturing process and the steel industry is certainly no exception. As a result the cross-sectional dimensions of steel members may vary somewhat from the values specified in the LRFD Manual. Maximum tolerances for the rolling of steel shapes are specifically set up by the American Society for Testing and Materials (ASTM) A6 Specification and are quoted in Part 1 of the Manual. As a result calculations can be made on the basis of the properties given in the Manual regardless of the manufacturer.

A few of the heavier W sections are not available from domestic producers. Nevertheless, there is no difficulty at the present time in obtaining any of these sections.

Through the years there have been changes in the sizes of steel sections. For instance, there may be insufficient demand to continue rolling a certain shape; an existing shape may be dropped because a similar sized but more efficient shape has been developed, and so forth. Occasionally the designer may need to know the properties of one of the discontinued shapes which are no longer listed in the latest edition of the Handbook or in other tables normally available to him

Structural shapes rolled at Bethlehem Steel Company's Bethlehem, Pa., plant. (Courtesy of Bethlehem Steel Corporation.)

or her. As an example, it may be desired to add another floor to an existing building which was constructed with shapes no longer rolled. In 1953 the AISC published a book entitled *Iron and Steel Beams 1873 to 1952* which gives a complete listing of iron and steel beams and their properties rolled in the United States during that period. Since that book was published there have been many additional shape changes. As a result the wise structural designer will carefully preserve old editions of the Manual so as to have them available when needed.

1-5 COLD-FORMED LIGHT-GAGE STEEL SHAPES

In addition to the hot-rolled steel shapes discussed in the previous section there are some cold-formed steel shapes. These are made by bending thin sheets of carbon or low-alloy steels into almost any desired cross section, such as the ones shown in Fig. 1-2.[5] These shapes—which may be used for light members in roofs, floors, and walls—vary in thickness from about 0.01 in. up to about 0.25 in. The thinner shapes are most often used for some types of panels.

---

[5] *Specification for the Design of Cold-formed Steel Structural Members* (Washington, D.C.: American Iron and Steel Institute, 1986).

Though cold-working does somewhat reduce ductility, it causes some strength increases. Under certain conditions, specifications will permit the use of these higher strengths.

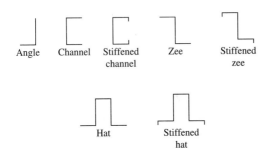

Figure 1-2 Cold-formed shapes.

Concrete and floor slabs are very commonly cast on formed steel decks which serve as economical forms for the wet concrete and which are left in place after the concrete hardens. Several types of this decking are available, some of which are shown in Fig. 1-3. The sections with the deeper cells have the useful feature that electrical and mechanical conduits can be placed in them. The use of steel decks for floor slabs is discussed in Chapter 16 of this text.

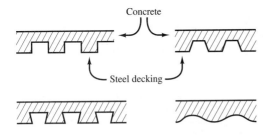

Figure 1-3 Some types of steel decks.

1-6 STRESS-STRAIN RELATIONSHIPS IN STRUCTURAL STEEL

In order to understand the behavior of steel structures, it is absolutely essential for the designer to be familiar with the properties of steel. Stress-strain diagrams present valuable information necessary to understand how steel will behave in a given situation. Satisfactory steel design methods cannot be developed unless complete information is available concerning the stress-strain relationships of the material being used.

If a piece of ductile structural steel is subjected to a tensile force it will begin to elongate. If the tensile force is increased at a constant rate, the amount of elongation will increase constantly within certain limits. In other words, elongation will double when the stress goes from 6000 to 12,000 psi (pounds per square inch). When the tensile stress reaches a value roughly equal to one-half of the ultimate strength of the steel, the elongation will begin to increase at a greater rate without a corresponding increase in the stress.

The largest stress for which Hooke's law applies or the highest point on the straight-line portion of the stress-strain diagram is the *proportional limit*. The largest stress that a material can withstand without being permanently deformed is called the *elastic limit*. This value is seldom actually measured and for most engineering materials including structural steel is synonymous with the propor-

Steel erection for the Chase Manhattan Bank Building, New York City. (Courtesy of Bethlehem Steel Corporation.)

Newport Bridge between Jamestown and Newport, R.I. (Courtesy of Bethlehem Steel Corporation.)

tional limit. For this reason the term *proportional elastic limit* is sometimes used.

The stress at which there is a decided increase in the elongation or strain without a corresponding increase in stress is said to be the *yield* stress. It is the first point on the stress-strain diagram where a tangent to the curve is horizontal. The yield stress is probably the most important property of steel to the designer, as so many design procedures are based on this value. Beyond the yield stress there is a range in which a considerable increase in strain occurs without increase in stress. The strain that occurs before the yield stress is referred to as the *elastic strain;* the strain that occurs after the yield stress, with no increase in stress, is referred to as the *plastic strain.* Plastic strains are usually from 10 to 15 times the elastic strains.

Yielding of steel without stress increase may be thought to be a severe disadvantage when in actuality it is a very useful characteristic. It has often performed the wonderful service of preventing failure due to omissions or mistakes on the designer's part. Should the stress at one point in a ductile steel structure reach the yield point, that part of the structure will yield locally without stress increase, thus preventing premature failure. This ductility allows the stresses in a steel structure to be readjusted. Another way of describing this phenomenon is to say that very high stresses caused by fabrication, erection, or loading will tend to equalize themselves. It might also be said that a steel structure has a reserve of plastic strain that enables it to resist overloads and sudden shocks. If it did not have this ability, it might suddenly fracture, like glass or other vitreous substances.

Following the plastic strain there is a range in which additional stress is necessary to produce additional strain. This is called *strain-hardening.* This portion of the diagram is not too important to today's designer because the strains are so large. A familiar stress-strain diagram for mild or low-carbon structural steel is shown in Fig. 1-4. Only the initial part of the curve is shown here because of the great deformation which occurs before failure. At failure in the mild steels the total strains are from 150 to 200 times the elastic strains. The curve will actually continue up to its maximum stress value and then "tail off" before failure. A sharp reduction in the cross section of the member takes place (called "necking") followed by failure.

The stress-strain curve of Fig. 1-4 is typical of the usual ductile structural steel and is assumed to be the same for members in tension or compression. (The compression members must be stocky because slender compression members subjected to compression loads tend to bend laterally, and their properties are greatly affected by the bending moments so produced.) The shape of the diagram varies with the speed of loading, the type of steel, and the temperature. One such variation is shown in the figure by the dotted line marked *upper yield.*

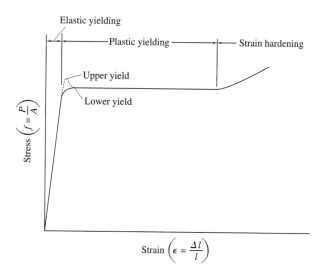

Figure 1-4 Typical stress-strain diagram for a mild or low-carbon structural steel at room temperature.

This shape stress-strain curve is the result when a mild steel has the load applied rapidly, while the *lower yield* is the case for slow loading.

You should note that the stress-strain diagram of Fig. 1-4 was prepared for mild steel at room temperature. Steels (particularly those with rather high carbon contents) may actually increase a little in strength as they are heated to a temperature of about 700°F. As temperatures are raised into the 800° to 1000°F range, steel strengths are drastically reduced, and at 1200°F they have little strength left.

Typical ratios of yield stresses at high temperatures to yield stresses at room temperatures are approximately 0.77 at 800°F, 0.63 at 1000°F, and 0.37 at 1200°F.[6] Temperatures in these ranges can be easily reached in steel members during fires, in localized areas of members when welding is being performed, in members in foundries over open flame, and so on.

When steel sections are cooled below 32°F their strengths will increase a little, but they will have substantial reductions in ductility and toughness.

A very important property of a structure which has not been stressed beyond its yield point is that it will return to its original length when the loads are removed. Should it be stressed beyond this point it will return only part of the way to its original position. This knowledge leads to the possibility of testing an existing structure by loading and unloading. If after the loads are removed the structure will not resume its original dimensions, it has been stressed beyond its yield point.

Steel is an alloy consisting almost entirely of iron (usually over 98 percent). It also contains small quantities of carbon, silicon, manganese, sulfur, phosphorus, and other elements. Carbon is the material that has the greatest effect on the properties of steel. The hardness and strength increase as the carbon percentage is increased, but unfortunately the resulting steel is more brittle and its weldability is adversely affected. A smaller amount of carbon will make the steel softer and more ductile, but also weaker. The addition of such elements as chromium, silicon, and nickel produces steels with considerably higher strengths. Though frequently quite useful, these steels, however, are appreciably more expensive and often are not as easy to fabricate.

A typical stress-strain diagram for a brittle steel is shown in Fig. 1-5. Unfortunately, low ductility or brittleness is a property usually associated with high strengths in steels (although not entirely confined to high-strength steels). As it is desirable to have both high strength and ductility, the designer may have to decide between the two extremes or to compromise. A brittle steel may fail suddenly without warning when overstressed, and during erection could possibly fail due to the shock of erection procedures.

---

[6] American Institute of Steel Construction, Volume I, *Manual of Steel Construction Load & Resistance Factor Design,* 2nd ed., (Chicago: AISC 1994), pp. 1-11 to 1-12.

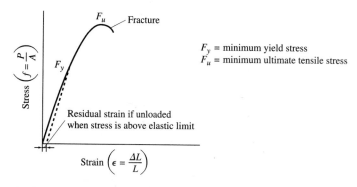

Figure 1-5 Typical stress-strain diagram for a brittle steel.

Brittle steels have a considerable range where stress is proportional to strain, but do not have clearly defined yield stresses. Yet, to apply many of the formulas given in structural steel design specifications, it is necessary to have definite yield stress values regardless of whether the steels are ductile or brittle.

If a mild steel member is strained beyond its elastic limit and then unloaded, it will not return to a condition of zero strain. As it is unloaded, its stress-strain diagram will follow a new path (shown by the dotted line in Fig. 1-5) parallel to the initial straight line. The result is a permanent or residual strain.

The yield stress for a brittle steel is usually defined as the stress at the point of unloading which corresponds to some arbitrarily defined residual strain (0.002 being the common value). In other words, we increase the strain by a designated amount and draw a line from that point, parallel to the straight-line portion of the stress-strain diagram, until the new line intersects the old. This intersection is the yield stress at that particular strain. If 0.002 is used, the intersection is usually referred to as the yield stress at 0.2 percent offset strain.

1-7 MODERN STRUCTURAL STEELS

The properties of steel can be greatly changed by varying the quantities of carbon present and by adding other elements such as silicon, nickel, manganese, and copper. A steel that has a significant amount of the latter elements is referred to as an alloy steel. Although these elements do have a great effect on the properties of steel, the actual quantities of carbon or other alloying elements are quite small. For instance, the carbon content of steel is almost always less than 0.5 percent by weight and is normally from 0.2 to 0.3 percent.

The chemistry of steel is extremely important in its effect on such properties of the steel as weldability, corrosion resistance, resistance to brittle fracture, and

One half of a 170-ft clear span roof truss for the Athletic and Convention Center, Lehigh University, Bethlehem, Pa. (Courtesy of Bethlehem Steel Corporation.)

so on. The ASTM specifies the exact maximum percentages of carbon, manganese, silicon, etc., which are permissible for a number of structural steels. Although the physical and mechanical properties of steel sections are primarily determined by their chemical composition, they are also influenced to a certain degree by the rolling process and by their stress history and heat treatment.

In the past few decades a structural carbon steel designated as A36 and having a minimum yield stress $F_y = 36$ ksi was the commonly used structural steel. Today (1995), however, most of the structural steel used in the United States is manufactured by melting scrap steel in electric furnaces. With this process a 50 ksi steel can be produced and sold at almost the same price as 36 ksi steel. *As a result, at the time of this writing 50 ksi steel produced by the electric furnace process is the commonly used structural steel in the United States.*

In recent decades the engineering and architecture professions have been continually requesting stronger steels, steels with more corrosion resistance, steels with better welding properties, and various other requirements. Research by the steel industry during this period has supplied several groups of new steels which satisfy many of the demands. Today there are quite a few structural steels designated by the ASTM and included in the LRFD Specification.

Structural steels are generally grouped into several major ASTM classifications: the *all-purpose carbon steel* (A36), the *structural carbon steel* (A529), the *high-strength low-alloy structural steel* (A572), the *atmospheric corrosion-resistant high-strength low-alloy structural steels* (A242 and A588) and *quenched and tempered alloy steel plate* (A514 and A852).

The paragraphs that follow include a few general remarks about these steel classifications. In addition, the seven ASTM steels mentioned here are listed in Table 1-1 together with a few general remarks concerning their uses and properties. (As you look at this table notice the phenomenon that the thinner steel is rolled the stronger it becomes. Thicker members tend to be more brittle, and their slower cooling rate causes a coarser microstructure of the steel.)

Carbon Steels

These steels have as their principal strengthening agents carefully controlled quantities of carbon and manganese. Carbon steels have their contents limited to the following maximum percentages: 1.7 percent carbon, 1.65 percent manganese, 0.60 percent silicon, and 0.60 percent copper. These steels are divided into four categories depending on carbon percentages as follows:

1. Low-carbon steel < 0.15 percent.
2. Mild steel 0.15 to 0.29 percent. (The structural carbon steels fall into this category.)
3. Medium-carbon steel 0.30 to 0.59 percent.
4. High-carbon steel 0.60 to 1.70 percent.

High-Strength Low-Alloy Steels

There are a large number of high-strength low-alloy steels, and they are included under several ASTM numbers. In addition to containing carbon and manganese, these steels obtain their higher strengths and other properties by the addition of one or more alloying agents such as columbium, vanadium, chromium, silicon, copper, and nickel. Included are steels with yield stresses as low as 40 ksi and as high as 70 ksi. These steels generally have much greater atmospheric corrosion resistance than the carbon steels.

The term *low-alloy* is used arbitrarily to describe steels for which the total of all the alloying elements does not exceed 5 percent of the total composition of the steel.

TABLE 1-1 PROPERTIES OF STRUCTURAL STEELS

ASTM designation	Type of steel	Forms	Recommended uses	Minimum yield stress F_y, KSI[a]	Specified minimum tensile strength F_u, KSI[b]
A36	Carbon	Shapes, bars, and plates	Bolted or welded buildings and bridges and other structural uses	36, but 32 if thickness > 8 in.	58–80
A529	Carbon	Shapes and plates to $\frac{1}{2}$ in.	Similar to A36	42–50	60–100
A572	High-strength low-alloy Columbium-Vanadium	Shapes, plates, and bars to 6 in.	Bolted or welded construction. Not for welded bridges for F_y grades 55 and above	42–65	60–80
A242	Atmospheric corrosion-resistant high-strength low-alloy	Shapes, plates, and bars to 5 in.	Bolted or welded construction; welding technique very important	42–50	63–70

TABLE 1-1 continued

			42–50	63–70
A588	Atmospheric corrosion-resistant high-strength low-alloy	Plates and bars to 4 in.	Bolted construction	
A852	Quenched and tempered alloy	Plates only to 4 in.	Welded or bolted construction, primarily for welded bridges and buildings. Welding technique of fundamental importance.	70	90–110
A514	Quenched and tempered low-alloy	Plates only $2\frac{1}{2}$ to 6 in.	Welded structures with great attention to technique, discouraged if ductility important	90–100	100–130

[a] F_y values vary with thickness and group (see Tables 1-1 and 1-2, Part 1, LRFD Manual.)

[b] F_u values vary by grade and type.

Atmospheric Corrosion-Resistant High-Strength Low-Alloy Structural Steels

When steels are alloyed with small percentages of copper, they become more corrosion-resistant. When exposed to the atmosphere, the surfaces of these steels oxidize and form a very tightly adherent film (sometimes referred to as a "tightly bound patina " or "a crust of rust") which prevents further oxidation and thus eliminates the need for painting. After this process takes place within 18 months to 3 years (depending on the type of exposure—rural, industrial, direct or indirect sunlight, etc.), the steel reaches a deep reddish-brown or black color.

Supposedly the first steel of this type was developed in 1933 by the U.S. Steel Corporation (now the USX Corporation) to provide resistance to the severe corrosive conditions of railroad coal cars.

You can see the many uses which can be made of such a steel, particularly for structures with exposed members that are difficult to paint—bridges, electrical transmission towers, and others. This steel is not considered to be satisfactory for use where it is frequently subject to saltwater sprays or fogs, or continually submerged in water (fresh or salt) or the ground, or where there are severe corrosive industrial fumes. It is also not satisfactory in very dry areas, as in some western parts of the United States. For the patina to form, the steel must be subjected to a wetting and drying cycle. Otherwise it will continue to look like unpainted steel.

Quenched and Tempered Alloy Steels

These steels have alloying agents in excess of those used in the carbon steels, and they are heat-treated by quenching and tempering to obtain strong and tough steels with yield strengths in the 70 to 110 ksi range. Quenching consists of rapid cooling of the steel with water or oil from temperatures of at least 1650°F down to about 300 or 400°F. In tempering, the steel is reheated to at least 1150°F and allowed to cool.

The quenched and tempered steels do not exhibit well-defined yield points, as do the carbon steels and the high-strength low-alloy steels. As a result their yield strengths are usually defined as the stress at 0.2 percent offset strain. The quenched and tempered steel plates and bars in Table 1-1 are A852 with a yield stress of 70 ksi and A514 with yield stresses of 90 or 100 ksi, depending on thickness.

In Section A3.1 of Part 6 of the LRFD Manual eight other ASTM grades of steel are listed (A53, A500, A501, A570, A606, A607, A618, and A709). These grades cover pipe, cold- and hot-formed tubing, sheet, strip, and structural steel for bridges.

Actual stress-strain curves for the three major types of steels described here (carbon, high-strength low-alloy, and quenched and tempered alloy) are shown

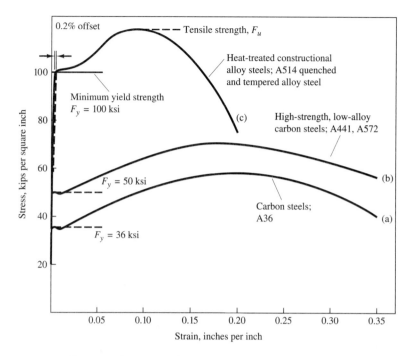

Figure 1-6 Stress-strain curves. (From C. G. Salmon and J. E. Johnson, *Steel Structures,* 3rd ed., Copyright © 1990, HarperCollins College Publishers, Inc.) p. 46.

in Fig. 1-6. The reader will be able to clearly see that the first two of these steels have well-defined yield points, while the quenched and tempered ones do not.

1-8 USES OF HIGH-STRENGTH STEELS

There are indeed other groups of high-strength steels such as the ultra-high-strength steels which have yield strengths from 160 to 300 ksi. These steels have not been included in the LRFD Manual because they have not been assigned ASTM numbers.

It is said that there are today more than 200 steels on the market that provide yield stresses in excess of 36 ksi. The steel industry is now experimenting with steels with yield stresses varying from 200 to 300 ksi, and this may be only the beginning. Many people in the steel industry feel that steels with 500 ksi yield strengths will be made available within a few years. The theoretical binding force between iron atoms has been estimated to be in excess of 4000 ksi.[7]

---

[7] L. S. Beedle et al., *Structural Steel Design* (New York: Ronald Press, 1964), p. 44.

Although the prices of steels increase with increasing yield stresses, the percentage of price increase does not keep up with the percentage of yield-stress increase. The result is that the use of the stronger steels will quite frequently be economical for tension members, beams, and columns. Perhaps the greatest economy can be realized with tension members (particularly those without bolt holes). They may provide a great deal of saving for beams if deflections are not important or if deflections can be controlled by methods described in later chapters. In addition considerable economy can frequently be achieved with high-strength steels for short- and medium-length stocky columns. Another application which can provide considerable saving is hybrid construction. In this type of construction two or more steels of different strengths are used, the weaker steels being used where stresses are smaller and the stronger steels where stresses are higher.

Among the other factors that might lead to the use of high-strength steels are the following:

1. Superior corrosion resistance.
2. Possible savings in shipping, erection, and foundation costs caused by weight saving.
3. Use of shallower beams permitting smaller floor depths.
4. Possible savings in fireproofing because smaller members can be used.

The first thought of most engineers in choosing a type of steel is the direct cost of the members. Such a comparison can be made quite easily, but determining which strength grade is most economical requires consideration of weights, sizes, deflections, maintenance, and fabrication. To make an accurate general comparison of the steels is probably impossible—rather, it is necessary to consider the specific job.

1-9 MEASUREMENT OF TOUGHNESS

The fracture toughness of steel is used as a general measure of its impact resistance or its ability to absorb sudden increases in stress at a notch. The more ductile steel is, the greater will be its toughness. On the other hand, the lower its temperature, the higher will be its brittleness.

There are several procedures available for estimating notch toughness, but the Charpy V-notch test is the most commonly used. Although this test (which is described in ASTM Specification A6) is somewhat inaccurate, it does help identify brittle steels. The energy required to fracture a small bar of rectangular cross section with a specified notch (see Fig. 1-7) is measured.

The bar is fractured with a pendulum swung from a certain height. The amount of energy needed to fracture the bar is determined from the height to which the pendulum rises after the blow. The test may be repeated for different temperatures and plotted as a graph as shown in Fig. 1-8. Such a graph clearly shows the relationship among temperature, ductility, and brittleness. The temperature at the point of steepest slope is the *transition temperature*.

Figure 1-7 Specimen for Charpy V-notch test.

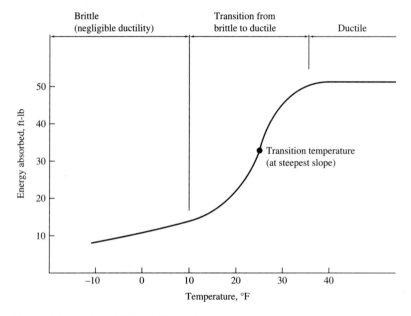

Figure 1-8 Results of Charpy V-notch test.

Different structural steels have different specifications for required absorbed energy levels (say 20 ft-lb at 20°F) depending on the temperature, stress, and loading conditions under which they are to be used. The topic of brittleness is continued in the next section.

1-10 JUMBO SECTIONS

In Table 1-2 of the LRFD Manual, steel shapes are grouped from 1 to 5 depending on their flange and web thicknesses. The heavy W sections in groups 4 and 5 (and structural tees cut from them) are often referred to as *jumbo sections.*

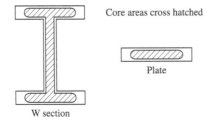
Core areas cross hatched

Plate

W section

Figure 1-9 Core areas where brittle failure may be a problem in thick heavy members.

Jumbo sections were originally developed for use as compression members and as such are quite satisfactory. However, designers have frequently used them for tension or flexural members. In these uses, flange and web areas have had some serious cracking problems where welding or thermal cutting has been used. These cracks may result in smaller load-carrying capacities and problems related to fatigue.[8]

Thick pieces of steel tend to be more brittle than thin ones. Some of the reasons for this are that the core areas of thicker shapes (shown in Fig. 1-9) are subject to less rolling, have higher carbon contents (needed to produce required yield stresses), and have higher tensile stresses from cooling. These topics are discussed in later chapters.

Jumbo sections spliced with welds can be satisfactorily used for axial tension or flexural situations if the procedures listed in Specification A3.1c of the Manual are carefully followed. Included among the requirements are the following:

1. The steel used must have certain absorbed energy levels as determined by the Charpy V-notch test (20 ft-lb at 70°F). It is absolutely necessary that the tests be made on specimens taken from the core areas (shown in Fig. 1-9), where brittle fracture has proved to be a problem.
2. Temperature must be controlled during welding, and the work must follow a certain sequence.
3. Special splice details are required.

1-11 LAMELLAR TEARING

The steel specimens used for testing and developing stress-strain curves usually have their longitudinal axes in the direction that the steel was rolled. Should specimens be taken with their longitudinal axes transverse to the rolling direc-

---

[8] R. Bjorhovde, "Solutions for the Use of Jumbo Shapes," *Proceedings 1988 National Steel Construction Conference,* AISC, Chicago, June 8–11, 1988, pp. 2-1 to 2-20.

tion "through the thickness" of the steel, the results will be lower ductility and toughness. Fortunately this is a matter of little significance for almost all situations. It can, however, be quite important when thick plates and heavy structural shapes are used in highly restrained welded joints. (It can be a problem for thin members, too, but it is much more likely to give trouble in thick ones.)

If a joint is highly restrained, the shrinkage of the welds in the through-the-thickness direction cannot be adequately redistributed, and the result can be a tearing of the steel called *lamellar tearing*. (*Lamellar* means "consisting of thin layers.") The situation is aggravated by the application of external tension. Lamellar tearing may show up as fatigue cracking after a number of cycles of load applications.

The lamellar tearing problem can be eliminated or greatly minimized with appropriate weld details and weld procedures. For example, the welds should be detailed so that shrinkage occurs as much as possible in the direction the steel was rolled. Several steel companies produce steels with enhanced through-the-thickness properties which provide much greater resistance to lamellar tearing. Even if such steels are used for heavy restrained joints, the special joint details mentioned above still are necessary.[9]

1-12 FURNISHING OF STRUCTURAL STEEL

The furnishing of structural steel consists of the rolling of the steel shapes, the fabrication of the shapes for the particular job (including cutting to the proper dimensions and punching the holes necessary for field connections), and their erection. Very rarely will a company perform all three of these functions, and the average company performs only one or two of them. For instance, many companies fabricate structural steel and erect it, while others may only be steel fabricators or steel erectors. There are approximately 400 to 500 companies in the United States which make up the fabricating industry for structural steel. Most of them do fabrication and erection.

Steel fabricators normally carry very little steel in stock because of the high interest and storage charges. When they get a job, they may order the shapes to certain lengths directly from the rolling mill, or they may obtain them from service centers. Service centers, which are an increasingly important factor in the supply of structural steel, buy and stock large quantities of structural steel which they buy at the best prices they can find anywhere in the world.

Structural steel is usually designed by an engineer in collaboration with an architectural firm. The designer makes design drawings that show member sizes, controlling dimensions, and any unusual connections. The company that is to

---

[9] "Commentary on Highly Restrained Welded Connections," *Engineering Journal*, AISC, 10, no. 3 (3d quarter, 1973), pp. 61–73.

fabricate the steel makes the detailed drawings subject to the engineer's approval. These drawings provide all the information necessary to fabricate the members correctly. They show the dimensions for each member, the locations of holes, the positions and sizes of connections, and the like. A part of a typical detail drawing for a bolted steel beam is shown in Fig. 1-10. There may be a few items included on this drawing that are puzzling to you since you have read only a few pages of this book. These items should become clear as you study the chapters to follow, however.

On actual detail drawings details probably will be shown for several members. Here, the author has shown only one member just to indicate the information needed so the shop can correctly fabricate the member. The darkened circles and rectangles indicate that the bolts are to be installed in the field, while the non-darkened ones show the connections which are to be made in the shop.

The erection of steel buildings is more a matter of assembly than nearly any other line of construction work. Each of the members is marked in the shop with letters and numbers to distinguish it from the other members to be used. The erection is performed in accordance with a set of erection plans. These plans are not detailed drawings, but are simple line diagrams showing the position of the various members in the building. The drawings show each separate piece or

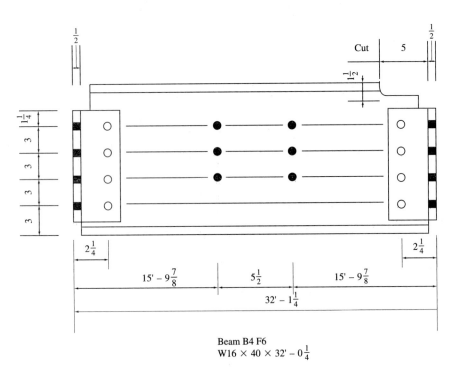

Beam B4 F6
W16 × 40 × 32' − 0$\frac{1}{4}$

Figure 1-10 Part of a detail drawing.

subassembly of pieces together with assigned shipping or erection marks so that the steelworkers can quickly identify and locate members in their correct positions in the structure. (Persons performing steel erection often are called *iron-workers,* which is a name held over from the days before structural steel.) Directions (north, south, east, or west) usually are painted on column faces.

Sometimes the erection drawing gives the sizes of the members, but this is not necessary. They may or may not be shown, depending on the practice of the particular fabricator.

Beams, girders, and columns will be indicated on the drawings by the letters B, G, or C followed by the number of the particular member as B5, G12, and so on. Often, there will be several members of these same designations where members are repeated in the building.

Multistory steel frames often will have several levels of identical or nearly identical framing systems. Thus, one erection plan may be used to serve several floors. For such situations the member designations for the columns, beams, and girders will have the level numbers incorporated in them. For instance, column C15(3–5) is column 15, third to 5th floors, while B4F6, or just B4(6), represents beam B4 for the sixth floor. A portion of a building erection drawing is shown in Fig. 1-11.

Figure 1-11 Part of an erection drawing showing where each member is to be located.

1-13 THE WORK OF THE STRUCTURAL DESIGNER

The structural designer arranges and proportions structures and their parts so that they will satisfactorily support the loads to which they may feasibly be subjected. It might be said that he or she is involved with the general layout of structures; studies of the possible structural forms that can be used; consideration of loading conditions; analysis of stresses, deflections, and so on; design of parts; and the preparation of design drawings. More precisely, the word *design* pertains to the proportioning of the various parts of a structure after the forces have been calculated, and it is this process which will be emphasized throughout the text, using structural steel as the material.

1-14 RESPONSIBILITIES OF THE STRUCTURAL DESIGNER

The structural designer must learn to arrange and proportion the parts of structures so that they can be practically erected and will have sufficient strength and reasonable economy. These items are discussed briefly below.

Safety

Not only must the frame of a structure safely support the loads to which it is subjected, but it must support them in such a manner that deflections and vibrations are not so great as to frighten the occupants or to cause unsightly cracks.

Cost

The designer needs to keep in mind the factors that can lower cost without sacrifice of strength. These items, which are discussed in more detail throughout the text, include the use of standard-size members, simple connections and details, and members and materials that will not require an unreasonable amount of maintenance through the years.

Practicality

Another objective is the design of structures that can be fabricated and erected without great problems arising. Designers need to understand fabrication methods and should try to fit their work to the fabrication facilities available.

Designers should learn everything possible about the detailing, fabrication, and field erection of steel. The more the designer knows about the problems,

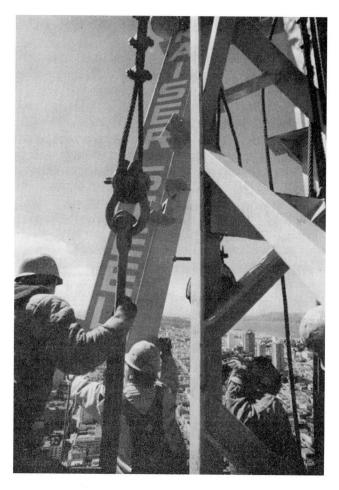

Steel erection for Transamerica Pyramid, San Francisco, Calif. (Courtesy of Kaiser Steel Corporation.)

tolerances, and clearances in shop and field the more probable it is that reasonable, practical, and economical designs will be produced. This knowledge should include information concerning the transportation of the materials to the job site (such as the largest pieces that can be transported practically by rail or truck), labor conditions, and the equipment available for erection. Perhaps the designer should ask, "Could I get this thing together if I were sent out to do it?"

Finally, he or she needs to proportion the parts of the structure so that they will not unduly interfere with the mechanical features of the structure (pipes, ducts, etc.) or the architectural effects.

1-15 ECONOMICAL DESIGN OF STEEL MEMBERS

The design of a steel member involves much more than a calculation of the properties required to support the loads and the selection of the lightest section providing these properties. Although at first glance this procedure would seem to give the most economical designs, many other factors need to be considered. Among these are:

1. The designer needs to select steel sections of sizes which are usually rolled. Steel beams and bars and plates of unusual sizes will be difficult to obtain during boom periods and will be expensive during any period. A little study on the designer's part will enable him or her to avoid these expensive shapes. Steel fabricators are constantly supplied with information from the steel companies and the steel warehousers as to the sizes and lengths of sections available. (Most structural shapes can be obtained in lengths from 60 to 75 ft depending on the producer, while it is possible under certain conditions to obtain some shapes up to 120 ft in length.)

2. A blind assumption that the lightest section is the cheapest one may be in considerable error. A building frame designed by the "lightest-section" procedure will consist of a large number of different shapes and sizes of members. Trying to connect these many-sized members and fit them in the building will be quite complicated, and the pound price of the steel will in all probability be rather high. A more reasonable approach would be to smooth out the sizes by selecting many members of the same sizes, although some of them may be slightly overdesigned.

3. The beams usually selected for the floors in buildings will normally be the deeper sections, because these sections for the same weights have the largest moments of inertia and the greatest resisting moments. As building heights increase, however, it may be economical to modify this practice. As an illustration, consider the erection of a 20-story building, for which each floor has a minimum clearance. It is assumed that the depths of the floor beams may be reduced by 6 in. without an unreasonable increase in beam weights. The beams will cost more, but the building height will be reduced by 20 × 6 in. = 120 in., or 10 ft, with resulting savings in walls, elevator shafts, column heights, plumbing, wiring, and footings.

4. The costs of erection and fabrication for structural steel beams are approximately the same for light or heavy members. Thus, beams should be spaced as far apart as possible to reduce the number of members which have to be fabricated and erected.[10]

5. Structural steel members should be painted only if so required by the applicable specification. You should realize that steel should not be painted if it

---

[10] H. Allison, "Low- and Medium-Rise-Steel Buildings" (Chicago: AISC, 1991) pp. 1–5.

Steel erection for Transamerica Pyramid, San Francisco, Calif. (Courtesy of Kaiser Steel Corporation.)

is to be in contact with concrete. Furthermore, the various fireproofing materials used for protecting steel members adhere better if the surfaces are unpainted.[11]

---

[11] Ibid, pp. 1–5.

6. It is very desirable to keep repeating the same section over and over again. Such a practice will reduce the detailing, fabrication, and erection costs.

7. For larger sections, particularly the built-up ones, the designer needs to have information pertaining to transportation problems. The desired information includes the greatest lengths and depths that can be shipped by truck or rail (see Section 1-17), clearances available under bridges and power lines leading to the project, and allowable loads on bridges. It may be possible to fabricate a steel roof truss in one piece, but is it possible to transport it to the job site and erect it in one piece?

8. Sections should be selected which are reasonably easy to erect and which have no conditions that will make them difficult to maintain. As an example, it is necessary to have access to all exposed surfaces of steel bridge members so that they may be periodically painted (unless one of the special corrosion-resistant steels is being used).

9. Buildings are often filled with an amazing conglomeration of pipes, ducts, conduits, and other items. Every effort should be made to select steel members that will fit in with the requirements made by these items.

10. The members of a steel structure are often exposed to the public, particularly in the case of steel bridges and auditoriums. Appearance may often be the major factor in selecting the type of structure, as where a bridge is desired that will fit in and actually contribute to the appearance of an area. Exposed members may be surprisingly graceful when a simple arrangement and perhaps curved members are used, but other arrangements may create a terrible eyesore. The student has certainly seen illustrations of each case. It is very interesting to know that beautiful structures in steel are usually quite reasonable in cost.

The question is often asked, *How do we achieve economy in structural steel design? The answer is simple: it lies in what the steel fabricator does not have to do.* (In other words, economy can be realized when fabrication is minimized.)

1-16 FAILURE OF STRUCTURES

Many people who are superstitious do not discuss flat tires or make their wills because they fear they would be tempting fate. These same people would probably not care to discuss the subject of engineering failures. Despite the prevalence of this superstition the author feels that an awareness of the items which have most frequently caused failures in the past is invaluable to experienced and inexperienced designers alike. Perhaps a study of past failures is more important than a study of past successes. Benjamin Franklin supposedly made the observation that "a wise man learns more from failures than from success."

The designer of little experience particularly needs to know where the most attention should be given and where outside advice is needed. The vast majority of designers, experienced and inexperienced, select members of sufficient size and strength. The collapse of structures is usually due to insufficient attention to the details of connections, deflections, erection problems, and foundation settlements. Rarely if ever do steel structures fail due to faults in the material, but rather due to its improper use.

A frequent fault displayed by designers is that after carefully designing the members of a structure, they carelessly select connections which may or may not be of sufficient size. They may even turn the job of selecting the connections over to drafters, who may not have sufficient understanding of the difficulties that can arise in connection design. Perhaps the most common mistake made in connection design is to neglect some of the forces acting on the connection, such as twisting moments. In a truss for which the members have been designed for axial forces only, the connections may be eccentrically loaded, resulting in moments that cause increasing stresses. These secondary stresses are occasionally so large that they need to be considered in design.

Another source of failure occurs where beams supported on walls have insufficient bearing or anchorage. Imagine a beam of this type supporting a flat roof on a rainy night when the roof drains are not functioning properly. As the water begins to form puddles on the roof, it tends to cause the beam to sag in the middle, causing a pocket to catch more rain, which will cause more beam sag, and so on. As the beam deflects, it pushes out against the walls, possibly causing collapse of walls or slippage of beam ends off the wall. Picture a 60-ft steel beam supported on a wall with only an inch or two of bearing when the temperature drops 50 or 60 degrees overnight. A collapse due to a combination of beam contraction, outward deflection of walls, and vertical deflection caused by precipitation loads is not difficult to visualize; furthermore, actual cases in engineering literature are not difficult to find.

Foundation settlements cause a large number of structural failures, probably more than any other factor. Most foundation settlements do not result in collapse, but they very often cause unsightly cracks and depreciation of the structure. If all parts of the foundation of a structure settle equally, the stresses in the structure theoretically will not change. The designer, usually not able to prevent settlement, has the goal of designing foundations in such a manner that equal settlements occur. Equal settlements may be an impossible goal, and consideration should be given to the stresses that would be produced if settlement variations occurred. The student's background in structural analysis will tell him or her that uneven settlements in statically indeterminate structures may cause extreme stress variations. Where foundation conditions are poor, it is desirable, if feasible, to use statically determinate structures whose stresses are not appreciably changed by support settlements. (The student will learn in subsequent discussions that the ultimate strength of steel structures is usually affected only slightly by uneven support settlements.)

Some structural failures occur because inadequate attention is given to deflections, fatigue of members, bracing against swaying, vibrations, and the possibility of buckling of compression members or the compression flanges of beams. The usual structure when completed is sufficiently braced with floors, walls, connections, and special bracing, but there are times during construction when many of these items are not present. As previously indicated, the worst conditions may well occur during erection, and special temporary bracing may be required.

1-17 HANDLING AND SHIPPING STRUCTURAL STEEL

The following general rules apply to the sizes and weights of structural steel pieces which can be fabricated in the shop, shipped to the job, and erected.

1. The maximum weights and lengths that can be handled in the shop and at a construction site are roughly 90 tons and 120 ft respectively.
2. Pieces as large as 8 ft high, 8 ft wide, and 60 ft long can be shipped on trucks with no difficulty (provided the axle or gross weights do not exceed the permissible values given by public agencies along the designated routes).
3. There are few problems in railroad shipment if pieces are no larger than 10 ft high, 8 ft wide, 60 ft long, and weigh no more than 20 tons.
4. Routes should be carefully studied and carriers consulted for weights and sizes exceeding the values given in 2 and 3 above.

1-18 CALCULATION ACCURACY

A most important point which many students with their superb pocket calculators and personal computers have difficulty in understanding is that structural design is not an exact science for which answers can confidently be calculated to eight places. Among the reasons for this fact are: the methods of analysis are based on partly true assumptions, the strengths of materials used vary appreciably, and maximum loadings can only be approximated. With respect to this last sentence, how many of the users of this book could estimate within 10 percent the maximum load in pounds per square foot that will ever occur on the building floor which they are now occupying? Calculations to more than two or three significant figures are obviously of little value, and may actually be harmful in that they mislead the student by giving him or her a fictitious sense of precision.

1-19 IMPACT OF COMPUTERS ON STRUCTURAL STEEL DESIGN

The availability of personal computers has drastically changed the way in which steel structures are analyzed and designed. In nearly every engineering school and office computers are routinely used to handle structural analysis problems.

Though they have been used much less for design work, the situation is rapidly changing as more and more software is developed and sold commercially.

Many calculations are involved in structural steel design, and many of these calculations are quite time consuming. With a computer the designer can tremendously reduce the time required for these calculations and thus supposedly have time to consider alternative designs.

Theoretically, the computer design of alternative systems for a few projects should substantially improve the designer's judgment in a short span of time. Without computers, the development of this same judgment will supposedly require the designer to work his or her way through quite a few projects.

Though computers do increase design productivity, they do undoubtedly tend at the same time to reduce the designer's "feel" for structures. This can be a special problem for young engineers with little previous design experience. Unless designers have this "feel," computer usage, though expediting the work and reducing math errors, may occasionally result in large mistakes.

It is interesting to note that up to the present time the feeling at most engineering schools has been that the best way to teach structural steel design is with chalk and blackboard.

1-20 COMPUTER-AIDED DESIGN IN THIS TEXT

Enclosed at the back of the book is a diskette containing a limited steel design software system called INSTEP. This software was prepared by Dr. James K. Nelson to assist the student in solving many of the problems presented in the text. It is limited in that many, but not all, of the checks necessary for complete designs are performed. Through use of INSTEP the student can easily study the effects of design changes and the significance of various design parameters.

INSTEP will operate on IBM and IBM-compatible microcomputers that have a hard disk and use the MS-DOS operating system. The contents of the diskette can be copied to a directory on the hard disk. To begin operation of INSTEP, simply change to that directory, enter INSTEP from the keyboard, and follow the instructions provided. INSTEP will operate successfully on microcomputers that have a Hercules, EGA, or VGA graphics monitor. If available, a mouse can be used to assist data entry.

A complete description of the INSTEP system, the assumptions and limitations of the procedures used, and information necessary to operate INSTEP can be obtained from within the system. During operation the designer is presented with various displays and choices. Data entry and design operations are performed through use of parameter and operation buttons in a user-oriented format. To the greatest extent possible, parameter names and notation have been chosen to be consistent with the textbook and the LRFD Manual.

Quite a few design examples are handled throughout this textbook using the INSTEP program. The first of these is Example 3-12 in Chapter 3.

Chapter **2**

Specifications, Loads, and Methods of Design

2-1 SPECIFICATIONS AND BUILDING CODES

The design of most structures is controlled by specifications. Even if not so controlled, the designer will probably refer to them as a guide. No matter how many structures a person has designed it is impossible for him or her to have encountered every situation. By referring to specifications, he or she is making use of the best available material on the subject. Engineering specifications which are developed by various organizations present the best opinion of those organizations as to what represents good practice.

Municipal and state governments concerned with the safety of the public have established building codes with which they control the construction of various structures within their jurisdiction. These codes, which are actually laws or ordinances, specify design loads, design stresses, construction types, material quality, and other factors. They vary considerably from city to city, a fact that causes some confusion among architects and engineers.

Several organizations publish recommended practices for regional or national use. Their specifications are not legally enforceable unless they are embodied in the local building code or made a part of a particular contract. Among these organizations are the AISC and AASHTO (American Association of State Highway and Transportation Officials). Nearly all municipal and state building codes have adopted the AISC Specification and nearly all state highway and transportation departments have adopted the AASHTO Specifications.

Readers should note that logical and clearly written codes are quite helpful to designers. Furthermore, there are far fewer structural failures in areas that have good codes that are strictly enforced.

South Fork Feather River Bridge in northern California, being erected by use of a 1626-ft-long cableway strung from 210-ft-high masts anchored on each side of the canyon. (Courtesy of Bethlehem Steel Corporation.)

Some people feel that specifications prevent engineers from thinking for themselves—and there may be some basis for the criticism. They say that the ancient engineers who built the great pyramids, the Parthenon, and the great Roman bridges were controlled by few specifications, which is certainly true. On the other hand, it should be said that only a few score of these great projects were built over many centuries, and they were apparently built without regard to cost of material, labor, or human life. They were probably built by intuition and by certain rules of thumb developed by observing the minimum size or strength of members which would fail only under given conditions. Their probably numerous failures are not recorded in history; only their successes endured.

Today, however, there are hundreds of projects being constructed at any one time in the United States which rival in importance and magnitude the famous structures of the past. It appears that if all engineers in our country were allowed to design projects such as these without restrictions there would be many disastrous failures. *The important thing to remember about specifications, therefore, is that they are not written for the purpose of restricting engineers but for the purpose of protecting the public.*

No matter how many specifications are written, it is impossible for them to cover every possible situation; however. As a result, no matter which code or specification is or is not being used, the ultimate responsibility for the design of a safe structure lies with the structural designer.

2-2 LOADS

Perhaps the most important and most difficult task faced by the structural designer is the accurate estimation of the loads which may be applied to a structure during its life. No loads that may reasonably be expected to occur may be overlooked. After loads are estimated, the next problem is to determine the worst possible combinations of these loads which might occur at one time. For instance, would a highway bridge completely covered with ice and snow be simultaneously subjected to fast-moving lines of heavily loaded trailer trucks in every lane and to a 90-mile lateral wind, or is some lesser combination of these loads more feasible?

The next two sections of this chapter provide a brief introduction to the types of loads with which the structural designer needs to be familiar. The purpose of these sections is not to discuss loads in great detail, but rather to give the reader a "feel" for the subject. As will be seen, loads are classed as being dead loads or live loads.

2-3 DEAD LOADS

Dead loads are loads of constant magnitude that remain in one position. They consist of the structural frame's own weight and other loads that are permanently attached to the frame. For a steel-frame building, the frame, walls, floors, roof, plumbing, and fixtures are dead loads.

To design a structure it is necessary for the weights or dead loads of the various parts to be estimated for use in the analysis. The exact sizes and weights of the parts are not known until the structural analysis is made and the members of the structure selected. The weights, as determined from the actual design, must be compared with the estimated weights. If large discrepancies are present, it will be necessary to repeat the analysis and design using better estimated weights.

Reasonable estimates of structure weights may be obtained by referring to similar types of structures or to various formulas and tables available in several locations. The weights of many materials are given in Part 7 of the LRFD Manual. Even more detailed information on dead loads is provided in Tables C1 and C2 of the American Society of Civil Engineers (ASCE) Standard 7.[1] An experienced designer can estimate very closely the weights of most materials and will spend little time repeating designs because of poor estimates.

---

[1] *American Society of Civil Engineers Minimum Design Loads for Buildings and Other Structures,* ASCE 7–93 (formerly ANSI A58.1) (New York, ASCE), pp. 49–50.

2-4 LIVE LOADS

Live loads are loads that may change in position and magnitude. Simply stated, all loads that are not dead loads are live loads. Live loads that move under their own power are said to be moving loads, such as trucks, people, and cranes, whereas those loads that may be moved are movable loads, such as furniture, warehouse materials, and snow. Other live loads include those caused by construction operations, wind, rain, earthquakes, blasts, soils, and temperature changes. A brief description of some of these loads follows:

1. *Floor loads.* The minimum gravity live loads to be used for building floors are usually clearly specified by the applicable building code. Unfortunately, however, the values given in these various codes vary from city to city and the designer must be sure that his or her designs meet the requirements of that locality. A few of the typical values for floor loadings are listed in Table 2-1. These values were adopted from the ASCE Standard 7.[2] In the absence of a governing code this seems to be an excellent one to follow.

TABLE 2-1   TYPICAL MINIMUM UNIFORM LIVE LOADS FOR DESIGN OF BUILDINGS

Type of building	LL (psf)
Apartment houses	
Apartments	40
Public rooms	100
Dining rooms and restaurants	100
Garages (passenger cars only)	50
Gymnasiums, main floors, and balconies	100
Office buildings	
Lobbies	100
Offices	50
Schools	
Classrooms	40
Corridors first floor	100
Corridors above 1st floor	80
Storage warehouses	
Light	125
Heavy	250
Stores (retail)	
First floor	100
Other floors	75

---

[2] Ibid, p. 4.

2. *Snow and ice.* In the colder states snow and ice loads are often quite important. One inch of snow is equivalent to approximately 0.5 psf (pounds per square foot), but it may be higher at lower elevations where snow is denser. For roof designs, snow loads of from 10 to 40 psf are used, the magnitude depending primarily on the slope of the roof and to a lesser degree on the character of the roof surface. The larger values are used for flat roofs and the smaller ones for sloped roofs. Snow tends to slide off sloped roofs, particularly those with metal or slate surfaces. A load of approximately 10 psf might be used for 45° (degree) slopes and a 40-psf load for flat roofs. Studies of snowfall records in areas with severe winters may indicate the occurrence of snow loads much greater than 40 psf, with values as high as 100 psf in northern Maine.

Snow is a variable load which may cover an entire roof or only part of it. There may be drifts against walls or buildup in valleys or between parapets. Snow may slide off one roof onto a lower one. The wind may blow it off one side of a sloping roof or the snow may crust over and remain in position even during very heavy winds.

Bridges are generally not designed for snow loads, since the weight of the snow is usually not significant in comparison with truck and train loads. In any case it is doubtful that a full load of snow and maximum traffic would be present at the same time. Bridges and towers are sometimes covered with layers of ice from 1 to 2 in. (inches) thick. The weight of the ice runs up to about 10 psf. Another factor to be considered is the increased surface area of the ice-coated members as it pertains to wind loads.

3. *Rain.* Though snow loads are a more severe problem than rain loads for the usual roof, the situation may be reversed for flat roofs, particularly those in warmer climates. If water on a flat roof accumulates faster than it runs off, the result is called *ponding* because the increased load causes the roof to deflect into a dish shape that can hold more water, which causes greater deflections, and so on. This process continues until equilibrium is reached or until collapse occurs. Ponding is a serious matter as illustrated by the large number of flat-roof failures that occur during rain storms every year in the United States.

Ponding will occur on almost any flat roof to a certain degree, even though roof drains are present. Roof drains may very well be used, but they may be inadequate during severe storms, or they may become partially or completely clogged. The best method of preventing ponding is to have an appreciable slope of the roof ($\frac{1}{4}$ in./ft or more) together with good drainage facilities. In addition to the usual ponding another problem may occur for very large flat roofs (with perhaps an acre or more of surface area). During heavy rainstorms strong winds frequently occur. If there is a great deal of water on the roof a strong wind may very well push a large quantity of it toward one end. The result can be a dangerous water depth as regards the load in psf on that end. For such situations *scuppers* are sometimes used. These are large holes or tubes in walls or parapets that enable water above a certain depth to quickly drain off the roof.

Cold-storage warehouse, Grand Junction, Colo. (Courtesy of the American Institute of Steel Construction, Inc.)

4. *Traffic loads for bridges.* Bridges are subjected to series of concentrated loads of varying magnitude caused by groups of truck or train wheels.

5. *Impact loads.* Impact loads are caused by the vibration of moving or movable loads. It is obvious that a crate dropped on the floor of a warehouse or a truck bouncing on uneven pavement of a bridge causes greater forces than would occur if the loads were applied gently and gradually. Impact loads are equal to the difference between the magnitude of the loads actually caused and the magnitude of the loads had they been dead loads.

The LRFD Specification (A4.2) requires that structures that are to support live loads that tend to cause impact are to have their assumed nominal loads for design increased by the following percentages if not otherwise specified:

For supports of elevators and elevator machinery	100%
For supports of light machinery, shaft or motor driven, not less than	20%
For supports of reciprocating machinery or power driven units, not less than	50%
For hangers supporting floors and balconies	33%
For cab-operated traveling crane support girders and their connections	25%
For pendant-operated traveling crane support girders and their connections	10%

6. *Lateral loads.* Lateral loads are of two main types: wind and earthquake. A survey of engineering literature for the past 150 years reveals many references to structural failures caused by wind. Perhaps the most infamous of these have been bridge failures such as those of the Tay Bridge in Scotland in 1879 (which caused the deaths of 75 persons) and the Tacoma Narrows Bridge (Tacoma, Washington) in 1940. But there have also been some disastrous building failures due to wind during the same period, such as that of the Union Carbide Building in Toronto in 1958. It is important to realize that a large percentage of building failures due to wind have occurred during their erection.[3]

(a) *Wind loads.* A great deal of research has been conducted in recent years on the subject of wind loads. Nevertheless, a great deal more work needs to be done as the estimation of these forces can by no means be classified as an exact science. The magnitudes of wind loads vary with geographical locations, heights above ground, types of terrain surrounding the buildings, including other nearby structures, and other factors.

Wind pressures are usually assumed to be uniformly applied to the windward surfaces of buildings and are assumed to be capable of coming from any direction. These assumptions are not very accurate because wind pressures are not very even over large areas, the pressures near the corners of buildings being probably greater than elsewhere due to wind rushing around the corners, and so on. From a practical standpoint, therefore, all of the possible variations cannot be considered in design, although today's specifications are becoming more and more detailed in their requirements.

When the designer working with large stationary buildings makes poor wind estimates the results are probably not too serious, but this is not the case when tall slender buildings (or long flexible bridges) are being considered. For many years the average designer ignored wind forces for buildings whose heights were not at least twice their least lateral dimensions. For such cases as these it was felt that the floors and walls provided sufficient lateral stiffnesses to eliminate the need for definite wind bracing systems. A better practice for designers to follow, however, is to consider all the possible loading conditions that a particular structure may have to resist. If one or more of these conditions (as perhaps wind loading) seem of little significance, they may then be neglected. Should buildings have their walls and floors constructed with modern lightweight materials and/or should they be subjected to unusually high wind loads (as in coastal or mountainous areas) they will probably have to be designed for wind loads even if the height/least lateral dimension ratios are less than 2.

Building codes do not usually provide for the estimated forces during tornadoes. The average designer considers the forces created directly in the paths of tornadoes to be so violent that it is not economically feasible to design buildings

---

[3] Wind Forces on Structures, Task Committee on Wind Forces. Committee on Loads and Stresses, Structural Division, ASCE, Final Report, *Transactions ASCE* 126, Part II (1961): 1124–1125.

Access bridge, Renton, Wash. (Courtesy of the Bethlehem Steel Corporation.)

to resist them. This feeling is undergoing some change, however, as it has been found that the wind resistance of structures (even for small buildings, including houses) can be greatly increased at very reasonable costs by using better practices of tying the parts of structures together from the roofs through the walls to the footings, and by making appropriate connections of window frames to walls and perhaps other parts of the structure.[4,5]

Wind forces act as pressures on vertical windward surfaces, pressures or suction on sloping windward surfaces (depending on the slope), and suction on flat surfaces and on leeward vertical and sloping surfaces (due to the creation of negative pressures or vacuums). The student may have noticed this definite suction effect where shingles or other roof coverings have been lifted from the leeward roof surfaces of buildings. Suction or uplift can easily be demonstrated by holding a piece of paper at two of its corners and blowing above it. For some common structures uplift may be as large as 20 to 30 psf or even more.

During the passing of a tornado or hurricane, a sharp reduction in atmospheric pressure occurs. This decrease in pressure is not reflected inside airtight buildings, and the inside pressures, being greater than the external pressures,

[4] P. R. Sparks, "Wind Induced Instability in Low-Rise Buildings," *Proceedings of the 5th U.S. National Conference on Wind Engineering,* Lubbock, Texas, November 6–18, 1985.

[5] P. R. Sparks, "The Risk of Progressive Collapse of Single-Story Buildings in Severe Storms," *Proceedings of the ASCE Structures Congress,* Orlando, Florida, August 17–20, 1987.

cause outward forces against the roofs and walls. Nearly everyone has heard stories of the walls of a building "exploding" outward during a storm.

The average building code in the United States makes no reference to wind velocities in the area or to shapes of the building or to other factors. It probably only requires the use of some specified wind pressure in design such as 20 psf on projected area in elevation up to 300 ft with an increase of 2.5 psf for each additional 100-ft increase in height. The values given in these codes are thought to be rather inaccurate for modern structural design.

For a period of several years the ASCE Task Committee on Wind Forces made a detailed study of existing information concerning wind forces. A splendid report of this information entitled *Wind Forces on Structures*[6] was presented by the committee in 1961. As stated in the report, its purpose was to provide a compact source of information which could be practically used by the design profession.

In the report much information is presented concerning wind-pressure coefficients for various types of structures, information concerning maximum wind velocities for particular geographical areas, and much additional data which can be of great value in realistically estimating wind forces.

The wind pressure on a building can be estimated with the expression to follow in which p is the pressure in pounds per square foot acting on vertical surfaces, C_s is a shape coefficient, and V is the basic wind velocity in miles per hour estimated from weather bureau records for that area of the country.

$$p = 0.002558C_s V^2$$

The coefficient C_s depends on the shape of the structure, primarily the roof. For box-type structures C_s is 1.3, of which 0.8 is for the pressure on the windward side and 0.5 is for suction on the leeward side. For such a building the total pressure on the two surfaces equals 20 psf for a wind velocity of 77.8 mph.

As described in the foregoing paragraphs, the accurate determination of the most critical wind loads on a building or bridge is an extremely involved problem; however, sufficient information is available today to permit satisfactory estimates on a reasonably simple basis.

(b) *Earthquake loads.* Many areas of the world fall in "earthquake territory," and in those areas it is necessary to consider seismic forces in design for all types of structures. Through the centuries there have been catastrophic failures of buildings, bridges, and other structures during earthquakes. It has been estimated that as many as 50,000 people lost their lives in the 1988 earthquake in Armenia.[7] The 1989 Loma Prieta and 1994 Northridge earthquakes in California caused many billions of dollars of property damage as well as considerable loss of life.

---

[6] Ibid., pp. 1124–1198.

[7] V. Fairweather, "The Next Earthquake," *Civil Engineering* (New York: ASCE, March 1990), pp. 54–57.

Fig. 2-1 shows seismic risk zones in various locations in the United States. These zones were established on the basis of past earthquakes.[8] This map indicates that the most severe risks are in the zones numbered 4 and the least severe risks are in those numbered 0. This somewhat dated map is presented to give the student a feeling for those areas of the country where earthquake risks are greatest. Much more detailed maps are used today to estimate potential seismic forces.

Steel structures can be economically designed and constructed to withstand the forces caused during most earthquakes. On the other hand, the cost of providing seismic resistance to existing structures (called *retrofitting*) can be extremely high. *Recent earthquakes have clearly shown that the average building or bridge that has not been designed for earthquake forces can be destroyed by an earthquake that is not particularly severe.*

During an earthquake there is an acceleration of the ground surface. This acceleration can be broken down into vertical and horizontal components. Usually the vertical component is assumed to be negligible, but the horizontal component can be severe. The structural analysis for the expected effects of an earthquake should include a study of the structure's response to the ground motion caused by the earthquake. However, it is common in design to approximate the effects of ground motion with a set of horizontal static loads acting at each level of the structure. Various formulas are used to change the earthquake accelerations into static forces that are dependent on the distribution of the mass of the structure, the type of framing, its stiffness, its location, and so on. Such an approximate approach is usually adequate for low-rise buildings, but it is not suitable for high-rises, where an overall dynamic analysis is necessary.

Some engineers seem to think that the seismic loads to be used in design are merely percentage increases of the wind loads. This surmise is incorrect, however, as seismic loads are different in their action and are not proportional to the exposed area of the building, but to the distribution of the mass of the building above the particular level being considered.

The forces due to the horizontal acceleration increase with the distance of the floor above the ground because of the "whipping effect" of the earthquake. Obviously, towers, water tanks, and penthouses on building roofs occupy precarious positions when an earthquake occurs.

Another factor to be considered in seismic design is the soil condition. Almost all of the structural damage and loss of life in the Loma Prieta earthquake occurred in areas that have soft clay soils. Apparently these soils amplified the motions of the underlying rock.[9]

---

[8] *American Society of Civil Engineers Minimum Design Loads for Buildings and Other Structures,* ASCE 7–88 (New York: ASCE), pp. 33–34.

[9] V. Fairweather, "The Next Earthquake," *Civil Engineering* (New York; ASCE, March 1990), pp. 54–57.

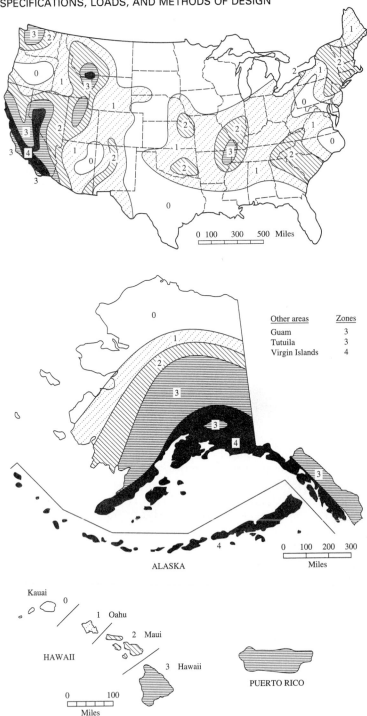

Figure 2-1 Map of seismic zones of the United States. The severity of risk ranges from a low in Zone 0 to a high in Zone 4. (From ASCE Standard 7-88, reprinted with the permission of the American Society of Civil Engineers.)

Using the LRFD Specification, the required design strengths for resisting earthquake loads are to be determined with the appropriate combination of factored loads provided in Sections 3 and 6 of the "Seismic Provisions for Structural Steel Buildings" given in Part 6 of the Manual. Detailed requirements are also presented there for the seismic design of members, connections, building frames and so on.

Of particular importance are the comments provided in the Commentary to the Seismic Provisions relating to drift. (*Drift* is defined as the movement or displacement of one story of a building with respect to the floor above or below.) Actually, the AISC Specification does not provide specific drift limits. It merely states that limits on story drift are to be used which are in accord with the governing code and shall not be so large as to impair the stability of the structure.

If structures are designed so that computed drifts during an earthquake of specified intensity are limited to certain maximum values, it will be necessary to provide added strength and stiffness to those structures. The result will be structures whose performance is substantially improved during earthquakes.

7. *Longitudinal loads.* Longitudinal loads are another type of load that needs to be considered in designing some structures. Stopping a train on a railroad bridge or a truck on a highway bridge causes longitudinal forces to be

Hungry Horse Dam and Reservoir, Rocky Mountains, in northwestern Montana. (Courtesy of the Montana Travel Promotion Division.)

applied. It is not difficult to imagine the tremendous longitudinal force developed when the driver of a 40-ton-trailer truck traveling 60 mph suddenly has to apply the brakes while crossing a highway bridge. There are other longitudinal load situations, such as ships running into docks, and the movement of traveling cranes that are supported by building frames.

8. *Other live loads.* Among the other types of live loads with which the structural designer will have to contend are *soil pressures* (such as the exertion of lateral earth pressures on walls or upward pressures on foundations); *hydrostatic pressures* (as water pressure on dams, inertia forces of large bodies of water during earthquakes, and uplift pressures on tanks and basement structures); *blast loads* (caused by explosions, sonic booms, and military weapons); *thermal forces* (due to changes in temperature resulting in structural deformations and resulting structural forces); and *centrifugal forces* (as those caused on curved bridges by trucks and trains or similar effects on roller coasters, etc.).

2-5 SELECTION OF DESIGN LOADS

To assist the designer in estimating the magnitudes of live loads with which he or she should proportion structures, various records have been assembled through the years in the form of building codes and specifications. These publications provide conservative estimates of live load magnitudes for various situations. One of the most widely used design load specifications for buildings is ASCE Standard 7. Some other commonly used specifications are:

1. For railroad bridges, American Railway Engineering Association (AREA).[10]
2. For highway bridges, American Association of State Highway and Transportation Officials (AASHTO).[11]
3. For buildings, the Uniform Building Code (UBC).[12]

These specifications will on many occasions clearly prescribe the loads for which structures are to be designed. Despite the availability of this information the designer's ingenuity and knowledge of the situation are often needed to predict what loads a particular structure will have to support in years to come. Over the past several decades insufficient estimates of future traffic loads by bridge designers have resulted in a great amount of replacement with wider and stronger structures.

---

[10]*Specifications for Steel Railway Bridges* [Washington, DC: American Railway Engineering Association (AREA), 1994].

[11]*AASHTO LRFD Bridge Design Specifications* [Washington, DC: American Association of State Highway and Transportation Officials (AASHTO), 1994].

[12]*Uniform Building Code* Whittier, Calif.: International Conference of Building Officials, (1994).

2-6 ELASTIC AND PLASTIC DESIGN METHODS DEFINED

Almost all existing steel structures were designed by the *elastic design* methods. The designer estimates the *working* or *service* loads, that is, the loads that the structure may feasibly have to support, and proportions the members on the basis of certain allowable stresses. These allowable stresses are usually some fraction of the specified minimum yield stress of the steel. Although the term "elastic design" is very commonly used to describe this method, the terms *allowable-stress design* or *working-stress design* are definitely more appropriate. Many of the provisions of the specifications for this method are actually based on plastic or ultimate-strength behavior and not on elastic behavior.

The ductility of steel has been shown to give it a reserve strength, and the realization of this fact is the theory behind *plastic design.* In this method the working loads are estimated and multiplied by certain load or overcapacity or safety factors and the members designed on the basis of collapse strengths. Another name for this method is *collapse design.*

The design profession has long been aware that the major portion of the stress-strain curve lies beyond the steel's elastic limit. Furthermore, tests through the years have made it clear that steels can resist stresses appreciably beyond their yield stresses, and that in cases of overload, statically indeterminate structures have the happy facility of spreading the load out due to the steel's ductility. On the basis of this information many plastic-design proposals have been made in recent decades. It is undoubtedly true that for certain types of structures, plastic design results in a more economical use of steel than does elastic design.

2-7 LOAD AND RESISTANCE FACTOR DESIGN

This book concerns a design method called *load and resistance factor design* (LRFD). However, because almost all structural steel design in the United States is today handled by the allowable stress design (ASD), the author feels that the student should have some familiarity with ASD. Rather than confusing readers by combining a discussion of ASD with LRFD, the author has presented an introduction to ASD in Appendix A. The material consists primarily of a series of example problems.

Despite the prevalence of ASD the design profession is gradually adopting the LRFD method. It seems that this trend will be accelerated in the next few years because of the various advantages of LRFD and because of the fact that nearly all of our college and university steel design classes are devoted almost exclusively to LRFD. This method includes many of the features of the design procedures commonly associated with ultimate design, plastic design, limit design, or collapse design.

Load and resistance factor design is based on a *limit states* philosophy. The term *limit state* is used to describe a condition at which a structure or some part of that structure ceases to perform its intended function. There are actually two categories of limit states, strength and serviceability.

Strength limit states are based on the safety or load-carrying capacity of structures and include plastic strengths, buckling, fracture, fatigue, overturning, and so on.

Serviceability limit states refer to the performance of structures under normal service loads and are concerned with the uses and/or occupancy of structures including such items as excessive deflections, slipping, vibrations, cracking and deterioration.

Not only must the structure be capable of supporting the design or ultimate loads, it must also be able to support the service or working loads in such a manner as to meet the requirements of the users or occupants of the structure. For instance, a high-rise building must be designed so that lateral deflections or drift are not excessive during ordinary storms; that is, the occupants must not become uncomfortable (frightened or seasick). As to the strength limit state, the frame will be designed so that it will safely resist the ultimate load occurring during that 50-year storm, although there may be some minor damage to the building and the occupants may suffer some discomfort.

The LRFD Specification concentrates on very specific requirements relating to the strength limit state and allows the designer some freedom of judgment in the serviceability area. This doesn't mean that the serviceability limit state is not significant, but by far the most important consideration (as in all structural specifications) is the life and property of the public. As a result public safety is not left to the judgment of the individual designer.

In LRFD the working or service loads (Q_i) are multiplied by certain load or safety factors (λ_i) almost always larger than 1.0 and the resulting "factored loads" used for designing the structure. The magnitudes of load factors vary depending on the type and combination of the loads and are discussed in detail in Section 2-8.

The structure is proportioned to have a design ultimate strength sufficient to support the factored loads. This strength is considered to equal the nominal or theoretical strength of the member (R_n) multiplied by a resistance or overcapacity factor (ϕ), which is normally less than 1.0, with which the designer attempts to account for the uncertainties in material strengths, dimensions, and workmanship. Furthermore these factors have been adjusted somewhat to ensure more uniform reliability in design as described in Section 2-11. Values of ϕ factors for different situations are discussed in detail in Section 2-9.

For a particular member the preceding information can be summarized as follows: (Sum of products of load effects and load factors) $\leq$ (resistance factor) (nominal resistance)

$$\sum \lambda_i Q_i \leq \phi R_n$$

The left-hand side of this expression refers to the load effects on the structure while the right-hand side refers to the resistance or capacity of the member.

In Canada a probability-based limit states design method has been in use for quite a few years for both hot-rolled and cold-formed steel structures. A uniform limit states steel design code is being adopted by nearly every European country. It is called Eurocode (EC) 3.[13] Similar codes have been prepared in Europe for reinforced concrete, timber, and masonry design, while much progress is being made in America toward similar codes.

At Washington University in St. Louis, Missouri, a research project was conducted on LRFD from 1969 through 1976 under the direction of T.V. Galambos and M. K. Ravindra. At the conclusion of the project, a document was published entitled "Proposed Criteria for Load and Resistance Factor Design of Steel Building Structures."[14, 15]

2-8 LOAD FACTORS

The purpose of load factors is to increase the loads to account for the uncertainties involved in estimating the magnitudes of dead or live loads. For instance, "How close in percent could you the reader estimate the worst wind or snow loads that will ever be applied to the building which you are now occupying?"

The value of the load factor used for dead loads is smaller than the one used for live loads because designers can estimate so much more accurately the magnitudes of dead loads than they can the magnitudes of live loads. In this regard the student will notice that loads which remain in place for long time periods will be less variable in magnitude, while those which are applied for brief periods, such as wind loads, will have larger variations. It is hoped that the LRFD design procedure will make the designer more conscious of load variations than he or she is with the allowable stress design method.

The LRFD Specification presents load factors and load combinations that were selected to be used with the recommended minimum loads given in ASCE 7-93 Standard.

The usual load combinations to be considered with the LRFD are given in their specification A4.1 by Equations A4-1 and A4-2. In these expressions the abbreviations used are D for dead loads, L for live loads due to gravity occupancy and movable equipment, L_r for roof live loads, S for snow loads, and R for initial rainwater or ice, not including a ponding contribution. The letter U represents the ultimate load.

$$U = 1.4D \qquad \text{(LRFD Equation A4-1)}$$

$$U = 1.2D + 1.6L + 0.5(L_r \text{ or } S \text{ or } R) \qquad \text{(LRFD Equation A4-2)}$$

[13] A. D. Weller, "An Introduction to EC 3." *The Structural Engineer, Journal of the Institution of Structural Engineers* 71, No.18 (London, September 21, 1993), pp. 326–331.

[14] M. K. Ravindra and T. V. Galambos, "Load and Resistance Factor Design for Steel," *ASCE Journal of Structural Division* 104, no. ST9 (September 1978), pp. 1337–1353.

[15] Research Report 45, C.E. Dept., Washington University, St. Louis, Mo., May 1976.

Impact loads will only be included in the second of these combinations (A4-2). Should wind (W) or earthquake (E) forces be involved, the following combinations need to be considered:

$$U = 1.2D + 1.6(L_r \text{ or } S \text{ or } R) \qquad \text{(LRFD Equation A4-3)}$$
$$+ (0.5L \text{ or } 0.8W)$$
$$U = 1.2D + 1.3W + 0.5L + 0.5(L_r \text{ or } S \text{ or } R) \quad \text{(LRFD Equation A4-4)}$$
$$U = 1.2D \pm 1.0E + 0.5L + 0.2S \qquad \text{(LRFD Equation A4-5)}$$

It is necessary to consider impact loading only in combination A4-3 of this group. There is a change in the value of the load factor for L for combinations A4-3, A4-4, and A4-5 for garages, public assembly areas, and all areas where the live load exceeds 100 psf. For such cases a load factor 1.0 is to be used and the load combinations become

$$U = 1.2D + 1.6(L_r \text{ or } S \text{ or } R) \qquad \text{(LRFD Equation A4-3')}$$
$$+ (1.0L \text{ or } 0.8W)$$
$$U = 1.2D + 1.3W + 1.0L + 0.5(L_r \text{ or } S \text{ or } R) \quad \text{(LRFD Equation A4-4')}$$
$$U = 1.2D \pm 1.0E + 1.0L \quad \text{ or } \quad 0.2S \qquad \text{(LRFD Equation A4-5')}$$

To account for possibilities of uplift, another load combination is given in the LRFD Specification. This condition is included to cover cases where tension forces develop owing to overturning moments. It will govern only for tall buildings where high lateral loads are present. In this combination the dead loads are reduced by 10 percent to take into account situations where they may have been overestimated.

The possibility that wind and earthquake forces may have either plus or minus signs need only be taken into account in this last equation A4-6. Thus in the preceding equations the signs used for W and E are the same as the signs used for the other items in those equations.

$$U = 0.9D \pm (1.3W \text{ or } 1.0E) \qquad \text{(LRFD Equation A4-6)}$$

The magnitudes of the loads (D, L, L_r, etc.) should be obtained from the governing building code or from ASCE 7-93. Wherever applicable the live loads used for design should be the reduced values specified for large floor areas, multistory buildings, and so on.

Examples 2-1 to 2-3 show the calculation of the factored loads using the applicable LRFD load combinations. The largest value obtained in each case is referred to as the *critical* or *governing load* and is to be used in design.

■ EXAMPLE 2-1

A floor system has W24 × 55 sections spaced 8 ft on-center supporting a floor dead load of 50 psf and a live load of 80 psf. Determine the governing load in lbs/ft which each beam must support.

Solution.   Noting that each foot of the beams must support itself (a dead load) plus $8 \times 1 = 8 \ ft^2$ of the building floor

$D = 55 + (8)(50) = 455$ lbs/ft

$L = (8)(80) = 640$ lbs/ft

Computing factored loads noting that as D and L are the only loads to be supported it is necessary to substitute only into LRFD equations A4-1 and A4-2.

$U = (1.4)(455) = 637$ lbs/ft   (A4-1)

$U = (1.2)(455) + (1.6)(640) = 1570$ lbs/ft $\leftarrow$ (A4-2)

Governing factored load $= 1570$ lbs/ft   ∎

■ EXAMPLE 2-2

A roof system with W16 $\times$ 40 sections spaced 9 ft on-center is to be used to support a dead load of 40 psf, a roof live or snow or rain load of 30 psf, and a wind load of 20 psf. Compute the governing factored load.

Solution.

$$D = 40 + (9)(40) = 400 \text{ lbs/ft}$$

$$L = 0$$

$$L_r \text{ or } S \text{ or } R = (9)(30) = 270 \text{ lbs/ft}$$

$$W = (9)(20) = 180 \text{ lbs/ft}$$

Substituting into the LRFD load combination expressions

$U = (1.4)(400) = 560$ lbs/ft   (A4-1)

$U = (1.2)(400) + 0 + (0.5)(270) = 615$ lbs/ft   (A4-2)

$U = (1.2)(400) + (1.6)(270) + (0.8)(180) = 1056$ lbs/ft   (A4-3)

$U = (1.2)(400) + (1.3)(180) + (.5)(270) = 849$ lbs/ft   (A4-4)

$U = (1.2)(400) + 0 + (0.2)(270) = 534$ lbs/ft   (A4-5)

$U = (0.9)(400) \pm (1.3)(180) = 594$ or 126 lbs/ft   (A4-6)

Governing factored loads $= 1056$ lbs/ft   ∎

■ EXAMPLE 2-3

The various axial loads for a building column have been computed according to the applicable building code with the following results: dead load $= 200$ k, load from roof $= 50$ k (roof live load), live load from floors (has been reduced as applicable for large floor area and multistory columns) = 250 k, wind = 80 k, and earthquake = 60 k. Determine the critical design load using the six LRFD load combinations.

Solution.

$$U = (1.4)(200) = 280 \text{ k} \hspace{3cm} \text{(A4-1)}$$
$$U = (1.2)(200) + (1.6)(250) + (0.5)(50) = 665 \text{ k} \leftarrow \hspace{0.8cm} \text{(A4-2)}$$
$$U = (1.2)(200) + (1.6)(50) + (0.5)(250) = 445 \text{ k} \hspace{1.2cm} \text{(A4-3(a))}$$
$$U = (1.2)(200) + (1.6)(50) + (0.8)(80) = 384 \text{ k} \hspace{1.3cm} \text{(A4-3(b))}$$
$$U = (1.2)(200) + (1.3)(80) + (0.5)(250) + (0.5)(50) = 494 \text{ k} \hspace{0.3cm} \text{(A4-4)}$$
$$U = (1.2)(200) \pm (1.0)(60) + (0.5)(250) = 425 \text{ or } 305 \text{ k} \hspace{0.6cm} \text{(A4-5)}$$
$$U = (0.9)(200) \pm (1.3)(80) = 284 \text{ or } 76 \text{ k} \hspace{2.1cm} \text{(A4-6(a))}$$
$$U = (0.9)(200) \pm (1.0)(60) = 240 \text{ or } 120 \text{ k} \hspace{1.8cm} \text{(A4-6(b))}$$

The critical factored load combination or design strength required for this column is 665 k as determined by LRFD Formula A4-2. It will be noted that the results of Formulas A4-5 and A4-6 do not indicate an uplift problem.   ∎

2-9 RESISTANCE FACTORS

To accurately estimate the ultimate strength of a structure it is considered necessary to take into account the uncertainties in material strengths, dimensions, and workmanship. With a resistance factor the designer attempts to show that the strength of a member cannot be computed exactly because of imperfections in analysis theory (remember the imperfect assumptions you made to analyze trusses and other members), variations in material properties, and imperfect dimensions of members.

This is done by multiplying the theoretical ultimate strength (called the *nominal strength* here) of each member by a resistance or overcapacity factor ϕ, which is almost always less than 1.0. These values are 0.85 for columns, 0.75 or 0.90 for tension members, 0.90 for bending or shear in beams, and so on.

Typical resistance factors from the LRFD Specification are shown in Table 2-2. Some terms used in this table (such as groove and fillet welds, slip-resistant bolts, and web crippling) will be defined in later chapters of this text. The magnitudes of the resistance factors given in the LRFD Specification are based on research recommendations from investigators at Washington University in St. Louis.

TABLE 2-2   TYPICAL RESISTANCE FACTORS

Resistance or ϕ factors	Situations
1.00	Bearing on the projected areas of pins, web yielding under concentrated loads, slip-resistant bolt shear values
0.90	Beams in bending and shear, fillet welds with stress parallel to weld axis, groove welds base metal, yielding on gross section of tension members
0.85	Columns, web crippling, edge distance and bearing capacity at holes
0.80	Shear on effective area of full-penetration groove welds, tension normal to effective area of partial-penetration groove welds
0.75	Bolts in tension, plug, or slot welds, fracture in the net section of tension members
0.65	Bearing on bolts (other than A307)
0.60	Bearing on concrete foundations

2-10 DISCUSSION OF SIZES OF LOAD AND RESISTANCE FACTORS

Students may feel that it is foolish and uneconomical to design structures with such large load factors and such small resistance factors. As the years go by, however, they will learn that these factors are subject to so many uncertainties that they may spend sleepless nights wondering if those they have used are sufficient (and they may join other designers in calling them "factors of ignorance"). Some of the uncertainties affecting these factors are:

1. Material strengths may initially vary appreciably from their assumed values and they will vary more with time due to creep, corrosion, and fatigue.

2. The methods of analysis are often subject to appreciable errors.

3. The so-called vagaries of nature or acts of God (hurricanes, earthquakes, etc.) cause conditions difficult to predict.

4. The stresses produced during fabrication and erection are often severe. Laborers in shop and field seem to treat steel shapes with reckless abandon. They drop them. They ram them. They force the members into position to line up the bolt holes. In fact, the stresses during fabrication and erection may exceed those which occur after the structure is completed. The floors for the rooms of apartment houses and office buildings are probably designed for service live loads varying from 40 to 80 psf (pounds per square foot). During the erection of such buildings the contractor may have 10 ft of bricks or concrete blocks or other construction materials or equipment piled up on some of the floors, causing loads of several hundred pounds per square foot. This discussion is not intended to criticize the practice (not that it is a good one), but rather to make the student aware of the things that happen during construction. (It is probable that the majority of steel structures are overloaded somewhere during construction, but hardly any of them fail. On many of these occasions the ductility of the steel has surely saved the day.)

5. There are technological changes that affect the magnitude of live loads. The constantly increasing traffic loads applied to bridges through the years is one illustration. The wind also seems to blow harder as the years go by, or at least building codes keep raising the minimum design wind pressures as more is learned about the subject.

6. Although the dead loads of a structure can usually be estimated quite closely, the estimate of the live loads is more inaccurate. This is particularly true in estimating the worst possible combination of live loads occurring at any one time.

7. Other uncertainties are the presence of residual stresses and stress concentrations, variations in dimensions of member cross sections, etc.

2-11 RELIABILITY AND THE LRFD SPECIFICATION

Probably fewer than 1 in 10 undergraduate engineering students take a course in statistics. Thus these statistically untrained students are often a little dismayed when they run into a statistically based subject such as this one. In the pages which follow these students will find that such worrying is unnecessary, as the LRFD theory and calculations are usually rather simple to follow. Furthermore, they may actually gain a little statistical knowledge along the way and thereby better understand the theory behind the development of LRFD.

The word *reliability* as used in this textbook refers to the estimated percentage of times that the strength of a structure will equal or exceed the maximum loading applied to that structure during its estimated life (say 50 years).

In this section the author briefly describes how:

1. LRFD investigators developed a procedure for estimating the reliability of given designs.
2. They set what to them were desirable reliability percentages for different situations.
3. They were able to adjust resistance or ϕ factors so steel designers are able to obtain the reliability percentages established in Item 2 above.

Before proceeding with this discussion a few comments are presented concerning the word *failure* as it is used in this discussion of reliability. Let us assume that a designer states that his or her designs are 99.7 percent reliable (and this is the approximate value obtained with most LRFD designs). This means that if this person were to design 1000 different structures, 3 of them would probably be overloaded at some time during their estimated 50-year lives and would fail. The reader probably and quite reasonably thinks this to be an unacceptably high rate of failure.

To the author, 99.7 percent reliability doesn't mean that 3 of 1000 structures are going to fall flat on the ground. To him it means that those structures
at some time will be loaded into the plastic range and perhaps the strain-
hardening range. As a result deformations may be quite large during the over-
loading and some slight damage may occur. It is not anticipated that any of these
structures will completely collapse. (The reader who is unfamiliar with statistics
might say he or she wants 100 percent reliability in design, but this is statistically
an impossible goal, as we will see in the paragraphs to follow.)

For this discussion it is assumed that we make a study of the reliability of
a large number of steel structures designed at various times and with different
past editions of the AISC Specification. To do this we will compute the resistance or strength, R, of each structure as well as the maximum loading, Q,
expected during the life of the structure. The structure will be deemed to be safe
if $R \geq Q$.

The actual values of R and Q are random variables, and it is therefore
impossible to say with 100 percent certainty that R is always equal to or larger
than Q for a particular structure. No matter how carefully a structure is designed
and constructed there will always be some little chance that Q will be greater
than R or that the strength limit state will be exceeded. The goal of the preparers
of the LRFD Specification was to keep this chance to a very small and consistent
percentage.

Thus the magnitudes of both resistances and loads are uncertain. If we were
to plot a curve of R/Q values for a large number of structures the result would
be a typical bell-shaped probability curve with mean values R_m and Q_m and a
standard deviation. If at any location $R < Q$ we will have exceeded the strength
limit state.

For convenience such a curve is plotted logarithmically and shown in Fig.
2-2. It is to be remembered that the logarithm of 1.0 is 0 and thus if $\ln R/Q < 0$
the strength limit state has been exceeded. Such a situation is represented by the

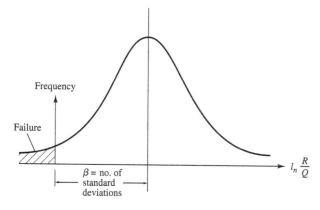

Figure 2-2 Definition of reliability index β.

shaded area on the diagram. Another way of expressing this fact is to say that the larger the number of standard deviations from the mean values to the shaded area the greater the reliability. In the figure the number of standard deviations is represented by β and is called the *reliability index*.

Even though the probable values of R and Q are not known very well, a formula has been developed with which values of β can be reasonably calculated.[16,17,18] The formula follows:

$$\beta = \frac{\ln(R_m/Q_m)}{\sqrt{V_R^2 + V_Q^2}}$$

In this expression R_m and Q_m are respectively the mean resistance and load effects while V_R and V_Q respectively the corresponding coefficients of variation.

As a result of the preceding work it is now possible to design a particular element in accordance with a certain edition of the AISC Specification and with the appropriate statistical information compute the value of β for the design. This process is called *calibration*.

The results of our study of the designs of these steel structures will show that the percentage of structures for which the design strengths equal or exceed the worst anticipated loading will vary as we examine designs made in accordance with the requirements of different editions of the AISC Specification. Furthermore our calculations will show that this reliability will vary for the designs of different types of members (such as columns and beams) made with the same edition of the AISC Specification.

Based upon the calculations of reliability described here the investigators decided to use consistent β values in this new specification. These were the values they selected:

1. $\beta = 3.00$ for members subject to gravity loading.
2. $\beta = 4.50$ for connections. (This value reflects the usual practice that connections should be stronger than the members they connect).
3. $\beta = 2.5$ for members subject to gravity plus wind loadings. (This value reflects an old practice that safety factors do not have to be as large for cases where lateral loads, which are of shorter duration, are involved.)
4. $\beta = 1.75$ for members subject to gravity loads plus earthquake loads.

---

[16] C. A. Cornell, "A Probability-Based Structural Code," *Journal, American Concrete Institute,* vol. 66, no. 12 (December 1969).

[17] B. Ellingwood, T. V. Galambos, J. G. MacGregor, and C. A. Cornell, "Development of a Probability-Based Load Criterion for American National Standard A58," National Bureau of Standards Special Publication 577 (June 1980).

[18] M. K. Ravindra and T. V. Galambos, "Load and Resistance Factor Design for Steel," *Journal of the Structural Division,* vol. 104, no ST9, (September 1978), pp. 1337–1353.

Then the values of ϕ for the various parts of the specification were adjusted so the β values shown here would be obtained in design. In effect this causes most LRFD designs to be almost identical with those obtained by the allowable stress method when the live to dead load ratio is 3.

2-12 ADVANTAGES OF LRFD

The average person looking at this material will ask "Will LRFD save money compared with allowable stress design (ASD)?" The answer is that it probably will, particularly if the live loads are small compared with the dead loads.

It should be noted, however, that the AISC has introduced LRFD not for the specific purpose of obtaining immediate economic advantages but because it helps provide a more uniform reliability for all steel structures whatever the loads, and it is written in a form that facilitates the introduction of the advances in knowledge that will occur through the years in structural steel design.

Despite this stated objective of the AISC, the author would like to continue for a little while with the topic of possible steel weight savings with LRFD. In allowable stress design the same safety factor was used for dead loads and live loads, whereas in LRFD a much smaller safety or load factor is used for dead loads (because they generally can be determined so much more accurately than can live loads). As a result of this fact a weight of steel comparison for ASD and LRFD designs will necessarily depend on the ratio of the live loads to dead loads.

For the usual building the live load to dead load ratio varies from approximately 0.25 to as high as 4.0 or even a little higher for some very light structures. Low-rise steel buildings and preengineered buildings generally will fall in the upper range of these ratios. In allowable stress design we use the same safety factors for dead and live loads regardless of their ratio to each other. Thus with ASD heavier members will result and the factor of safety will increase as the live load to dead load ratio decreases.

It can be shown that for the lower range of L to D values, that is, less than 3, there can be steel weight savings with LRFD perhaps as much as one-sixth for tension members and columns and perhaps as much as one-tenth for beams. On the other hand, if we have very high L and D ratios there is almost no increase in steel weight with LRFD compared with ASD.[19]

PROBLEMS

2-1. Structural steel beams are to be placed 12 ft on-center under a reinforced concrete floor slab. If they are to support a dead working load $D = 90$ psf of floor area and a live working load $L = 80$ psf of floor area, determine the factored uniform load per foot which each beam must support. (*Ans.* 2832 lb/ft)

---

[19] J. A. Edinger, "Introduction to the Proposed AISC Load and Resistance Factor Design Specification," *Engineering Journal*, AISC, 21 no. 1 (first quarter, 1984), pp. 62–65.

2-2. For the design of a roof the estimated working or service loads are: dead load D = 20 psf, snow S = 30 psf, and wind W = 20 psf. Calculate the factored loads in psf to be used for design.

2-3. A column is to support the following service or working loads: D = 50 k axial compression dead load, L = 40 k floor axial compression live load, and W = 30 k axial compression or tension wind load. Compute the required design strength of the member. (*Ans.* 124 k)

2-4. Determine the required design strength for a column for which dead load D = 80 k, floor live load L = 60 k, roof live load L_r = 20 k, and wind W = 55 k.

2-5. The estimated service or working axial loads and bending moments for a beam-column are as follows: D = 80 k, L = 10 k, M_D = 18 ft-k and M_L = 8 ft-k. Compute the axial load and moment values which must be used in design. (*Ans.* axial load = 112 k, moment = 34.4 ft-k)

2-6. The following axial service or working loads have been estimated for the design of a particular column: dead load D = 60 k, floor live load L = 45 k, roof live load L_r = 15 k, and wind load W = 40 k. Determine the required design strength for the member.

2-7. The working or service loads have been determined for a particular building column as required by the applicable building code with the following results: D = 200 k, L = 150 k, L_r or S or R from roof = 25 k, W = 100 k and E = 40 k. Determine the critical factored design load for the column. (*Ans.* 492.5 k)

Chapter 3

Analysis of Tension Members

3-1 INTRODUCTION

Tension members are found in bridge and roof trusses, towers, bracing systems, and in situations where they are used as tie rods. The selection of a section to be used as a tension member is one of the simplest problems encountered in design. As there is no danger of buckling, the designer needs only to compute the factored force to be carried by the member and divide that force by a design stress to determine the effective cross-sectional area required. Then it is necessary to select a steel section that provides the required area. Though these introductory calculations for tension members are quite simple, they do serve the important tasks of getting students started with design ideas and getting their "feet wet" regarding the massive LRFD Manual.

One of the simplest forms of tension members is the circular rod, but there is some difficulty in connecting it to many structures. The rod has been used frequently in the past, but has only occasional uses today in bracing systems, light trusses, and in timber construction. One important reason rods are not popular with designers is that they have been used improperly so often in the past that they have a bad name; however, if designed and installed correctly, they are satisfactory for many situations.

The average size rod has very little stiffness and may quite easily sag under its own weight, injuring the appearance of the structure. The threaded rods formerly used in bridges often worked loose and rattled. Another disadvantage of rods is the difficulty of fabricating them with the exact lengths required and the consequent difficulties of installation.

When rods are used in wind bracing it is a good practice to produce initial tension in them, as this will tighten up the structure and reduce rattling and swaying. To obtain initial tension the members may be detailed shorter than their required lengths, a method that gives the steel fabricator very little trouble. A common rule of thumb used is to detail the rods about $\frac{1}{16}$ in. short for each 20 ft. of length. (Approximate stress $f = \epsilon E = [\frac{1}{16}/(12)(20)](29 \times 10^6) = 7550$ psi.) Another very satisfactory method involves tightening the rods with some sort of sleeve nut or turnbuckle. Part 8 of the LRFD Manual provides detailed information for these devices.

The preceding discussion on rods should illustrate why rolled shapes such as angles have supplanted rods for most applications. In the early days of steel structures, tension members consisted of rods, bars, and perhaps cables. Today, although the use of cables is increasing for suspended-roof structures, tension members usually consist of single angles, double angles, tees, channels, W sections, or sections built up from plates or rolled shapes. These members look better than the old ones, are stiffer, and are easier to connect. Another type of tension section often used is the welded tension plate or flat bar, which is very satisfactory for use in transmission towers, signs, foot bridges, and similar structures.

The tension members of steel roof trusses may consist of single angles as small as $2\frac{1}{2} \times 2 \times \frac{1}{4}$ for minor members. A more satisfactory member is made from two angles placed back to back with sufficient space between them to permit the insertion of plates (called gusset plates) for connection purposes. Where steel sections are used back-to-back in this manner, they should be connected every 4 or 5 ft to prevent rattling, particularly in bridge trusses. Single angles and double angles are probably the most common types of tension members in use. Structural tees make very satisfactory chord members for welded trusses because web members can conveniently be connected to them.

For bridges and large roof trusses tension members may consist of channels, W or S shapes, or even sections built up from some combination of angles, channels, and plates. Single channels are frequently used, as they have little eccentricity and are conveniently connected. Although, for the same weight, W sections are stiffer than S sections, they may have a connection disadvantage in their varying depths. For instance, the W12 × 79, W12 × 72, and W12 × 65 all have slightly different depths (12.38 in., 12.25 in., and 12.12 in., respectively), while the S sections of a certain nominal size all have the same depths. For instance, the S12 × 50, the S12 × 40.8, and the S12 × 35 all have 12.00-in. depths.

Although single structural shapes are a little more economical than built-up sections, the latter are occasionally used when the designer is unable to obtain sufficient area or rigidity from single shapes. Where built-up sections are used it is important to remember that field connections will have to be made and paint applied; therefore, sufficient space must be available to accomplish these things.

Members consisting of more than one section need to be tied together. Tie plates (also called tie bars) located at various intervals or perforated cover plates serve to hold the various pieces in their correct positions. These plates serve to correct any unequal distribution of loads between the various parts. They also

Erection of Federal Reserve bank building, Minneapolis, Minn. (Courtesy of IR Construction Products Co.)

keep the slenderness ratios (to be discussed) of the individual parts within limitations and they may permit easier handling of the built-up members. Long individual members such as angles may be inconvenient to handle due to flexibility, but when four angles are laced together into one member, as shown in Fig. 3-1, the member has considerable stiffness. None of the intermittent tie

Figure 3-1 Types of tension members.

Federal Reserve bank building, Minneapolis, Minn. (Courtesy of IR Construction Products Co.)

plates may be considered to increase the effective areas of the sections. As they do not theoretically carry portions of the force in the main sections, their sizes are usually governed by specifications and perhaps by some judgment on the designer's part. Perforated cover plates (see Fig. 6-9) are an exception to this rule, as part of their areas can be considered as being effective in resisting axial load.

A few of the various types of tension members in general use are illustrated in Fig. 3-1. In this figure the dotted lines represent the intermittent tie plates or bars used to connect the shapes.

Steel cables are made with special steel alloy wire ropes which are cold-drawn to the desired diameter. The resulting wires with strengths of about 200,000 to 250,000 psi can be economically used for suspension bridges, cable supported roofs, ski lifts, and other similar applications.

Normally, to select a cable tension member the designer uses a manufacturer's catalog. From the catalog the yield stress of the steel and the cable size required for the design force are determined. It is also possible to select clevises or other devices to use for connectors at the cable ends.

3-2 DESIGN STRENGTH OF TENSION MEMBERS

A ductile steel member without holes subject to a tensile load can resist without fracture a load larger than its gross cross-sectional area times its yield stress because of strain hardening. However, a tension member loaded until strain hardening is reached will lengthen a great deal before fracture, a fact that will in all probability take away the member's usefulness and may even cause failure of the structural system of which the member is a part.

If, on the other hand, we have a tension member with bolt holes, it can possibly fail by fracture at the net section through the holes. This failure load may very well be smaller than the load required to yield the gross section away from the holes. It is to be realized that the portion of the member where we have a reduced cross-sectional area due to the presence of holes normally is very short compared with the total length of the member. Though the strain-hardening situation is quickly reached at the net section portion of the member, yielding there may not really be a limit state of significance because the overall change in length of the member due to yielding in this small part of the member length may be negligible.

As a result of the preceding information the LRFD Specification (D1) states that the design strength of a tension member, $\phi_t P_n$, is to be the smaller of the values obtained by substituting into the following two expressions:

For the limit state of yielding in the gross section (which is intended to prevent excessive elongation of the member)

$$P_n = F_y A_g \qquad\qquad \text{(LRFD Equation D1-1)}$$
$$P_u = \phi_t F_y A_g \quad \text{with} \quad \phi_t = 0.90$$

For fracture in the net section where bolt or rivet holes are present

$$P_n = F_u A_e \qquad\qquad \text{(LRFD Equation D1-2)}$$
$$P_u = \phi_t F_u A_e \quad \text{with } \phi_t = 0.75$$

In the preceding expression F_u is the specified minimum tensile stress and A_e is the effective net area that can be assumed to resist tension at the section through the holes. This area may be somewhat smaller than the actual net area, A_n, because of stress concentrations and other factors that are discussed in Section 3-5 of this chapter. (See Table 1-1 in Manual for values of F_y and F_u.)

The design strengths presented here are not applicable to threaded steel rods or to members with pin holes (as in eyebars). These situations are discussed in Sections 4-3 and 4-4.

It is not likely that stress fluctuations will be a problem in the average building frame because the changes in load in such structures usually occur only occasionally and produce relatively minor stress variations. Full design wind or earthquake loads occur so infrequently that they are not considered in fatigue design. Should there, however, be frequent variations or even reversals in stress the matter of fatigue must be considered. This subject is presented in Section 4-5.

3-3 NET AREAS

The presence of a hole obviously increases the unit stress in a tension member, even if the hole is occupied by a bolt. (When fully tightened high-strength bolts are used there may be some disagreement with this statement under certain conditions.) There is less area of steel to which the load can be distributed and there will be some concentration of stress along the edges of the hole.

Tension is assumed to be uniformly distributed over the net section of a tension member, although photoelastic studies show there is a decided increase in stress intensity around the edges of holes, sometimes equaling several times the stresses if the holes were not present. For ductile materials, however, a uniform stress distribution assumption is reasonable when the material is loaded beyond its yield stress. Should the fibers around the holes be stressed to their yield stress they will yield without further stress increase, with the result that there is a redistribution or balancing of stresses. At ultimate load it is reasonable to assume a uniform stress distribution. The importance of ductility on the strength of bolted tension members has been clearly demonstrated in tests. Tension members (with bolt holes) made from ductile steels have proved to be as much as one-fifth to one-sixth stronger than similar members made from brittle steels with the same strengths. We have already shown in Chapter 1 that it is possible for steel to lose its ductility and become subject to brittle fracture. Such a condition can be created by fatigue-type loads and by very low temperatures.

This initial discussion is applicable only for tension members subjected to relatively static loading. Should tension members be designed for structures subjected to fatigue-type loadings, considerable effort should be made to minimize the items causing stress concentrations such as points of sudden change of cross section and sharp corners. In addition, as described in Section 4-5, the members may have to be enlarged.

The term "net cross-sectional area" or simply "net area" refers to the gross cross-sectional area of a member minus any holes, notches, or other indentations. In considering the area of such items as these, it is important to realize that it is usually necessary to subtract an area a little larger than the actual hole. For instance, in fabricating structural steel which is to be connected with bolts, the holes are usually punched $\frac{1}{16}$ in. larger than the diameter of the bolt. Furthermore, the punching of the hole is assumed to damage or even destroy $\frac{1}{16}$ in. (1.6 mm) more of the surrounding metal; therefore, the area of the holes subtracted is $\frac{1}{8}$ in. (3 mm) larger than the diameter of the bolt. The area of the holes subtracted is rectangular and equals the diameter of the hole times the thickness of the metal. (Should the holes be slotted as described in Chapter 12, the usual practice is to add $\frac{1}{16}$ in. to the actual width of the holes.)

For steel much thicker than bolt diameters it is difficult to punch out the holes to the full sizes required without excessive deformation of the surrounding material. These holes may be subpunched (with diameters $\frac{3}{16}$ in. undersized) and then reamed out to full size after the pieces are assembled. Very little material is damaged by this quite expensive process, as the holes are even and smooth and it is considered unnecessary to subtract the $\frac{1}{16}$ in. for damage to the sides. Sometimes when very thick pieces are being connected the holes may be drilled to the diameter of the bolts or rivets plus $\frac{1}{32}$ in. This is a very expensive process, however, and should be avoided if possible.

It may be necessary to have an even greater latitude in meeting dimensional tolerances during erection and for high-strength bolts larger than $\frac{5}{8}$ in. in diameter. For such a situation, holes larger than the standard ones may be used without reducing the performance of the connections. These oversized holes can be short-slotted or long-slotted as described in Section 12-9.

Example 3-1 illustrates the calculations necessary for determining the net area of a plate type of tension member.

■ EXAMPLE 3-1

Determine the net area of the $\frac{3}{8} \times 8$-in. plate shown in Fig. 3-2. The plate is connected at its end with two lines of $\frac{3}{4}$-in. bolts.

Figure 3-2

Solution.

$$\text{Net area} = A_n = \left(\frac{3}{8}\right)(8) - (2)\left(\frac{3}{4} + \frac{1}{8}\right)\left(\frac{3}{8}\right) = 2.34 \text{ in.}^2 \ (1510 \text{ mm}^2) \ ■$$

The connections of tension members should be arranged so that no eccentricity is present. (An exception to this rule is permitted by the LRFD Specification for certain bolted and welded connections as described in Chapter 14.) If this arrangement is possible the stress is assumed to be spread uniformly across the net section of a member. Should the connections have eccentricities, moments will be produced that will cause additional stresses in the vicinity of the connection. Unfortunately it is often quite difficult to arrange connections without eccentricity. Although specifications cover some situations, the designer may have to give consideration to eccentricities in some cases by making special estimates.

The lines of action of truss members meeting at a joint are assumed to coincide. Should they not coincide, ecentricity is present and secondary stresses are the result. The centers of gravity (c.g.s) of truss members are assumed to

Two International Place, Boston, Mass. (Courtesy of Owen Steel Company, Inc.)

coincide with the lines of action of their respective forces. No problem is present in a symmetrical member as its center of gravity is at its center line, but for unsymmetrical members the problem is a little more difficult. For these members the center line is not the center of gravity, but the usual practice is to arrange the members at a joint so their gage lines coincide. If a member has more than one gage line, the one closest to the actual center of gravity of the member is used in detailing. Figure 3-3 shows a truss joint in which the c.g.s coincide.

Figure 3-3 Lining up c.g.s. of members.

3-4 EFFECT OF STAGGERED HOLES

Should there be more than one row of bolt holes in a member it is often desirable to stagger them in order to provide as large a net area as possible at any one section to resist the load. In the preceding paragraphs tensile members have been assumed to fail transversely as along line *AB* in either Fig. 3-4(a) or 3-4(b). Figure 3-4(c) shows a member in which a failure other than a transverse one is possible. The holes are staggered, and failure along section *ABCD* is possible unless the holes are a large distance apart.

To determine the critical net area in Fig. 3-4(c), it might seem logical to compute the area of a section transverse to the member (as *AE*) less the area of one hole and then the area along section *ABCD* less two holes. The smallest value obtained along these sections would be the critical value. This method is at fault, however. Along the diagonal line from *B* to *C* there is a combination of direct stress and shear and a somewhat smaller area should be used. The strength of the member along section *ABCD* is obviously somewhere between the strength obtained by using a net area computed by subtracting one hole from the trans- verse cross-sectional area and the value obtained by subtracting two holes from section *ABCD*.

Tests on joints show that little is gained by using complicated theoretical formulas to consider the staggered-hole situation, and the problem is usually handled with an empirical equation. The LRFD Specification (B2) and other specifications use a very simple method for computing the net width of a tension

Figure 3-4 Possible failure sections in plates.

TABLE 3-1   USUAL GAGES FOR ANGLES, IN INCHES

Leg	8	7	6	5	4	3½	3	2½	2	1¾	1½	1⅜	1¼	1
g	4½	4	3½	3	2½	2	1¾	1⅜	1⅛	1	⅞	⅞	¾	⅝
g_1	3	2½	2¼	2										
g_2	3	3	2½	1¾										

member along a zigzag section.[1] The method is to take the gross width of the member regardless of the line along which failure might occur, subtract the diameter of the holes along the zigzag section being considered, and add for each inclined line the quantity given by the expression $s^2/4g$.

In this expression s is the longitudinal spacing (or pitch) of any two holes and g is the transverse spacing (or gage) of the same holes. The values of s and g are shown in Fig. 3-4(c). There may be several paths, any one of which may be critical at a particular joint. Each possibility should be considered and the one giving the least value should be used. The smallest net width obtained is multiplied by the plate thickness to give the net area, A_n. Example 3-2 illustrates the method of computing the critical net area of a section which has three lines of bolts. (For angles, the gage for holes in opposite legs is considered to be the sum of the gages from the back of the angle minus the thickness of the angle.)

Holes for bolts and rivets are normally punched in steel angles at certain standard locations. These locations or gages are dependent on the angle-leg widths and on the number of lines of holes. Table 3-1, which is taken from Part 9 of the LRFD Manual, shows these gages. It is unwise for the designer to require different gages from those given in the table unless unusual situations are present, because of the appreciably higher fabrication costs that will result.

■ EXAMPLE 3-2

Determine the critical net area of the ½-in.-thick plate shown in Fig. 3-5 using the LRFD Specification (Section B2). The holes are punched for ¾-in. bolts.

Figure 3-5

---

[1] V. H. Cochrane, "Rules for Riveted Hole Deductions in Tension Members," *Engineering News-Record* (New York, November 16, 1922), pp. 847–848.

Solution.   The critical section could possibly be *ABCD*, *ABCEF*, or *ABEF*. Hole diameters to be subtracted are $\frac{3}{4} + \frac{1}{8} = \frac{7}{8}$ in. The net widths for each case are as follows:

$$ABCD = 11 - (2)\left(\frac{7}{8}\right) = 9.25 \text{ in.}$$

$$ABCEF = 11 - (3)\left(\frac{7}{8}\right) + \frac{(3)^2}{(4)(3)} = 9.125 \text{ in.}   \text{(controls)}$$

$$ABEF = 11 - (2)\left(\frac{7}{8}\right) + \frac{(3)^2}{(4)(6)} = 9.625 \text{ in.}$$

The reader should note it is a waste of time to check route *ABEF* for this plate. Two holes need to be subtracted for routes *ABCD* and *ABEF*. As *ABCD* is a shorter route it obviously controls over *ABEF*.)

$$A_n = (9.125)\left(\frac{1}{2}\right) = 4.56 \text{ in.}^2 \qquad \blacksquare \quad Ans.$$

The problem of determining the minimum pitch of staggered bolts such that no more than a certain number of holes need be subtracted to determine the net section is handled in Example 3-3.

■ **EXAMPLE 3-3**
For the two lines of bolt holes shown in Fig. 3-6, determine the pitch which will give a net area *DEFG* equal to the one along *ABC*. The problem may also be stated as follows: determine the pitch that will give a net area equal to the gross area less one bolt hole. The holes are punched for $\frac{3}{4}$-in. bolts.

Figure 3-6

Solution.

$$ABC = 6 - (1)\left(\frac{7}{8}\right) = 5.125 \text{ in.}$$

$$DEFG = 6 - (2)\left(\frac{7}{8}\right) + \frac{s^2}{(4)(2)} = 4.25 + \frac{s^2}{8}$$

$$ABC = DEFG$$

$$5.125 = 4.25 + \frac{s^2}{8}$$

$$s = 2.65 \text{ in.} \qquad\qquad\qquad\qquad\qquad\qquad ■ \quad Ans.$$

The $s^2/4g$ rule is merely an approximation or simplification of the complex stress variations which occur in members with staggered arrangements of bolts. Steel specifications can only provide minimum standards and designers will have to logically apply such information to complicated situations which the specifications could not cover in their attempts at brevity and simplicity. The next few paragraphs present a discussion and numerical examples of the $s^2/4g$ rule applied to situations not specifically addressed in the LRFD Specification.

The LRFD Specification does not include a method to be used for determining the net widths of sections other than plates and angles. For channels, W sections, S sections, and others the web and flange thicknesses are not the same. As a result it is necessary to work with net areas rather than net widths. If the holes are placed in straight lines across such a member the net area can simply be obtained by subtracting the cross-sectional areas of the holes from the gross area of the member. If the holes are staggered it is necessary to multiply the $s^2/4g$ values by the applicable thickness to change it to an area. Such a procedure is illustrated for a W section in Example 3-4 where bolts pass through the web only.

■ EXAMPLE 3-4
Determine the net area of the W12 × 16 ($A_g = 4.71$ in.^{2}) shown in Fig. 3-7 assuming the holes are for 1-in. bolts.

Figure 3-7

Solution.   Net areas:

$$ABDE = 4.71 - (2)\left(1\frac{1}{8}\right)(0.220) = 4.21 \text{ in.}^2$$

$$ABCDE = 4.71 - (3)\left(1\frac{1}{8}\right)(0.220) + (2)\frac{(2)^2}{(4)(3)}(0.220) = 4.11 \text{ in.}^2 \quad \blacksquare$$

If the zigzag line goes from a web hole to flange hole the thickness changes at the junction of the flange and web. In Example 3-5 the author has computed the net area of a channel that has bolt holes staggered in its flanges and web. The channel is assumed to be flattened out into a single plate as shown in parts (b) and (c) of Fig. 3-8. The net area along route *ABCDEF* is determined by taking the area of the channel minus the area of the holes along the route in the flanges and web plus the $s^2/4g$ values for each zigzag line times the appropriate thickness. For line *CD*, $s^2/4g$ has been multiplied by the thickness of the web. *For lines BC and DE (which run from holes in the web to holes in the flange) an approximate procedure has been used in which the $s^2/4g$ values have been multiplied by the average of the web and flange thicknesses.*

■ **EXAMPLE 3-5**
Determine the net area along route *ABCDEF* for the C15 × 33.9 ($A_g = 9.96$ in.^{2}) shown in Fig. 3-8. Holes are for $\frac{3}{4}$-in. bolts.

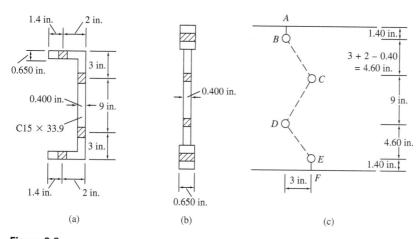

(a)   (b)   (c)

Figure 3-8

Solution.

$$\text{Approximate net } A \text{ along } ABCDEF = 9.96 - (2)\left(\frac{7}{8}\right)(0.650)$$
$$- (2)\left(\frac{7}{8}\right)(0.400)$$
$$+ \frac{(3)^2}{(4)(9)}(0.4000)$$
$$+ (2)\frac{(3)^2}{(4)(4.60)}\left(\frac{0.650 + 0.400}{2}\right)$$
$$= 8.736 \text{ in.}^2 \quad \blacksquare$$

3-5 EFFECTIVE NET AREAS

When a member other than a flat plate or bar is loaded in axial tension until failure occurs across its net section its actual tensile failure stress will probably be less than the coupon tensile strength of the steel *unless all of the various elements which make up the section are connected so stress is transferred uniformly across the section.*

If the forces are not transferred uniformly across a member cross section there will be a transition region of uneven stress running from the connection out into the member for some distance. This is the situation shown in Fig. 3-9(a) where a single angle tension member is connected by one leg only. At the connection more of the load is carried by the connected leg, and it takes the transition distance shown in part (b) of the figure for the stress to spread uniformly across the whole angle.

In the transition region the stress in the connected part of the member may very well exceed F_y and go into the strain-hardening range. Unless the load is reduced the member may fracture prematurely. The further we move out from the connection the more uniform the stress becomes. In the transition region the shear transfer has "lagged" and the phenomenon is referred to as *shear lag.*

In such a situation the flow of tensile stress between the full member cross section and the smaller connected cross section is not 100 percent effective. As

(a) Angle connected by one leg only   (b) Stress in transition range $> F_y$

Figure 3-9 Shear lag.

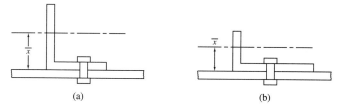

Figure 3-10 Reducing shear lag by reducing length of unconnected leg and and thus $\bar{x}$.

a result the LRFD Specification (B3) states that the effective net area, A_e, of such a member is to be determined by multiplying an area A (which is the net area or the gross area or the directly connected area as described in the next few pages) by a reduction factor U. The use of a factor such as U accounts for the nonuniform stress distribution in a simple manner. An explanation of the way in which U factors are determined follows.

$$A_e = AU \qquad \text{(LRFD Equation B3-1)}$$

The angle shown in Fig. 3-10(a) is connected at its ends to only one leg. You can easily see that its area effective in resisting tension can be appreciably increased by shortening the width of the unconnected leg and lengthening the width of the connected leg as shown in Fig. 3-10(b).

Investigators have found that a convenient measure of the effectiveness of a member such as an angle connected by one leg is the distance $\bar{x}$ measured from the plane of the connection to the centroid of the area of the whole section.[2,3] The smaller the value of $\bar{x}$ the larger the effective area of the member. The specification in effect reduces the length L of a connection with shear lag to a shorter effective length L'. The value of U then equals L'/L or $1 - \bar{x}/L$. Several values of $\bar{x}$ are shown in Fig. 3-11. The remainder of this section is devoted to the determination of the effective areas of various bolted, and welded tension members.

Bolted Members

Should a tension load be transmitted by bolts, A equals the net area A_n of the member and U is computed as follows:

$$U = 1 - \frac{\bar{x}}{L} \le 0.9 \qquad \text{(LRFD Equation B3-2)}$$

The length L used in this expression is equal to the distance between the first and last bolts in the line. When there are two or more lines of bolts, it is the length of the line with the maximum number of bolts. Should the bolts be staggered, it is the out-to-out dimension between the extreme bolts. You will

[2] E. H. Gaylord, Jr., and C. N. Gaylord, *Design of Steel Structures*, 2d ed. (New York: McGraw-Hill Book Company, 1972), pp. 119–123.

[3] W. H. Munse and E. Chesson, Jr., "Riveted and Bolted Joints: Net Section Design," *Journal of the Structural Division*, ASCE, 89, STI (February 1963).

Figure 3-11 Values of $\bar{x}$ for different shapes.

note that the longer the connection (L) becomes the larger U will become as will the effective area of the member. (On the other hand we will learn in the connection chapters of this text that the effectiveness of connectors is somewhat reduced if very long connections are used.) Insufficient data are available for the case in which only one bolt is used in each line. It is thought that a conservative approach for this case is to let $A_e = A_n$ of the connected element.

In order to calculate U for a W section connected by its flanges only, we will assume that the section is split into two structural tees. Then the value of $\bar{x}$ used will be the distance from the outside edge of the flange to the c.g. of the structural tee as shown in part (c) of Fig. 3-11. Parts (b) and (c) of Fig. C-B3.1 of the LRFD Commentary illustrate the recommended procedures for calculating $\bar{x}$ values for channel and I-shaped sections where the loads are transferred by means of bolts passing through the member webs only.

The LRFD Specification permits the designer to use larger values of U than obtained from the equation if such values can be justified by tests or other rational criteria.

Section B3 of the LRFD Commentary provides suggested $\bar{x}$ values for use in the equation for U for several situations not addressed in the Specification. Included are values for W and C sections bolted only through their webs. Also considered are single angles with two lines of staggered bolts in one of their legs. The basic idea for computing $\bar{x}$ for these cases is presented in the following paragraph.[4]

---

[4] W. S. Easterling and L. G. Giroux, "Shear Lag Effects in Steel Tension Members," *Engineering Journal*, AISC, no. 3 (3rd Quarter, 1993), pp. 77–89.

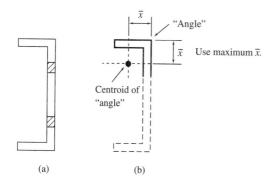

Figure 3-12 Computing $\bar{x}$ for a channel bolted through its web.

The channel of Fig. 3-12(a) is connected with two lines of bolts through its web. The "angle" part of this channel above the center of the top bolt is shown darkened in part (b) of the figure. This part of the channel is unconnected. For shear lag purposes we can compute the vertical distance from the center of the top bolt to the centroid of the "angle" above and the horizontal distance from the outside face of the web to the "angle" centroid. The larger value will represent the worst situation and will be the $\bar{x}$ used in the equation. It is felt that with this idea in mind the reader will be able to understand the values shown in the Commentary for other sections.

Example 3-6 illustrates the calculations necessary for determining the effective net area of a bolted W section connected only to its flanges. In addition the design strength of the member is computed.

■ EXAMPLE 3-6

Determine the tensile design strength of a W10 × 45 with two lines of $\frac{3}{4}$-in.-diameter bolts in each flange using A572 Grade 50 steel with $F_y = 50$ ksi and $F_u = 65$ ksi and the LRFD Specification. There are assumed to be at least three bolts in each line 4 in. on center, and the bolts are not staggered with respect to each other.

Solution.   Using a W10 × 45 ($A_g = 13.3$ in.², $d = 10.10$ in., $b_f = 8.020$ in., $t_f = 0.620$ in.)

(a) $P_u = \phi_t F_y A_g = (0.90)(50)(13.3) = 598.5$ k

(b) $A_n = 13.3 - (4)\left(\frac{7}{8}\right)(0.620) = 11.13$ in.² $= A$

Referring to the tables for half of a W10 × 45 (or to a WT5 × 22.5) we find $\bar{x} = 0.907$ in. Then

$$U = 1 - \frac{\bar{x}}{L} = 1 - \frac{0.907}{8} = 0.89$$

$$A_e = UA = (0.89)(11.13) = 9.91 \text{ in.}^2$$

$$P_u = \phi_t F_u A_e = (0.75)(65)(9.91) = 483.1 \text{ k}$$

Design strength $P_u = 483.1$ k ■

The 1986 LRFD Specification presented a set of standard U values that could be used for bolted members instead of substituting into the $1 - \frac{\bar{x}}{L}$ expression. These values which are listed in Table 3-2 are still acceptable. (For Example 3-6 note that U from the table is 0.90 as $b_f/d = 8.020/10.10 > 2/3$.) The tabulated values are particularly useful for situations in which initial design size selections are being made and we have insufficient information to calculate U values. We will encounter this situation in the first two examples of Chapter 4.

TABLE 3-2 PERMISSIBLE U VALUES FOR BOLTED CONNECTIONS

(a) W, M, or S shapes with flange widths not less than two-thirds the depth, and structural tees cut from these shapes, provided the connection is to the flanges and has no fewer than three fasterners per line in the direction of stress, $U = 0.90$

(b) W, M, or S shapes not meeting the conditions of subparagraph a, structural tees cut from these shapes, and all other shapes including built-up cross sections, provided the connection has no fewer than three fasteners per line in the direction of stress, $U = 0.85$.

(c) All members having only two fasteners per line in the direction of stress, $U = 0.75$.

When a tension load is transmitted by fillet welds to some but not all elements of a cross section, the weld strength will control.

Welded Members

When tension loads are transferred by welds the following rules from LRFD Specification B.3 are to be used to determine values for A and U. (A_e as for bolted connections $= AU$.)

1. Should the load be transmitted only by longitudinal welds to other than a plate member, or by longitudinal welds in combination with transverse welds, A is to equal the gross area of the member A_g.
2. Should a tension load be transmitted only by transverse welds, A is to equal the area of the directly connected elements and U is to equal 1.0.
3. Tests have shown that when flat plates or bars connected by longitudinal fillet welds (a term to be described in Chapter 14) are used as tension members, they may fail prematurely by shear lag at the corners if the welds are too far apart. Therefore, the LRFD Specification states that when such situations are encountered the length of the welds may not be less than the width of the plates or bars. The letter A represents the area

of the plate and UA is the effective net area. For such situations the following values of U are to be used (LRFD Specification B3.2(d)).

When $l \geq 2w$	$U = 1.0$
When $2w > l \geq 1.5w$	$U = 0.87$
When $1.5w > l \geq w$	$U = 0.75$

where $l =$ weld length, in.
        $w =$ plate width (distance between welds), in.

For combinations of longitudinal and tranverse welds l is to be used equal to the length of the longitudinal weld because the transverse weld has little or no effect on the shear lag (that is, it does little to get the load into the unattached parts of the member).

In cases in which fillet welds are used to transmit tension loads to some but not all of the elements of a cross section the weld strength will control.

Example 3-7 illustrates the calculation of the effective area and the design strength of a welded member.

■ EXAMPLE 3-7

The 1 × 6 in. plate shown in Fig. 3-13 is connected to a 1 × 10 in. plate with longitudinal fillet welds to support a tensile load. Determine the design strength P_u of the member if $F_y = 50$ ksi and $F_u = 65$ ksi.

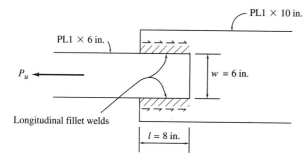

Figure 3-13

Solution.   Considering the smaller PL

(a) $P_u = \phi_t F_y A_g = (0.90)(50)(1 \times 6) = 270$ k
(b) $A = A_g = 1 \times 6 = 6$ in.^{2}
      $1.5w = 9$ in. $> l = 8$ in. $> w = 6$ in.
      $\therefore U = 0.75$
      $A_e = AU = (6.0)(0.75) = 4.50$ in.^{2}
      $P_u = \phi_t F_u A_e = (0.75)(65)(4.50) = 219.4$ k ←
      Design strength $P_u = 219.4$ k

Figure 3-14 Angle welded to 8-in. leg only.

■ EXAMPLE 3-8

Compute the design strength P_u for the angle shown in Fig. 3-14. It is welded on the ends and sides of the 8 in. leg only and $F_y = 50$ ksi and $F_u = 70$ ksi.

Solution.   As only one leg of the angle is connected, a reduced effective area needs to be computed.

(a) $P_u = \phi_t F_y A_g = (0.9)(50)(9.94) = 447.3$ k

(b) $U = 1 - \dfrac{\bar{x}}{L} = 1 - \dfrac{1.56}{6.00} = 0.74$

$A_e = AU = (9.94)(0.74) = 7.36$ in.^{2}

$P_u = \phi_t F_u A_e = (0.75)(70)(7.36) = 386.4$ k $\leftarrow$

Design strength $P_u = 386.4$ k   ■

3-6 CONNECTING ELEMENTS FOR TENSION MEMBERS

When splice or gusset plates are used as statically loaded tensile connecting elements their strength shall be determined as follows:

For yielding of welded or bolted connection elements

$\phi = 0.90$

$R_n = A_g F_y$ (LRFD Equation J5-1)

For fracture of bolted connection elements

$\phi = 0.75$

$R_n = A_n F_u$ with $A_n \leq 0.85 F_y$ (LRFD Equation J5-2)

The net area A_n to be used in the second of these expressions may not exceed 85 percent of A_g. Tests have shown for decades that bolted tension connection elements rarely have an efficiency greater than 85 percent, even if the holes represent a very small percentage of the gross area of the elements. In Example 3-9 the strength of a pair of tensile connecting plates is computed.

■ EXAMPLE 3-9

The tension member ($F_y = 50$ ksi and $F_u = 65$ ksi) of Example 3-6 is assumed to be connected at its ends with two $\frac{3}{8} \times 12$-in. plates as shown in Fig. 3-15. If two lines of $\frac{3}{4}$-in. bolts are used in each plate, determine the design tensile force which the plates can transfer.

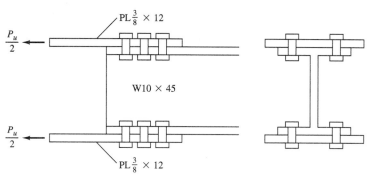

Figure 3-15

Solution.

$$P_u = \phi_t F_y A_g = (0.9)(50)\left(2 \times \frac{3}{8} \times 12 \right) = 405 \text{ k}$$

$$A_n \text{ of 2 plates} = \left(\frac{3}{8} \times 12 - \frac{7}{8} \times 2 \times \frac{3}{8} \right)2 = 7.69 \text{ in.}^2$$

$$0.85A_g = (0.85)\left(2 \times \frac{3}{8} \times 12 \right) = 7.65 \text{ in.}^2 = A_n$$

$$P_u = \phi_t F_u A_n = (0.75)(65)(7.65) = 372.9 \text{ k} \leftarrow$$

$$\underline{\underline{P_u = 372.9 \text{ k}}}$$

3-7 BLOCK SHEAR

The design strength of a tension member is not always controlled by $\phi_t F_y A_g$ or $\phi_t F_u A_e$ or by the strength of the bolts or welds with which the member is connected. It may instead be controlled by its *block shear* strength as described in this section.

The failure of a member may occur along a path involving tension on one plane and shear on a perpendicular plane as shown in Fig. 3-16, where several

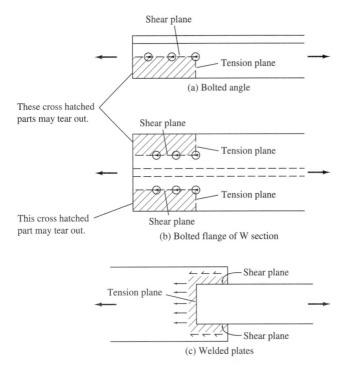

These cross hatched parts may tear out.

This cross hatched part may tear out.

Figure 3-16 Block shear.

possible block shear failures are shown. For these situations it is possible for a "block" of steel to tear out.

When a tensile load applied to a particular connection is increased the fracture strength of the weaker plane will be approached. That plane will not fail then because it is restrained by the stronger plane. The load can be increased until the fracture strength of the stronger plane is reached. During this time the weaker plane is yielding. The total strength of the connection equals the fracture strength of the stronger plane plus the yield strength of the weaker plane.[5] Thus it is not realistic to add the fracture strength of one plane to the fracture strength of the other plane to determine the block shear resistance of a particular member. *You can see that block shear is a tearing or rupture situation and not a yielding situation.*

The member shown in Fig. 3-17(a) has a large shear area and a small tensile area, thus the primary resistance to a block shear failure is shearing and not tensile. The LRFD Specification states that it is logical to assume that when a shear fracture occurs on this large shear-resisting area the small tensile area has yielded.

---

[5] L. B. Burgett, "Fast Check for Block Shear," *Engineering Journal,* AISC, 29, no. 4 (4th Quarter, 1992, pp. 125–127.

Large shear
area

Small tensile area

(a)
Shear fracture and tension yielding

(b)
Free body of "block" that tends to
shear out in angle of part (a)

Small shear
area

Large tensile
area

(c)
Tensile fracture and shear yielding

Figure 3-17 Block shear.

Part (b) of Fig. 3-17 shows a free body of the block that tends to tear out of the angle of part (a). You can see in this sketch that the block shear is caused by the bolts bearing on the back of the bolt holes.

In part (c) of Fig. 3-17 a member is shown which, so far as block shear goes, has a large tensile area and a small shear area. The LRFD feels that for this case the primary resisting force against a block shear failure will be tensile and not shearing. Thus a block shear failure cannot occur until the tensile area fractures. At that time it seems logical to assume that the shear area has yielded.

Based on the preceding discussion the LRFD Specification (J4.3) states that the block shear design strength of a particular member is to be determined by (1) computing the tensile fracture strength on the net section in one direction and adding to that value the shear yield strength on the gross area on the perpendicular segment and (2) computing the shear fracture strength on the gross area subject to tension and adding it to the tensile yield strength on the net area subject to shear on the perpendicular segment.

Test results show that this procedure gives good results. Furthermore, it is consistent with the calculations previously used for tension members where gross areas are used for one limit state of yielding $(\phi_t F_y A_g)$ and net area for the fracture limit state $(\phi_t F_y A_e)$. The LRFD Specification (J4.3) states that the block shear rupture design strength is to be determined as follows:

1. If $F_u A_{nt} \geq 0.6 F_u A_{nv}$ we will have shear yielding and tension fracture and thus should use the equation to follow:

$$\phi R_n = \phi[0.6 F_y A_{gv} + F_u A_{nt}]$$ (LRFD Equation J4-3a)

2. If $0.6F_uA_{nv} > F_uA_{nt}$ we will have tension yielding and shear fracture and thus should use the following equation:

$$\phi R_n = \phi[0.6 \, F_u A_{nv} + F_y A_{gt}] \qquad \text{(LRFD Equation J4-3b)}$$

in which $\phi = 0.75$

A_{gv} = gross area subjected to shear
A_{gt} = gross area subjected to tension
A_{nv} = net area subjected to shear
A_{nt} = net area subjected to tension

Examples 3-10 and 3-11 illustrate the determination of the block shear strengths for two members. This topic of block shear is continued in the connection chapters of this text, where we will find that it is absolutely necessary to check beam connections where the top flange of the beams are coped or cut back as illustrated in Figs. 10-3(c) and 15-6(b).

■ **EXAMPLE 3-10**

The A572 Grade 50 tension member shown in Fig. 3-18 is connected with three $\frac{3}{4}$-in. bolts. Determine the block shearing strength of the member and its tensile strength.

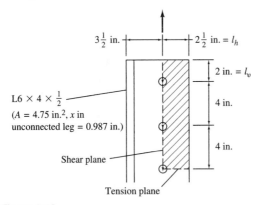

Figure 3-18

Solution.

$$A_{gv} = (10)\left(\frac{1}{2}\right) = 5.0 \text{ in.}^2$$

$$A_{gt} = (2.5)\left(\frac{1}{2}\right) = 1.25 \text{ in.}^2$$

$$A_{nv} = \left(10 - 2.5 \times \frac{7}{8}\right)\left(\frac{1}{2}\right) = 3.91 \text{ in.}^2$$

$$A_{nt} = \left(2.50 - \frac{1}{2} \times \frac{7}{8}\right)\left(\frac{1}{2}\right) = 1.03 \text{ in.}^2$$

$$F_u A_{nt} = (65)(1.03) = 66.9 \text{ k} < 0.6F_u A_{nv} = (0.6)(65)(3.91) = 152.5 \text{ k}$$

$$\therefore \text{Use LRFD Equation J4-3b}$$

$$\phi R_n = 0.75 \, [(0.6)(65)(3.91) + (50)(1.25)] = 161.2 \text{ k}$$

Tensile strength of angle

(a) $P_u = \phi_t F_y A_g = (0.9)(5c)(4.75) = 213.7$ k ←
(b) $A_n = 4.75 - (1)(\tfrac{7}{8})(\tfrac{1}{2}) = 4.31$ in.^{2} = A

$$U = 1 - \frac{0.987}{8} = 0.88$$

$A_e = uA = (0.88)(4.31) = 3.79$ in.^{2}
$P_u = \phi_t F_u A_e = (0.75)(65)(3.79) = 184.8$ k
P_u for member = smaller of $\phi R_n = 161.2$ k or of $P_u = 184.8$ k
$\underline{\underline{P_u = 161.2 \text{ k}}}$ ■

Tables are available in Part 8 of the LRFD Manual with which the block shear strengths of coped W beams can be determined. In Table 8-47(a) values of $\phi F_u A_{nt}$ are tabulated per inch of material thickness, and then in Table 8-47(b) values of $\phi(0.6 F_y A_{gv})$ per inch of material thickness are given.

■ **EXAMPLE 3-11**
Determine the block shear design strength of the A36 welded member shown in Fig. 3-19.

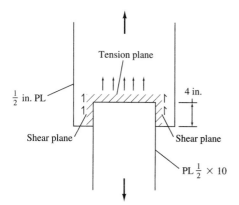

Figure 3-19

Solution.

$$A_{gv} = \left(\frac{1}{2}\right)(8) = 4.0 \text{ in.}^2$$

$$A_{gt} = \left(\frac{1}{2}\right)(10) = 5.0 \text{ in.}^2$$

$$A_{nv} = \left(\frac{1}{2}\right)(8) = 4.0 \text{ in.}^2$$

$$A_{nt} = \left(\frac{1}{2}\right)(10) = 5.0 \text{ in.}^2$$

$$F_u A_{nt} = (58)(5.0) = 290 \text{ k} > 0.6 F_u A_{nv} = (0.6)(58)(4.0) = 139.2 \text{ k}$$

$$\therefore \text{ Must use LRFD Equation J4-3a}$$

$$\phi R_{bs} = \overline{0.75[(0.6)(36)(4.0) + (58)(5.0)]} = 282.3 \text{ k}$$

Tensile strength of plate

$$P_u = \phi F_y A_g = (0.9)(36)\left(\frac{1}{2} \times 10\right) = 162.0 \text{ k} \leftarrow$$

Design strength of plate = 162.0 k ⬛

Sometimes cases are encountered where it is not altogether clear what sections should be considered for block shear calculations. For such situations designers will have to use their own judgment. One such case is shown in Fig. 3-20. In part (a) of the figure it is first assumed that web tear-out will occur along the lines *abcdef*. An alternate tear-out possibility for the same member along lines *abdef* is shown in part (b) of the figure. For this connection it is assumed that the load to be resisted is distributed equally among the five bolts. Thus, when web tear out is considered for case (b), we will assume only $\frac{4}{5}P_u$ is carried by the section in question, because one of the bolts is outside of the tear-out area.

(a)   (b)

Figure 3-20

Notice that the total block shear strength of the member will equal the block shear strength along path *abdef* plus the strength of bolt C, as it also must fail. To compute the width of the tension planes *abc* and *abd* for these cases, it seems logical to make use of the $s^2/4g$ expression presented in Section 3-4.

3-8 COMPUTER EXAMPLES

In this section the design strength of a W section is determined using the computer diskette enclosed with the book. When you finish using this particular option of INSTEP you will probably think the whole thing was a waste of time because you could have solved it just as quickly using your pocket claculator. This example, however, is presented primarily to show you how to apply the program. Subsequent applications for computing design loads; for analyzing and designing beams, columns, connections, base plates; and so on will result in appreciable time and work savings.

Axial tension	Bending beams	Bolted joints	Column base PL
Columns	Beam columns	Welded joints	Bearing beam PL
Design loads	Built-up members	System help	Quit

Figure 3-21 Problems that can be addressed by the INSTEP program.

To run the program from a disk drive (say, the A drive) the user needs to insert the disk in that drive, type INSTEP and press the ENTER key. The computer screen will then show a list of options available as shown in Fig. 3-21. The user will select the option desired using the keyboard or the mouse. It will be necessary to input only the information requested by screen prompts. As each bit of information is supplied, the user must press the RETURN key. After the last requested item is supplied the computer will show the results of the desired computations.

In the next several paragraphs the use of the computer program INSTEP is described in detail for a person using a keyboard. (The mouse can be used just as easily.) Once the list of problems that can be addressed with this diskette is displayed, as shown in Fig. 3-21, the user selects the one desired. The top horizontal row of options from left to right are Axial Tension, Bending Beams, and so on. The user presses F1 for the first option, F2 for the second, and on to F4 for Column Base Plates.

For the second row it is necessary to press the SHIFT key and F1 simultaneously for the first option, the SHIFT key and F2 for the second option, and so on to the fourth option on that row. Finally, for the bottom row it is necessary to press the CONTROL key simultaneously with F1 for the first option and so on. *A complete description of the use of INSTEP is available with the SYSTEM HELP option on the bottom row.* Nevertheless, a brief introduction to the program is presented in the next few paragraphs. Each of the other computer examples in this text are preceded with only a brief remark or two about the particular problem in question.

In the example problem to follow (Example 3-12) the INSTEP program is used to determine the design tensile strength of a W section. This very simple problem will not only show the reader how to use this particular option, but should enable him or her to make use of any of the other options.

After the axial tension option is selected with the F1 key, the input data needed to solve this type of problem will appear on the left half of the screen. A letter is shown by each item. The first of these requests the steel grade. The letter next to it—C, in this case—is entered and a list of steel grades appears on the screen. The reader should type the letter appearing by the steel grade to be used. In the example to follow the author typed H to represent A572 Grade 50 steel. For the second input item the letter D is typed and then a beam size is specified (here it was a W10 × 45) and so on for the other input items. (By the way, any time a steel shape is specified, you can have all of its properties listed on the screen by pressing the F10 key.) *Notice that the units used for inputting data are inches and kips.*

The last input item listed for the Axial Tension option is the factored load. You can put in 0 or any other number—the computer will compute the design strength of the member and compare it to the number you input. After all of the data needed are furnished, press the F1 key and the solution will appear on the right-hand side of the screen. Sometimes the solution will require more than one screen. By pressing the F1 key you can move on to the next page of the solution. After the problem is completed, return to the options list by pressing the F4 key.

With the INSTEP program you can quickly revise the data for a particular problem in any way you see fit. You then press the F1 key and new answers will be provided. In this way the program can be used to study the effect on the design of changing one or more input items.

■ EXAMPLE 3-12

Using the program INSTEP, determine the design tensile strength of a 12-ft long W12 × 136 consisting of A572 Grade 50 steel if the net area is assumed to be 35.52 in.^{2} and $U = 0.9$.

Solution.

INPUT

STEEL AND TRIAL MEMBER SIZE

$\boxed{C}$ Type Steel: A572 Grade 50
$\boxed{D}$ Trial Size: W12 × 136

TENSION DESIGN PARAMETERS

$\boxed{G}$ L = 144
$\boxed{H}$ A_n = 35.52
$\boxed{I}$ U = .9

FACTORED DESIGN LOAD

$\boxed{L}$ P_u = 0

OUTPUT

YIELDING IN THE GROSS SECTION

P_n = 39.9(50) = 1995.
$0.9P_n$ = 1795.5 > 0 OK

FRACTURE IN THE NET SECTION

P_n = (.9)35.52(65) = 2077.92
$0.75P_n$ = 1558.44 > 0 OK

LIMITING SLENDERNESS RATIO

$$L/r = \frac{144}{3.16} = 45.57 < 300 \quad OK$$

This Member is Acceptable ■

One of the options of the computer program INSTEP is the calculation of the design or factored loads for a particular set of service loads. This option, which was shown as *Design Loads* in Fig. 3-21 can save the user the appreciable amount of time required to tediously substitute into LRFD Equations A4-1 to A4-6. Example 3-13 illustrates this type of calculation.

■ EXAMPLE 3-13

Using the computer program INSTEP compute the governing factored loads for the following: $D = 200$ k, $L_r = 50$ k, $L = 250$ k, $W = 80$ k, and $E = 60$ k. These were the values previously considered in Example 2-3.

Solution.

```
                         INPUT

                  COMPONENTS OF LOAD

              F   Dead Load : 200
              G   Live Load : 250
              H   Roof Load : 50
              I   Snow Load : 0
              J   Rain Load : 0
              K   Wind Load : 80
              L   Seismic   : 60

                       OUTPUT
```

DEAD LOAD ONLY (A4-1)

U = 1.4(200) = 280

DEAD + LIVE + ROOF LOADS (A4-2)

U = 1.2(200) + 1.6(250) + 0.5(50) = 665

DEAD + LIVE + SNOW LOADS (A4-2)

U = 1.2(200) + 1.6(250) + 0.5(0) = 640

DEAD + LIVE + RAIN LOADS (A4-2)

U = 1.2(200) + 1.6(250) + 0.5(0) = 640

DEAD + ROOF + LIVE LOADS (A4-3)

U = 1.2(200) + 1.6(50) + 0.5(250) = 445

DEAD + SNOW + LIVE LOADS (A4-3)

U = 1.2(200) + 1.6(0) + 0.5(250) = 365

DEAD + RAIN + LIVE LOADS (A4-3)

U = 1.2(200) + 1.6(0) + 0.5(250) = 365

DEAD + ROOF + WIND LOADS (A4-3)

U = 1.2(200) + 1.6(50) + 0.8(80) = 384

DEAD + SNOW + WIND LOADS (A4-3)

U = 1.2(200) + 1.6(0) + 0.8(80) = 304

DEAD + RAIN + WIND LOADS (A4-3)

U = 1.2(200) + 1.6(0) + 0.8(80) = 304

DEAD + WIND + LIVE + ROOF LOADS (A4-4)

U = 1.2(200) + 1.3(80) + 0.5(250 + 50) = 494

DEAD + WIND + LIVE + SNOW LOADS (A4-4)

U = 1.2(200) + 1.3(80) + 0.5(250 + 0) = 469

DEAD + WIND + LIVE + RAIN LOADS (A4-4)

U = 1.2(200) + 1.3(80) + 0.5(250 + 0) = 469

DEAD + SEISMIC + LIVE + SNOW LOADS (A4-5)

U = 1.2(200) + 1.0(60) + 0.5(250) + 0.2(0) = 425
U = 1.2(200) − 1.0(60) + 0.5(250) + 0.2(0) = 305

DEAD + WIND LOADS (A4-6)

U = 0.9(200) + 1.3(80) = 284
U = 0.9(200) − 1.3(80) = 76

DEAD + SEISMIC LOADS (A4-6)

U = 0.9(200) + 1.0(60) = 240
U = 0.9(200) − 1.0(60) = 120

SUMMARY OF EXTREME VALUES

Maximum: 665 kips
Minimum: 76 kips ∎

PROBLEMS (*Use standard-size bolt holes for all problems*)

3-1 to 3-18. *Compute the net area of each of the given members.*

Problem 3-1   (*Ans.* 9.84 in.²)

Problem 3-2

Problem 3-3   (*Ans.* 14.12 in.²)

Problem 3-4

3-5. An L8 × 4 × $\frac{9}{16}$ with one line of $\frac{7}{8}$-in. bolts in each leg. (*Ans.* 5.30 in.²)

3-6. A pair of 7 × 4 × $\frac{1}{2}$ Ls with two rows of $\frac{3}{4}$-in. bolts in the long legs and one row in the short legs.

3-7. A W21 × 93 with two holes in each flange and two in the web, all for $\frac{7}{8}$-in. bolts. (*Ans.* 22.42 in.²)

3-8. The $\frac{3}{4}$ × 12 plate shown in the illustration. The holes are for $\frac{7}{8}$-in. bolts.

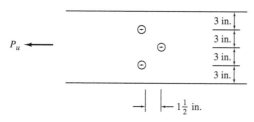

Problem 3-8

3-9. The $\frac{1}{2}$ × 9 plate shown in the illustration. The holes are for $\frac{7}{8}$-in. bolts. (*Ans.* 3.67 in.²)

Problem 3-9

3-10. The $\frac{7}{8}$ × 12 plate shown in the illustration. The holes are for 1-in. bolts.

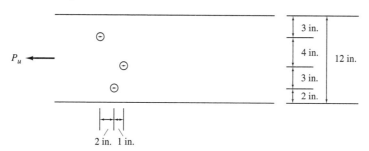

Problem 3-10

3-11. The 8 × 6 × $\frac{9}{16}$ angle shown has one line of $\frac{3}{4}$-in. bolts in each leg. The bolts are 3 in. on center in each line and are staggered $1\frac{1}{2}$ in. with respect to each other. (*Ans.* 6.62 in.)

Problem 3-11

3-12. For the plate shown find the minimum pitch s for which only two bolts need be subtracted at any one section in calculating the net area. Bolts are $\frac{3}{4}$ in.

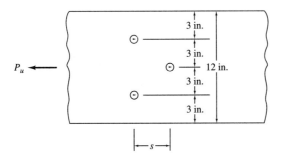

Problem 3-12

3-13. Find the minimum pitch s in the plate of Prob. 3-12 so that only two and one-half bolts need be subtracted at any one section. (*Ans.* 1.62 in.)

3-14. An L8 × 8 × $\frac{5}{8}$ is used as a tension member with one gage line of 1-in. bolts in each leg at the usual gage location. (See Table 3-1.) What is the minimum amount of stagger necessary so that only one bolt need be subtracted from the gross area of the angle? Compute the net area of this member if the holes are staggered at 2 in.

3-15. An L7 × 4 × $\frac{3}{4}$ is shown. Two rows of $\frac{7}{8}$-in. bolts are used in the long leg and one in the short leg. Determine the minimum stagger (or pitch s in the figure) necessary so that only two holes need be subtracted in determining the net area. (*Ans.* 2.65 in.)

Problem 3-15

3-16. A $8 \times 6 \times \frac{3}{4}$ angle has one line of holes in each leg for $\frac{3}{4}$-in. bolts. Determine the minimum pitch so that only one and one-half holes need to be deducted to obtain the net area. (Use the usual gage for angles as given in Table 3-1 of this chapter.)

3-17. Determine the effective net cross-sectional area of the C15 $\times$ 40 shown. Holes are for $\frac{3}{4}$-in bolts. (*Ans.* 10.05 in.²)

Problem 3-17

3-18. Compute the effective net area of the built-up section shown if the holes are punched for $\frac{7}{8}$-in bolts. Assume at least three bolts in each line.

Problem 3-18

3-19 to 3-22. *Determine the effective net areas of the sections shown using the U values given in Table 3-2 of this chapter. Assume three bolts in a line.*

Problem 3-19. (*Ans.* 3.66 in.²)

3-20. Determine the effective net area of the MC12 × 45 shown. Assume the holes are for 1-in. bolts.

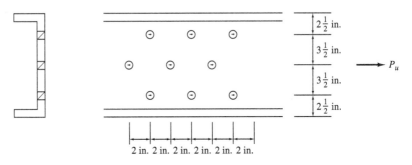

Problem 3-20

3-21. A C12 × 20.7 is connected through its web with three gage lines of $\frac{3}{4}$-in. bolts. The gage lines are 3 in. on centers, and the bolts are spaced 4 in. on centers along the gage line. If the center row of bolts is staggered with respect to the outer row, determine the effective net cross-sectional area of the channel. Assume there are three bolts in each line. (*Ans.* 4.67 in.²)

Problem 3-22

Problem 3-23 to 3-36. *Use the formula for U.*

3-23. Repeat Prob. 3-19. (*Ans.* 3.79 in.²)

3-24. Repeat Prob. 3-20.

3-25. Repeat Prob. 3-22. (*Ans.* 7.65 in.²)

3-26. Using A36 steel determine the tensile design strength P_u of a W12 × 45 with two lines of $\frac{3}{4}$-in. bolts in each flange (three bolts in each line 3 in. on center). Neglect block shear strength.

3-27. Determine the tensile design strength P_a of a W18 × 119 (A572 Grade 50) if it has two lines of 1-in. bolts in each flange (at least four bolts in each line 3 in. on center). Neglect block shear. (*Ans.* 1138.3 k)

3-28. A single-angle tension member (7 × 4 × $\frac{3}{4}$) has two gage lines in its long leg and one in the short leg for $\frac{3}{4}$-in. bolts arranged as shown. Determine the tensile strength P_u of this member if A572 Grade 50 steel is used and if block shear is neglected.

Problem 3-28

3-29. Compute the design strength of the bolted connection shown neglecting block shear. The angle is made of A36 steel and the bolts are $\frac{3}{4}$ in. (*Ans.* 110.8 k)

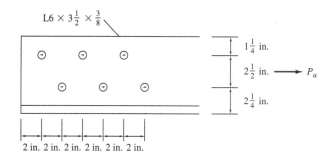

Problem 3-29

3-30. Determine the tensile design strength P_u of the pair of $6 \times 6 \times \frac{3}{4}$ angles shown, made from A242 Grade 50 steel. Standard gages are to be used as determined from Table 3-1 or the LRFD Manual for the $\frac{3}{4}$-in. bolts. Neglect block shear.

Problem 3-30

3-31. The $7 \times 4 \times \frac{3}{8}$ angle shown is connected with three 1-in. bolts. If the angle consists of A36 steel, determine its block shear strength. Compare the results with the tensile design strength of the member. (*Ans.* 100.8 k)

Problem 3-31

3-32. A W12 × 53 is connected at its ends with the plates shown. Determine the block shear strength of the member if A36 steel is used and if it is connected with six $\frac{7}{8}$-in. bolts in each flange as shown. Compare the result with the tensile design strength of the member. Do not check plate strength.

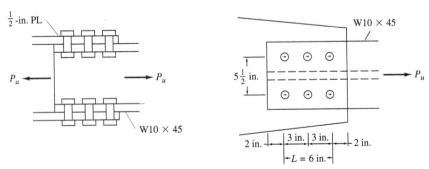

Problem 3-32

3-33. Repeat Prob. 3-26 if A242 Grade 50 steel is used and block shear is included. (*Ans.* 475.9 k)

Problem 3-33

3-34. Compute the tensile design strength of the $6 \times 6 \times \frac{1}{2}$ angle shown if it consists of a steel with $F_y = 50$ ksi and $F_u = 65$ ksi. Consider block shear as well as the tensile strength of the angle.

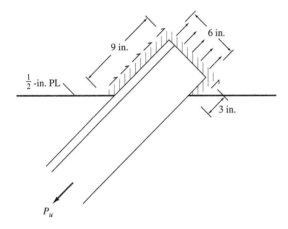

Problem 3-34

3-35. A W14 × 82 has two lines of $\frac{7}{8}$-in. bolts (three in a line 4 in. on center) in each flange. If it consists of A572 Grade 50 steel, determine the maximum dead and live service tensile loads (P_D and P_L) it can support if 30 % of the service load is dead and 70% is live. Use the U value given in LRFD Spec. B.3. (*Ans.* $P_D = 169.6$ k, $P_L = 395.7$ k)

3-36. Repeat Prob. 3-35 if the member is a C12 × 30 with three lines of $\frac{3}{4}$-in. bolts (four in a line 3 in. on center) in the web.

3-37. An A572 Grade 50 WT15 × 62 has transverse welds to its flange only at its ends. Determine its design tensile strength P_u using the LRFD Spec. B.3 expression for U. (*Ans.* 476.8 k)

3-38. The two MC13 × 50s shown have transverse welds along their webs only. Compute the design tensile strength P_u of the member if A36 steel is used.

Problem 3-38

3-39 to 3-40. *Using the enclosed computer program solve the following problems. In each case select U from Table 3-2 in this chapter.*

3-39. Determine the design tensile strength of a 30-ft long W12 × 58 with two lines of four $\frac{7}{8}$-in. bolts 4 in. on center in each flange. Use A572 Grade 50 steel. (*Ans.* 633.6 k)

3-40. Same question and data as for Prob. 3.39 except the member is a W14 × 90.

3-41. Repeat Problem 3.40 if A36 steel is used. (*Ans.* 858.6 k)

3-42. Determine the design tensile strength of a 24-ft long MC12 × 40 with one line of three $\frac{3}{4}$-in. bolts in each flange 4 in. on center. Use A572 Grade 50 steel.

3-43. Repeat Prob. 3.42 using A572 Grade 60 steel. (*Ans.* 452 k)

Chapter 4

Design of Tension Members

4-1 SELECTION OF SECTIONS

The determination of the design strengths of various tension members was presented in Chapter 3. In this chapter the selection of members to support given tension loads is described. Although the designer has considerable freedom in the selection, the resulting members should have the following properties: (a) compactness, (b) dimensions that fit into the structure with reasonable relation to the dimensions of the other members of the structure, and (c) connections to as many parts of the sections as possible to minimize shear lag.

The choice of member type is often affected by the type of connections used for the structure. Some steel sections are not very convenient to bolt together with the required gusset or connection plates, while the same sections may be welded together with little difficulty. Tension members consisting of angles, channels, and W or S sections will probably be used when the connections are made with bolts, while plates, channels, and structural tees might be used for welded structures.

Various types of sections are selected for tension members in the examples to follow, and in each case where bolts are used some allowance is made for holes. Should the connections be made *entirely* by welding, no holes have to be added to the net areas to give the required gross area. *The student should realize, however, that very often welded members may have holes punched in them for temporary bolting during field erection before the permanent field welds are made. These holes need to be considered in design.* It is also to be remembered that in LRFD Equation D1-2 ($P_n = F_u A_e$) the value of A_e may be less than A_g even though there are no holes, depending on the arrangement of welds and on whether all of the parts of the members are connected.

The slenderness ratio of a member is the ratio of its unsupported length to its least radius of gyration. Steel specifications give preferable maximum values of slenderness ratios for both tension and compression members. The purpose of such limitations for tension members is to ensure the use of sections with sufficient stiffness to prevent undesirable lateral deflections or vibrations. Although tension members are not subject to buckling under normal loads, stress reversal may occur during shipping and erection and perhaps due to wind or earthquake. The specifications recommend that slenderness ratios be kept below certain maximum values in order that some minimum compressive strengths be provided in the members. For tension members other than rods LRFD Specification B7 recommends maximum slenderness ratios of 300. Members whose design is controlled by tension loads, but which may be subjected to some compression due to other loading conditions, are not required to meet the preferable maximum slenderness ratio requirement for compression members, which is 200. (For slenderness ratios greater than 200, design compressive stresses will be very small—in fact, less than 5.33 ksi. This will be discussed further later in the text.)

It should be noted that out-of-straightness does not affect the strength of tension members very much because the tension loads tend to straighten the members. (The same statement cannot be made for compression members.) For this reason the LRFD specification is a little more liberal in its consideration of tension members, including those subject to some compressive forces due to transient loads such as wind or earthquake.

The *recommended* maximum slenderness ratio of 300 is not applicable to tension rods. Maximum L/r values for rods are left to the designer's judgment. If a maximum value of 300 was specified for them, they would seldom be used because of their extremely small radii of gyration.

The 1989 AASHTO Specifications provide mandatory maximum slenderness ratios of 200 for main tension members and 240 for secondary members. (A *main member* is defined by the AASHTO as one in which stresses result from dead and/or live loads, while *secondary members* are those used to brace structures or to reduce the unbraced length of other members—main or secondary.) *No such distinction is made in the LRFD Specification between main and secondary members.*

Example 4-1 illustrates the design of a bolted tension member with a W section, while Example 4-2 illustrates the selection of a bolted single-angle tension member. In both cases the LRFD Specification is used. The design strength P_u is the lesser of (a) $\phi_t F_y A_g$ or (b) $\phi_t F_u A_e$.

(a) To satisfy the first of these expressions the minimum gross area must be at least equal to the following

$$\min A_g = \frac{P_u}{\phi_t F_y} \tag{1}$$

(b) To satisfy the second expression the minimum value of A_e must be at least

$$\min A_e = \frac{P_u}{\phi_t F_u}$$

Transfer truss, 150 Federal Street, Boston, Mass. (Courtesy Owen Steel Company, Inc.)

And since $A_e = UA_n$ for a bolted member the minimum value of A_n is

$$\min A_n = \frac{\min A_e}{U} = \frac{P_u}{\phi_t F_u U}$$

Then the minimum A_g for the second expression must at least equal the minimum value of A_n plus the estimated hole areas.

$$\min A_g = \frac{P_u}{\phi_t F_u U} + \text{estimated hole areas} \tag{2}$$

The designer can substitute into Equations (1) and (2), taking the larger value of A_g so obtained for an initial size estimate. It is, however, well to notice that the maximum preferable slenderness ratio L/r is 300. From this value it is easy to compute the least preferable value of r for a particular design, that is, the value of r for which the slenderness ratio will be exactly 300. It is undesirable to consider a section whose least r is less than this value because its slenderness ratio would exceed the preferable maximum value of 300.

$$\min r = \frac{L}{300} \tag{3}$$

For the two examples to follow the author has substituted into the two ordinary gravity load expressions:

$$P_u = 1.4D$$

$$P_u = 1.2D + 1.6L$$

As the first of these expressions will not control unless the dead load is more than 8 times as large as the live load, the first expression is omitted for the remaining problems in this text (unless $D > 8L$).

In Example 4-1 a W section is selected for a given set of tensile loads. For this first application of the tension design formulas the author has narrowed the

Bridge over Allegheny River at Kittaning, PA. (Courtesy American Bridge Company.)

problem down to one series of W shapes so the reader can concentrate on the application of the formulas and not become lost in considering W8s, W10s, W14s, and so on. Exactly the same procedure can be used for trying these other series as is used here for the W12.

■ EXAMPLE 4-1

Select a 30-ft-long W12 section of A572 Grade 50 steel to support a tensile service dead load P_D = 130 k and a tensile service live load P_L = 110 k. As shown in Fig. 4-1, the member is to have two lines of bolts in each flange for $\frac{7}{8}$-in. bolts (at least three in a line 4 in. on center).

 Figure 4-1 Cross section of member for Example 4-1.

Solution.   Considering the two ordinary load conditions

$$P_u = 1.4P_D = (1.4)(130) = 182 \text{ k}$$

$$P_u = 1.2P_D + 1.6P_L = (1.2)(130) + (1.6)(110) = 332 \text{ k} \leftarrow$$

Computing the minimum A_g required

1. min $A_g = \dfrac{P_u}{\phi_t F_y} = \dfrac{332}{(0.90)(50)} = 7.38 \text{ in.}^2$

2. min $A_g = \dfrac{P_u}{\phi_t F_u U}$ + estimated hole areas

Assume U = 0.90 from Table 3-2 and assume flange thickness is about 0.380 in. after looking at W12 sections in the LRFD Manual which have areas of 7.38 in.² or more.

$$\text{min } A_g = \frac{332}{(0.75)(65)(0.90)} + (4)(1.00)(0.380) = 9.0 \text{ in.}^2 \leftarrow$$

3. Preferable min $r = \dfrac{L}{300} = \dfrac{(12)(30)}{300} = 1.2 \text{ in.}$

Try W12 × 35 (A_g = 10.3 in.², d = 12.50 in., b_f = 6.56 in., t_f = 0.520 in., r_y = 1.54 in.)

Checking

1. $P_u = \phi_t F_y A_g = (0.90)(50)(10.3) = 463.5 \text{ k} > 332 \text{ k}$ (OK)
2. $\bar{x}$ for half of W12 × 35 or that is a WT6 × 17.5 = 1.30 in.
    $L = (2)(4) = 8 \text{ in.}$

$$U = 1 - \frac{\overline{X}}{L} = 1 - \frac{1.30}{8} = 0.84 \text{ (from Table 3-2 } U = 0.85$$

$$A_n = 10.3 - (4)(1.00)(0.520) = 8.22 \text{ in.}^2$$

$$P_u = \phi_t F_u A_e = (0.75)(6.5)(0.84 \times 8.22) = 336.6 \text{ k} > 332 \text{ k} \quad \text{(OK)}$$

3. $\dfrac{L}{r} = \dfrac{(12)(30)}{1.54} = 234 < 300$ (OK)

Use W12 × 35 ∎

In Example 4-2 a broader example is presented in that the lightest satisfactory angle in the LRFD Manual is selected for a given set of tensile loads.

∎ EXAMPLE 4-2

Design a 9-ft single-angle tension member to support a dead tensile working load of 30 k and a live tensile working load of 40 k. The member is to be connected to one leg only with $\frac{7}{8}$-in. bolts (at least three in a line 3 in. on center). Assume that only one bolt is to be located at any one cross section. Use A36 steel.

Discussion.   There are many different angles listed in the LRFD Manual which will support the load for the conditions described. As a result it may seem rather difficult to arrive at the absolutely lightest satisfactory section. It is possible, however, to set up a table with which the various possible sections may be methodically considered for various angle thicknesses. In the table presented with this solution the author has listed various possible thicknesses of angles which will provide sufficient areas. He has then found the lightest angle for each thickness which provides the estimated area needed. Finally, a glance at the completed table reveals the lightest angle or the one with the smallest cross-sectional area. A similar process could be followed for designing tension members from other steel sections.

Solution.   This problem is one which can be conveniently handled with the use of a table, and such a procedure is followed here for a few angle sizes. The limiting values are calculated as follows.

$$P_u = (1.2)(30) + (1.6)(40) = 100 \text{ k}$$

1. min A_g required $= \dfrac{P_u}{\phi_t F_y} = \dfrac{100}{(0.90)(36)} = 3.09 \text{ in.}^2$
2. From Table 3-2 assume $U = 0.85$

   min A_n required $= \dfrac{P}{\phi_t F_u U} = \dfrac{100}{(0.75)(58)(0.85)} = 2.70 \text{ in.}^2$
3. min $r = \dfrac{L}{300} = \dfrac{(12)(9)}{300} = 0.36 \text{ in.}$

Angle t (in.)	Area of one 1-in. bolt hole (in.^{2})	Gross area required = larger of $P_u/\phi_t F_y$ or $P_u/\phi_t F_u U$ + est. hole area (in.^{2})	Lightest angles available, their areas (in.^{2}) and least radii of gyration (in.)
$\frac{5}{16}$	0.312	3.09	$6 \times 6 \times \frac{5}{16}$ $(A = 3.65, r = 1.20)$
$\frac{3}{8}$	0.375	3.09	$6 \times 3\frac{1}{2} \times \frac{3}{8}$ $(A = 3.42, r = 0.767)$
$\frac{7}{16}$	0.438	3.14	$5 \times 3 \times \frac{7}{16}$ $(A = 3.31, r = 0.651)$
			$4 \times 4 \times \frac{7}{16}$ $(A = 3.31, r = 0.785)$
$\frac{1}{2}$	0.500	3.20	$4 \times 3 \times \frac{1}{2}$ $(A = 3.25, r = 0.639)$ ←
			$3\frac{1}{2} \times 3\frac{1}{2} \times \frac{1}{2}$ $(A = 3.25, r = 0.683)$ ←
$\frac{5}{8}$	0.625	3.33	$4 \times 3 \times \frac{5}{8}$ $(A = 3.98, r = 0.637)$

Use L4 $\times$ 3 $\times \frac{1}{2}$ or L3$\frac{1}{2} \times$ 3$\frac{1}{2} \times \frac{1}{2}$ ■

4-2 BUILT-UP TENSION MEMBERS

Sections D2 and J3.5 of the LRFD Specification provide a definite set of rules describing how the different parts of built-up tension members are to be connected together.

1. When a tension member is built up from elements in continuous contact with each other, such as a plate and a shape, or two plates, the longitudinal spacing of connectors between those elements must not exceed 24 times the thickness of the thinner plate, or 12 in. if the member is to be painted or if it is not to be painted and not to be subjected to corrosive conditions.
2. Should the member consist of unpainted weathering steel elements in continuous contact and be subject to atmospheric corrosion, the maximum permissible connector spacings are 14 times the thickness of the thinner plate, or 7 in.
3. Should a tension member be built up from two or more shapes separated by intermittent fillers, the shapes must be connected to each other at intervals such that the slenderness ratio of the individual shapes between the fasteners does not exceed 300.
4. The distance from the center of any bolts to the nearest edge of the connected part under consideration may not be larger than 12 times the thickness of the connected part, or 6 in.

Example 4-3 illustrates the review of a tension member that is built up from two channels that are separated from each other. Included in the problem is the design of tie plates or tie bars to hold the channels together as shown in Fig. 4-2(b). These plates which are used to connect the parts of built-up mem-

Figure 4-2 Built-up section for Example 4-3.

bers on their open sides result in more uniform stress distribution among the various parts. Section D-2 of the LRFD Specification provides empirical rules for their design. (Perforated cover plates may also be used.) The rules are based on many decades of experience with built-up tension members.

In Fig. 4-2 the location of the bolts, that is, the usual gage for these channels, is shown as being $1\frac{3}{4}$ in. from the back of the channels. The LRFD Manual does not provide the usual gages except for angles, and those are given in Part 9. For other rolled shapes such as Cs, Ws, and Ss, it is necessary to refer to a manufacturer's catalog or to old AISC Steel Manuals. The gages necessary to solve problems in this text are furnished by the author. (Though rather standard values of these gages are used by the steel industry for the various rolled shapes, they are not specified in the LRFD Manual in order to give steel fabricators more freedom in placing the holes.)

In Fig. 4-2 the distance between the lines of bolts connecting the tie plates to the channels can be seen to equal 8.50 in. The LRFD Specification (D2) states that the length of tie plates (lengths are always measured parallel to the long direction of the members in this text) may not be less than two-thirds the distance between the lines of connectors. Furthermore, their thickness may not be less than one-fiftieth of this distance.

The minimum permissible width of tie plates (not mentioned in the specification) is the width between the lines of connectors plus the necessary edge distance on each side to keep the bolts from splitting the plate. For this example this minimum edge distance is taken as $1\frac{1}{2}$ in. from Table J3.4 of the LRFD Specification. (Detailed information concerning edge distances for bolts is provided in Chapter 12.) The plate dimensions are rounded off to agree with the plate sizes available from the steel mills as given in the Bars and Plates section of Part 1 of the LRFD Manual. It is much cheaper to select standard thicknesses and widths rather than picking odd ones that will require cutting or other operations.

The LRFD Specification (D2) provides a maximum spacing between tie plates by stating that the L/r of each individual component of a built-up member running along by itself between tie plates should preferably not exceed 300. If the designer substitutes into this expression ($L/r = 300$), the least r of an individual component of the built-up member, the value of L may be computed. This will be the maximum spacing of the tie plates preferred by the LRFD Specification for this member.

■ EXAMPLE 4-3

Two C12 × 30s (see Fig. 4-2) have been selected to support a dead tensile working load of 120 k and a 240-k live tensile working load. The member is 30 ft long, consists of A36 steel, and has one line of at least three $\frac{7}{8}$-in. bolts in each channel flange 3 in. on center. Using the LRFD Specification, determine whether the member is satisfactory and design the necessary tie plates. Assume centers of bolt holes are $1\frac{3}{4}$ in. from the backs of the channels.

Solution.   Using C12 × 30s ($A_g = 8.82$ in.² each, $t_f = 0.501$ in., $I_x = 162$ in.⁴ each, $I_y = 5.14$ in.⁴ each, y axis 0.674 in. from back of C, $r_y = 0.763$ in.)

Load to be resisted

   $P_u = (1.2)(120) + (1.6)(240) = 528$ k

Design strengths

   $P_u = \phi_t F_y A_g = (0.90)(36)(2 \times 8.82) = 571.5$ k > 528 k   (OK)

   $A_n = [8.82 - (2.0)(0.501)] 2 = 15.64$ in.²

   $U = 0.85$ from Table 3-2

   $P_u = \phi_t F_u A_n U = (0.75)(58)(15.64)(0.85) = 578.3$ k > 528 k   (OK)

Slenderness ratio

$$I_x = (2)(162) = 324 \text{ in.}^4$$

$$I_y = (2)(5.14) + (2)(8.82)(5.33)^2 = 511 \text{ in.}^4$$

$$r_x = \sqrt{\frac{324}{17.64}} = 4.29 \text{ in.}$$

$$\frac{L}{r} = \frac{(12 \times 30)}{4.29} = 83.9 < 300$$

Design of tie plates (LRFD Specification D2)

Distance between lines of bolts $= 12.00 - (2)(1\frac{3}{4}) = 8.50$ in.

Minimum length of tie plates $= (\frac{2}{3})(8.50) = 5.67$ in. (say 6 in.)

Minimum thickness of tie plates $= (\frac{1}{50})(8.50) = 0.17$ in. (say $\frac{3}{16}$ in.)

Minimum width of tie plates $= 8.50 + (2)(1\frac{1}{2}) = 11.5$ in. (say 12 in.)

Maximum preferable spacing of tie plates
least r of one $C = 0.763$ in.

Maximum preferable $\dfrac{L}{r} = 300$

$$\frac{(12)(L)}{0.763} = 300$$

$$L = 19.08 \text{ ft}$$

Use $\frac{3}{16} \times 6 \times 1$ ft 0 in. tie plates 15 ft 0 in. o.c. (on center)   ■

4-3 RODS AND BARS

When rods and bars are used as tension members they may be simply welded at their ends, or they may be threaded and held in place with nuts. The LRFD nominal tensile design stress for threaded rods is given in their Table J3.2 and equals $\phi 0.75F_u$ and is to be applied to the gross area of the rod A_D computed with the major thread diameter, that is, the diameter to the outer extremity of the thread. The area required for a particular tensile load can then be calculated from the following expression:

$$A_D = \frac{P_u}{\phi 0.75F_u} \quad \text{with } \phi = 0.75$$

In Table 8-7 of the Manual entitled "Threading Dimensions for High-Strength and Non-High-Strength Bolts" properties of standard threaded rods are presented. Example 4-4 which follows illustrates the selection of a rod using this table. It will be noted that the LRFD Specification (Part 6, Section J1.7) states that the factored load P_u used for connection design may not be less than 10 k except for lacing, sag rods, or girts.

■ EXAMPLE 4-4

Using A36 steel and the LRFD Specification, select a standard threaded rod of A36 steel to support a tensile working dead load of 10 k and a tensile working live load of 20 k.

Solution.

$$P_u = (1.2)(10) + (1.6)(20) = 44 \text{ k} \leftarrow$$

$$A_D = \frac{P_u}{\phi 0.75 F_u} = \frac{44}{(0.75)(0.75)(58)} = 1.35 \text{ in.}^2$$

Use $1\frac{3}{8}$-in.-diameter rod with 6 threads per in. ($A_D = 1.49$ in.^{2}) ■

As shown in Fig. 4-3, upset rods sometimes are used, where the rod ends are made larger than the regular rod and the threads are placed in the upset ends. Threads obviously reduce the cross-sectional area of a rod. If a rod is upset and the threads are placed in that part of the rod, the result will be a larger cross-sectional area at the root of the thread than we would have if the threads were placed in the regular part of the rod.

Table J3.2 in the LRFD Specification states that the nominal tensile strength of the threaded portion of the upset ends is equal to $0.75 F_u A_D$, where A_D is the cross-sectional area of the rod at its major thread diameter. This value must be larger than the nominal body area of the rod (before upsetting) times F_y.

Upsetting permits the designer to use the entire area of the regular part of the bar for strength calculations. Nevertheless, the use of upset rods probably is not economical and should be avoided unless a large order is being made.

One situation in which tension rods are sometimes used is in steel-frame industrial buildings with purlins running between their roof trusses to support the roof surface. These types of buildings will also frequently have girts running between the columns along the vertical walls. (Girts are horizontal beams used on the sides of buildings, usually industrial, to resist lateral bending due to wind. They also are often used to support corrugated or other types of siding.) Sag rods may be required to provide support for the purlins parallel to the roof surface and vertical support for the girts along the walls. For roofs with steeper slopes than 1 vertically to 4 horizontally, sag rods are often considered necessary to provide lateral support for the purlins, particularly where the purlins consist of steel channels. Steel channels are commonly used as purlins, but they have very little resistance to lateral bending. Although the resisting moment needed parallel to the roof surface is small, an extremely large channel is required to provide such

Figure 4-3 A round upset rod.

New Albany Bridge crossing the Ohio River between Louisville, Ky., and New Albany, Ind. (Courtesy of the Lincoln Electric Company.)

a moment. The use of sag rods for providing lateral support to purlins made from channels usually is economical because of the bending weakness of channels about their y axes. For light roofs (as where trusses support corrugated steel roofs) sag rods will probably be needed at the one-third points if the trusses are more than 20 ft on centers. Sag rods at the midpoints are usually sufficient if the trusses are less than 20 ft on centers. For heavier roofs such as those made of slate, cement tile, or clay tile, sag rods will probably be needed at closer intervals. The one-third points will probably be necessary if the trusses are spaced at greater intervals than 14 ft and the midpoints will be satisfactory if truss spacings are less than 14 ft. Some designers assume that the load components parallel to the roof surface can be taken by the roof, particularly if it consists of corrugated steel sheets, and that tie rods are unnecessary. This assumption, however, is open to some doubt and definitely should not be followed if the roof is very steep.

Designers have to use their own judgment in limiting the slenderness values for rods, as they will usually be several times the limiting values mentioned for other types of tension members. A common practice of many designers is to use rod diameters no less than 1/500th of their lengths to obtain some rigidity, even though design calculations may permit smaller sizes.

It is usually desirable to limit the minimum size of sag rods to $\frac{5}{8}$ in. because smaller rods than these are often injured during construction. The threads on smaller rods are quite easily injured by overtightening, which seems to be a frequent habit of construction workers. Sag rods are designed for the purlins of a roof truss in Example 4-5. The rods are assumed to support the simple beam reactions for the components of the gravity loads (roofing, purlins, snow and ice) parallel to the roof surface. Wind forces are assumed to act perpendicular to the roof surfaces and theoretically will not affect the sag rod forces. The maximum force in a sag rod will occur in the top sag rod because it must support the sum of the forces in the lower sag rods. It is theoretically possible to use smaller rods for the lower sag rods, but this reduction in size will probably be impractical.

■ EXAMPLE 4-5

Design the sag rods for the purlins of the truss shown in Fig. 4-4. Purlins are to be supported at their one-third points between the trusses which are spaced 21 ft on centers. Use A36 steel and assume a minimum size rod of $\frac{5}{8}$ in. is permitted. A clay tile roof weighing 16 psf (0.77 kN/m²) of roof surface is used and supports a snow load of 20 psf (0.96 kN/m²) of horizontal projection of roof surface. Details of the purlins and the sag rods and their connections are shown in Figs. 4-4 and 4-5 on page 114. In these figures the dotted lines represent a common practice of using ties and struts in the end panels in the plane of the roof to give greater resistance to loads located on one side of the roof (a loading situation which might occur when snow is blown off one side of the roof during a severe windstorm).

Solution.   Gravity loads in psf of roof surface are as follows.

Average weight in psf of the 7 purlins on each side of the roof

$$= \frac{(7)(11.5)}{37.9} = 2.1 \text{ psf}$$

$$\text{Snow} = 20\left(\frac{3}{\sqrt{10}}\right) = 19.0 \text{ psf}$$

$$\text{Tile roofing} = 16.0 \text{ psf}$$

$$w_u = (1.2)(2.1 + 16.0) + (0.5)(19.0) = 31.22 \text{ psf}$$

$$w_u = (1.2)(2.1 + 16.0) + (1.6)(19.0) = 52.12 \text{ psf} \leftarrow$$

Component of loads parallel to roof surface $= \left(\frac{1}{\sqrt{10}}\right)(52.12) = 16.48 \text{ psf}$

You can see in Figs. 4-4 and 4-5 that half of the load component parallel to the roof surface between the top two purlins on each side of the truss is carried directly to the horizontal sag rod between the purlins. In this example there are seven purlins (with six spaces between them) on each side of the truss. Thus, 1/12th of the total inclined load goes directly to the horizontal sag rod and 11/12ths goes to the top inclined sag rod.

Load on top inclined sag rod $= \left(\frac{11}{12}\right)(37.9)(7)(16.48) = 4008 \text{ lb}$

$$A_D = \frac{P_u}{\phi 0.75 F_u} = \frac{4.008}{(0.75)(0.75)(58)} = 0.123 \text{ in.}^2$$

Use $\frac{5}{8}$-in. rod (minimum practical size),
11 threads per in. ($A_D = 0.307 \text{ in.}^2$)

Force in tie rod between ridge purlins

$$T = (37.9)(7)(16.48)\frac{\sqrt{10}}{3} = 4608 \text{ lb}$$

Figure 4-4 Plan of two bays of roof.

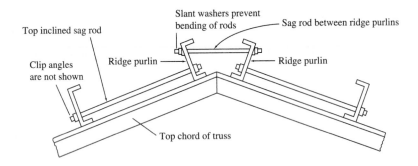

Figure 4-5 Details of sag rod connections.

or it equals $(4.008)\left(\dfrac{\sqrt{10}}{3}\right) + \left(\dfrac{1}{12}\right)(37.9)(7)(16.48) = 4608$ lb

$$A_D = \frac{4.608}{(0.75)(0.75)(58)} = 0.141 \text{ in.}^2$$

Use $\frac{5}{8}$-in. rod ■

4-4 PIN-CONNECTED MEMBERS

Until the early years of the twentieth century nearly all bridges in the United States were pin-connected, but today pin-connected bridges are used infrequently because of the advantages of bolted and welded connections. One trouble with the old pin-connected trusses was the wearing of the pins in the holes, which caused looseness of the joints.

An eyebar is a special type of pin-connected member whose ends where the pin holes are located are enlarged as shown in Fig. 4-6. Though just about obsolete today, eyebars at one time were very commonly used for the tension members of bridge trusses.

Pin-connected eyebars are used occasionally today as tension members for long-span bridges and as hangars for some types of bridges and other structures where they are normally subjected to very large dead loads. As a result the eyebars are usually prevented from rattling and wearing under live loads.

Eyebars are generally made not by forging them, but by thermally cutting them from plates. As stated in the LRFD Commentary (D3), extensive testing has shown that thermally cut members result in more balanced designs. The heads of eyebars are specially shaped so as to provide optimum stress flow around the holes. These proportions are based on long experience and testing with forged eyebars, and the results are rather conservative for today's thermally cut members.

The LRFD Specification (D3) provides detailed requirements for pin-connected members as to strength and proportions of the pins and plates. The design strength of such a member is the lowest value obtained from the following equations. Reference is made to Fig. 4-7.

Figure 4-6 An eyebar.

(a) Tensile strength on net section

(b) Shear design strength on effective area

(c) Bearing strength of surface (this is bearing on projected rectangular area behind bolt)

(d) Tensile strength of gross section

Figure 4-7 Strength of pin-connected tension members.

1. Tensile strength on the effective net area. See Fig. 4-7(a).

$$\phi = \phi_t = 0.75$$

$$P_n = 2tb_{\text{eff}}F_u \qquad \text{(LRFD Equation D3-1)}$$

In which t = plate thickness and $b_{\text{eff}} = 2t + 0.63$, but may not exceed the distance from the hole edge to the edge of the part measured perpendicular to the line of force.

2. Shear design strength on the effective area. See Fig. 4-7(b).

$$\phi = \phi_{\text{sf}} = 0.75$$

$$P_n = 0.6A_{\text{sf}}F_u \qquad \text{(LRFD Equation D3-2)}$$

In which $A_{\text{sf}} = 2t(a + d/2)$, where a is the shortest distance from the edge of the pin hole to the member edge measured parallel to the force.

3. Strength of surfaces in bearing. See Fig. 4-7(c).

$$\phi = 0.75$$

$$P_n = 1.8F_y A_{\text{pb}} \qquad \text{(LRFD Equation J8-1)}$$

In which A_{pb} = projected bearing area = dt. Notice that LRFD Equation J8-1 applies to milled surfaces, pins in reamed, drilled, or bored holes, and ends of fitted bearing stiffeners. (LRFD Specification J8 also provides other equations for determining the bearing strength for expansion rollers and rockers.)

4. Tensile strength of the gross section. See Fig. 4-7(d).

$\phi = 0.90$

$$P_n = F_y A_g \hspace{4cm} \text{(LRFD Equation D1-1)}$$

The LRFD Specification (D3) says that thicknesses $< \frac{1}{2}$ in. for both eyebars and pin-connected plates are only permissible when external nuts are provided to tighten the pin plates and filler plates into snug contact. The bearing design strength of such plates are provided in LRFD Specification J8.

In addition to the other requirements mentioned, LRFD Specification D3 specifies certain proportions between the pins and the eyebars. These values are based on long experience in the steel industry, and experimental work by B. G. Johnston.[1] It has been found that when eyebars and pin-connected members are made from steels with yield stresses greater than 70 kips per square inch (ksi) there may be a possibility of "dishing" (a complicated inelastic stability failure where the head of the eyebar tends to curl laterally into a dish-like shape). For this reason the LRFD Specification requires stockier member proportions for such situations (hole diameter not to exceed five times the plate thickness and the width of the eyebar reduced accordingly).

4-5 DESIGN FOR FATIGUE LOADS

It is not likely that fatigue stresses will be a problem in the average building frame because the changes in load in such structures usually occur only occasionally and produce relatively minor stress variations. Should there, however, be frequent variations or even reversals of stress, the matter of fatigue must be considered. Fatigue can be a problem in buildings when crane runway girders or heavy vibrating or moving machinery or equipment are supported.

If steel members are subjected to loads that are applied and then removed or changed significantly many thousands of times, cracks may occur, and they may spread so much as to cause failure. (The steel must be subjected to either stress reversals or to variations in tension stress because fatigue problems occur only when tension is present.) The result of fatigue loading is that steel members may fail at stresses well below the stresses at which they would fail if they were subject to static loads. The fatigue strength of a particular member is dependent on the number of cycles of load change, the range of load change, and the size of flaws.

---

[1] B. G. Johnston, "Pin-Connected Plate Links," *Transactions ASCE* 104 (1939).

Through the years many thousands of fatigue tests have been conducted, and the results have been discussed in innumerable papers on the subject. Despite all of this work, however, a clear understanding of fatigue behavior is not yet available to the steel design profession. As a result, steel design for fatigue situations is based almost entirely on test results.

One common method of running fatigue tests is the axial load method, in which members are subjected to alternating axial stress and the results are shown on an S-N curve. In this curve the data are shown with the maximum stress (S) plotted as the ordinate and the number of cycles of loading required to cause failure (N) plotted as the abscissa. A typical S-N curve is sketched in Fig. 4-8. Obviously, the values will change with different grades of steel and different temperatures.

The member specimens are tested at different stress levels and subjected to a sufficient number of cycles of loading to cause failure. You can see in Fig. 4-8 that the fatigue life of a member increases when the maximum stress is decreased. Furthermore, at low stress values the fatigue life becomes very large. In fact, there is a stress at which life seems to be infinite. This stress is usually called the *endurance limit*. Its value is obviously quite important for a material subject to many millions of cycles of loading, as, say, some type of rotating machinery.

The LRFD Specification in its Appendix K provides a simple design method for considering repeated loads. With this procedure the number of stress cycles, the expected range of stress (that is, the difference between the maximum and minimum estimated stresses), and the type and location of the member are considered. Based on this information a maximum allowable stress range is given *for service or working loads.*

In detail the maximum calculated stress in a member by the LRFD Specification may not exceed the basic nominal stress in that type of member, nor may its maximum range of stress exceed the permissible stress range provided in Appendix K of the Specification.

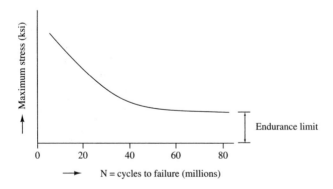

Figure 4-8 A typical S-N curve.

If it is anticipated that there will be fewer than 20,000 cycles of loading, no consideration needs to be given to fatigue. If the number of loading cycles exceeds 20,000, a permissible stress range is determined as follows.

1. The loading condition is determined from Table A-K3.1 of Appendix K of the LRFD Specification. For instance, if it is anticipated that there will be no fewer than 100,000 cycles of loading (that is, roughly 10 applications a day for 25 years) and no more than 500,000 cycles, loading condition 2 is to be used as determined from the table.
2. The type and location of material is selected from Fig. A-K3.1 of the Appendix. If a tension member consists of two angles fillet-welded to a plate it should be classified as illustrative example 17 in the figure. (The term *fillet weld* will be defined in Chapter 14. In such a weld one member is lapped over another and they are welded together.)
3. From Table A-K3.2 the stress category A, B, C, D, E, or F is selected. For a fillet-welded tension connection classified as illustrative example 17, the stress category is E.
4. Finally, from Table A-K3.3 of the Appendix the allowable stress range for the service loads for stress category E and loading condition 2 is $F_{sr} = 13$ ksi.

Example 4-6 presents the design of a two-angle tension member subjected to fluctuating loads using Appendix K of the LRFD Specification. Stress fluctuations and reversals are an everyday problem in the design of bridge structures. The 1989 AASHTO Specifications in their Article 10.3 provide allowable stress ranges determined in a manner very similar to those of the LRFD Specification.

■ EXAMPLE 4-6

An 18-ft member is to consist of a pair of equal leg angles with fillet-welded end connections. The dead load working tensile force is 30 k, while it is estimated that the working live load force may be applied 250,000 times and may vary from a compression of 12 k to a tension of 65 k. Select the angles using A36 steel and the LRFD Specification.

Solution.   Referring to Appendix K of the LRFD Specification and selecting the following values:
From Table A-K3.1—loading condition 2
From Fig. A-K3.1 and Table A-K3.2—illustrative example 17 and stress category E
Therefore, from Table A-K3.3—allowable stress range is 13 ksi
Range of P_u
For maximum tension

$$P_u = (1.2)(30) + (1.6)(65) = 140 \text{ k}$$

For compression

$$P_u = (1.4)(30) = 42 \text{ k}$$

$$P_u = (1.2)(30) + (1.6)(-12) = +16.8 \text{ k}$$

Therefore, member remains in tension

Estimated section size

$$A_g = \frac{P_u}{\phi_t F_y} = \frac{140}{(0.90)(36)} = 4.32 \text{ in.}^2$$

Try 2Ls 4 × 4 × $\frac{5}{16}$ ($A = 4.80$ in.^{2}, $r = 1.24$ in.)

Max service load tension $f_t = \dfrac{30 + 65}{4.80} = 19.79$ ksi

Min service load tension $f_t = \dfrac{30 - 12}{4.80} = \underline{3.75 \text{ ksi}}$

Actual stress range = 16.04 ksi

> 13 ksi   NG (means no good)

Try 2Ls 4 × 4 × $\frac{1}{2}$ ($A = 7.50$ in.^{2}, $r = 1.22$ in.)

Max service load $f_t = \dfrac{30 + 65}{7.50} = 12.67$ ksi

Min service load $f_t = \dfrac{30 - 12}{7.50} = \underline{2.40 \text{ ksi}}$

Actual stress range = 10.27 ksi < 13 ksi   (OK) ∎

4-6 COMPUTER EXAMPLE

In Example 4-7 the author presents the selection of a W12 section using the INSTEP program. You can select a series as W14, W12 or whatever, assume a value for U, and the computer will select the lightest satisfactory section in that series. You can then type that section in as the trial size, and the value of ϕP_n and $\dfrac{L}{r}$ will be shown.

∎ EXAMPLE 4-7

Using the computer diskette STEEL, select a 12-ft W12 section of A572 Grade 50 steel to support a tensile load = 1440 k. Assume that the effective net area of the member equals 90 percent of its gross area.

Solution.

INPUT

STEEL AND TRIAL MEMBER SIZE

- [C] Type Steel: A572 Grade 50
- [D] Trial Size: W12

TENSION DESIGN PARAMETERS

G L = 144
H A_n = 0
I U = .9

FACTORED DESIGN LOAD

L P_u = 1440

OUTPUT

The lightest acceptable W12 shape is being
determined. The net area is being assumed
to be 90% of the gross area of the member.
This assumption must be verified!

Suggested Trial Size: W12 × 136   ■

PROBLEMS

4-1 to 4-8. *Select sections for the conditions described using A36 steel unless otherwise noted, and neglecting block shear. $U = 1 - \dfrac{\bar{x}}{L}$ except for Prob. 4-8.*

4-1. Select the lightest W14 section available to support working tensile loads of $P_D = 220$ k and $P_L = 250$ k. The member is to be 30 ft long and is assumed to have two lines of holes for 1-in. bolts in each flange. There will be at least three bolts in each line 4 in. on center. (*Ans.* W14 × 74)

4-2. Repeat Prob. 4-1 selecting a W12 section.

4-3. Select the lightest W12 available to support a factored tensile load P_u of 380 k. Assume there are two lines of $\frac{3}{4}$-in. bolts in each flange (at least three bolts in each line 4 in. on center). The member is to be 28 ft long. (*Ans.* W12 × 45)

4-4. Select the lightest American Standard channel that will safely support the service tensile loads $P_D = 80$ k and $P_L = 110$ k. The member is to be 18 ft long and is assumed to have one line of holes for 1-in. bolts in each flange. Assume there are at least three holes in each line 4 in. on center.

4-5. Repeat Prob. 4-4 if an MC is to be used. (*Ans.* MC10 × 33.6)

4-6. Select the lightest available S section to support the service tensile loads $P_D = 80$ k and $P_L = 100$ k. The member is to be 20 ft long, and is assumed to have two lines of $\frac{7}{8}$-in. bolts in each flange (4 in a line 3 in. on center).

4-7. Repeat Prob. 4-6 if the member is to consist of A572 Grade 50 steel (4 bolts in a line 3 in. on center in the web only). (*Ans.* S12 × 31.8)

4-8. A welded tension member is to support a design load $P_u = 650$ k and is to consist of two channels placed 12 in. back-to-back with the flanges turned in. Select the lightest standard channels available. Assume $U = 0.87$. The member is to be 30 ft long.

4-9 to 4-16.   *Select the lightest section for each of the situations described. Assume bolts are 4 in. on center. Do not consider block shear. Determine U from LRFD Specification B.3 except for Prob. 4-11.*

Prob. no.	Section	P_D (kips)	P_L (kips)	Length (ft)	Steel	End connection	
4-9	W12	100	150	22	A572 Grade 50	Two lines of $\frac{3}{4}$-in. bolts (3 in a line) each flange	(Ans. W12 × 40)
4-10	W14	200	240	24	A572 Grade 50	Two lines of $\frac{7}{8}$-in. bolts (3 in a line) each flange	
4-11	W10	80	60	18	A572 Grade 50	Longitudinal welds to flanges only with $U = 0.87$.	(Ans. W10 × 17)
4-12	W12	400	100	28	A36	Two lines of $\frac{3}{4}$-in. bolts (2 in a line) each flange	
4-13	MC	70	90	20	A36	Transverse welds to flanges only	(Ans. MC12 × 35)
4-14	S	50	80	18	A572 Grade 50	One line of 1-in. bolts (3 in a line) each flange	
4-15	WT7	100	60	16	A36	Transverse welds to flange only	(Ans. WT7 × 26.5)
4-16	WT6	80	120	15	A242 Grade 46	Longitudinal welds to flange only	

4-17. A36 steel is to be used in selecting a single angle member to resist service tensile loads of $P_D = 70$ k and $P_L = 80$ k. The member is to be 20 ft long and is assumed to be connected with one line of four $\frac{7}{8}$-in. bolts in the longer leg if an unequal leg angle is used. Neglect block shear. $U = 1 - \frac{\bar{x}}{L}$. (*Ans.* L5 × 5 × $\frac{3}{4}$ or 6 × 4 × $\frac{3}{4}$)

4-18. Select a pair of C8s for the condition shown. Use A572 Grade 50 steel and assume transverse welds at the ends along the webs only. $L = 24$ ft, $P_u = 300$ k. Do not consider block shear.

10 in.

Problem 4-18

4-19. Repeat Prob. 4-17 using a pair of angles. long legs b. to b. Assume angle legs are touching and assume one hole for a $\frac{7}{8}$-in. bolt is to be taken out of each angle. Also assume $U = 0.85$. (*Ans.* 2Ls 6 × $3\frac{1}{2}$ × $\frac{3}{8}$)

4-20. Design member $L_2 L_3$ of the truss shown in the accompanying illustration. It is to consist of a pair of angles with a $\frac{3}{8}$-in. gusset plate between the angles at each joint. Use A36 steel and assume one line of three $\frac{3}{4}$-in. bolts in each vertical angle leg 4 in. on center. Consider only the angles shown in the double-angle tables of the Manual. For each load $P_D = 20$ k and $P_r = 12$ k (roof load). Do not consider block shear.

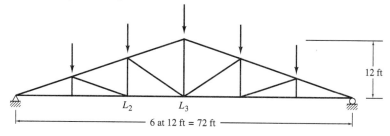

L_2 L_3

6 at 12 ft = 72 ft

12 ft

Problem 4-20

4-21. Select a single-angle tension member to resist the service loads $P_D = 80$ k and $P_L = 70$ k. The member is to be 18 ft long and is to be connected in the longer leg with one line of four $\frac{7}{8}$-in. bolts 4 in. on center. Assume $F_y = 40$ ksi and $F_u = 60$ ksi. Neglect block shear. $U = 1 - \frac{\bar{x}}{L}$. (*Ans.* L8 × 4 × $\frac{9}{16}$ or L6 × 6 × $\frac{9}{16}$)

4-22. Repeat Prob. 4-8 assuming that one line of $\frac{7}{8}$-in. bolts will be used in each flange with at least three bolts in each line 4 in. on center. Also design tie plates. Assume distance or gage from back of channel to center of line of bolts is 2 in. U is to be determined from LRFD Specification B3.

4-23. A tension member is to consist of four equal leg angles arranged as shown in the accompanying illustration to support the service loads $P_D = 180$ k and $P_L = 320$ k. The member is assumed to be 30 ft long and is to have one line of three $\frac{7}{8}$-in. bolts in each leg. Design the member with A572 Grade 50 steel including the necessary tie plates. Neglect block shear. (*Ans.* 4 Ls 5 × 5 × $\frac{1}{2}$)

18 in.

├─── 18 in. ───┤   **Problem 4-23**

4-24. Select a standard threaded round rod as a hanger to resist the service loads $P_D = 10$ k and $P_L = 12$ k using A36 steel.

4-25. Select a standard threaded rod to resist service loads $P_D = 12$ k and $P_L = 15$ k using A572 Grade 50 steel. (*Ans.* $1\frac{1}{4}$-in. rod 7 threads per in.)

4-26. The horizontal thrust at the base of the three-hinged arch shown in the accompanying illustration is to be resisted by a tie rod of A36 steel. What size threaded round tie rod should be used if the arch supports the working loads shown in the figure?

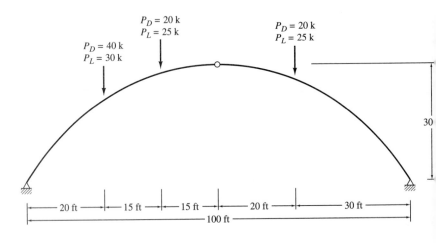

Problem 4-26

4-27. The roof trusses for a particular industrial building are spaced 21 ft on centers, have a roof covering weighing an estimated 6 psf of roof surface, and have purlins spaced as shown in the accompanying illustration and weighing an estimated 3 psf of roof surface. Design sag rods with A36 steel assuming a snow load of 30 psf of horizontal roof surface. The sag rods are to be used at the one-third points. (*Ans.* $\frac{5}{8}$-in. rod 11 threads per in.)

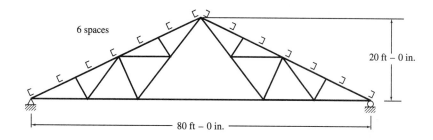

6 spaces

20 ft – 0 in.

80 ft – 0 in.

Problem 4-27

Chapter 5

Introduction to Axially Loaded Compression Members

5-1 GENERAL

There are several types of compression members, the column being the best known. Among the other types are the top chords of trusses, bracing members, the compression flanges of rolled beams and built-up beam sections, and members that are subjected simultaneously to bending and compressive loads. Columns are usually thought of as being straight vertical members whose lengths are considerably greater than their thicknesses. Short vertical members subjected to compressive loads are often called struts or simply compression members; however, the terms *column* and *compression member* will be used interchangeably in the pages that follow.

There are three general modes by which axially loaded columns can fail. These are flexural buckling, local buckling and torsional buckling. These modes of buckling are briefly defined below.

1. *Flexural buckling* (also called Euler buckling) is the primary type of buckling discussed in this chapter. Members are subject to flexure or bending when they become unstable.
2. *Local buckling* occurs when some part or parts of the cross section of a column are so thin that they buckle locally in compression before the other modes of buckling can occur. The susceptibility of a column to local buckling is measured by the width-thickness ratios of the parts of its cross section. This topic is addressed in Section 5-7 of this chapter.
3. *Torsional buckling* may occur in columns that have certain cross-sectional configurations. These columns fail by twisting (torsion) or by a combination of torsional and flexural buckling. This topic is initially addressed in Section 6-8 of this text.

The longer a column becomes for the same cross section the greater becomes its tendency to buckle and the smaller becomes the load it will support.

The tendency of a member to buckle is usually measured by its *slenderness ratio*, which has previously been defined as the ratio of the length of the member to its least radius of gyration. The tendency to buckle is also affected by such factors as the types of end connections, eccentricity of load application, imperfection of column material, initial crookedness of columns, residual stresses from manufacture, etc.

The loads supported by a building column are applied by the column section above and by the connections of other members directly to the column. The ideal situation is for the loads to be applied uniformly across the column, with the center of gravity of the loads coinciding with the center of gravity of the column. Furthermore, it is desirable for the column to have no flaws, to consist of a homogeneous material, and to be perfectly straight, but these situations are obviously impossible to achieve.

Loads that are exactly centered over a column are referred to as *axial* or *concentric loads*. The dead loads may or may not be concentrically placed over an interior building column and the live loads may never be centered. For an outside column the load situation is probably even more eccentric, as the center of gravity of the loads will usually fall well on the inner side of the column. In other words, it is doubtful that a perfect axially loaded column will ever be encountered in practice.

The other desirable situations are also impossible to achieve because of the following: imperfections of cross-sectional dimensions, residual stresses, holes punched for bolts, erection stresses, and transverse loads. It is difficult to take into account all of these imperfections in a formula.

Slight imperfections in tension members and beams can be safely disregarded as they are of little consequence. On the other hand, slight defects in columns may be of major significance. A column that is slightly bent at the time it is put in place may have significant bending moments equal to the column load times the initial lateral deflection. (Fig. 3 of Section 7-11 of the LRFD Commentary on the Code of Standard Practice for Steel Buildings and Bridges, located in Part 6 of the LRFD Manual, shows that the maximum out-of-straightness permitted in columns is $L/1000$ where L is the distance between braced points. Section E2 of the Commentary on the LRFD Specification states that average out-of-straightness values of $L/1500$ were used in developing the LRFD column formulas.)

Obviously, a column is a more critical member in a structure than is a beam or tension member because minor imperfections in materials and dimensions mean a great deal. This fact can be illustrated by a bridge truss that has some of its members damaged by a truck. The bending of tension members probably will not be serious as the tensile loads will tend to straighten those members; but the bending of any compression members is a serious matter, as compressive loads will tend to magnify the bending in those members.

The preceding discussion should clearly show that column imperfections cause them to bend and the designer must consider stresses due to those moments as well as due to axial loads. Chapters 5 to 7 are limited to a discussion of axially loaded columns, while members subjected to a combination of axial loads and bending loads are discussed in Chapter 11.

Two International Place, Boston, Mass. (Courtesy of Owen Steel Company, Inc.)

The spacing of columns in plan establishes what is called a *bay*. For instance, if the columns are 20 ft on center in one direction and 25 ft in the other direction the bay size is 20 ft × 25 ft. Larger bay sizes increase the user's flexibility in space planning. As to economy, a detailed study by John Ruddy[1]

---

[1] J. L. Ruddy, "Economics of Low-Rise Steel-Framed Structures," *Engineering Journal,* AISC, 20, no. 3 (3d quarter, 1983), pp. 107–118.

indicates that when shallow spread footings are used, bays with length-to-width ratios of about 1.25 to 1.75 and areas of about 1000 sq ft are the most cost efficient. When deep foundations are used, his study shows that larger bay areas are more economical.

5-2 RESIDUAL STRESSES

Research at Lehigh University has shown that residual stresses and their distribution are very important factors affecting the strength of axially loaded steel columns. These stresses are of particular importance for columns with slenderness ratios varying from approximately 40 to 120, a range that includes a very large percentage of practical columns. A major cause of residual stress is the uneven cooling of shapes after hot-rolling. For instance, in a W shape the outer tips of the flanges and the middle of the web cool quickly, while the areas at the intersection of the flange and web cool more slowly.

The quicker cooling parts of the sections when solidified resist further shortening, while those parts that are still hot tend to shorten further as they cool. The net result is that the areas that cooled more quickly have residual compressive stresses, while the slower cooling areas have residual tensile stresses. The magnitude of these stresses varies from about 10 to 15 ksi (69 to 103 MPa) although some values greater than 20 ksi (138 MPa) have been found.

When rolled-steel column sections with their residual stresses are tested, their proportional limits are reached at P/A values of only a little more than half of their yield stresses and the stress-strain relationship is nonlinear from there up to the yield stress. Because of the early localized yielding occurring at some points of the column cross sections, buckling strengths are appreciably reduced. Reductions are greatest for columns with slenderness ratios varying from approximately 70 to 90 and may possibly be as high as 25 percent.[2]

As a column load is increased, some parts of the column will quickly reach the yield stress and go into the plastic range because of residual compression stresses. The stiffness of the column will be reduced and become a function of the part of the cross section that is still elastic. A column with residual stresses will behave as though it has a reduced cross section. This reduced section or elastic portion of the column will change as the applied stresses change. The buckling calculations for a particular column with residual stresses can be handled by using an effective moment of inertia I_e of the elastic portion of the cross section or by using the tangent modulus. For the usual sections used as columns the two methods give fairly close results.

Welding can produce severe residual stresses in columns that actually can approach the yield point in the vicinity of the weld. Another important fact is that columns may actually be appreciably bent by the welding process, decid-

---

[2] L. S. Beedle and L. Tall, "Basic Column Strength," *Proc. ASCE* 86 (July 1960), pp. 139–173.

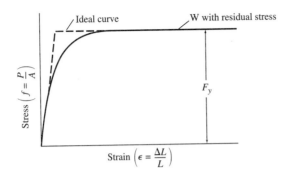

Figure 5-1 Effect of residual stresses on column stress-strain diagram.

edly affecting their load-carrying ability. Fig. 5-1 shows the effect of residual stresses (due to cooling and fabrication) on the stress-strain diagram for a hot-rolled W shape.

The welding together of built-up shapes frequently causes even higher residual stresses than those caused by the uneven cooling of hot-rolled H-shaped sections.

Residual stresses may also be caused during fabrication when *cambering* is performed by cold bending and due to cooling after welding. Cambering is the bending of a member in one direction so that it won't look so bad when the service loads bend it in the other direction. For instance, we may bend a beam initially upward   so that it will be approximately straight when its normal gravity loads are applied.

5-3 SECTIONS USED FOR COLUMNS

Theoretically, an endless number of shapes can be selected to safely resist a compressive load in a given structure. From a practical viewpoint, however, the number of possible solutions is severely limited by such considerations as sections available, connection problems, and type of structure in which the section is to be used. The paragraphs that follow are intended to give a brief résumé of the sections which have proved to be satisfactory for certain conditions. These sections are shown in Fig. 5-2 and the letters in parentheses in the paragraphs to follow refer to the parts of this figure.

The sections used for compression members usually are similar to those used for tension members with certain exceptions. The exceptions are caused by the fact that the strengths of compression members vary in some inverse relation to the slenderness ratios and stiff members are required. Individual rods, bars, and plates usually are too slender to make satisfactory compression members unless they are very short and lightly loaded.

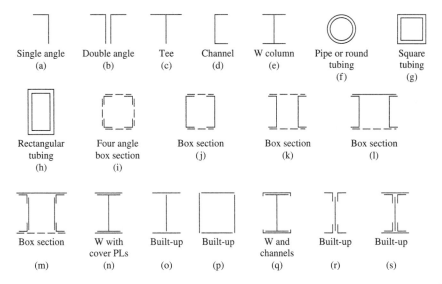

Figure 5-2 Types of compression members.

Single-angle members (a) are satisfactory for bracing and compression members in light trusses. Equal-leg angles may be more economical than unequal-leg angles because their least r values are greater for the same area of steel. The top chord members of bolted roof trusses might consist of a pair of angles back to back (b). There will often be a space between them for the insertion of a gusset or connection plate at the joints necessary for connections to other members. An examination of this section will show that it is probably desirable to use unequal-leg angles with the long legs back-to-back to give a better balance between the r values about the x and y axes.

If roof trusses are welded, gusset plates may be unnecessary and structural tees (c) might be used for the top chord compression members because the web members can be welded directly to the stems of the tees. Single channels (d) are not satisfactory for the average compression member because of their almost negligible r values about their web axes. They can be used if some method of providing extra lateral support in the weak direction is available. The W shapes (e) are the most common shapes used for building columns and for the compression members of highway bridges. Their r values, although far from being equal about the two axes, are much more nearly balanced than for channels.

Several famous bridges constructed during the nineteenth century (such as the Firth of Forth Bridge in Scotland and Ead's Bridge in St. Louis) made extensive use of tube-shaped members. Their use, however, declined due to connection problems and manufacturing costs, but with the development of economical welded tubing their use is again increasing (although the tube shapes of today are very small compared with the giant ones used in those early steel bridges).

Hollow structural sections (HSS) or structural tubing is today a very valuable part of the steel which is used for buildings and bridges and other structures. These clean, neat-looking sections are easily fabricated and erected. For small and medium loads the round or pipe sections (f) are quite satisfactory. They are often used as columns in long series of windows, as short columns in warehouses, as columns for the roofs of covered walkways, in the basements and garages of residences, etc. Pipe columns have the advantage of being equally rigid in all directions and usually are very economical unless moments are large. The LRFD Handbook furnishes the sizes of these sections and classifies them as being standard, extra strong, and double extra strong.

Square and rectangular tubing (g) and (h) have not been used to a great extent for columns until recently. In fact, for many years only a few steel mills manufactured steel tubing for structural uses. Perhaps the major reason tubing was not used to a great extent was the difficulty of making connections with rivets or bolts. This problem has been fairly well eliminated, however, by the advent of modern welding. The use of tubing for structural purposes by architects and engineers in the years to come will probably be greatly increased for several reasons. These include

1. The most efficient compression member is one that has a constant radius of gyration about its centroid, a property available in round tubing. Square tubing is the next most efficient compression member.
2. Four-sided and round structural tubing are much easier to paint than are the six-sided open W, S, and M sections. Furthermore, the rounded corners make it easier to apply paint or other coatings uniformly around the sections.
3. They have less surface area to paint or fireproof.

Bents fabricated from tubing sections in New Jersey. (Courtesy of Bethlehem Steel Corporation.)

4. They have excellent torsional resistance.

5. The surfaces of tubing are quite attractive.

6. When exposed the wind resistance of round tubing is only about two-thirds of that of flat surfaces of the same width.

7. If cleanliness is important, hollow structural tubing doesn't have the problem of dirt collecting between the flanges of open structural shapes.

A slight disadvantage that comes into play in certain cases is that the ends of tubing that are subject to corrosive atmospheres may have to be sealed to protect their inaccessible inside surfaces from corrosion. Although making very attractive exposed members for beams, tubing is at a definite weight disadvantage as compared with W sections, which have so much larger resisting moments for the same weights.

For many column situations, the weight of square or rectangular tube sections—usually referred to as hollow structural sections (HSS)—can be less than one half the weights required for open-profile sections (W, M, S, channel, and angle sections). It is true that tubes may cost perhaps 25 percent more per pound than open sections, but this still allows the possibility of up to 20 percent savings in many cases.[3]

Hollow structural sections are available with yield strengths up to 50 ksi and can be obtained with enhanced atmospheric corrosion resistance. Today they make up about 3 percent of the structural steel fabricated for buildings and bridges in the United States. In Japan and Europe the values are respectively 15 percent and 25 percent and still growing. Thus it is probable that their use will continue to rise in the United States in the years to come.[4] Detailed information including various tables on hollow structural sections can be obtained from the American Institute of Hollow Steel Sections, 929 McLaughlin Run Road, Suite 8, Pittsburgh, PA 15017.

Where compression members are designed for very large structures it may be necessary to use built-up sections. Built-up sections are needed where the members are long and support very heavy loads and/or when there are connection advantages. Generally speaking, a single shape such as a W section is more economical than a built-up section having the same cross-sectional area. With heavy column loads high-strength steels can frequently be used with very economical results if their increased strength permits the use of W sections rather than built-up members.

When built-up sections are used they must be connected on their open sides with some type of lacing (also called lattice bars) to hold the parts together in their proper positions and to assist them in acting together as a unit. The ends of these members are connected with tie plates (also called batten plates or stay

---

[3] "High Design, Low Cost", *Modern Steel Construction*" (Chicago: AISC, March-April 1990), pp. 32–34.

[4] Ibid., p. 34.

plates). Several types of lacing for built-up compression members are shown in Fig. 6-9.

The dashed lines in Fig. 5-2 represent lacing or discontinuous parts and the solid lines represent parts that are continuous for the full length of the members. Four angles are sometimes arranged as shown in (i) to produce large r values. This type of member may often be seen in towers and in crane booms. A pair of channels (j) is sometimes used as a building column or as a web member in a large truss. It will be noted that there is a certain spacing for each pair of channels at which their r values about the x and y axes are equal. Sometimes the channels may be turned out as shown in (k).

A section well suited for the top chords of bridge trusses is a pair of channels with a cover plate on top (l) and with lacing on the bottom. The gusset or connection plates at joints are conveniently connected to the insides of the channels and may also be used as splices. When the largest channels available will not produce a top chord member of sufficient strength a built-up section of the type shown in (m) may be used.

When the rolled shapes do not have sufficient strength to resist the column loads in a building or the loads in a very large bridge truss, their areas may be increased by adding plates to the flanges (n). In recent years it has been found that for welded construction a built-up column of the type shown in part (o) is a more satisfactory shape than a W with welded cover plates (n). It seems that in bending (as where a beam frames into the flange of a column) it is difficult to efficiently transfer tensile force through the cover plate to the column without pulling the plate away from the column. For very heavy column loads a welded box section of the type shown in (p) has proved to be quite satisfactory. Some other built-up sections are shown in parts (q) through (s). The built-up sections shown in parts (n) through (q) have an advantage over those shown in parts (i) through (m) in that they do not require the expense of the lattice work necessary in the former. Lateral shearing forces are negligible for the single column shapes and for the nonlatticed built-up sections, *but they are definitely not negligible for the built-up latticed columns.*

Today *composite columns* are being increasingly used. These columns usually consist of steel pipe and structural tubing filled with concrete or of W shapes encased in concrete usually square or rectangular in cross section. These columns are discussed in Chapter 17.

5-4 DEVELOPMENT OF COLUMN FORMULAS

The use of columns goes back before the dawn of history, but it was not until 1729 that a paper was published on the subject by Pieter van Musschenbroek, a Dutch mathematician.[5] He presented an empirical column formula for estimat-

---

[5] L. S. Beedle et al., *Structural Steel Design* (New York: Ronald Press, 1964), p. 269.

ing the strength of rectangular columns. A few years later in 1757 Leonhard Euler, a Swiss mathematician, wrote a paper of great value concerning the buckling of columns. He was probably the first person to realize the significance of buckling. The Euler formula, the most famous of all column expressions, is derived in Appendix B of this text. This formula, which is discussed in the next section, marked the real beginning of theoretical and experimental investigation of columns.

Engineering literature is filled with formulas developed for ideal column conditions, but these conditions are not encountered in actual practice. Consequently, practical column design is based primarily on formulas that have been developed to fit with reasonable accuracy test-result curves. The reasoning behind this procedure is simply the fact that the independent derivation of column expressions does not yield formulas that give results comparing closely with test-result curves for all slenderness ratios.

The testing of columns with various slenderness ratios results in a scattered range of values such as those shown by the broad band of dots in Fig. 5-3. The dots will not fall on a smooth curve even if all of the testing is done in the same laboratory because of the difficulty of exactly centering the loads, lack of perfect uniformity of the materials, varying dimensions of the sections, residual stresses, end restraint variations, etc. The usual practice is to attempt to develop formulas which give results represented by an approximate average of the test results. The student should also realize that laboratory conditions are not field conditions and column tests probably give the limiting values of column strengths.

The magnitudes of the yield stresses of the sections tested are quite important for short columns as their failure stresses are close to those yield stresses. For columns with intermediate slenderness ratios the yield stresses are of lesser importance in their effect on failure stresses and they are of no significance for long slender columns. For intermediate range columns residual stresses have more effect on the results while the failure stresses for long slender columns are

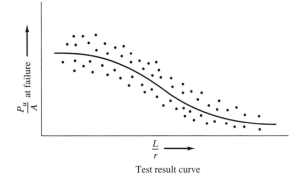

Test result curve

Figure 5-3 Test result curve.

very sensitive to end support conditions. In addition to residual stresses and nonlinearity of material, another dominant factor in its effect on column strength is member out-of-straightness.

5-5 THE EULER FORMULA

The stress at which a column buckles obviously decreases as the column becomes longer. After it reaches a certain length, that stress will have fallen to the proportional limit of the steel. For that length and greater lengths the buckling stress will be elastic.

For a column to buckle elastically it will have to be long and slender. Its buckling load P can be computed with the Euler formula which follows:

$$P = \frac{\pi^2 EI}{r^2}$$

This formula usually is written in a slightly different form which involves the column's slenderness ratio. Since $r = \sqrt{I/A}$ we can say that $I = Ar^2$. Substituting this value into the formula the Euler buckling or critical stress is obtained. It is referred to as F_e in the LRFD Manual.

$$\frac{P}{A} = \frac{\pi^2 E}{(L/r)^2} = F_e$$

Example 5-1 illustrates the application of the Euler formula to a steel column. If the value obtained for a particular column exceeds the steel's proportional limit the elastic Euler formula is not applicable.

■ **EXAMPLE 5-1**

 (a) A W10 × 22 is used as a 15-ft pin-connected column. Using the Euler expression determine the column's critical or buckling load. Assume the steel has a proportional limit of 36 ksi.
 (b) Repeat part (a) if the length is changed to 8 ft.

Solution.

 (a) Using a W10 × 22 ($A = 6.49$ in.², $r_x = 4.27$ in., $r_y = 1.33$ in.)

$$\text{Minimum } r = r_y = 1.33 \text{ in.}$$

$$\frac{L}{r} = \frac{(12)(15)}{1.33} = 135.34$$

Critical or buckling stress $F_e = \dfrac{(\pi^2)(29 \times 10^3)}{(135.34)^2}$

$= 15.63$ ksi $<$ the proportional
limit of 36 ksi

OK column is in elastic range

Critical or buckling load $= (15.63)(6.49) = 101.4\text{k}$

(b) Using an 8-ft W10 × 22

$$\frac{L}{r} = \frac{(12)(8)}{1.33} = 72.18$$

Critical or buckling stress $F_e = \dfrac{(\pi^2)(29 \times 10^3)}{(72.18)^2}$

$= 54.94$ ksi > 36 ksi

$\therefore$ column is in inelastic range and Euler equation not applicable ■

450 Lexington Ave., New York City. (Courtesy of Owen Steel Company, Inc.)

The student should carefully note that the buckling load determined from the Euler equation is independent of the strength of the steel used.

The equation is only useful if the end support conditions are carefully considered. The results obtained by application of the formula to specific examples compare very well with test results for centrally loaded, long slender columns with rounded ends. Designers, however, do not encounter perfect columns of this type. The columns with which they work do not have rounded ends and are not free to rotate because their ends are bolted or welded to other members. These practical columns have different amounts of restraint against rotation, varying from slight restraint to almost fixed conditions. For the actual cases encountered in practice where the ends are not free to rotate, different length values can be used in the formula and more realistic buckling stresses will be obtained.

To successfully use the Euler equation for practical columns the value of L should be the distance between points of inflection in the buckled shape. This distance is referred to as the *effective length* of the column. For a pinned-end column (whose ends can rotate but cannot translate) the points of inflection or zero moment are located at the ends a distance L apart. For columns with different end conditions the effective lengths may be entirely different. Effective lengths are discussed extensively in the next section.

5-6 END RESTRAINT AND EFFECTIVE LENGTHS OF COLUMNS

End restraint and its effect on the load-carrying capacity of columns is a very important subject indeed. Columns with appreciable end restraint can support considerably more load than those with little end restraint as at hinged ends.

The effective length of a column was defined in the last section as the distance between points of zero moment in the column, that is, the distance between its inflection points. In steel specifications the effective length of a column is referred to as KL where K is the *effective length factor*. K is the number that must be multiplied by the length of the column to obtain its effective length. Its size depends on the rotational restraint supplied at the ends of the column and upon the resistance to lateral movement provided.

The concept of effective lengths is simply a mathematical method of taking a column, whatever its end and bracing conditions, and replacing it with an equivalent pinned-end braced column. A complex buckling analysis could be made for a frame to determine the critical stress in a particular column. The K factor is determined by finding the pinned-end column with an equivalent length that provides the same critical stress. The K factor procedure is a method of making simple solutions for complicated frame buckling problems.

Columns with different end conditions have entirely different effective lengths. For this initial discussion it is assumed that no sidesway or joint translation is possible. Sidesway or joint translation means that one or both ends of a column can move laterally with respect to each other. Should a column be

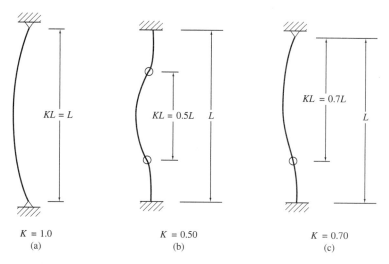

Figure 5-4 Effective lengths for columns in braced frames (sidesway prevented).

connected with frictionless hinges as shown in part (a) of Fig. 5-4, its effective length would be equal to the actual length of the column and K would equal 1.0. If there were such a thing as a perfectly fixed-ended column, its points of inflection (or points of zero moment) would occur at its one-fourth points and its effective length would equal $L/2$ as shown in part (b) of Fig. 5-4. As a result its K value would equal 0.50.

Obviously, the smaller the effective length of a particular column, the smaller its danger of lateral buckling and the greater its load-carrying capacity. In part (c) of Fig. 5-4 a column is shown with one end fixed and one end pinned. The K for this column is theoretically 0.70.

Actually there are neither perfect pin connections nor any perfect fixed ends, and the usual column falls in between the two extremes. This discussion would seem to indicate that column effective lengths always vary from an absolute minimum of $L/2$ to an absolute maximum of L, but there are exceptions to this rule. An example is given in Fig. 5-5(a) where a simple bent is shown. The base of each of the columns is pinned and the other end is free to rotate and move laterally (called sidesway). Examination of this figure will show that the effective length will exceed the actual length of the column as the elastic curve will theoretically take the shape of the curve of a pinned-end column of twice its length and K will theoretically equal 2.0. Notice in part (b) of the figure how much smaller the lateral deflection of column AB would be if it were pinned both top and bottom so as to prevent sidesway.

Structural steel columns serve as parts of frames, and these frames are sometimes *braced* and sometimes *unbraced*. A braced frame is one for which sidesway or joint translation is prevented by means of bracing, shear walls, or lateral support from adjoining structures. An unbraced frame does not have any

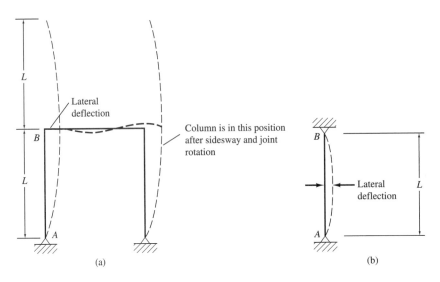

Figure 5-5

of these types of bracing supplied and must depend on the stiffness of its own members to prevent lateral buckling. For braced frames K values can never be greater than 1.0, but for unbraced frames the K values will always be greater than 1.0 because of sidesway.

Table C-C2.1 of the "Commentary on the LRFD Specification" gives recommended effective length factors when ideal conditions are approximated. This table is reproduced here as Table 5-1 with the permission of the AISC. Two sets of K values are provided in the table, one being the theoretical values and the other being the recommended design values based on the fact that perfectly pinned and fixed conditions are not possible. If the ends of the column of Fig. 5-4(b) were not quite fixed, the column would be a little freer to bend laterally and its points of inflection would be farther apart. The recommended design K given is 0.65, while the theoretical value is 0.5. As no column ends are perfectly fixed or perfectly hinged, the designer may wish to interpolate between the values given in the table, the interpolation to be based on his or her judgment of the actual restraint conditions.

The values in Table 5-1 are very useful for preliminary designs. When using this table we will almost always apply the design values and not the theoretical values. In fact, the theoretical values should be used only for those very rare situations where fixed ends are really almost perfectly fixed and/or when simple supports are almost perfectly frictionless (this means almost never).

You will note in the table that for cases (a), (b), (c), and (e) the design values are greater than the theoretical values, but that is not the situation for cases (d) and (f), where the values are the same. The reason for this in each of these two latter cases is that if the pinned conditions are not perfectly frictionless the K values will become smaller, not larger. Thus, by making the design values the same as the theoretical ones, we are staying on the safe side.

TABLE 5-1   COLUMN EFFECTIVE LENGTHS

	(a)	(b)	(c)	(d)	(e)	(f)
Buckled shape of column is shown by dashed line						
Theoretical K value	0.5	0.7	1.0	1.0	2.0	2.0
Recommended design value when ideal conditions are approximated	0.65	0.80	1.2	1.0	2.10	2.0

End condition code	Rotation fixed and translation fixed
	Rotation free and translation fixed
	Rotation fixed and translation free
	Rotation free and translation free

Source: Load and Resistance Factor Design Specification for Structural Steel Buildings, December 1, 1993, (Chicago: AISC, 1994), p. 6-184 in the LRFD Manual.

The K values in Table 5-1 are probably very satisfactory to use for designing isolated columns, but for the columns in continuous frames they are probably only satisfactory for making preliminary or approximate designs. Such columns are restrained at their ends by their connections to various beams and the beams themselves are connected to other columns and beams at their other ends and thus also restrained. These connections can appreciably affect the K values. As a result, for most situations the values in Table 5-1 are not sufficient for final designs.

For continuous frames it is necessary to use a more accurate method for computing K values. Usually this is done by using the alignment charts which will be presented in the first section of Chapter 7. There we will find charts for determining K values for columns in frames braced against sidesway and for frames not braced against sidesway. These charts should almost always be used for final column designs.

5-7 STIFFENED AND UNSTIFFENED ELEMENTS

Up to this point in the text the author has only considered the overall stability of members and yet it is entirely possible for the thin flanges or webs of a column or beam to buckle locally in compression well before the calculated buckling strength of the whole member is reached. When thin plates are used to carry compressive stresses they are particularly susceptible to buckling about their weak axes due to the small moments of inertia in those directions.

The LRFD Specification (Section B5) provides limiting values for the width-thickness ratios of the individual parts of compression members and for the parts of beams in their compression regions. The student is quite well aware of the lack of stiffness of thin pieces of cardboard or metal or plastic with free edges. If, however, one of these elements is folded or restrained, its stiffness is appreciably increased. For this reason two categories are listed in the LRFD Manual: *stiffened elements* and *unstiffened elements.*

An unstiffened element is a projecting piece with one free edge parallel to the direction of the compression force, while a stiffened element is supported along the two edges in that direction. These two types of elements are illustrated in Fig. 5-6. In each case the width, *b*, and the thickness, *t*, of the elements in question are shown.

Depending on the ranges of different width-thickness ratios for compression elements and depending on whether the elements are stiffened or unstiffened, the elements will buckle at different stress situations.

For establishing width-thickness ratio limits for the elements of compression members, the LRFD Specification divides members into three classifications as follows: compact sections, noncompact sections, and slender com-

(a) Unstiffened elements

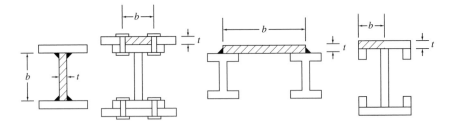

(b) Stiffened elements

Figure 5-6

pression elements. These classifications, which decidedly affect the design compression stresses to be used for columns, are discussed in the paragraphs to follow.

Compact Sections   A compact section is one that has a sufficiently stocky profile so that it is capable of developing a fully plastic stress distribution before buckling. The term *plastic* means stressed throughout to the yield stress and is discussed at length in Chapter 8. For a compression member to be classified as compact its flanges must be continuously connected to its web or webs and the width-thickness ratios of its compression elements may not be greater than the limiting ratios λ_P given in Table 5-2 on pages 144–145 (Table B5.1 of Part 6 of the LRFD Manual).

Noncompact Sections   A noncompact section is one for which the yield stress can be reached in some but not all of its compression elements before buckling occurs. It is not capable of reaching a fully plastic stress distribution. In Table 5-2 the noncompact sections are those which have width-thickness ratios greater than λ_P but not greater than λ_r.

Slender Compression Elements   A slender element with a cross section that does not satisfy the width-thickness requirements of Table 5-2 can still be used as a column, but the procedure for doing so is very complex. Furthermore, the reduction in design stress is severe. As a result it is usually more economical to thicken members to take them out of the slender range.

Almost all of the W, M, and S shapes listed in the LRFD Manual are compact for 36 and 50 ksi yield stress steels. A few of them are noncompact (and are so indicated in the column and beam tables of the Manual). None of them are classified as being slender for these two yield stresses.

Should the width-thickness limits for noncompact sections be exceeded, Appendix B5.3 of the LRFD Specification must be consulted. The formulas presented there are so complex and tedious to apply that we would be wise to almost never use members falling in this classification. Slender compression elements are discussed in detail and a numerical example is presented in Appendix C of this book.

5-8 LONG, SHORT, AND INTERMEDIATE COLUMNS

A column subject to an axial compression load will shorten in the direction of the load. If the load is increased until the column buckles, the shortening will stop and the column will suddenly bend or deform laterally and may at the same time twist in a direction perpendicular to its longitudinal axis.

The strength of a column and the manner in which it fails are greatly dependent on its effective length. A very short stocky steel column may be loaded until the steel yields and perhaps on into the strain hardening range. As a result it can support about the same load in compression that it can in tension.

TABLE 5-2 LIMITING WIDTH-THICKNESS RATIOS FOR COMPRESSION ELEMENTS

	Description of Element	Width-Thick-ness Ratio	Limiting Width-Thickness Ratios	
			λ_p (compact)	λ_r (noncompact)
Unstiffened Elements	Flanges of I-shaped rolled beams and channels in flexure	b/t	$65/\sqrt{F_y}\,[c]$	$141/\sqrt{F_y - 10}$
	Flanges of I-shaped hybrid or welded beams in flexure	b/t	$65/\sqrt{F_{yf}}$	$\dfrac{162}{\sqrt{(F_{yf} - 16.5)/k_e}}[f]$
	Flanges projecting from built-up compression members	b/t	NA	$109/\sqrt{F_y/k_e}\,[f]$
	Outstanding legs of pairs of angles in continuous contact, flanges of channels in axial compression; angles and plates projecting from beams or compression members	b/t	NA	$95/\sqrt{F_y}$
	Legs of single angle struts; legs of double angle struts with separators; unstiffened elements, i.e., supported along one edge	b/t	NA	$76/\sqrt{F_y}$
	Stems of tees	d/t	NA	$127/\sqrt{F_y}$
Stiffened Elements	Flanges of square and rectangular box and hollow structural sections of uniform thickness subject to bending or compression; flange cover plates and diaphragm plates between lines of fasteners or welds	b/t	$190/\sqrt{F_y}$	$238/\sqrt{F_y}$

TABLE 5-2 (Continued)

Description of Element		Width-Thick-ness Ratio	Limiting Width-Thickness Ratios	
			λ_p (compact)	λ_r (noncompact)
Stiffend Elements	Unsupported width of cover plates perforated with a succession of access holes [b]	b/t	NA	$317/\sqrt{F_y}$
	Webs in flexural compression [a]	h/t_w	$640/\sqrt{F_y}$ [c]	$970/\sqrt{F_y}$ [g]
	Webs in combined flexural and axial compression	h/t_w	for $P_u/\phi_b P_y \leq 0.125$ [c] $\dfrac{640}{\sqrt{F_y}}\left(2.33 - \dfrac{2.75\,P_u}{\phi_b P_y}\right)$ for $P_u/\phi_b P_y > 0.125$ [c] $\dfrac{191}{\sqrt{F_y}}\left(2.33 - \dfrac{P_u}{\phi_b P_y}\right) \geq \dfrac{253}{\sqrt{F_y}}$	[g] $\dfrac{970}{\sqrt{F_y}}\left(1 - 0.74\dfrac{P_u}{\phi_b P_y}\right)$
	All other uniformly compressed stiffened elements, i.e., supported along two edges	b/t h/t_w	NA	$253/\sqrt{F_y}$
	Circular hollow sections in axial compression	D/t	NA [d]	$3,300/F_y$
	In flexure		$2,070/F_y$	$8,970/F_y$

[a] For hybrid beams, use the yield strength of the flange F_{yf} instead of F_y.

[b] Assumes net area of plate at widest hole.

[c] Assumes an inelastic rotation capacity of 3. For structures in zones of high seismicity, a greater rotation capacity may be required.

[d] For plastic design use $1,300/F_y$.

[e] F_r = compressive residual stress in flange
   = 10 ksi for rolled shapes
   = 16.5 ksi for welded shapes

[f] $k_c = \dfrac{4}{\sqrt{h/t_w}}$ but not less than $0.35 \leq k_c \leq 0.763$

[g] For members with unequal flanges, see Appendix B 5.1. F_y is the specified minimum yield stress of the type of steel being used.

As the effective length of a column increases, its buckling stress will decrease. If the effective length exceeds a certain value the buckling stress will be less than the proportional limit of the steel. Columns in this range are said to fail *elastically*.

As previously shown in Section 5-5, very long steel columns will fail at loads that are proportional to the bending rigidity of the column (EI) and independent of the strength of the steel. For instance, a long column constructed with a 36 ksi yield stress steel will fail at just about the same load as one constructed with a 100 ksi yield stress steel.

Columns are sometimes classed as being long, short, or intermediate. A brief discussion of each of these classifications is presented in the paragraphs to follow.

Long Columns   The Euler formula predicts very well the strength of long columns where the axial buckling stress remains below the proportional limit. Such columns will buckle *elastically*.

Short Columns   For very short columns the failure stress will equal the yield stress and no buckling will occur. (For a column to fall into this class it would have to be so short as to have no practical application. Thus no further reference is made to them here.)

Intermediate Columns   For intermediate columns some of the fibers will reach the yield stress and some will not. The members will fail by both yielding and buckling and their behavior is said to be *inelastic*. Most columns fall into this range. (For the Euler formula to be applicable for such columns it would have to be modified according to the reduced modulus concept or the tangent modulus concept to account for the presence of residual stresses.)

In Section 5-9 formulas are presented with which the LRFD estimates the strength of columns in these different ranges.

5-9 COLUMN FORMULAS

The LRFD Specification provides one equation (it's the Euler equation) for long columns with inelastic buckling and an empirical parabolic equation for short and intermediate columns. With these equations a critical or buckling stress, F_{cr}, is determined for a compression member. Once this stress is computed for a particular compression member it is multiplied by the cross-sectional area of the member to obtain the member's nominal strength. The design strength of the member can then be determined as follows:

$$P_n = A_g F_{cr} \qquad \text{(LRFD Equation E2-1)}$$

$$P_u = \phi_c A_g F_{cr} \quad \text{with } \phi = 0.85$$

One LRFD equation for F_{cr} is for inelastic buckling and the other is for elastic buckling. In both equations λ_c in easily remembered form is $\sqrt{F_y/F_e}$

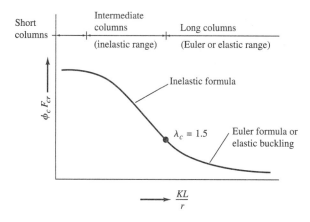

Figure 5-7

where F_e is the Euler stress $\pi^2 E/(KL/r)^2$. Substituting this value for F_e we get the form of λ_c given in the LRFD Specification.

$$\lambda_c = \frac{KL}{r\pi} \sqrt{\frac{F_y}{E}} \qquad \text{(LRFD Equation 2-4)}$$

Both equations for F_{cr} include the estimated effects of residual stresses and initial out-of-straightness of the members. The inelastic formula that follows is an empirical or test result formula.

$$F_{cr} = (0.658^{\lambda_c^2})\, F_y \text{ for } \lambda_c \leq 1.5 \qquad \text{(LRFD Equation E2-2)}$$

The other equation is for elastic or Euler buckling and is the familiar Euler equation multiplied by 0.877 to estimate the effect of out-of-straightness.

$$F_{cr} = \left(\frac{0.877}{\lambda_c^2}\right) F_y \text{ for } \lambda_c > 1.5 \qquad \text{(LRFD Equation E2-3)}$$

These equations are represented graphically in Fig. 5-7.

After looking at these equations the reader might think that their use would be very tedious and time-consuming with a pocket calculator. Such calculations, however, rarely have to be made because the LRFD Manual provides computed $\phi_c F_{cr}$ values for steels with $F_y = 36$ ksi and 50 ksi for KL/r values from 1 to 200 and has shown the results in Tables 3-36 and 3-50 of Part 6 of the LRFD Manual. Furthermore the Manual has another table (Table 4) from which the user may obtain values for steels with any F_y values.

5-10 MAXIMUM SLENDERNESS RATIOS

In Part 6, Section B7, the LRFD states that compression members *preferably* should be designed with KL/r ratios not exceeding 200. The reader might note from LRFD Tables 3-36 and 3-50 that design stresses $\phi_c F_{cr}$ for KL/r values of

200 are both 5.33 ksi. Should slenderness ratios larger than these be used the $\phi_c F_{cr}$ values will be very small and will require the user to substitute into the column formulas provided in Section 5-9.

5-11 EXAMPLE PROBLEMS

In this section three simple numerical column problems are presented. In each case the design strength of a column is calculated. In Example 5-2(a) the author determines the strength of a W section. The value of K is determined as described in Section 5-6, the effective slenderness ratio is computed, and the design stress of the member $\phi_c F_{cr}$ is selected from the appropriate LRFD table and multiplied by the cross-sectional area of the column.

It will be noted that the LRFD Manual in its Part 3 has further simplified the calculations required by computing the column design strength $\phi_c F_{cr} A_g$ for each of the steel shapes normally used as columns for F_y values of 36 and 50 ksi for most of the commonly used effective lengths or KL values given in feet. The use of these tables is illustrated in Example 5-2(b).

■ **EXAMPLE 5-2**

(a) Using the column design stress values shown in Table 3-50, Part 6 of the LRFD Manual, determine the design strength $(P_u = \phi_c P_n)$ of the $F_y = 50$ ksi axially loaded column shown in Fig. 5-8.

(b) Repeat the problem using the column tables of Part 2 of the Manual.

Figure 5-8

Solution.

(a) Using a W12 × 72 ($A = 21.1$ in.^{2}, $r_x = 5.31$ in., $r_y = 3.04$ in.)

$K = 0.80$ from Table 5-1

Obviously $(KL/r)_y > (KL/r)_x$ and thus controls

$$\left(\frac{KL}{r}\right)_y = \frac{(0.80)(12 \times 15)}{3.04} = 47.37$$

$\phi_c F_{cr} = 36.07$ ksi from Table 3-50, Part 6, LRFD Manual

$$P_u = \phi_c P_n = \phi_c F_{cr} A_g = (36.07)(21.1) = 761.1 \text{ k}$$

(b) Entering column tables Part 2 of Manual with $K_y L_y$ in feet.

$$K_y L_y = (0.80)(15) = 12 \text{ ft}$$
$$P_u = \phi_c P_n = 761 \text{ k} \qquad \blacksquare$$

In Example 5-3 the author illustrates the computations necessary to determine the design strength of a built-up column section. Several special requirements for built-up column sections are described in Chapter 6.

■ **EXAMPLE 5-3**
Determine the design strength $\phi_c P_n$ of the axially loaded column shown in Fig. 5-9 if $KL = 19$ ft and 50 ksi steel is used.

Figure 5-9

Solution.

$$A = (20)(\tfrac{1}{2}) + (2)(12.6) = 35.2 \text{ in.}^2$$

$$\bar{y} \text{ from top} = \frac{(10)(0.25) + (2)(12.6)(9.50)}{35.2} = 6.87 \text{ in.}$$

$$I_x = (2)(554) + (25.2)(2.63)^2 + (\tfrac{1}{12})(20)(\tfrac{1}{2})^3 + (10)(6.62)^2$$
$$= 1721 \text{ in.}^4$$
$$I_y = (2)(14.4) + (12.6)(6.877)^2(2) + (\tfrac{1}{12})(\tfrac{1}{2})(20)^3 = 1554 \text{ in.}^4$$

$$\text{Least } r = r_y = \sqrt{\frac{1554}{35.2}} = 6.64 \text{ in.}$$

$$\left(\frac{KL}{r}\right)_y = \frac{(12)(19)}{6.64} = 34.34$$

$$\phi_c F_{cr} = 38.99 \text{ ksi}$$

$$\phi_c P_n = P_u = (38.99)(35.2) = 1372.4 \text{ k}$$

To determine the design compression stress to be used for a particular column it is theoretically necessary to compute both $(KL/r)_x$ and $(KL/r)_y$. The reader will notice, however, that for most of the steel sections used for columns r_y will be much less than r_x. As a result only $(KL/r)_y$ is calculated for most columns and used in the applicable column formulas.

For some columns, particularly the long ones, bracing is supplied perpendicular to the weak axis thus reducing the slenderness or the length free to buckle in that direction. This may be accomplished by framing braces or beams into the sides of a column. For instance, horizontal members called *girts* running parallel to the exterior walls of a building frame may be framed into the sides of columns. The result is stronger columns and ones for which the designer needs to calculate both $(KL/r)_x$ and $(KL/r)_y$. The larger ratio obtained for a particular column indicates the weaker direction and will be used for calculating the design stress $\phi_c F_{cr}$ for that member.

Bracing members must be capable of providing the necessary lateral forces without buckling themselves. The forces to be taken are quite small and are often conservatively estimated to equal 0.02 times the column design loads. These members can be selected as are other compression members. A bracing member must be connected to other members that can transfer the horizontal force by shear to the next restrained level. If this is not done little lateral support will be provided for the original column in question.

If the lateral bracing were to consist of a single bar or rod (⊥) it would not prevent twisting and torsional buckling of the column (see Chapter 6). As torsional buckling is a difficult problem to handle, we should provide lateral bracing that prevents lateral movement and twist.[6]

---

[6] J. A. Yura, "Elements for Teaching Load and Resistance Factor Design" (New York: AISC, August 1987), p. 20.

Steel columns may also be built into substantial masonry walls in such a manner that they are substantially supported in the weaker direction. The designer, however, should be quite careful in assuming complete lateral support parallel to the wall, because a poorly built wall will not provide 100 percent lateral support.

Example 5-4 illustrates the calculations necessary to determine the design strength of a column with two unbraced lengths.

■ EXAMPLE 5-4

(a) Using Table 3-50 of Part 6 of the LRFD Manual, determine the design strength $\phi_c P_n$ of the 50 ksi axially loaded W14 × 90 shown in Fig. 5-10. Because of its considerable length this column is braced perpendicular to its weak or y axis at the points shown in the figure. These connections are assumed to permit rotation of the member but to prevent translation or sidesway.

(b) Repeat part (a) using the column tables of Part 2 of the Manual.

Figure 5-10

Solution.

(a) Using W14 × 90 ($A = 26.5$ in.², $r_x = 6.14$ in., $r_y = 3.70$ in.)
   Determining effective lengths

$$K_x L_x = (0.80)(32) = 25.6 \text{ ft}$$
$$K_y L_y = (1.0)(10) = 10 \text{ ft} \leftarrow$$
$$K_y L_y = (0.80)(12) = 9.6 \text{ ft}$$

Computing slenderness ratios

$$\left(\frac{KL}{r}\right)_x = \frac{(12)(25.6)}{6.14} = 50.03 \leftarrow$$

$$\left(\frac{KL}{r}\right)_y = \frac{(12)(10)}{3.70} = 32.43$$

$$\phi_c F_{cr} = 35.39 \text{ ksi}$$

$$\phi_c P_n = (35.39)(26.5) = \underline{\underline{937.8 \text{ k}}}$$

(b) Noting from part (a) solution that there are two different KL values

$$K_x L_x = 25.6 \text{ ft}$$

$$K_y L_y = 10 \text{ ft}$$

Controlling $K_y L_y$ for use in tables is either 10 ft or $\dfrac{K_x L_x}{r_x/r_y}$

$\dfrac{r_x}{r_y}$ for W14 × 90 from column tables = 1.66

$$\frac{K_x L_x}{r_x/r_y} = \frac{25.6}{1.66} = 15.42 \text{ ft} \leftarrow$$

From column tables with $K_y L_y = 15.42$ ft we find by interpolation

$$\phi_c P_n = \underline{\underline{937.8 \text{ k}}}$$ ■

5-12 COMPUTER EXAMPLE

The axial design compression strength of a W section is determined in Example 5-5 with the computer program INSTEP.

■ **EXAMPLE 5-5**
Using the enclosed diskette determine the design compression strength of a W14 × 132 which consists of A572 Grade 50 steel. $K_x = K_y = 1.0$ and $L_x = L_y = 20$ ft.

Solution.   The author specifies the design load P_u to be equal to 0. Regardless of the value specified, the computer is going to print out the design capacity of the column.   ■

INPUT

STEEL AND TRIAL MEMBER SIZE

B Type Steel: A572 Grade 50
C Trial Size: W14 × 132

UNBRACED LENGTH PARAMETERS

F $L_x = 240$
G $K_x = 1$
H $L_y = 240$
I $K_y = 1$

FACTORED DESIGN LOAD

L $P_u = 0$

OUTPUT

CHECK WIDTH/THICKNESS RATIOS

$$\left[\frac{b}{t}\right] = 14.725/2/1.03$$
$$= 7.14806 < 95/\sqrt{50} \quad \text{OK}$$

$$\left[\frac{h}{t}\right] = (14.66 - 2(1.6875))/.645$$
$$= 17.4961 < 253/\sqrt{50} \quad \text{OK}$$

FIND CONTROLLING KL/r RATIO

$$\left[\frac{KL}{r}\right]_x = 1(240)/6.28$$
$$= 38.2166 < 200 \quad \text{OK}$$

$$\left[\frac{KL}{r}\right]_y = 1(240)/3.76$$
$$= 63.8298 < 200 \quad \text{OK}$$

COMPUTE DESIGN LOAD

$$\lambda_c = \frac{63.8298}{3.14159}\sqrt{\frac{50}{29000}} = .843645$$

$F_{cr} = (0.658)^{.84365^2} (50) = 37.119$
$P_n = (38.8)37.119 = 1440.21$
$0.85P_n = 1224.18 > P_u \quad \text{OK}$

This Member is Acceptable   ■

PROBLEMS

5-1 to 5-4. *Determine the critical buckling load for each of the columns using the Euler equation. $E = 29 \times 10^6$ psi. Proportional limit = 30,000 psi. Assume simple ends and maximum permissible $L/r = 200$.*

5-1. A solid square bar 1.0 in. × 1.0 in.
   (a) $L = 3$ ft 0 in. (*Ans.* 18.44 k)
   (b) $L = 4$ ft 6 in. (*Ans.* 8.20 k)

5-2. The pipe section shown.
   (a) $L = 36$ ft 0 in.
   (b) $L = 24$ ft 0 in.
   (c) $L = 12$ ft 0 in.

$\frac{1}{2}$ in.

8 in.

Problem 5-2

5-3. A W12 × 50. $L = 20$ ft 0 in. (*Ans.* 280.6 k)

5-4. The two C12 × 30s shown for $L = 36$ ft 0 in.

C12 × 30

12 in.

Problem 5-4

5-5. Using 50 ksi steel and the LRFD specification determine the design strength $\phi_c P_n$ for each of the compression members.
   (a) A W14 × 90 with $KL = 12$ ft (*Ans.* 1008 k)
   (b) A W12 × 72 with $KL = 22$ ft (*Ans.* 516.7 k)
   (c) An S10 × 35 with $KL = 12$ ft (*Ans.* 86.0 k)

5-6. Using the LRFD Specification determine the design strength $\phi_c P_n$ for each of the following columns.
   (a) A W14 × 68 with fixed ends, $L = 18$ ft 6 in., $F_y = 50$ ksi.
   (b) A W12 × 40 with pinned ends, $L = 22$ ft 0 in., $F_y = 50$ ksi.
   (c) A W10 × 49 with one end fixed and the other pinned, $L = 22$ ft 6 in., $F_y = 50$ ksi.
   (d) A W14 × 193 with fixed ends, $L = 26$ ft 0 in., $F_y = 50$ ksi.

(e) A pipe 12 × strong with pinned ends, $L = 20$ ft 0 in., A36 steel.

(f) Two 6 × 6 × 1 Ls separated $\frac{3}{8}$ in. (for gusset PLs at ends), pinned ends, $L = 24$ ft 6 in., $F_y = 50$ ksi.

5-7. A W12 × 79 with a $\frac{3}{4}$ × 16-in. cover plate bolted to each flange is to be used for a column with $KL = 20$ ft. Find its design strength $\phi_c P_n$ if $F_y = 50$ ksi (*Ans. $P_u = 1526.9$ k*).

5-8. Using $F_y = 50$ ksi determine the design strength $\phi_c P_n$ of the axially loaded compression members shown.

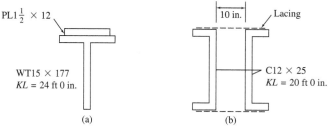

Problem 5-8

5-9. Using $F_y = 50$ ksi determine the design strength $\phi_c P_n$ of the axially loaded compression members shown.

Problem 5-9

5-10. Using a steel with $F_y = 50$ ksi calculate the design strength $\phi_c P_n$ of the concentrically loaded columns shown.

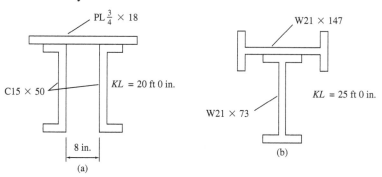

Problem 5-10

5-11. Using $F_y = 50$ ksi determine $\phi_c P_n$ for the axially loaded columns shown.

PL1 × 22

W18 × 71

12 in.

20 ft

(a) (Ans. 2524.3 k)

PL$\frac{3}{4}$ × 12

4Ls 8 × 4 × $\frac{5}{8}$
(long legs
b. to b.)

28 ft

(b) (Ans. 212.7 k)

Problem 5-11

5-12. Determine the design strength $\phi_c P_n$ for each of the axially loaded columns shown. $F_y = 50$ ksi.

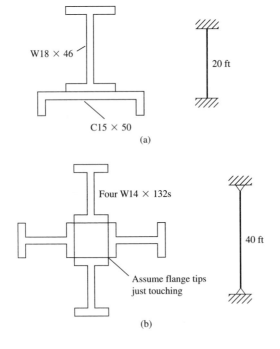

W18 × 46

C15 × 50

20 ft

(a)

Four W14 × 132s

Assume flange tips
just touching

40 ft

(b)

Problem 5-12

5-13. A 27-ft axially loaded W 12 × 65 column is laterally supported perpendicular to its y axis at middepth. Determine $\phi_c P_n$ for this column if $F_y = 50$ ksi and if effective length factors $= 1.0$ are assumed in each case. (*Ans.* 616.5 k)

5-14. Determine the maximum service live load that the column shown can support if the load is 1/3 dead load and 2/3 live load. $K_x L_x = 24$ ft and $K_y L_y = 12$ ft. $F_y = 50$ ksi.

PL$\frac{3}{4}$ × 14

PL$\frac{3}{4}$ × 20

PL$\frac{3}{4}$ × 14

Problem 5-14

5-15. Compute the maximum total service live load that can be applied to the A572, Grade 50 section shown if $K_x L_x = 15$ ft and $K_y L_y = 10$ ft. Assume the service load is 30 percent dead and 70 percent live. (*Ans.* $P_L = 287.2$k)

PL1 × 12

PL1 × 8

Problem 5-15

5-16 to 5-18. *Using the computer diskette enclosed with this book repeat the following problems.*

5-16. Prob. 5-5.

5-17. Parts (a) to (d) of Prob. 5-6 (*Ans.* 661 k, 135 k, 361 k, 2014 k)

5-18. Prob. 5-13.

Chapter 6

Design of Axially Loaded Compression Members

6-1 INTRODUCTION

In this chapter the designs of several axially loaded columns are presented. Included are the selections of single shapes, W sections with cover plates, and built-up sections constructed with channels. Also included are the designs of sections whose unbraced lengths in the x and y directions are different and the sizing of lacing and tie plates for built-up sections with open sides. Another topic considered is the flexural-torsional buckling of sections.

The design of columns with formulas involves a trial-and-error process. The design stress $\phi_c F_{cr}$ is not known until a column size is selected and vice versa. Once a trial section is assumed, the r values for that section can be obtained and substituted into the appropriate column equation to determine its design stress. Examples 6-1, 6-3, and 6-4 illustrate this procedure.

The designer may assume a design stress, divide that stress into the factored column load to give an estimated column area, select a column section which has approximately that area, determine its design stress, multiply that stress by the cross-sectional area of the section to obtain the member's design strength to see if the section selected is over- or underdesigned, and if appreciably so, try another size. The student may feel that he or she does not have sufficient background or knowledge to make reasonable initial design stress assumptions. If this student will read the information contained in the next few paragraphs, however, he or she will immediately be able to make excellent estimates.

The effective slenderness ratio (KL/r) for the average column of 10- to 15-ft length will generally fall between about 40 and 60. If for a particular column a KL/r somewhere in this approximate range is assumed and substituted into the appropriate column equation (in the LRFD this usually means looking into the

Georgia Railroad Bank and Trust Company Building, Atlanta, Ga. (Courtesy of Bethlehem Steel Corporation.)

tables where the design stresses have already been calculated for KL/r values from 0 to 200) the result will usually be a very satisfactory design stress estimate.

In Example 6-1 a column with $KL = 10$ ft is selected using the LRFD formulas. An effective slenderness ratio of 50 is assumed, the design stress for that value is determined from Table 3-50 of Part 6 of the Manual, and the resulting stress is divided into the factored column load to obtain an estimated column area. After a trial section is selected with approximately that area, its actual slenderness ratio and design strength are determined. The first estimated size in Example 6-1, though quite close, is a little too small, but the next larger section in that series of shapes is tried and found to be satisfactory.

To estimate the effective slenderness ratio for a particular column the designer may estimate a value a little higher than 40 to 60 if the column is appreciably longer than the 10- to 15-ft range and vice versa. A very heavy factored column load, say in the 750- or 1000-k range or higher will require a rather large column for which the radii of gyration will be larger and the designer may estimate a little smaller value of KL/r. For lightly loaded bracing members he or she may estimate high slenderness ratios perhaps over 100.

■ **EXAMPLE 6-1**

Using $F_y = 36$ ksi select the lightest W14 available for the service column loads $P_D = 100$ k and $P_L = 160$ k. $KL = 10$ ft.

Solution.

$$P_u = (1.2)(100) + (1.6)(160) = 376 \text{ k}$$

Assume $\dfrac{KL}{r} = 50$

$\phi_c F_{cr}$ from Table 3-36 (Part 6 of Manual) $= 26.83$ ksi

$$A \text{ required} = \frac{376}{26.83} = 14.01 \text{ in.}^2$$

Try W14 × 48 ($A = 14.1$ in.², $r_x = 5.85$ in., $r_y = 1.91$ in.)

Obviously $(KL/r)_y > (KL/r)_x$ and ∴ controls

$$(KL/r)_y = \frac{(12)(10)}{1.91} = 62.83$$

$\phi_c F_{cr}$ from Table 3-36 $= 24.86$ ksi

$$\phi_c P_n = (24.86)(14.1) = 350 \text{ k} < 376 \text{ k} \qquad (\text{NG})$$

∴ try next larger W14.

Try W14 × 53 ($A = 15.6$ in.², $r_x = 5.89$ in., $r_y = 1.92$ in.)

$$(KL/r)_y = \frac{(12)(10)}{1.92} = 62.5$$

$\phi_c F_{cr} = 24.91$ ksi

$$\phi_c P_n = (24.91)(15.6) = 388.6 \text{ k} > 376 \text{ k} \qquad (\text{OK})$$

Use W14 × 53   ■

6-2 LRFD DESIGN TABLES

For Example 6-2, Part 3 of the LRFD Manual is used to select various column sections from tables without the necessity of using a trial-and-error process. These tables provide axial design strengths ($\phi_c P_n$) for various practical effective lengths of the steel sections commonly used as columns (Ws, pipes, tubes, pairs of angles, and structural tees). The values are given with respect to the least radius of gyration for steels with $F_y = 36$ ksi and 50 ksi (except that the pipe strengths are given for 36 ksi steel only, while the strengths for the square and rectangular tubes are for 46 ksi steel).

For most columns consisting of single steel shapes the effective slenderness ratio with respect to the y axis $(KL/r)_y$ is larger than the effective slenderness ratio with respect to the x axis $(KL/r)_x$. As a result the controlling or smaller design stress is for the y axis. Because of this, the LRFD tables provide design strengths of columns with respect to their y axes. We will learn in the pages to follow how to handle situations in which $(KL/r)_x$ is larger than $(KL/r)_y$.

The resulting tables are very simple to use. The designer takes the KL value for the weaker direction in feet, enters the table in question from the left-hand side, and moves horizontally across the table. Under each section is listed the design strength $\phi_c P_n$ for that KL and the steel yield stress. (The 50 ksi column values are shaded in the tables.) As an illustration it is assumed that we have a factored design load $P_u = \phi_c P_n = 1200$ k, $K_y L_y = 12$ ft, and we want to select the lightest available W14 section using 50 ksi steel. We enter the tables with $KL = 12$ ft in the left column and read from left to right for 50 ksi steel the numbers 7240, 8530, 7760, 7030 k, and so on until several pages later where the consecutive values 1220 and 1110 k are found. The 1110 k value is not sufficient, and we go back to the 1220 k which falls under the W14 × 109. A similar procedure can be followed for the other available shapes.

Example 6-2 which follows illustrates the selection of W sections as well as pipes and tubes. It is possible to support a given column load with a standard pipe column; or with an extra strong pipe column (× strong) which has a smaller diameter but thicker walls, thus is heavier and more expensive; or with a double extra strong pipe column (×× strong) which has an even smaller diameter and even thicker walls and heavier weight.

■ **EXAMPLE 6-2**

Using the LRFD column tables of Part 3 of the Manual,

    (a) Select the lightest W section available for the loads, steel, and KL of Example 6-1.

    (b) Select the lightest standard, extra strong, and double extra strong pipe columns for the situation of part (a) of this problem.

    (c) Select the lightest square and rectangular tubes satisfactory for the situation of part (a), except use an F_y of 46 ksi.

Solution.

(a) Enter tables with $K_y L_y = 10$ ft and $P_u = \phi_c P_n = 376$ k
   Lightest suitable section in each W series

   W14 × 53 ($\phi_c P_n = 389$ k)
   W12 × 53 ($\phi_c P_n = 422$ k)
   W10 × 49 ($\phi_c P_n = 392$ k) ←
   W8 × 58 ($\phi_c P_n = 441$ k)

   Use W10 × 49

(b) Pipe columns

   Pipe 12 std ($\phi_c P_n = 429$ k) wt = 49.56 lb/ft
   Pipe 10 × strong ($\phi_c P_n = 465$ k) wt = 54.74 lb/ft
   Pipe 6 × strong ($\phi_c P_n = 399$ k) wt = 53.16 lb/ft

Eversharp, Inc., Building at Milford, Conn. (Courtesy of Bethlehem Steel Corporation.)

(c) Square and rectangular tubing ($F_y = 46$ ksi)

$$8 \times 8 \times \tfrac{3}{8}\,(\phi_c P_n = 392 \text{ k}) \text{ wt} = 37.60 \text{ lb/ft}$$
$$12 \times 10 \times \tfrac{1}{4}\,(\phi_c P_n = 390 \text{ k}) \text{ wt} = 36.03 \text{ lb/ft} \qquad ■$$

An axially loaded column is laterally restrained in its weak direction in Fig. 6-1. Example 6-3 illustrates the design of such a column with its different unsupported lengths in the x and y directions. The student can easily solve this problem by trial and error. A trial section can be selected as described in Section 6-1 of this chapter, the slenderness values $(KL/r)_x$ and $(KL/r)_y$ computed, $\phi_c F_{cr}$ determined for the larger value and multiplied by A_g to obtain $\phi_c P_n$. Then, if necessary, another size can be tried, and so on.

For this discussion it is assumed that K is the same in both directions. Then if we are to have equal strengths about the x and y axes the following relation must hold

$$\frac{L_x}{r_x} = \frac{L_y}{r_y}$$

For L_y to be equivalent to L_x we would have

$$L_x = L_y \frac{r_x}{r_y}$$

If $L_y(r_x/r_y)$ is less than L_x then L_x controls; if greater than L_x then L_y controls.

This brace must be a section which prevents lateral movement and twisting of the column. A rod or bar is not satisfactory.

Figure 6-1 A column laterally restrained at mid-depth in its weaker direction.

Based on the preceding information the LRFD Manual provides a method with which a section can be selected from its tables with little trial and error when the unbraced lengths are different. The designer enters the appropriate table with $K_y L_y$, selects a shape, takes the r_x/r_y value given in the table for that shape, and multiplies it by L_y. If the result is larger than $K_x L_x$, then $K_y L_y$ controls and the shape initially selected is the correct one. If the result of the multiplication is less than $K_x L_x$, then $K_x L_x$ controls and the designer will reenter the tables with a larger $K_y L_y$ equal to $K_x L_x/(r_x/r_y)$ and select the final section.

Example 6-3 illustrates the two procedures described here for selecting a W section which has different effective lengths in the x and y directions.

■ **EXAMPLE 6-3**

Select the lightest satisfactory W12 for the following conditions: $F_y = 50$ ksi, $P_u = 900$ k, $K_x L_x = 26$ ft, and $K_y L_y = 13$ ft.

(a) By trial and error
(b) Using LRFD tables

Solution.

(a) Using trial and error

$$\text{Assume } \frac{KL}{r} = 50$$

$$\phi_c F_{cr} = 35.40 \text{ ksi}$$

$$A \text{ required} = \frac{900}{35.40} = 25.42 \text{ in.}^2$$

Try W12 × 87 ($A = 25.6$ in.^{2}, $r_x = 5.38$ in., $r_y = 3.07$ in.)

$$\left(\frac{KL}{r}\right)_x = \frac{(12)(26)}{5.38} = 57.99 \leftarrow$$

$$\left(\frac{KL}{r}\right)_y = \frac{(12)(13)}{3.07} = 50.81$$

$$\phi_c F_{cr} = 33.23 \text{ ksi}$$

$$\phi_c P_n = (33.23)(25.6) = 850.6 \text{ k} < 900 \text{ k} \qquad \text{(NG)}$$

A subsequent check of the next larger W section (W12 × 96) shows it will work.

$$\underline{\underline{\text{Use W12 × 96}}}$$

(b) Using LRFD tables

Enter tables with $K_y L_y = 13$ ft

$$\text{Try W12 × 87}\left(\frac{r_x}{r_y} = 1.75\right) \quad \text{with } \phi_c P_n \text{ based on } K_y L_y$$

$$(K_yL_y)\left(\frac{r_x}{r_y}\right) = (13)(1.75) = 22.75 < K_xL_x$$

Therefore, K_xL_x controls.

Reenter tables with new $K_yL_y = \dfrac{K_xL_x}{r_x/r_y} = \dfrac{26}{1.75} = 14.86$

Use W12 × 96 ∎

6-3 COLUMN SPLICES

For multistory buildings column splices are desirably placed 4 feet above finished floors to permit the attachment of safety cables to the columns as may be required at floor edges or openings. This offset also enables us to keep the splices from interfering with the beam and column connections.

Typical column splices are shown in Fig. 6-2. The column ends are usually milled so they can be placed firmly in contact with each other for purposes of load transfer. When the contact surfaces are milled, a large part of the axial compression (if not all) can be transferred through the contacting areas. It is obvious that splice plates are necessary even though full contact is made between the columns and only axial loads are involved. For instance, the two column sections need to be held together during erection and afterward. What is necessary to hold them together is primarily based on the experience and judgment of the designer. Splice plates are even more necessary when consideration is given to the shears and moments existing in practical columns subjected to off-center loads, lateral forces, moments, and so on.

There is obviously a great deal of difference between tension splices and compression splices. In tension splices all load has to be transferred through the splice, whereas in splices for compression members a large part of the load can be transferred directly in bearing between the columns. The splice material is then needed to transfer only the remaining part of the load.

The amount of load to be carried by the splice plates is difficult to estimate. Should the column ends not be milled, the plates should be designed to carry 100 percent of the load. When the surfaces are milled and axial loads only are involved, the amount of load to be carried by the plates might be estimated to be from 25 to 50 percent of the total load. If bending is involved perhaps 50 to 75 percent of the total load may have to be carried by the splice material.

The bridge specifications spell out very carefully splice requirements for compression members, but the LRFD Specification does not. A few general requirements are given in Section J1.4 of the LRFD Specification. Such splices and a detailed discussion is presented in Section 11 of the Manual.

Fig. 6-2(a) shows a splice that may be used for columns with substantially the same nominal depths. The student will notice in the LRFD Manual that the

(a)

(b)

Figure 6-2 Column splices. (a) Columns from same W series with total depths close to each other (d_{lower} < 2 in. greater than d_{upper}). (b) Columns from different W series.

W shapes of a given nominal size can be divided into groups that are rolled with the same set of rolls. Because of the fixed dimensions of each set of rolls the clear distances between the flanges are constant for each shape in that group, although their total depths may vary greatly. For instance, the inside distance for each of the 28 shapes (running from the W14 × 61 to the W14 × 730) is 12.60 in., although their total depths vary from 13.89 in. to 22.42 in.

It is most economical to use the simple splices of Fig. 6-2(a). This can easily be accomplished by using one series of shapes for as many stories of a building as possible. For instance, we may select a particular W14 column section for the top story or top two stories of a building and then keep selecting heavier and heavier W14s for that column as we come down in the building. We may also switch to higher strength steel columns as we move down in the building, thus enabling us to stay with the same W series for even more floors. It will be necessary to use filler plates between the splice plates and the upper column if the upper column has a total depth significantly less than that of the lower column.

Fig. 6-2(b) shows a type of splice which can be used for columns of equal or different nominal depths. For this type of splice the butt plate is shop-welded to the lower column, and the clip angles used for field erection are shop-welded to the upper column. In the field the erection bolts shown are installed, and the upper column is field-welded to the butt plate. The horizontal welds on this plate resist shears and moments in the columns.

Welded column splice for Colorado State Building, Denver, Colo. (Courtesy of Lincoln Electric Company.)

Sometimes splices are used on all four sides of columns. The web splices are bolted in place in the field and field-welded to the column webs. The flange splices are shop-welded to the lower column and field-welded to the top column. The web plates may be referred to as *shear plates* and the flange plates as *moment plates*.

For multistory buildings the columns may be fabricated for one, two, or more stories. Theoretically, column sizes can be changed at each floor level so that the lightest total column weight is used. The splices needed at each floor will be quite expensive, however, and as a result it is usually more economical to use the same column sizes for at least two stories, even though the total steel weight will be higher. Seldom are the same sizes used for as many as three stories because three-story columns are so difficult to erect. The two-story heights work out very well most of the time.

6-4 BUILT-UP COLUMNS

As previously described in Section 5-3, compression members may be constructed with two or more shapes built-up into a single member. They may consist of parts in contact with each other, such as cover-plated sections $\underline{\text{I}}$, or they may consist of parts in near contact with each other, such as pairs of angles][that may be separated by a small distance from each other equal to the thickness of the end connections or gusset plates between them. They may consist of parts that are spread well apart, such as pairs of channels [ ] or four angles $\ulcorner \urcorner$, and so on.

Two-angle sections probably are the most common type of built-up member. (For example, they frequently are used as the members of light trusses.) When a pair of angles are used as a compression member they need to be fastened together so they will act as a unit. Welds may be used at intervals (with a spacer bar between the parts if the angles are separated) or they may be connected with "stitch bolts." When the connections are bolted, washers or "ring fills" are placed between the parts to keep them at the proper spacing if the angles are to be separated.

For long columns it may be convenient to use built-up sections where the parts of the columns are spread out or widely separated from each other. Before heavy W sections were made available, such sections were very commonly used in both buildings and bridges. Today these types of built-up columns are commonly used for crane booms and for the compression members of various kinds of towers. The widely spaced parts of these types of built-up members must be carefully laced or tied together.

Sections 6-5 and 6-6 of this chapter concern compression members that are built up from parts in direct contact (or nearly so) with each other. Section 6-7 addresses built-up compression members whose parts are spread widely apart.

6-5 BUILT-UP COLUMNS WITH COMPONENTS IN CONTACT WITH EACH OTHER

Should a column consist of two equal size plates, as shown in Fig. 6-3, and should those plates not be connected together, each plate will act as a separate column and each will resist approximately half of the total column load. In other words, the total moment of inertia of the column will equal two times the moment of inertia of one plate. The two "columns" will act the same and have equal deformations, as shown in part (b) of the figure.

Should the two plates be connected together sufficiently to prevent slippage on each other, as shown in Fig. 6-4, they will act as a unit. Their moment of inertia may be computed for the whole built-up section as shown in the figure and will be four times as large as it was for the column of Fig. 6-3, where slipping between the plates was possible. The reader should also notice that the plates of the column of Fig. 6-4 will deform different amounts as the column bends laterally.

Should the plates be connected in a few places, it would appear that the strength of the resulting column would be somewhere in between the two cases just described.

Reference to Fig. 6-3(b) shows that the greatest displacement between the two plates tends to occur at the ends and the least displacement tends to occur at middepth. As a result connections placed at column ends which will prevent slipping between the parts have the greatest strengthening effect, while those placed at middepth have the least effect.

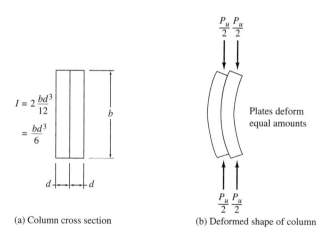

(a) Column cross section

(b) Deformed shape of column

Figure 6-3 Column consisting of two plates not connected to each other.

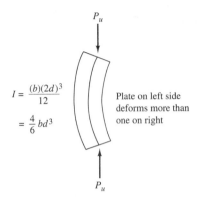

$$I = \frac{(b)(2d)^3}{12}$$

$$= \frac{4}{6}bd^3$$

Plate on left side
deforms more than
one on right

Figure 6-4 Column consisting of two plates
fully connected to each other.

Should the plates be fastened together at their ends with slip-resistant connectors, those ends will deform together and the column will take the shape
shown in Fig. 6-5. As the plates are held together at the ends the column will
bend in an S shape as shown in the figure.

If the column were to bend in the S shape shown, its K factor would
theoretically equal 0.5 and its KL/r value would be the same as the one for the
continuously connected column of Fig. 6-4.[1]

$$\frac{KL}{r} \text{ for the column of Fig. 6-4} = \frac{(1)(L)}{\sqrt{\frac{4}{6}bd^3/2bd}} = 1.732L$$

$$\frac{KL}{r} \text{ for the end-fastened column of Fig. 6-5} = \frac{(0.5)(L)}{\sqrt{\frac{1}{6}bd^3/2bd}} = 1.732L$$

Thus the design stresses are equal for the two cases and the columns would
carry the same loads. This is true for the particular case described here, but is
not applicable for the common case where the parts of Fig. 6-5 begin to separate.

Slip-resistant connection

Slip-resistant connection

Figure 6-5 Column consisting of two plates connected at its ends only.

---

[1] J. A. Yura, "Elements for Teaching Load and Resistance Factor Design" (Chicago: AISC, July
1987), pp. 17–19.

6-6 CONNECTION REQUIREMENTS FOR BUILT-UP COLUMNS WHOSE COMPONENTS ARE IN CONTACT WITH EACH OTHER

Several requirements concerning built-up columns are presented in LRFD Specification E4. When such columns consist of different components which are in contact with each other and which are bearing on base plates or milled surfaces, they must be connected at their ends with bolts or welds. If welds are used the weld lengths must at least equal the maximum width of the member. If bolts are used they may not be spaced longitudinally more than four diameters on center and the connection must extend for a distance at least equal to $1\frac{1}{2}$ times the maximum width of the member.

The LRFD Specification also requires the use of welded or bolted connections between the end ones described in the last paragraph. These must be sufficient to provide for the transfer of calculated stresses. If it is desired to have a close fit over the entire faying surfaces between the components it may be necessary to place the connectors even closer than is required for shear transfer.

When the component of a built-up column consists of an outside plate, the LRFD Specification provides specific maximum spacings for fastening. If intermittent welds are used along the edges of the components or if bolts are provided along all gage lines at each section, their maximum spacing may not be greater than $127/\sqrt{F_y}$ times the thickness of the thinner outside plate or 12 in. Should these fasteners be staggered on each gage line, however, they may not be spaced farther apart on each gage line than $190/\sqrt{F_y}$ times the thickness of the thinner part or 18 in.

In Chapter 12 high-strength bolts are referred to as being *snug-tight* or *slip-critical*. Snug-tight bolts are those that are tightened until all the plies of a connection are in firm contact with each other. This usually means the tightness obtained by the full manual effort of a worker with a spud wrench, or the tightness obtained after a few impacts with a pneumatic wrench.

Slip-critical bolts are tightened much more than are snug-tight bolts. They are tightened until their bodies or shanks have very high tensile stresses (approaching the lower bound of their yield stress). Such bolts clamp the connected parts of a connection so tightly together between the bolt and nut heads that loads are resisted by friction and slippage is nil. (We will see in Chapter 12 that where slippage is potentially a problem slip-critical bolts should be used. For example, they should be used if the working or service loads cause a large number of stress changes resulting in a possible fatigue situation in the bolts.)

In the following discussion the letter *a* represents the distance between connectors and r_i is the least radius of gyration of an individual component of the column. If compression members consisting of two or more shapes are used they must be connected together at intervals such that the effective slenderness ratio Ka/r_i of each of the component shapes between the connectors is not larger than 3/4 times the governing or controlling slenderness ratio of the whole built-up member. The end connections must be made with welds or slip-critical bolts with clean mill scale or blasted cleaned faying surfaces with Class A coatings. (A Class A coating is one that has a mean slip coefficient not less than 0.33. See Section 5(b) in "Specification for Structural Joints Using A325 or A490 Bolts" in Part 6 of the LRFD Manual.) Class A surfaces are typical in structural steel work. Snug-tight bolts may be used for interior bolts.

The design strength of compression members built up from two or more shapes in contact with each other will be determined with the usual applicable LRFD Equations (E2-1, E2-2, and E2-3) with one exception. Should the column tend to buckle in such a manner that relative deformations in the different parts cause shear forces in the connectors between the parts, it will be necessary to modify the KL/r value for that axis of buckling. This modification is required by Section E4 of the LRFD Specification.

Reference is made here to the cover-plated column of Fig. 6-6. If this section tends to buckle about its y axis the connectors between the W shape and the plates are not subjected to any calculated load. If, on the other hand, it tends to buckle about its x axis the connectors are subjected to shearing forces. The flanges of the W section and the cover plates will have different stresses and thus different deformations. The result will be shear in the connection between these parts and $(KL/r)_x$ will have to be modified by LRFD Equation E4-1 or E4-2 as described below. Equation E4-1 is based upon test results that supposedly account for shear deformations in the connectors. Equation 4-2 is based upon theory and was checked by means of tests.

(a) For intermediate connectors that are snug-tight bolted:

$$\left(\frac{KL}{r}\right)_m = \sqrt{\left(\frac{KL}{r}\right)_0^2 + \left(\frac{a}{r_i}\right)^2} \qquad \text{(LRFD Equation E4-1)}$$

Note that it is important to remember that the design strength of a built-up column will be reduced if the spacing of connectors is such that one of the components of the column can buckle before the whole column buckles.

(b) For intermediate connectors that are welded or have fully tensioned bolts as required for slip-critical joints:

$$\left(\frac{KL}{r}\right)_m = \sqrt{\left(\frac{KL}{r}\right)_0^2 + 0.82 \frac{\alpha^2}{(1 + \alpha^2)} \left(\frac{a}{r_{ib}}\right)^2} \qquad \text{(LRFD Equation E4-2)}$$

$x \;-\!\cdot\!-\!\!-\!-\!-\!-\!-\!\cdot\!-\!\cdot\,x$

Figure 6-6

In these two equations

$\left(\dfrac{KL}{r}\right)_0 =$ unmodified slenderness of the whole built-up member acting as a unit

$\left(\dfrac{KL}{r}\right)_m =$ modified slenderness of built-up member

$a =$ distance between connectors, in.

$r_i =$ minimum radius of gyration of individual component, in.

$r_{ib} =$ radius of gyration of individual component relative to its centroidal axis parallel to the member axis of buckling, in.

$h =$ distance between centroids of individual components perpendicular to the member axis of buckling, in.

$\alpha =$ separation ratio $= \dfrac{h}{2r_{ib}}$

For the case in which the column tends to buckle about an axis such as to cause shear in the connection between the column parts, it will be necessary to compute a modified slenderness ratio $(KL/r)_m$ for that axis and to check to see whether that value will cause a change in the design strength of the member. If it does, it may be necessary to revise sizes and repeat the steps just described.

Example 6-4 illustrates the design of a column consisting of a W section with cover plates bolted to its flanges as shown in Fig. 6-7. Even though snug-tight bolts are used for this column, you should realize that LRFD Specification E4 states that the end bolts must be slip-critical or the ends must be welded. This is required so the parts of the built-up section will not slip with respect to each other and thus will act as a unit in resisting loads. (As a practical note, the usual steel company required to tighten the end bolts to a slip-critical condition will probably just go ahead and tighten them all to that condition.)

As this type of built-up section is not shown in the column tables of the LRFD Manual, it is necessary to use a trial-and-error design procedure. An effective slenderness ratio is assumed. Then, $\phi_c F_{cr}$ for that slenderness ratio is determined and divided into the column design load to estimate the column area required. The area of the W section is subtracted from the estimated total area to obtain the estimated cover plate area. Cover plate sizes are then selected to provide the required estimated area.

13.12 in.

W12 × 120 (A = 35.3 in.^{2}, d = 13.12 in., b_f = 12.320 in., I_x = 1070 in.^{4}, I_y = 345 in.^{4})

Figure 6-7 W section used as column with cover plates.

Next the properties of the trial section are calculated. You will note that it is necessary to compute a modified value of $(KL/r)_x$. In this case the section tends to buckle about its y axis—thus longitudinal shear is not developed between the W section and the cover plates.

■ EXAMPLE 6-4

It is desired to design a column for $P_u = 2375$ k using $F_y = 50$ ksi and $KL = 14$ ft. A W12 × 120 (for which $\phi_c P_n = 1220$ k from the Part 3 tables of the Manual) is on hand. Design cover plates to be snug-tight bolted at 6 in. spacings to the W section shown in Fig. 6-7 to enable the column to support the required load.

Solution.

$$\text{Assume } \frac{KL}{r} = 50$$

$$\phi_c F_{cr} = 35.40 \text{ ksi}$$

$$A \text{ required} = \frac{2375}{35.40} = 67.09 \text{ in.}^2$$

$$-A \text{ of W12} \times 120 = -35.30$$

$$\text{Estimated } A \text{ of 2 plates} = 31.79 \text{ in}^2. \text{ or } 15.90 \text{ in.}^2 \text{ each}$$

Try 1 PL1 × 16 each flange

$$A = 35.30 + (2)(1)(16) = 67.30 \text{ in.}^2$$

$$I_x = 1070 + (2)(16)(7.06)^2 = 2665 \text{ in.}^4$$

$$r_x = \sqrt{\frac{2665}{67.30}} = 6.29 \text{ in.}$$

$$\left(\frac{KL}{r}\right)_x = \frac{(12)(14)}{6.29} = 26.71$$

$$I_y = 345 + (2)\left(\frac{1}{12}\right)(1)(16)^3 = 1027.7 \text{ in.}^4$$

$$r_y = \sqrt{\frac{1027.7}{67.30}} = 3.91 \text{ in.}$$

$$\left(\frac{KL}{r}\right)_y = \frac{(12)(14)}{3.91} = 42.97$$

$$P_u = (37.14)(67.30) = 2500 \text{ k} > 2375 \text{ k} \tag{OK}$$

Use W12 × 120 with 1 cover plate 1 × 16 each flange.   (Note: Many other plate sizes could have been selected.)   ■

6-7 BUILT-UP COLUMNS WITH COMPONENTS NOT IN CONTACT WITH EACH OTHER

Example 6-5 presents the design of a member built up from two channels that are not in contact with each other. The parts of such members need to be connected or laced together across their open sides. The design of lacing is discussed immediately after this example and is illustrated in Example 6-6.

■ EXAMPLE 6-5

Select a pair of 12-in. standard channels for the column and load shown in Fig. 6-8 using A36 steel. For connection purposes the back-to-back distance of the channels is to be 12 in.

$P_u = 450$ k

20 ft

$P_u = 450$ k

— 12 in. —

Figure 6-8 Column built up from two channels.

Solution.

Assume $\dfrac{KL}{r} = 50$

$$\phi_c F_{cr} = 26.83 \text{ ksi}$$

$$A \text{ required} = \frac{450}{26.83} = 16.77 \text{ in.}^2$$

Try 2C12 × 30s (For each channel $A = 8.82$ in.^{2}, $I_x = 162$ in.^{4}, $I_y = 5.14$ in.^{4}, $\bar{x} = 0.674$ in.)

$$I_x = (2)(162) = 324 \text{ in.}^4$$

$$I_y = (2)(5.14) + (2)(8.82)(5.326)^2 = 511 \text{ in.}^4$$

$$r_x = \sqrt{\frac{324}{(2)(8.82)}} = 4.29 \text{ in.}$$

$$KL = (1.0)(20) = 20 \text{ ft}$$

$$\frac{KL}{r} = \frac{(12)(20)}{4.29} = 55.94$$

$$\phi_c F_{cr} = 25.95 \text{ ksi}$$

$$\phi_c P_n = (25.95)(17.64) = 457.8 \text{ k} > 450 \text{ k}$$

Use 2C12 × 30s   ∎

The open sides of compression members that are built up from plates or shapes may be connected together with continuous cover plates with perforated holes for access purposes, or they may be connected together with lacing and tie plates.

The purpose of the perforated cover plates and the lacing or lattice work is to hold the various parts parallel and the correct distance apart and to equalize the stress distribution between the various parts. The student will understand the necessity for lacing if he or she considers a built-up member consisting of several sections (such as the four-angle member of Fig. 5-2) which supports a heavy compressive load. Each of the parts will tend to individually buckle laterally unless they are tied together to act as a unit in supporting the load. In addition to lacing, it is necessary to have tie plates (also called stay plates or batten plates) as near the ends of the member as possible and at intermediate points if the lacing is interrupted. Parts (a) and (b) of Fig. 6-9 show arrangements of tie plates and lacing. Other possibilities are shown in parts (c) and (d) of the same figure.

The failure of several structures in the past has been attributed to inadequate lacing of built-up compression members. Perhaps the best-known example was the failure of the Quebec Bridge in 1907. Following its collapse the general opinion was that the lattice work of the compression chords was too weak and resulted in failure.

If continuous cover plates perforated with access holes are used to tie the members together, the LRFD Specification E4 states that (a) they must comply with the limiting width-thickness ratios specified for compression elements in Section B5.1 of the LRFD Specification; (b) the ratio of the access hole length (in the direction of stress) to the hole width may not exceed 2; and (c) the clear distance between the holes in the direction of stress may not be less than the transverse distance between the nearest lines of connecting fasteners or welds. Stress concentrations and secondary bending stresses are usually neglected, but lateral shearing forces must be checked as they are for other types of lattice work. (The unsupported width of such plates at access holes is assumed to contribute to the design strength $\phi_c P_n$ of the member if the conditions as to sizes,

Figure 6-9 Lacing and perforated cover plates.

width-thickness ratios, etc., described in LRFD Specification E4 are met.) Perforated cover plates are attractive to many designers because of several advantages they possess.

1. They are easily fabricated with modern gas cutting methods.
2. Some specifications permit the inclusion of their net areas in the effective section of the main members, provided the holes are made in accordance with their empirical requirements which have been developed on the basis of extensive research.
3. Painting of the members is probably simplified as compared with ordinary lacing bars.

Dimensions of tie plates and lacing are usually controlled by specifications. Section E4 of the LRFD Specification states that tie plates shall have a thickness at least equal to one-fiftieth of the distance between the connection lines of welds or other fasteners.

Lacing may consist of flat bars, angles, channels, or other rolled sections. These pieces must be so spaced that the individual parts being connected will not have L/r values between connections which exceed the governing value for the entire built-up member. (The governing value is KL/r for the whole built-up section.) Lacing is assumed to be subjected to a shearing force normal to the member equal to not less than 2 percent of the compression design strength $\phi_c P_n$ of the member. The LRFD column formulas are used to design the lacing in the usual manner. Slenderness ratios are limited to 140 for single lacing and 200 for double lacing. Double lacing or single lacing made with angles should preferably be used if the distance between connection lines is greater than 15 in.

Example 6-6 illustrates the design of lacing and end tie plates for the built-up column of Example 6-5. Bridge specifications are somewhat different in their lacing requirements from the LRFD, but the design procedures are much the same.

■ EXAMPLE 6-6

Using the LRFD Specification design bolted single lacing for the column of Example 6-5. Reference is made to Fig. 6-10. Assume $\frac{3}{4}$-in. bolts.

Figure 6-10 Two-channel column section with lacing.

Solution.   Distance between lines of bolts is 8.5 in. < 15 in.; therefore, single lacing OK.

Assume lacing bars are inclined at 60° with axis of member. Length of channels between lacing connections is 8.5/cos 30° = 9.8 in., and L/r of 1 channel between connections is $9.8/0.763 = 12.9 < 55.94$, which is L/r of main member previously determined in Example 6-5.

Force on lacing bar:

$V_u = 0.02$ times design compressive strength of member (from Example 6-5)

$V_u = (0.02)(457.8) = 9.16 \text{ k}$

$\frac{1}{2}V_u = 4.58 \text{ k} = $ shearing force on each plane of lacing

Force in bar (with reference to bar dimensions in Fig. 6-10):

$$\left(\frac{9.8}{8.5}\right)(4.58) = 5.28 \text{ k}$$

Properties of flat bar:

$I = \frac{1}{12}bt^3$

$A = bt$

$r = \sqrt{\dfrac{\frac{1}{12}bt^3}{bt}} = 0.289t$

Design of bar:

Assume $\dfrac{L}{r} = $ maximum value of 140

$\dfrac{9.8}{0.289t} = 140$

$t = 0.242 \text{ in. (try } \frac{1}{4}\text{-in. flat bar)}$

$\dfrac{L}{r} = \dfrac{9.8}{(0.289)(0.250)} = 136$

$\phi_c F_{cr} = 11.54 \text{ ksi}$

Area required $= \dfrac{5.28}{11.54} = 0.458 \text{ in.}^2$ $(1.83 \times \frac{1}{4}$ needed$)$

Minimum edge distance if $\frac{3}{4}$-in. bolt used $= 1\frac{1}{4}$ in.

$\therefore$ Minimum length of bar $= 9.8 + (2)(1\frac{1}{4}) = 12.3$ in. say 14 in.

Use $\frac{1}{4} \times 2\frac{1}{2} \times 1$ ft 2 in. bars

Design of end tie plates:

Minimum length $= 8.5$ in.

Minimum $t = (\frac{1}{50})(8.5) = 0.17$ in.

Minimum width $= 8.5 + (2)(1\frac{1}{4}) = 11$ in.

Use $\frac{3}{16} \times 8\frac{1}{2} \times 0$ ft 12 in. end tie plates   ■

6-8 INTRODUCTORY REMARKS CONCERNING FLEXURAL-TORSIONAL BUCKLING OF COMPRESSION MEMBERS

Axially loaded compression members can theoretically fail in three different fashions: by flexural buckling, by torsional buckling, or by flexural-torsional buckling.

Flexural buckling (also called Euler buckling) is the situation considered up to this point in our column discussions where we have computed slenderness ratios for the principal column axes and determined $\phi_c F_{cr}$ for the highest ratios so obtained. Doubly symmetrical column members (such as W sections) are subject only to flexural buckling and torsional buckling.

As torsional buckling can be very complex, it is very desirable to prevent its occurrence. This may be done by careful arrangements of the members and by providing bracing to prevent lateral movement and twisting. If sufficient end supports and intermediate lateral bracing are provided, flexural buckling will always control. The column design strengths given in the LRFD column tables for W, M, S, tube, and pipe sections are based on flexural buckling.

Open sections such as Ws, Ms, and channels have little torsional strength, but box beams have a great deal. Thus if a torsional situation is encountered, it may be well to use box sections or to make box sections out of W sections by adding welded side plates $\left(\underline{\boxed{}}\right)$. Another way in which torsional problems can be reduced is to shorten the lengths of members that are subject to torsion.

For a singly symmetrical section such as a tee or double angle, Euler buckling may occur about the x or y axis. For equal-leg single angles Euler buckling may occur about the z axis. For all these sections flexural-torsional buckling is definitely a possibility and may control. (It will always control for unequal-leg single-angle columns.) The values given in the LRFD column load tables for double-angle and structural tee sections were computed for buckling about the weaker of the x or y axis and for flexural-torsional buckling.

The average designer does not consider the torsional buckling of symmetrical shapes or the flexural-torsional buckling of unsymmetrical shapes. The feeling is that these conditions don't control the critical column loads, or at least don't affect them very much. Should we have unsymmetrical columns or even symmetrical columns made up of thin plates, however, we will find that torsional buckling or flexural-torsional buckling may significantly reduce column capacities.

A flexural-torsional example is presented in Appendix D of this book.

6-9 SINGLE-ANGLE COMPRESSION MEMBERS

You will note that the author has not discussed the design of single-angle compression members up to this point. The AISC has long been concerned about the problems involved in loading such members fairly concentrically. It can be done rather well if the ends of the angles are milled and if the loads are applied through bearing plates. In practice, however, single-angle columns are often used in such a manner that rather large eccentricities of load applications are present. The sad result is that it is somewhat easy to greatly underdesign such members.

In Part 6 of the LRFD Manual a special specification is provided for the design of single-angle compression members. Though this specification includes information for tensile, shear, compressive, flexural, and combined loadings, the present discussion is concerned only with the compression case. Some rather complicated formulas are presented in the specification for computing the axial design strengths of single angles.

These formulas were developed to account for three different limit states that might occur in single-angle columns. These are flexural buckling, local buckling of thin angle legs, and flexural torsional buckling. Using these expressions the design strengths of concentrically loaded single angles are computed for 36 ksi and 50 ksi steels for a range of KL values. These strengths are shown at the end of Part 3 of the LRFD Manual.

Just before the single-angle column tables are presented in Part 3 of the Manual, a very practical example problem is presented in which a single-angle strut loaded with an estimated eccentricity is analyzed.

6-10 COMPUTER EXAMPLE

In Example 6-7 the computer diskette is used to select a column for an axial load. For the trial size the user specifies a series as W12, W10, etc. The computer will print out the lightest suitable section in the series selected. To try another series it is necessary only to revise the trial size and press the F1 key again.

■ EXAMPLE 6-7

Select the lightest W12 section to support an axial compression design load $P_u = 720$ k. Use A36 steel and assume $K_x = K_y = 1.0$ and $L_x = 20$ ft and $L_y = 10$ ft.

Solution.

INPUT

STEEL AND TRIAL MEMBER SIZE

| B | Type Steel: A36 |
| C | Trial Size: W12 |

UNBRACED LENGTH PARAMETERS

F	L_x = 240
G	K_x = 1
H	L_y = 120
I	K_y = 1

FACTORED DESIGN LOAD

| L | P_u = −720 |

OUTPUT

The lightest acceptable W12 shape is being determined. The adequacy of the suggested section must be verified by the design engineer!

Suggested Trial Size: W12 × 96   ■

PROBLEMS

All of the columns for these problems are assumed to be in frames that are braced against sidesway.

6-1 to 6-3. *Use a trial-and-error procedure in which a KL/r value is estimated, a value of $\phi_c F_{cr}$ is determined from the appropriate LRFD table (3-36 or 3-50 in Part 6 of the Manual), the estimated column area is determined, a trial section is selected, its P_u is computed, and then another size is tried if necessary, and so on.*

6-1. Select the lightest available W12 section to support the axial compression loads P_D = 140 k and P_L = 120 k if KL = 14 ft and A572 Grade 50 steel is used. (*Ans.* W12 × 50)

6-2. Select the lightest available W14 section to support the axial loads P_D = 200 k and P_L = 300 k if KL = 12 ft and F_y = 50 ksi.

6-3. Repeat Prob. 6-2 if F_y = 36 ksi. (*Ans.* W14 × 90)

6-4 to 6-20. *Take advantage of all available column tables in the Manual, particularly those in Part 3.*

6-4. Repeat Prob. 6-1.

6-5. Repeat Prob. 6-2. (*Ans.* W14 × 74)

6-6. Repeat Prob. 6-3.

6-7. Several building columns are to be designed using A572 Grade 50 steel and the LRFD Specification. Select the lightest available W sections for these columns that are described as follows:

(a) $P_u = 600$ k, $L = 14$ ft, pinned end supports. (*Ans.* W12 × 65)

(b) $P_u = 400$ k, $L = 12$ ft, fixed end supports. (*Ans.* W10 × 39)

(c) $P_u = 800$ k, $L = 18$ ft 6 in., fixed at bottom, pinned at top. (*Ans.* W12 × 87)

(d) $P_u = 1600$ k, $L = 17$ ft, pinned end supports. (*Ans.* W14 × 159)

6-8. Select the lightest W section available in 50 ksi steel to serve as a pinned end column to support the following axial loads: $P_D = 200$ k and P_W due to wind $= 400$ k. Assume KL is 15 ft.

6-9. A W section is to be selected to support an axial compressive load $P_u = 1800$ k. The member, which is to be 24 ft long and is to be pinned top and bottom, has lateral support (pinned) supplied in the weak direction at middepth. Select the lightest W12 or 14 section using A572 Grade 50 steel. (*Ans.* W14 × 176, W12 × 190)

6-10. Repeat Prob. 6-9 if $F_y = 36$ ksi.

6-11. Repeat Prob. 6-9 if lateral support is provided in the weak direction at the one-third points and the total column length is changed to 33 ft. Use A36 steel. (*Ans.* W14 × 257, W12 × 252)

6-12. A 27-ft column is laterally supported in the weak direction at its middepth. Select the lightest W section that can adequately support the axial gravity loads $P_D = 200$ k and $P_L = 150$ k using $F_y = 50$ ksi. Assume all Ks are 1.0.

6-13. A 14-ft column is to be built into a wall in such a manner that it will be continuously braced in its weak direction but not in its strong direction. If the member is to consist of A36 steel and is assumed to have pinned ends, select the lightest satisfactory W12 section available using the LRFD Specification. $P_u = 1000$ k. (*Ans.* W12 × 120)

6-14. Repeat Prob. 6-13 if $F_y = 50$ ksi.

6-15. A W14 section of 50 ksi steel is to be selected to support the axial compressive loads $P_D = 300$ k and $P_L = 350$ k. The member which is to be 30 ft long is to be fixed top and bottom and is to have lateral support at its one-third points perpendicular to the y axis (pinned). (*Ans.* W14 × 90)

6-16. Using A36 steel (except $F_y = 46$ ksi for square and rectangular tubing) select the lightest available rolled sections (W, HP, square, rectangular, or round tubing) that are adequate for the following situations:

   (a) $P_u = 300$ k, $L = 14$ ft, pinned ends.
   (b) $P_u = 420$ k, $L = 15$ ft, fixed ends.
   (c) $P_u = 740$ k, $L = 20$ ft, one end pinned and the other fixed.

6-17. Assuming axial loads only, select W sections for an interior column of the laterally braced frame shown in the accompanying illustration. Use $F_y = 50$ ksi and the LRFD Specification. Each column section can be used for one or two stories before it is spliced, whichever seems advisable. Miscellaneous data: concrete weighs 150 lb/ft³. LL on roof = 30 psf. Roofing = 6 psf. LL on interior floors = 80 psf. Partition load on interior floors = 15 psf. All joints are assumed to be pinned so as to simplify calculations. Frames 30 ft on center. (*Ans.* W14 × 53 top two stories and W14 × 90 bottom two stories)

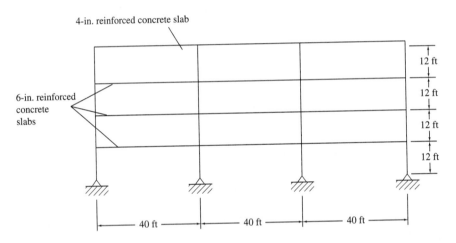

Problem 6-17

6-18. It is desired to design a column for $P_u = 2875$ k using A572 Grade 50 steel with $KL = 12$ ft. A W14 × 145 is on hand with a good supply of $\frac{3}{4}$-in. thick plates. Design cover plates to be welded to the flanges of the W section to enable the column to support the required load.

6-19. Determine the design compressive strength of the section shown if snug-tight bolts 4 ft on center are used to connect the A572 Grade 50 angles. $K_x L_x = K_y L_y = 24$ ft. (*Ans.* 111.2 k)

2Ls 8 × 6 × $\frac{1}{2}$
(short legs b. to b.)

$\frac{3}{8}$ in.   **Problem 6-19**

6-20. Repeat Prob. 6-19 if the angles are welded together with their long legs back to back at intervals of 6 ft.

6-21. Four $4 \times 4 \times \frac{1}{2}$ angles are used to form the member shown in the accompanying illustration. The member is 30 ft long, has pinned ends, and consists of A572 Grade 50 steel. Determine the design compressive strength of the member. Design single lacing and end tie plates assuming connections are made to the angles with $\frac{3}{4}$-in. bolts. (*Ans. P_u = 547.9 k; end tie plates $\frac{9}{32} \times 13 \times 1$ ft 4 in.; single lacing at 60° $\frac{3}{8} \times 2\frac{1}{2} \times 1$ ft $5\frac{1}{2}$ in.*)

18 in.

18 in.

Problem 6-21

6-22. Select a pair of standard channels to support an axial compressive load of P_u = 925 k. The member is to be 24 ft long with both ends pinned and is to be arranged as shown in the accompanying illustration. Use F_y = 50 ksi, and design single lacing and end tie plates assuming $\frac{3}{4}$-in. bolts are to be used for connections. Assume bolts located $2\frac{1}{4}$ in. from back of channels.

8 in.

Problem 6-22

6-23 to 6-28. *Using the enclosed computer diskette repeat the following problems:*

6-23. Prob. 6-1 (*Ans. W12 × 50*)

6-24. Prob. 6-3 using A572 Grade 50 steel.

6-25. Prob. 6-7 (*Ans.* W12 × 65, W10 × 39, W12 × 87, W14 × 159)

6-26. Prob. 6-9

6-27. Prob. 6-11 using A572 Grade 50 steel. (*Ans.* W14 × 193, W12 × 210)

Chapter 7

Design of Axially Loaded Compression Members Continued

7-1 FURTHER DISCUSSION OF EFFECTIVE LENGTHS

The subject of effective lengths was introduced in Chapter 5, and some suggested K factors were presented in Table 5-1. These factors were developed for columns with certain idealized conditions of end restraint which may be very different from practical design conditions. The table values are usually quite satisfactory for preliminary designs and for situations in which sidesway is prevented by bracing. Should the columns be part of a continuous frame subject to sidesway, however, it is often advantageous to make a more detailed analysis as described in this section. To a lesser extent this is also desirable for columns in frames braced against sidesway.

Perhaps a few explanatory remarks should be made at this point, defining sidesway as it pertains to effective lengths. For this discussion sidesway refers to a type of buckling. In statically indeterminate structures sidesway occurs where the frames deflect laterally due to the presence of lateral loads or unsymmetrical vertical loads or where the frames themselves are unsymmetrical. Sidesway also occurs in columns whose ends can move transversely when they are loaded until buckling occurs.

Should frames with diagonal bracing or rigid shear walls be used, the columns will be prevented from sidesway and provided with some rotational

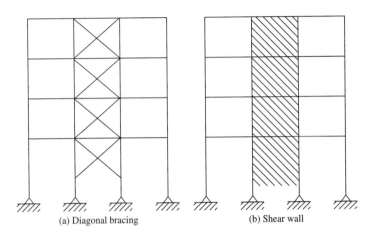

(a) Diagonal bracing   (b) Shear wall

Figure 7-1 Sidesway inhibited.

restraint at their ends. For these situations, pictured in Fig. 7-1, the K factors will fall somewhere between cases (a) and (d) of Table 5-1.

The LRFD Specification (C2) states that $K = 1.0$ should be used for columns in frames with sidesway inhibited unless an analysis shows that a smaller value can be used. $K = 1.0$ is often quite conservative, and an analysis made as described herein may result in some savings.

The effective length of a column is a property of the whole structure of which the column is a part. In many existing buildings it is probable that the masonry walls provide sufficient lateral support to prevent sidesway. When light curtain walls are used, however, as they often are in modern buildings, there is probably little resistance to sidesway. Sidesway is also present in tall buildings in appreciable amounts unless a definite diagonal bracing system or shear walls are used. For these cases it seems logical to assume that resistance to sidesway is primarily provided by the lateral stiffness of the frame alone.

Theoretical mathematical analyses may be used to determine effective lengths, but such procedures are usually too lengthy and perhaps too difficult for the average designer. The usual procedure is to use either Table 5-1, interpolating between the idealized values as the designer feels appropriate, or the alignment charts that are described in this section.

The charts shown in Fig. 7-2 present a practical method for estimating K values.[1,2] They were developed from a slope-deflection analysis of the frames that includes the effect of column loads. One chart was developed for columns braced against sidesway and one for columns subject to sidesway. Their use

---

[1] O. G. Julian and L. S. Lawrence, "Notes on J and L Monograms for Determination of Effective Lengths" (1959).

[2] Structural Stability Research Council, *Guide to Stability Design Criteria for Metal Structures*, 4th ed. T. V. Galambos, ed. (New York: Wiley, 1988).

(a) Sidesway prevented

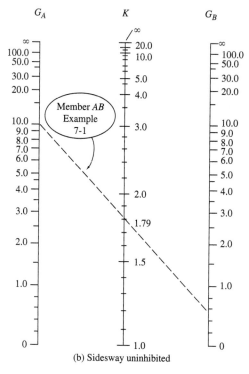

(b) Sidesway uninhibited

The subscripts A and B refer to the joints at the two ends of the column section being considered. G is defined as

$$G = \frac{\Sigma \dfrac{I_c}{L_c}}{\Sigma \dfrac{I_g}{L_g}}$$

in which Σ indicates a summation of all members rigidly connected to that joint and lying in the plane in which buckling of the column is being considered, I_c is the moment of inertia and L_c the unsupported length of a column section, and I_g is the moment of inertia and L_g the unsupported length of a girder or other restraining member. I_c and I_g are taken about axes perpendicular to the plane of buckling being considered.

"For column ends supported by but not rigidly connected to a footing or foundation, G is theoretically infinity, but, unless actually designed as a true friction free pin, may be taken as '10' for practical designs. If the column end is rigidly attached to a properly designed footing, G may be taken as 1.0. Smaller values may be used if justified by analysis."

From American Institute of Steel Construction, *Manual of Steel Construction Load & Resistance Factor Design*, Volume I, 2nd. ed. (Chicago: AISC, 1994) p. 6-186.

Figure 7-2 Alignment charts for effective lengths of columns in continuous frames.

Shearson Lehman/American Express Information Services Center, New York City. (Courtesy of Owen Steel Company, Inc.)

enables the designer to obtain good K values without struggling through lengthy trial-and-error procedures with the buckling equations.

To use the alignment charts it is necessary to have preliminary sizes for the girders and columns framing into the column in question before the K factor can be determined for that column. In other words, before the chart can be used, we have to either assume some member sizes or carry out a preliminary design.

When we say sidesway is *inhibited* we mean there is something present other than just columns and girders to prevent sidesway or the horizontal translation of the joints. That means we have a definite system of lateral bracing or we have shear walls. If we say that sidesway is *uninhibited* we are saying that resistance to horizontal translation is supplied only by the bending strength and stiffness of the girders and beams of the frame in question with its continuous joints.

The resistance to rotation furnished by the beams and girders meeting at one end of a column is dependent on the rotational stiffnesses of those members. The moment needed to produce a unit rotation at one end of a member if the other end of the member is fixed is referred to as its *rotational stiffness*. From our structural analysis studies this works out to be equal to $4EI/L$ for a homogeneous member of constant cross section. Based on the preceding we can say that the rotational restraint at the end of a particular column is proportional to the ratio of the sum of the column stiffnesses to the girder stiffnesses meeting at that joint.[3]

$$G = \frac{\sum \dfrac{4EI}{L} \text{ for columns}}{\sum \dfrac{4EI}{L} \text{ for girders}} = \frac{\sum \dfrac{I_c}{L_c}}{\sum \dfrac{I_g}{L_g}}$$

To determine a K value for a particular column, the following steps are taken:

1. Select the appropriate chart (sidesway prevented or sidesway uninhibited).
2. Compute G at each end of the column and label the values G_A and G_B as desired.
3. Draw a straight line on the chart between the G_A and G_B values and read K where the line hits the center K scale.

When G factors are being computed for a rigid frame structure (rigid in both directions) the torsional resistance of the girders is generally neglected in the calculations. With reference to Fig. 7-3 it is assumed that we are calculating G for the joint shown for buckling in the plane of the paper. For such a case the torsional resistance of the girder shown, which is perpendicular to the plane being considered, is probably neglected.

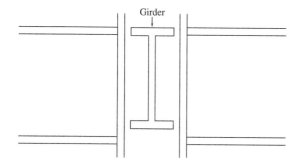

Girder

Figure 7-3

---

[3] W. T. Segui, *LRFD Steel Design* (Boston: PWS Publishing Company, 1994) pp. 96–97.

The Structural Stability Research Council (SSRC) makes several recommendations concerning the use of the alignment charts.

1. For pinned columns G is theoretically infinite, as where a column is connected to a footing with a frictionless hinge. Recognizing that such a connection is not frictionless, it is recommended that G be made equal to 10 where such nonrigid supports are used.
2. For rigid connections of columns to footings G theoretically approaches zero, but from a practical standpoint a value of 1.0 is recommended as no connections are perfectly rigid.
3. If a beam or girder is rigidly attached to a column, its stiffness I/ℓ should be multiplied by the appropriate factor given in Table 7-1, depending on the condition at the far end of the member.

TABLE 7-1   MULTIPLIERS FOR RIGIDLY ATTACHED MEMBERS

Condition at Far End of Girder	Sidesway Prevented, Multiply by:	Sidesway Uninhibited, Multiply By:
Pinned	1.5	0.5
Fixed against rotation	2.0	0.67

If the girders at a joint are very stiff (that is, they have very large I/L values) the value of $G = \Sigma\,(I_c/L_c)/\Sigma(I_g/L_g)$ will approach zero and the K factors will be small. If G is very small the column moments cannot rotate the joint very much, thus the joint is close to a fixed end situation. Usually, however, G is appreciably larger than zero, resulting in significantly larger values of K.

The effective lengths of each of the columns of a frame are estimated with the alignment charts in Example 7-1. (When sidesway is possible, it will be found that the effective lengths are always greater than the actual lengths as is illustrated in this example. When frames are braced in such a manner that sidesway is not possible, K will be less than 1.0.) An initial design has provided preliminary sizes for each of the members in the frame. After the effective lengths are determined, each column can be redesigned. Should the sizes change appreciably, new effective lengths can be determined, the column designs repeated, and so on. Several tables are used in the solution of this example. These should be self-explanatory after the clear directions given on the alignment chart are examined.

■ EXAMPLE 7-1
Determine the effective lengths of each of the columns of the frame shown in Fig. 7-4 if the frame is not braced against sidesway. Use the alignment charts of Fig. 7-2(b) and assume the far ends of the girders are fixed against rotation.

Figure 7-4

Solution.   Stiffness factors:

Member	Shape	I	L	I/L
AB	W8 × 24	82.8	144	0.575
BC	W8 × 24	82.8	120	0.690
DE	W8 × 40	146	144	1.014
EF	W8 × 40	146	120	1.217
GH	W8 × 24	82.8	144	0.575
HI	W8 × 24	82.8	120	0.690
BE	W18 × 50	800	240	3.333
CF	W16 × 36	448	240	1.867
EH	W18 × 97	1750	360	4.861
FI	W16 × 57	758	360	2.106

G factors for each joint:

Joint	$\Sigma(I_c/L_c)/\Sigma(I_g/L_g)$	G
A	See Fig. 7-2(b)	10.0
B	$\dfrac{0.575 + 0.690}{(0.67)(3.333)}$	0.566
C	$\dfrac{0.690}{(0.67)(1.867)}$	0.552
D	See Fig. 7-2(b)	10.0
E	$\dfrac{1.014 + 1.217}{(0.67)(3.333 + 4.861)}$	0.406
F	$\dfrac{1.217}{(0.67)(1.867 + 2.106)}$	0.457
G	See Fig. 7-2(b)	10.0
H	$\dfrac{0.575 + 0.690}{(0.67)(4.861)}$	0.388
I	$\dfrac{0.690}{(0.67)(2.106)}$	0.489

Column K factors from chart (Fig. 7-2(b)):

Column	G_A	G_B	K^*
AB	10.0	0.566	1.79
BC	0.566	0.522	1.17
DE	10.0	0.406	1.77
EF	0.406	0.457	1.13
GH	10.0	0.388	1.76
HI	0.388	0.489	1.14

*It is a little difficult to read the charts to the 3 places shown by the author. He has used a larger copy of Fig. 7-2 for his work. For all practical design purposes the K values can be read to two places which can easily be accomplished with this figure.

For most buildings the values of K_x and K_y should be examined separately. The reason for such individual study lies in the different possible framing conditions in the two directions. Many multistory frames consist of rigid frames in one direction and conventionally connected frames with sway bracing in the other. In addition the points of lateral support may often be entirely different in the two planes.

There is available a set of rather simple equations for computing effective length factors. On some occasions the designer may find these expressions very convenient to use as compared to the alignment charts just described. Perhaps the most useful situation is for computer programs. You can see that it would be rather inconvenient to stop occasionally in the middle of a computer design to read K factors from the charts and input them to the computer. The equations, however, can easily be included in the programs, eliminating the necessity of using alignment charts.[4]

The alignment chart of Fig. 7-2(b) for frames with sidesway uninhibited always indicates that $K \geq 1.0$. In fact calculated K factors of 2.0 to 3.0 are common, and even larger values are occasionally obtained. To many designers such large factors seem completely unreasonable. If the designer obtains seemingly high K factors, he or she should carefully review the numbers used to enter the chart (that is, the G values) as well as the basic assumptions used in preparing the charts. These assumptions are discussed in detail in Section 7-2.

7-2 STIFFNESS REDUCTION FACTORS

The alignment charts were developed on the basis of a set of idealized conditions which are seldom if ever completely met in a real structure. A complete list of these assumptions is shown in Section C2 of the Commentary on the LRFD

---

[4] P. Dumonteil, "Simple Equations for Effective Length Factors," *Engineering Journal,* AISC, 29, no. 3 (3rd quarter, 1992), pp. 111–115.

Specification. Among these are: column behavior is purely elastic, all columns buckle simultaneously, all members have constant cross sections, all joints are rigid, and so on.

If the actual conditions are different from these assumptions, unrealistically high K factors may be obtained from the charts, and overconservative designs may result. A large percentage of columns will fall in the inelastic range, but the alignment charts were prepared assuming elastic failure. This situation, previously discussed in Chapter 5, is illustrated in Fig. 7-5. For such cases the chart K values are too conservative and should be corrected as described in this section.

In the elastic range the stiffness of a column is proportional to EI where $E = 29{,}000$ ksi, while in the inelastic range its stiffness is more accurately proportional to $E_T I$ where E_T is a reduced or tangent modulus.

The buckling strength of columns in framed structures was shown in the alignment charts to be related to

$$G = \frac{\text{column stiffness}}{\text{girder stiffness}} = \frac{\Sigma(EI/L) \text{ columns}}{\Sigma(EI/L)\text{girders}}$$

If the columns behave elastically, the modulus of elasticity will be canceled from the preceding expression for G. If the column behavior is inelastic, however, the column stiffness factor will be smaller and will equal $E_T I/L$. As a result the G factor used to enter the alignment chart will be smaller, and the K factor selected from the chart will be smaller.

Though the alignment charts were developed for elastic column action, they may be used for an inelastic column situation if the G value is multiplied by a correction factor called the *stiffness reduction factor* (SRF). This reduction factor equals the tangent modulus over the elastic modulus (E_T/E) and is approximately equal to $F_{cr \text{ inelastic}}/F_{cr \text{ elastic}} \approx (P_u/A)/F_{cr \text{ elastic}}$. Values of this correction factor are shown for various P_u/A values in Table 7-2 which is Table 3-1 in the

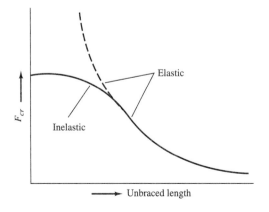

Figure 7-5

TABLE 7-2 STIFFNESS REDUCTION FACTORS (SRF) FOR COLUMNS

P_u/A ksi	F_y		P_u/A ksi	F_y	
	36 ksi	50 ksi		36 ksi	50 ksi
42	—	0.03	26	0.38	0.82
41	—	0.09	25	0.45	0.85
40	—	0.16	24	0.52	0.88
39	—	0.21	23	0.58	0.90
38	—	0.27	22	0.65	0.93
37	—	0.33	21	0.70	0.95
36	—	0.38	20	0.76	0.97
35	—	0.44	19	0.81	0.98
34	—	0.49	18	0.85	0.99
33	—	0.53	17	0.89	1.00
32	—	0.58	16	0.92	↓
31	—	0.63	15	0.95	
30	0.05	0.67	14	0.97	
29	0.14	0.71	13	0.99	
28	0.22	0.75	12	1.00	
27	0.30	0.79	11	↓	

−Indicates not applicable.

From American Institute of Steel Construction, *Manual of Steel Construction Load & Resistance Factor Design,* Volume I, 2nd. ed. (Chicago, AISC, 1994) pp. 3–7. Reprinted with the permission of AISC.

LRFD Manual. A direct design method for considering inelastic buckling is presented in the Manual. It involves the following steps:

1. Calculate P_u and select a trial column size.
2. Calculate P_u/A and pick the SRF from Table 7-2. (If P_u/A is less than the values given in the table the column is in the elastic range and no reduction needs to be made.)
3. The value of G_{elastic} is computed and multiplied by the SRF and K is picked from the chart.
4. The effective slenderness ratio KL/r is computed, and $\phi_c F_{cr}$ obtained from the Manual is multiplied by the column area to obtain P_u. If this value is appreciably different from the value computed in Step 1, another trial column size is attempted and the four steps are repeated.

Example 7-2 which follows illustrates these steps for the design of a column in a frame subject to sidesway. *In this example, note that the author has only considered in-plane behavior and only bending about the x axis.* As a result of inelastic behavior, the effective length factor is appreciably reduced.

■ **EXAMPLE 7-2**

Select a W12 section for column *AB* of the frame shown in Fig. 7-6 assuming (a) elastic column behavior and (b) inelastic column behavior. $P_u = 1210$ k and

A36 steel is used . The columns above and below AB are assumed to be approximately the same size as AB. Consider only in-plane behavior. The far ends of the girders are fixed against rotation.

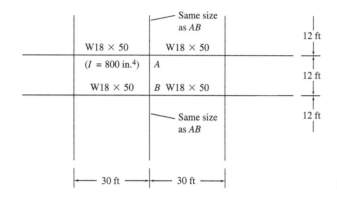

Figure 7-6

Solution.

(a) Assuming column in elastic range and selecting a trial section based on an estimated $K_y L_y = 12$ ft

Try W12 × 170 ($A = 50.0$ in.², $I_x = 1650$ in.⁴, $r_x = 5.74$ in.)

$$G_A = G_B = \frac{\Sigma(I_c/L_c)}{\Sigma(I_g/L_g)} = \frac{(2)(1650/12)}{(2)(800/30)(0.67)} = 7.70$$

$K = 2.65$ from Fig. 7-1 alignment chart

$$\left(\frac{KL}{r}\right)_x = \frac{(2.65)(12 \times 12)}{5.74} = 66.48$$

$\phi_c F_{cr} = 24.25$ ksi

$P_u = (24.25)(50.0) = 1212 \text{ k} > 1210 \text{ k}$ (OK)

Use W12 × 170

(b) Inelastic solution

Try the next lightest section W12 × 152 ($A = 44.7$ in.²,

$I_x = 1430$ in.⁴, $r_x = 5.66$ in.)

$$\frac{P_u}{A} = \frac{1210}{44.7} = 27.07 \text{ ksi}$$

SRF = 0.294 from Table 3-1 in Manual

∴ Column is in inelastic range

$$G_A = G_B = \frac{\Sigma(I_c/L_c)}{\Sigma(I_g/L_g)} (\text{SRF})$$

$$= \frac{(2)(1430/12)}{(2)(800/30)(0.67)} (0.294) = 1.96$$

$K = 1.57$ from Fig. 7-2(b) alignment chart

$$\frac{KL}{r} = \frac{(1.57)(12 \times 12)}{5.66} = 39.94$$

$\phi_c F_{cr} = 28.14$ ksi

$P_u = (28.14)(44.7) = 1258 \text{ k} > 1210 \text{ k}$ (OK)

Use W12 × 152 ∎

7-3 COLUMNS LEANING ON EACH OTHER FOR IN-PLANE DESIGN

When we have an unbraced frame with beams rigidly attached to columns, it is safe to design each column individually using the sidesway uninhibited alignment chart to obtain the K factors (which will probably be appreciably larger than 1.0).

A column cannot buckle by sidesway unless all of the columns on that story buckle by sidesway. One of the assumptions on which the alignment chart of Fig. 7-2(b) was prepared was that all of the columns on the story in question would buckle at the same time. If this assumption is correct the columns cannot support or brace each other because if one gets ready to buckle they all supposedly are ready to buckle.

In some situations, however, certain columns in a frame have some excess buckling strength. If, for instance, the buckling loads of the exterior columns of the unbraced frame of Fig. 7-7 have not been reached when the buckling loads of the interior columns are reached, the frame will not buckle. The interior columns in effect will lean against the exterior columns, that is, the exterior columns will brace the interior ones. For this situation shear resistance is provided in the exterior columns which resist the sidesway tendency.[5]

A pin-ended column which does not help provide lateral stability to a structure is referred to as a "leaning" column. Such a column depends on the other parts of the structure to provide lateral stability. Section C2.2 of the LRFD Specification states that the effects of gravity loaded "leaning" columns shall be included in the design of moment frame columns.

There are many practical situations in which some columns have excessive buckling strength. This might happen when the designs of different columns on

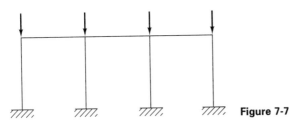

Figure 7-7

[5] J. A. Yura, "The Effective Length of Columns in Unbraced Frames," *Engineering Journal*, AISC, 8, no. 2 (second quarter, 1971), pp. 37–42.

a particular story are controlled by different loading conditions. For such cases failure of the frame will occur only when the gravity loads are increased sufficiently to offset the extra strength of the lightly loaded columns. As a result the critical loads for the interior columns of Fig. 7-7 are increased and, in effect, their effective lengths are decreased. In other words, if the exterior columns are bracing the interior ones against sidesway, the K factors for those interior columns are approaching 1.0. Yura[6] says that the effective length of some of the columns in a frame subject to sidesway can be reduced to 1.0 in this type of situation even though there is no apparent bracing system present.

The net effect of the information presented here is that the total gravity load that an unbraced frame can support equals the sum of the strength of the individual columns. In other words, the total gravity load that will cause sidesway buckling in a frame can be split up among the columns in any proportion so long as the maximum load applied to any one column does not exceed the maximum load that column could support if it were braced against sidesway with $K = 1.0$.

For this discussion the unbraced frame of Fig. 7-8(a) is considered. It is assumed that each column has a $K = 2.0$ and will buckle under the loads shown.

When sidesway occurs the frame will lean to one side as shown in part (b) of the figure and $P\Delta$ moments equal to 200Δ and 700Δ will be developed.

Suppose that we load the frame with 200 k on the left-hand column and 500 k on the right-hand column (or 200 k less than we had before). We know that for this situation, which is shown in part (c) of the figure, the frame will not buckle by sidesway until we reach a moment of 700Δ at the right-hand column base. This means that our right-hand column can take an additional moment of 200Δ. Thus, as Yura says, the right-hand column has a reserve of strength that can be used to brace the left-hand column and prevent its sidesway buckling.

Obviously the left-hand column is now braced against sidesway, and sidesway buckling will not occur until the moment at its base reaches 200Δ. Therefore, it can be designed with a K factor less than 2.0 and can support an

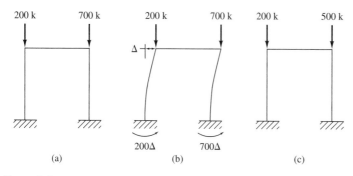

(a)   (b)   (c)

Figure 7-8

[6] Ibid., pp. 39–40.

additional load of 200 k, giving it a total load of 400 k—*but this load must not be greater than would be its capacity if it were braced against sidesway with K = 1.0.* It should be mentioned that the total load the frame can carry still is 900 k, as in part (a) of the figure.

The advantage of the frame behavior described here is illustrated in Example 7-3 where it is assumed that the interior columns of a frame are braced against sidesway by the exterior columns. As a result the interior columns are assumed to each have K factors equal to 1.0. They are designed for the factored loads shown (660 k each). Then the K factors are determined for the exterior columns with the sidesway uninhibited chart of Fig. 7-1, and they are each designed for column loads equal to 440 + 660 = 1100 k.

■ EXAMPLE 7-3

For the frame of Fig. 7-9, which consists of A36 steel, beams are rigidly connected to the exterior columns while all other connections are simple. The columns are braced top and bottom against sidesway out of the plane of the frame so that $K = 1.0$ in that direction. Sidesway is possible in the plane of the frame. Design the interior columns assuming $K = 1.0$ and the exterior columns with K as determined from the alignment chart and $P_u = 1100$ k. (With this approach to column buckling, the interior columns could carry no load at all since they appear to be unstable under sidesway conditions.)

Figure 7-9

Solution. Design of interior columns

Assume $K = 1.0$, $KL = (1.0)(15) = 15$ ft, $P_u = 660$ k
$$\text{Use W14} \times 90$$

Design of exterior columns

Out of plane $K_y = 1.0$, $P_u = 440$ k

In plane $P_u = 440 + 660 = 1100$ k, K_x to be determined from alignment chart. Estimating a column size a little larger than would be required for $P_u = 1100$ k. Try W14 × 159 ($A = 46.7$ in.², $I_x = 1990$ in.⁴, $r_x = 6.38$ in.)

$$G_{\text{top}} = \frac{1900/15}{2100/30 \times 0.5} = 3.62$$

(noting that girder stiffness is multiplied by 0.5 since sidesway is permitted and far end of girder is hinged.)

Figure 7-10

$$G_{\text{bottom}} = 10$$

$$K_x = 2.40$$

$$\frac{K_x L_x}{r_x} = \frac{(2.40)(12 \times 15)}{6.38} = 67.71 \qquad \text{(OK)}$$

$$\phi_c F_{cr} = 24.04 \text{ ksi}$$

$$P_u = (24.04)(46.7) = 1122.7 \text{ k} > 1100 \text{ k}$$

$$\underline{\underline{\text{Use W14} \times 159}} \qquad \blacksquare$$

Space is not taken in this example to apply stiffness reduction factors (as described in Section 7-2) to the columns of this frame. Section C2.2 of the LRFD Specification states that *stiffness reduction adjustment due to column inelasticity is permitted when the "leaning" column theory has been applied.*

It is rather frightening to think of additions to existing buildings and the leaner column theory. If we have a building (represented by the solid lines in Fig. 7-10) and we decide to add onto it (indicated by the dashed lines in the same figure), we may think that we can use the old frame to brace the new one and that we can keep expanding laterally with no effect on the existing building. Sadly, we may be in for quite a surprise. The leaning of the new columns may cause one of the old ones to fail.

7-4 BASE PLATES FOR CONCENTRICALLY LOADED COLUMNS

The design compressive stress in a concrete or other type of masonry footing is much smaller than it is in a steel column. When a steel column is supported by a footing it is necessary for the column load to be spread over a sufficient area to keep the footing from being overstressed. Loads from steel columns are transferred through a steel base plate to a fairly large area of the footing below. (It will be noted that a footing performs a related function in that it spreads the load over an even larger area so that the underlying soil will not be overstressed.)

The base plates for steel columns can be welded directly to the columns or they can be fastened by means of some type of bolted or welded lug angles. These connection methods are illustrated in Fig. 7-11. A base plate welded directly to the column is shown in part (a) of the figure. For small columns these plates are probably shop-welded to the columns, but for larger columns it may be necessary to ship the plates separately and set them to the correct elevations. For this second case the columns are connected to the footing with anchor bolts that pass through the lug angles which have been shop-welded to the columns.

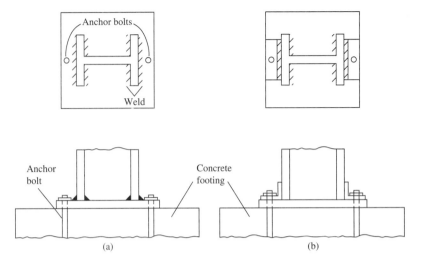

Figure 7-11 Column base plates.

This type of arrangement is shown in part (b) of the figure. Some designers like to use lug angles on both flanges and web.

A critical phase in the erection of a steel building is the proper positioning of column base plates. If they are not located at their correct elevations, serious stress changes may occur in the beams and columns of the steel frame. One of the following three methods is used for preparing the site for the erection of a column to its proper elevation: leveling plates, leveling nuts, or preset base plates. An article by Ricker[7] describes these procedures in considerable detail.

For small-to-medium base plates (up to 20 to 22 in.), approximately $\frac{1}{4}$-in.-thick leveling plates with the same dimensions as the base plates (or a little larger) are shipped to the job and carefully grouted in place to the proper elevations. Then the columns with their attached base plates are set on the leveling plates.

As these leveling plates are very light and can be handled manually they are set by the foundation contractor. This is also true for the lighter base plates. On the other hand large base plates which have to be lifted with a derrick or crane are usually set by the steel erector.

For larger base plates up to about 36 in., some types of leveling nuts are used to adjust the base plates up or down. To ensure stability during erection these nuts must be used on at least four anchor bolts.

If the base plates are larger than about 36 in., the columns with the attached base plates are so heavy and cumbersome that it is difficult to ship them together.

For such cases the base plates are shipped to the job and placed in advance of the steel erection. They can be leveled with shims or wedges.

For tremendously large base plates weighing several tons or more, angle frames may be built to support the plates. They are carefully leveled and filled with concrete, which is screeded off to the correct elevations and the base plates are set directly on the concrete.

A column transfers its load to the supporting pier or footing through the base plate. Should the supporting concrete area A_2 be larger than the plate area A_1, the concrete strength will be higher. In that case the concrete surrounding the contact area supplies appreciable lateral support to the directly loaded part, with the result that the loaded concrete can support more load. This fact is reflected in the design stresses.

The lengths and widths of column base plates are usually selected in multiples of even inches and their thicknesses in multiples of $\frac{1}{8}$ in. up to $1\frac{1}{4}$ in. and in multiples of $\frac{1}{4}$ in. thereafter. To make sure that column loads are spread uniformly over their base plates, it is essential to have good contact between the two. The surface preparation of these plates is governed by Section M2.8 of the LRFD Specification. In that section it is stipulated that bearing plates of 2 in. or less in thickness may be used without milling if satisfactory contact is obtained. (Milled surfaces have been accurately sawed or finished to a true plane.) Plates over 2 in. thick up through 4 in. may be straightened by pressing, or they may be milled at the option of the steel fabricator. Plates thicker than 4 in. must be milled if they are beyond the flatness tolerances specified in Tables 1-19 and 1-20 of Part 1 of the LRFD Manual entitled "Permissible Variations from Flatness."

Column base plate with anchor bolts at each corner. Column welded to base plate. Grout placed below base plate to transfer compression load evenly to reinforced concrete foundation. (Courtesy American Institute of Steel Construction.)

If the bottom surfaces of the plates are to be in contact with cement grout to ensure full bearing contact on the foundation, they do not have to be milled. Furthermore, the top surfaces of plates thicker than 4 in. do not have to be milled if full-penetration welds (to be described in Chapter 14) are used. Notice that when finishing is required as described here the plates will have to be ordered a little thicker than required for their final dimensions to allow for the cuts.

Initially, columns will be considered that support average-size loads. Should the loads be very small, so that base plates are very small, the design procedure will have to be revised as described later in this section.

The LRFD Specification does not specify a particular method for designing column base plates. The method presented here is based upon example problems presented in Part 11 of the LRFD Manual.

To analyze the base plate shown in Fig. 7-12, the column is assumed to apply a total load equal to P_u to the base plate. Then the load P_u is assumed to be transmitted uniformly through the plate to the footing below with a pressure equal to P_u/A where A is the area of the base plate. The footing will push back with a pressure of P_u/A and will tend to curl up the cantilevered parts of the base plate outside of the column, as shown in the figure. This pressure will also tend to push up the base plate between the flanges of the column.

With reference to Fig. 7-12, Part 11 of the LRFD Manual suggests that maximum moments in a base plate occur at distances $0.80\ b_f$ and $0.95\ d$ apart. The bending moment can be calculated at each of these sections and the larger value used to determine the plate thickness needed. This method of analysis is only a rough approximation of the true conditions because the actual plate stresses are caused by a combination of bending in two directions.

Plate Area   The design strength of the concrete in bearing beneath the base plate must at least equal the load to be carried. When the base plate covers the entire area of the concrete support this strength equals ϕ_c (0.60 for bearing on concrete) times the nominal strength of the concrete $0.85\ f_c'$ times A_1 (where f_c' is the 28-day compression strength of the concrete in ksi and A_1 is the area of the plate).

$$P_u = \phi_c P_p = \phi_c(0.85\ f_c'A_1) \qquad\qquad \text{(from LRFD Equation J9-1)}$$

$$A_1 = \frac{P_u}{\phi_c\ 0.85\ f_c'}$$

Should the full area of the concrete support not be covered by the plate, the concrete underneath the plate, surrounded by concrete outside, will be somewhat stronger. For this situation the LRFD Specification (J9b) permits the design strength above ($\phi 0.85\ f_c'A_1$) to be increased by multiplying it by $\sqrt{A_2/A_1}$. In the resulting expression A_2 is the maximum area of the portion of the supporting concrete which is geometrically similar to and concentric with the loaded

Figure 7-12

area. The value of $\sqrt{A_2/A_1}$ is limited to a maximum value of 2 as shown in the expression below.

$$\phi_c P_P = \phi_c(0.85 f_c' A_1) \sqrt{\frac{A_2}{A_1}} \quad \text{with} \quad \sqrt{\frac{A_2}{A_1}} \le 2 \qquad \text{(from LRFD Equation J9-2)}$$

Then

$$A_1 = \frac{P_u}{\phi_c(0.85 f_c' A_1) \sqrt{\frac{A_2}{A_1}}} \quad \text{where} \quad \sqrt{\frac{A_2}{A_1}} \text{ may not be} > 2$$

Furthermore, A_1 may not be less than the depth of the column times its flange width.

$$A_1 = b_f d$$

After the controlling value of A_1 is determined as described above, the plate dimensions B and N (shown in Fig. 7-12) are selected to the nearest 1 or 2 in. so that the values of m and n shown in the figure are roughly equal. Such a procedure will make the cantilever moments in the two directions approximately the same. This will enable us to keep the plate thicknesses to a minimum. The condition $m = n$ can be approached if the following equation is satisfied.

$$N \approx \sqrt{A_1} + \Delta$$

where A_1 = area of plate = BN

$$\Delta = 0.5 \, (0.95 \, d - 0.80b_f)$$

$$B \approx \frac{A_1}{N}$$

Plate Thickness   To determine the required plate thickness, moments are taken in the two directions as though the plate is cantilevered out by the dimensions m and n. Reference is again made here to Fig. 7-12. These moments are respectively $(P_u/BN)(m)(m/2) = P_u m^2/2BN$ and $(P_u/BN)(n)(n/2) = P_u n^2/2BN$ both computed for a 1-in. width of plate.

The design moment strength of the plate per in. of width must at least equal the larger of these two moments. In the next few chapters we will learn to compute this value. For a plate it will equal $\phi F_y(t^2/4)$ where t is the plate thickness and $\phi = 0.9$. Equating this expression to the maximum computed moment the required value of t can be determined as follows

$$\phi F_y \frac{t^2}{4} = \frac{P_u m^2}{2BN} \quad \text{or} \quad \frac{P_u n^2}{2BN}$$

$$t = m \sqrt{\frac{2P_u}{0.9F_y BN}} \quad \text{or} \quad n \sqrt{\frac{2P_u}{0.9F_y BN}}$$

If lightly loaded base plates, as for the columns of low-rise buildings and pre-engineered metal buildings, are designed by the procedure just described they will have quite small areas. They will, as a result, extend very little outside the edges of the columns and the computed moments and the resulting plate thicknesses will be very small, perhaps so small as to be impractical.

Several procedures for handling this problem have been proposed. In 1990 W. A. Thornton[8] combined three of the methods into a single procedure applicable to either heavily loaded or lightly loaded base plates. This modified method

---

[8] W. A. Thornton, "Design of Base Plates for Wide Flange Columns—A Concatenation of Methods," *Engineering Journal*, AISC, 27, no. 4 (4th quarter, 1990) pp. 173–174.

is used for the example base plate problems in Part 11 of the LRFD Manual as well as for the example problems in this text.

Thornton proposed that the thickness of the plates be determined using the largest of m, n, or $\lambda n'$. He called this largest value ℓ.

$$\ell = \max (m, n, \text{ or } \lambda n')$$

To determine $\lambda n'$ it is necessary to substitute into the following expressions which are developed in his paper.

$$\phi_c P_p = \phi_c 0.85 f'_c A_1 \quad \text{for plates covering the full area of the concrete support}$$

$$\phi_c P_p = \phi_c f'_c A_1 \sqrt{\frac{A_2}{A_1}} \quad \text{where } \sqrt{\frac{A_2}{A_1}} \text{ must be} \le 2 \text{ for plates not covering}$$

the entire area of concrete support

$$X = \left[\frac{4db_f}{(d + b_f)^2} \right] \frac{P_u}{\phi_c P_p}$$

$$\lambda = \frac{2\sqrt{X}}{1 + \sqrt{1 - X}} \le 1$$

$$\lambda n' = \frac{\lambda \sqrt{db_f}}{4}$$

Finally the thickness of the plate is determined with

$$t = \ell \sqrt{\frac{2P_u}{0.9 F_y BN}}$$

Three example base plate designs are presented in the next few pages. Example 7-4 illustrates the design of a base plate supported by a large reinforced concrete footing with A_2 many times as large as A_1. In Example 7-5 a base plate is designed which is supported by a concrete pedestal where the plate covers the entire concrete area.

Finally in Example 7-6 a base plate is selected for a column that is to be supported on a pedestal 4 in. wider on each side than the plate. This means A_2 cannot be determined until the plate area (A_1) is computed. This puts us into the following trial and error procedure:

1. Compute $A_1 = \dfrac{P_u}{\phi_c 0.85 f'_c}$

2. Then $A_2 = (B + 4)(N + 4)$

3. Now A_1 can be determined $= \dfrac{P_u}{\phi_c 0.85 f'_c \sqrt{\dfrac{A_2}{A_1}}} = \dfrac{\text{above } A_1}{\sqrt{\dfrac{A_2}{A_1}}}$ and values

selected for B and N.

4. The value of A_2 determined in Step 2 has now changed so we continue through Steps 2 to 3 until A_1 changes very little.

■ **EXAMPLE 7-4**

Design a base plate of A36 steel for a W12 × 65 column (F_y = 50 ksi) which supports a factored axial load P_u = 720 k. The concrete has a specified compressive strength f_c' = 3 ksi. Assume footing size is 9 ft × 9 ft.

Solution.   Using a W12 × 65 column (d = 12.12 in., b_f = 12.00 in.)

$$A_2 = (12 \times 9)(12 \times 9) = 11{,}664 \text{ in.}^2$$

Determine required base plate area

Assuming the area of the supporting concrete will be far greater than the area of the base plate such that $\sqrt{A_2/A_1} \geq 2$.

$$A_1 = \frac{P_u}{\phi_c(0.85\,f_c')\sqrt{\dfrac{A_2}{A_1}}} = \frac{720}{(0.6)(0.85)(3)(2)} = 235.3 \text{ in.}^2 \leftarrow$$

Checking

$$\sqrt{\frac{A_2}{A_1}} = \sqrt{\frac{11{,}664}{235.3}} = 7.04 > 2 \qquad \text{(OK)}$$

The base plate should be at least as large as the column.

$$A_1 = db_f = (12.12)(12.00) = 145.44 \text{ in.}^2$$

Optimizing base-plate dimensions

$$\Delta = \frac{0.95d - 0.8b_f}{2} = \frac{(0.95)(12.12) - (0.8)(12.00)}{2} = 0.957 \text{ in.}$$

$$N = \sqrt{A_1} + \Delta = \sqrt{235.3} + 0.957 = 16.30 \text{ in.} \quad \text{Say 16 in.}$$

$$B = \frac{A_1}{N} = \frac{235.3}{16} = 14.71 \text{ in.} \quad \text{Say 15 in.}$$

Computing required base-plate thickness

$$m = \frac{N - 0.95d}{2} = \frac{16 - (0.95)(12.12)}{2} = 2.24 \text{ in.}$$

$$n = \frac{B - 0.8b_f}{2} = \frac{15 - (0.8)(12.00)}{2} = 2.70 \text{ in.}$$

$$\phi_c P_p = 0.6(0.85\,f_c'A_1)\sqrt{\frac{A_2}{A_1}} = (0.6)(0.85)(3)(16)(15)(2) = 734.4 \text{ k}$$

$$X = \frac{4db_f}{(d + b_f)^2}\frac{P_u}{\phi_c P_p} = \frac{(4)(12.12)(12.00)}{(12.12 + 12.00)^2}\frac{720}{734.4} = 0.980$$

$$\lambda = \frac{2\sqrt{X}}{1 + \sqrt{1 - X}} = \frac{2\sqrt{0.980}}{1 + \sqrt{1 - 0.980}} = 1.73 > 1.0 \quad \text{Use 1.0}$$

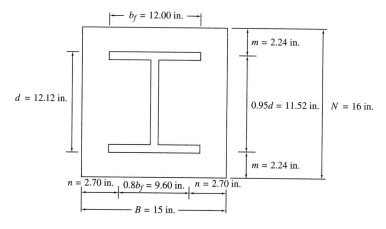

Figure 7-13

$$\lambda n' = \frac{\lambda \sqrt{d b_f}}{4} = \frac{1\sqrt{(12.12)(12.00)}}{4} = 3.01 \text{ in.}$$

$$\ell = \max{(m, n, \lambda n')} = 3.01 \text{ in.}$$

$$t_{reqd} = \ell \sqrt{\frac{2P_u}{0.9 F_y B N}} = 3.01 \sqrt{\frac{(2)(720)}{(0.9)(36)(15)(16)}} = 1.30 \text{ in.}$$

Use PL$1\frac{1}{2}$ × 15 × 1 ft 4 in. ■

■ EXAMPLE 7-5

A base plate is to be designed for a W12 × 152 column ($F_y = 50$ ksi) supporting a factored axial load $P_u = 960$ k. Determine the dimensions for the plate (A36 steel) and for a concrete pedestal ($f'_c = 3$ ksi) if the plate is to cover the entire concrete area.

Solution.   Using a W12 × 152 ($d = 13.71$ in., $b_f = 12.48$ in.)
Determine required base-plate area

$$A_1 = \frac{P_u}{\phi_c(0.85 f'_c)} = \frac{960}{(0.6)(0.85)(3)} = 627.5 \text{ in.}^2 \leftarrow$$

$$A_1 = d b_f = (13.71)(12.48) = 171.1 \text{ in.}^2$$

Optimizing base-plate dimensions

$$\Delta = \frac{0.95d - 0.8 b_f}{2} = \frac{(0.95)(13.71) - (0.8)(12.48)}{2} = 1.52 \text{ in.}$$

$$N = \sqrt{A_1} + \Delta = \sqrt{627.5} + 1.52 = 26.57 \text{ in.}   \underline{\text{Say 27 in.}}$$

$$B = \frac{A_1}{N} = \frac{627.5}{27} = 23.24 \text{ in.}   \underline{\text{Say 24 in.}}$$

Computing required base-plate thickness

$$m = \frac{N - 0.95d}{2} = \frac{27 - (0.95)(13.71)}{2} = 6.99 \text{ in.}$$

$$n = \frac{B - 0.8b_f}{2} = \frac{24 - (0.8)(12.48)}{2} = 7.01 \text{ in.}$$

$$\phi_c P_p = 0.6(0.85 f_c' A_1) = (0.6)(0.85)(3)(27)(24) = 991.4 \text{ k}$$

$$X = \frac{4db_f}{(d + b_f)^2} \frac{P_u}{\phi_c P_p} = \frac{(4)(13.71)(12.48)}{(13.71 + 12.48)^2} \frac{960}{991.4} = 0.966$$

$$\lambda = \frac{2\sqrt{X}}{1 + \sqrt{1 - X}} = \frac{(2)\sqrt{0.966}}{1 + \sqrt{1 - 0.966}} = 1.66 > 1.0 \quad \underline{\underline{\text{Use 1.0}}}$$

$$\lambda n' = \frac{\lambda \sqrt{db_f}}{4} = \frac{1.0\sqrt{(13.71)(12.48)}}{4} = 3.27$$

$$\ell = \max (m, n, \lambda n') = 7.01 \text{ in.}$$

$$t_{reqd} = \ell \sqrt{\frac{2P_u}{0.9F_y BN}} = 7.01 \sqrt{\frac{(2)(960)}{(0.9)(36)(24)(27)}} = 2.12 \text{ in.}$$

Use PL2$\frac{1}{4}$ × 24 × 2 ft 3 in. with 24 × 27 pedestal ∎

■ EXAMPLE 7-6
Repeat Example 7-5 if the column is to be supported by a concrete pedestal 4 in. wider on each side than the base plate.

Solution.   Using a W12 × 152 ($d = 13.71$ in., $b_f = 12.48$ in.)
Determining required base-plate area
   If base plate had same area as pedestal, A_1 could be determined as follows:

$$A_1 = \frac{P_u}{\phi_c(0.85 f'_c)} = \frac{960}{(0.6)(0.85)(3)} = 627.5 \text{ in.}^2$$

$$A_1 = db_f = (13.71)(12.48) = 171.1 \text{ in.}^2$$

If we try a plate 25 × 26 ($A_1 = 650$ in.^{2}) the pedestal area A_2 will equal $(25 + 4)(26 + 4) = 870$ in.^{2} Then

$$\sqrt{\frac{A_2}{A_1}} = \sqrt{\frac{870}{650}} = 1.16$$

And

$$A_1 = \frac{P_u}{\phi_c(0.85 f'_c) \sqrt{\dfrac{A_2}{A_1}}} = \frac{960}{(0.6)(0.85)(3)(1.16)} = 540.9 \text{ in.}^2$$

Trying a plate 23 × 24 (552 in.^{2}) the pedestal area will be $(23 + 4) × (24 + 4) = 756$ in.^{2} and $\sqrt{A_2/A_1} = \sqrt{756/540.9} = 1.18$. Thus no appreciable change in A_1.

Optimizing base-plate dimensions

$$\Delta = \frac{0.95d - 0.8b_f}{2} = \frac{(0.95)(13.71) - (0.8)(12.48)}{2} = 1.52 \text{ in.}$$

$$N = \sqrt{A_1} + \Delta = \sqrt{540.9} + 1.52 = 24.78 \text{ in.} \quad \text{Say 25 in.}$$

$$B = \frac{540.9}{25} = 21.6 \quad \underline{\text{Say 22 in.}}$$

Use pedestal 26 × 29

$$\sqrt{\frac{A_2}{A_1}} = \sqrt{\frac{(26)(29)}{(22)(25)}} = 1.17 \quad \therefore \underline{\text{No change}}$$

Computing required base-plate thickness

$$m = \frac{N - 0.95d}{2} = \frac{25 - (0.95)(13.71)}{2} = 5.99 \text{ in.}$$

$$n = \frac{B - 0.8b_f}{2} = \frac{22 - (0.8)(12.48)}{2} = 6.01 \text{ in.}$$

$$\phi_c P_p = 0.6(0.85 f'_c A_1)\sqrt{\frac{A_2}{A_1}} = (0.6)(0.85)(3)(22 \times 25)\sqrt{\frac{26 \times 29}{22 \times 25}}$$

$$= 984.6 \text{ k}$$

$$X = \frac{4db_f}{(d + b_f)^2}\frac{P_u}{\phi_c P_p} = \frac{(4)(13.71)(12.48)}{(13.71 + 12.48)^2}\frac{960}{984.6} = 0.973$$

$$\lambda = \frac{2\sqrt{X}}{1 + \sqrt{1 - X}} = \frac{2\sqrt{0.973}}{1 + \sqrt{1 - 0.973}} = 1.69 > 1.0 \quad \underline{\text{Use 1.0}}$$

$$\lambda n' = \frac{\lambda\sqrt{db_f}}{4} = \frac{1\sqrt{(13.71)(12.48)}}{4} = 3.27 \text{ in.}$$

$$\ell = \max{(m, n, \lambda n')} = 6.01 \text{ in.}$$

$$t_{reqd} = \ell\sqrt{\frac{2P_u}{0.9F_y BN}} = 6.01\sqrt{\frac{(2)(960)}{(0.9)(36)(22)(25)}} = 1.97 \text{ in.}$$

Use base PL2 × 22 × 2 ft 1 in. and 26 × 29 pedestal ■

Moment Resisting Column Bases

The designer will often be faced with the need for moment resisting column bases. Before such a topic is introduced, the author feels that the student needs to be familiar with the design of welds (Chapter 14) and moment resisting connections between members (Chapters 14 and 15). As a result, the subject of moment resisting base plates has been placed in Appendix E.

7-5 COMPUTER EXAMPLE

In Example 7-7 a column base plate is designed for the information used in Example 7-4. To use the program it is necessary to input trial dimensions for the plate. Any reasonable values can be specified. The computer will then calculate the dimensions required based on those trial dimensions. The user will then see if the initial assumed dimensions are satisfactory. If not, some very accurate trial dimensions can be assumed from the results provided in the first solution.

■ **EXAMPLE 7-7**

Repeat Example 7-4 if $P_u = 548$ k using the INSTEP diskette if $P_u = 548$ k.

Solution.

```
                    INPUT

        STEEL & COLUMN PARAMETERS
        [B]  Type Steel: A36
        [C]  Column: W12×65
        [D]  Design Load: −720

        CONCRETE FOOTING PARAMETERS
        [G]  fc′ = 3
        [H]  Bc = 108
        [I]  Nc = 108

        TRIAL BASE PLATE SIZE
        [L]  B = 15
        [M]  N = 16
        [N]  t = 0

                   OUTPUT

        EVALUATION OF TRIAL BASE PLATE

        COMPUTE ACTUAL AREAS

        Ap = 15(16) = 240 sq. in.
        Ac = 108(108) = 11664 sq. in.
        Ac/Ap = 11664/240 = 48.6
        Use Ac/Ap = 4.0
```

COMPUTE BEARING CAPACITY

$0.6P_n = 0.85(3)240[4]^{0.5} = 734.4$ kips

COMPUTE REQUIRED PLATE THICKNESS

$m = 0.5[16 - 0.95(12.12)] = 2.243$ inches
$n = 0.5[16 - 0.80(12)] = 2.7$ inches
$n' = 0.25[12.12(12)]^{0.5} = 3.01496$ inches
$X = .980368$
$£ = 1.7369$ Use $£ = 1$
$£n' = 3.01496$ inches
$t_r = 1.29743$ inches   ■

PROBLEMS

7-1. Select a W12 section for column AB shown in the accompanying illustration, which is an elevation view; $P_u = 1100$ k. Use 50 ksi steel. The columns above and below AB are to be approximately the same size as AB. Consider only in-plane behavior. Sidesway is uninhibited and the far ends of the girders are fixed against rotation. (a) Assume elastic column behavior. (b) Assume inelastic column behavior. (*Ans.* W12 × 106, W12 × 96)

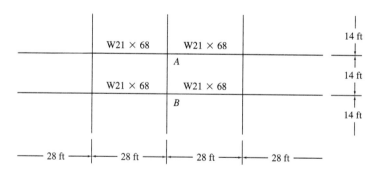

Problem 7-1

7-2. Repeat Prob. 7-1 if F_y is 36 ksi.

7-3. Repeat Prob. 7-1. if a W14 is used. (*Ans.* W14 × 109, W14 × 99)

7-4.   Select a W14 section for column CD shown in the accompanying illustration. Consider only in-plane behavior. (a) Assume elastic column behavior. (b) Assume inelastic column behavior. $P_u = 1330$ k and $F_y = 50$ ksi. The columns above and below are assumed to be approximately the same size as CD. Sidesway is uninhibited and far ends of girders fixed against rotation.

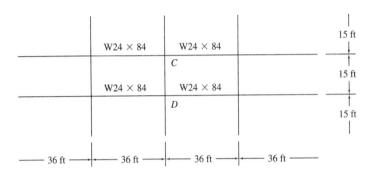

Problem 7-4

7-5.   Design W12 columns for the bent shown in the accompanying illustration with 50 ksi steel using the inelastic K-factor procedure. The columns are braced top and bottom against sidesway out of the plane of the frame so that $K = 1.0$ in that direction. Sidesway is possible in the plane of the frame. Design the right-hand column using $K = 1.0$ and the left-hand column with K as determined from the alignment chart and $P_u = 1900$ k. The beam is rigidly connected to the left column but has only a simple connection to the right column. (*Ans.* W12 × 96, W12 × 210)

Problem 7-5

7-6.   Repeat Prob. 7-5 if the P_u loads are 1000 k each and A36 steel is used.

7-7.   The columns for the frames shown in the accompanying illustration are braced top and bottom against sidesway out of the plane of the frame so that $K = 1.0$ in that direction. Sidesway is possible in the plane of the frame. Design the interior column assuming $K = 1.0$ and the exterior columns with K as determined from the alignment chart and $P_u = 2000$ k. Use $F_y = 50$ ksi and a W14 section. (*Ans.* W14 × 193, W14 × 233)

Problem 7-7

7-8.   Repeat Prob. 7-7 using $F_y = 36$ ksi.

7-9.   For the frame shown in the accompanying illustration the beams are rigidly connected to the exterior columns while all other connections are simple. The columns are braced top and bottom against sidesway out of the plane of the frame so that $K = 1.0$ in that direction. Sidesway is possible in the plane of the frame. Design W14 interior columns of 50 ksi steel assuming $K = 1.0$ and W14 exterior columns with K as determined from the alignment chart and $P_u = 1625$ k. (*Ans.* W14 × 90, W14 × 193)

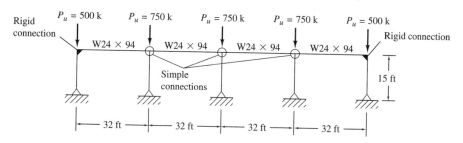

Problem 7-9

7-10.   Repeat Prob. 7.9 using $F_y = 36$ ksi.

7-11.   Design a base plate with A36 steel for a W14 × 82 column with a dead working load of 120 k and a working live load of 460 k. The concrete 28-day strength is 3 ksi. Footing size is 11 by 11 ft. (*Ans.* PL1$\frac{1}{2}$ × 15 × 1 ft 8 in.)

7-12. Design a base plate with A36 steel for a W12 × 106 column with a dead working load of 100 k and a live working load of 420 k. The concrete 28-day strength is 4 ksi. Footing size is 12 by 12 ft.

7-13. Repeat Prob. 7.12 if the column is supported by a 28- by 28-in. pedestal. (*Ans.* PL1$\frac{1}{2}$ × 13 × 1 ft 4 in.)

7-14. Design a column base plate of A36 steel for a W14 × 120 column supporting an axial load $P_u = 960$ k. Footing size is 10 by 10 ft and f'_c is 3 ksi.

7-15. Design a column base plate for a W14 × 90 column with $P_u = 650$ k if the steel is A36 and $f'_c = 3$ ksi for the concrete. Assume the column is to be supported by a pedestal 6 in. wider on each side than the plate. (*Ans.* PL1$\frac{1}{4}$ × 18 × 1 ft 6 in., 24 × 24 pedestal)

7-16 to 7-17. *Repeat the problems indicated using the enclosed LRFD computer diskette.*

7-16. Prob. 7-12.

7-17. Prob. 7-14. (*Ans.* PL1$\frac{3}{4}$ × 18 × 1 ft 6 in.)

Chapter **8**

Introduction to Beams

8-1 TYPES OF BEAMS

Beams are usually said to be members that support transverse loads. They are probably thought of as being used in horizontal positions and subjected to gravity or vertical loads; but there are frequent exceptions—rafters, for example.

Among the many types of beams are joists, lintels, spandrels, stringers, and floor beams. *Joists* are the closely spaced beams supporting the floors and roofs of buildings, while *lintels* are the beams over openings in masonry walls such as windows and doors. A *spandrel beam* supports the exterior walls of buildings and perhaps part of the floor and hallway loads. The discovery that steel beams as a part of a structural frame could support masonry walls (together with the development of passenger elevators) is said to have permitted the construction of today's "skyscrapers." *Stringers* are the beams in bridge floors running parallel to the roadway, whereas *floor beams* are the larger beams in many bridge floors which are perpendicular to the roadway of the bridge and are used to transfer the floor loads from the stringers to the supporting girders or trusses. The term *girder* is rather loosely used but usually indicates a large beam and perhaps one into which smaller beams are framed. These and other types of beams are discussed in the sections to follow.

8-2 SECTIONS USED AS BEAMS

The W shapes will normally prove to be the most economical beam sections and they have largely replaced channels and S sections for beam usage. Channels are sometimes used for beams subjected to light loads, such as purlins, and in places where clearances available require narrow flanges. They have very little resistance to lateral forces and need to be braced as illustrated by the sag rod problem in Chapter 4. The W shapes have more steel concentrated in their flanges than do S beams and thus have larger moments of inertia and resisting moments for

Harrison Avenue Bridge, Beaumont, Tex. (Courtesy of Bethlehem Steel Corporation.)

the same weights. They are relatively wide and have appreciable lateral stiffness. (The small amount of space devoted to S beams in the LRFD Manual clearly shows how much their use has decreased from former years. They are today used primarily for special situations as where narrow flange widths are desirable, or where shearing forces are very high, or where the greater flange thickness next to the web may be desirable where lateral bending occurs as perhaps with crane rails.)

Another common type of beam section is the open web joist or bar joist which is discussed at length in Chapter 19. This type of section which is commonly used to support floor and roof slabs is actually a light shop-fabricated parallel chord truss. It is particularly economical for long spans and light loads.

8-3 BENDING STRESSES

For an introduction to bending stresses the rectangular beam and stress diagrams of Fig. 8-1 are considered. (For this initial discussion the beam's compression flange is assumed to be fully braced against lateral buckling. Lateral buckling is discussed at length in Chapter 9). If the beam is subjected to some bending moment the stress at any point may be computed with the usual flexure formula $f_b = Mc/I$. *It is to be remembered, however, that this expression is only applicable when the maximum computed stress in the beam is below the elastic limit.* The formula is based on the usual elastic assumptions: stress is propor-

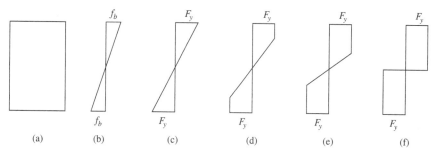

Figure 8-1

tional to strain, a plane section before bending remains a plane section after bending, etc. The value of I/c is a constant for a particular section and is known as the *section modulus* (S). The flexure formula may then be written as follows:

$$f_b = \frac{M}{S}$$

Initially, when the moment is applied to the beam the stress will vary linearly from the neutral axis to the extreme fibers. This situation is shown in part (b) of Fig. 8-1. If the moment is increased there will continue to be a linear variation of stress until the yield stress is reached in the outermost fibers, as

Erection of steel beams with a truck crane. (Courtesy of the American Institute of Steel Construction, Inc.)

shown in part (c) of the figure. The *yield moment* of a cross section is defined as the moment that will just produce the yield stress in the outermost fiber of the section.

If the moment in a ductile steel beam is increased beyond the yield moment the outermost fibers that had previously been stressed to their yield stress will continue to have the same stress but will yield, and the duty of providing the necessary additional resisting moment will fall on the fibers nearer to the neutral axis. This process will continue with more and more parts of the beam cross section stressed to the yield stress as shown by the stress diagrams of parts (d) and (e) of the figure, until finally a full plastic distribution is approached as shown in part (f). Note that the variation of strain from the neutral axis to the outer fibers remains linear for all of these cases. When the stress distribution has reached this stage a *plastic hinge* is said to have formed because no additional moment can be resisted at the section. Any additional moment applied at the section will cause the beam to rotate with little increase in stress.

The *plastic moment* is the moment that will produce full plasticity in a member cross section and create a plastic hinge. The ratio of the plastic moment M_p to the yield moment M_y is called the *shape factor*. The shape factor equals 1.50 for rectangular sections and varies from about 1.10 to 1.20 for standard rolled-beam sections.

8-4 PLASTIC HINGES

This section is devoted to a description of the development of a plastic hinge in the simple beam shown in Fig. 8-2. The load shown is applied to the beam and increased in magnitude until the yield moment is reached and the outermost fiber is stressed to the yield stress. The magnitude of the load is further increased with the result that the outer fibers begin to yield. The yielding spreads out to the other fibers away from the section of maximum moment as indicated in the figure. The length in which this yielding occurs away from the section in question is dependent on the loading conditions and the member cross section. For a concentrated load applied at the center line of a simple beam with a rectangular cross section, yielding in the extreme fibers at the time the plastic hinge is formed will extend for one-third of the span. For a W shape in similar circumstances yielding will extend for approximately one-eighth of the span. During

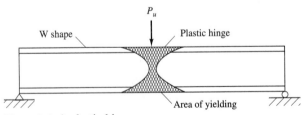

Figure 8-2 A plastic hinge.

this same period the interior fibers at the section of maximum moment yield gradually until nearly all of them have yielded and a plastic hinge is formed as shown in Fig. 8-2.

Although the effect of a plastic hinge may extend for some distance along the beam, it is assumed to be concentrated at one section for analysis purposes. For the calculation of deflections and for the design of bracing, the length over which yielding extends is quite important.

When steel frames are loaded to failure, the points where rotation is concentrated (plastic hinges) become quite visible to the observer before collapse occurs.

8-5 ELASTIC DESIGN

Until recent years almost all steel beams were designed on the basis of the elastic theory. The maximum load that a structure could support was assumed to equal the load that first caused a stress somewhere in the structure to equal the yield stress of the material. The members were designed so that computed bending stresses for service loads did not exceed the yield stress divided by a safety factor (say 1.5 to 2.0). Engineering structures have been designed for many decades by this method with satisfactory results. The design profession, however, has long been aware that ductile members do not fail until a great deal of yielding occurs after the yield stress is first reached. This means that such members have greater margins of safety against collapse than the elastic theory would seem to indicate.

8-6 THE PLASTIC MODULUS

The yield moment M_y equals the yield stress times the elastic modulus. The elastic modulus equals I/c or $bd^2/6$ for a rectangular section and the yield moment equals $\frac{F_y bd^2}{6}$. This same value can be obtained by considering the resisting internal couple shown in Fig. 8-3.

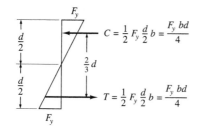

$$C = \frac{1}{2} F_y \frac{d}{2} b = \frac{F_y bd}{4}$$

$$T = \frac{1}{2} F_y \frac{d}{2} b = \frac{F_y bd}{4}$$

Figure 8-3

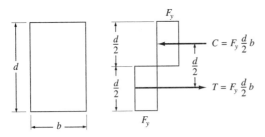

Figure 8-4

The resisting moment equals T or C times the lever arm between them, as follows:

$$M_y = \left(\frac{F_y bd}{4}\right)\left(\frac{2}{3}d\right) = \frac{F_y bd^2}{6}$$

The elastic section modulus can again be seen to equal $bd^2/6$ for a rectangular beam.

The resisting moment at full plasticity can be determined in a similar manner. The result is the so-called plastic moment, M_p. It is also the nominal moment of the section, M_n. This plastic or nominal moment equals T or C times the lever arm between them. For the rectangular beam of Fig. 8-4 we have

$$M_p = M_n = T\frac{d}{2} = C\frac{d}{2} = \left(F_y\frac{bd}{2}\right)\left(\frac{d}{2}\right) = F_y\frac{bd^2}{4}$$

The plastic moment is said to equal the yield stress times the plastic modulus. From the foregoing expression for a rectangular section, the plastic modulus Z can be seen to equal $bd^2/4$. The shape factor, which equals M_n/M_y, F_yZ/F_yS, or Z/S, is $(bd^2/4)/(bd^2/6) = 1.50$ for a rectangular section. A study of the plastic modulus determined here shows that it equals the statical moment of the tension and compression areas about the neutral axis. Unless the section is symmetrical, the neutral axis for the plastic condition will not be in the same location as for the elastic condition. The total internal compression must equal the total internal tension. As all fibers are considered to have the same stress (F_y) in the plastic condition, the areas above and below the neutral axis must be equal. This situation does not hold for unsymmetrical sections in the elastic condition. Example 8-1 illustrates the calculations necessary to determine the shape factor for a tee beam and the nominal uniform load w_n that the beam can theoretically support.

■ EXAMPLE 8-1

Determine M_y, M_n, and Z for the steel tee beam shown in Fig. 8-5. Also calculate the shape factor and the nominal load (w_n) which can be placed on the beam for a 12-ft simple span. $F_y = 36$ ksi.

Figure 8-5

Solution.   Elastic calculations:

$$A = (8)\left(1\frac{1}{2}\right) + (6)(2) = 24 \text{ in.}^2$$

$$\bar{y} = \frac{(12)(0.75) + (12)(4.5)}{24} = 2.625 \text{ in. from top flange}$$

$$I = \left(\frac{1}{3}\right)(2)(1.125^3 + 4.875^3) + \left(\frac{1}{12}\right)(8)\left(1\frac{1}{2}\right)^3 + (12)(1.875)^2$$

$$= 122.4 \text{ in.}^4$$

$$S = \frac{I}{c} = \frac{122.4}{4.875} = 25.1 \text{ in.}^3$$

$$M_y = F_y S = \frac{(36)(25.1)}{12} = 75.3 \text{ ft-k}$$

Plastic calculations: Neutral axis is at base of flange.

$$Z = (12)(0.75) + (12)(3) = 45 \text{ in.}^3$$

$$M_n = F_y Z = \frac{(36)(45)}{12} = 135 \text{ ft-k}$$

$$\text{Shape factor} = \frac{M_n}{M_y} \quad \text{or} \quad \frac{Z}{S} = \frac{45}{25.1} = 1.79$$

$$M_n = \frac{w_n L^2}{8}$$

$$w_n = \frac{(8)(135)}{(12)^2} = 7.5 \text{ k/ft} \qquad \blacksquare$$

The values of the plastic moduli for the standard steel beam sections are tabulated in Part 4 of the LRFD Manual in the "Load Factor Design Selection Table for Shapes Used as Beams" as well as being listed for each shape in the "Dimensions and Properties" section of the Handbook (Part 1). These Z values will frequently be used throughout the text.

8-7 THEORY OF PLASTIC ANALYSIS

The basic plastic theory has been shown to be a major change in the distribution of stresses after the stresses at certain points in a structure reach the yield stress. The theory is that those parts of the structure that have been stressed to the yield stress cannot resist additional stresses. They instead will yield the amount required to permit the extra load or stresses to be transferred to other parts of the structure where the stresses are below the yield stress and thus in the elastic range and able to resist increased stress. Plasticity can be said to serve the purpose of equalizing stresses in cases of overload.

As early as 1914 the Hungarian, Dr. Gabor Kazinczy, recognized that the ductility of steel permitted a redistribution of stresses in an overloaded statically indeterminate structure.[1] In the United States Prof. J. A. Van den Broek introduced his plastic theory which he called limit design. This theory was published in a paper entitled "Theory of Limit Design" in February 1939 in the *Proceedings of the ASCE.*

For this discussion the stress-strain diagram is assumed to have the idealized shape shown in Fig. 8-6. The yield stress and the proportional limit are assumed to occur at the same point for this steel, and the stress-strain diagram is assumed to be a perfectly straight line in the plastic range. Beyond the plastic range there is a range of strain hardening. This latter range could theoretically permit steel members to withstand additional stress, but from a practical standpoint the strains occuring are so large that they cannot be considered. Furthermore, inelastic buckling will limit the ability of a section to develop a moment greater than M_n even if strain hardening is significant.

Figure 8-6

---

[1] Lynn S. Beedle, *Plastic Design of Steel Frames* (New York: Wiley, 1958), p. 3.

8-8 THE COLLAPSE MECHANISM

A statically determinate beam will fail if one plastic hinge develops. To illustrate this fact, the simple beam of constant cross section loaded with a concentrated load at midspan shown in Fig. 8-7(a) is considered. Should the load be increased until a plastic hinge is developed at the point of maximum moment (underneath the load in this case) an unstable structure will have been created as shown in part (b) of the figure. Any further increase in load will cause collapse. P_n represents the nominal or theoretical maximum load that the beam can support.

For a statically indeterminate structure to fail it is necessary for more than one plastic hinge to form. The number of plastic hinges required for failure of statically indeterminate structures will be shown to vary from structure to structure, but may never be less than two. The fixed-end beam of Fig. 8-8 cannot fail unless the three plastic hinges shown in the figure are developed.

Although a plastic hinge may have formed in a statically indeterminate structure, the load can still be increased without causing failure if the geometry of the structure permits. The plastic hinge will act like a real hinge insofar as

Figure 8-7

Figure 8-8

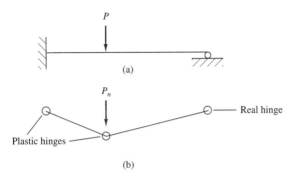

Figure 8-9

increased loading is concerned. As the load is increased there is a redistribution of moment because the plastic hinge can resist no more moment. As more plastic hinges are formed in the structure there will eventually be a sufficient number of them to cause collapse. Actually, some additional load can be carried after this time before collapse occurs as the stresses go into the strain hardening range, but the deflections that would occur are too large to be permissible.

The propped beam of Fig. 8-9 is an example of a structure that will fail after two plastic hinges develop. Three hinges are required for collapse, but there is a real hinge on the right end. In this beam the largest elastic moment caused by the design concentrated load is at the fixed end. As the magnitude of the load is increased a plastic hinge will form at that point.

The load may be further increased until the moment at some other point (here it will be at the concentrated load) reaches the plastic moment. Additional load will cause the beam to collapse. The arrangement of plastic hinges and perhaps real hinges which permit collapse in a structure is called the *mechanism.* Parts (b) of Figs. 8-7, 8-8, and 8-9 show mechanisms for various beams.

After observing the large number of fixed-end and propped beams used for illustration in this text, the student may form the mistaken idea that he or she will frequently encounter such beams in engineering practice. These types of beams are difficult to find in actual structures but are very convenient to use in illustrative examples. They are particularly convenient for introducing plastic analysis before continuous beams and frames are considered.

8-9 THE VIRTUAL-WORK METHOD

One very satisfactory method used for the plastic analysis of structures is the *virtual-work method.* The structure in question is assumed to be loaded to its nominal capacity, M_n, and is then assumed to deflect through a small additional displacement after the ultimate load is reached. The work performed by the external loads during this displacement is equated to the internal work absorbed by the hinges. For this discussion the *small-angle theory* is used. By this theory

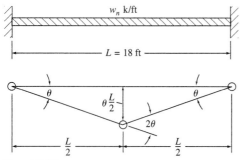

Figure 8-10

the sine of a small angle equals the tangent of that angle and also equals the same angle expressed in radians. In the pages to follow the author uses these values interchangeably because the small displacements considered here produce extremely small rotations or angles.

As a first illustration the uniformly loaded fixed-ended beam of Fig. 8-10 is considered. This beam and its collapse mechanism are shown in the figure. Owing to symmetry, the rotations at the end plastic hinges are equal and they are represented by θ in the figure; thus the rotation at the middle plastic hinge will be 2θ.

The work performed by the total external load $(w_n L)$ is equal to $w_n L$ times the average deflection of the mechanism. The average deflection equals one-half the deflection at the center plastic hinge ($\frac{1}{2} \times \theta \times L/2$). The external work is equated to the internal work absorbed by the hinges or to the sum of M_n at each plastic hinge times the angle through which it works. The resulting expression can be solved for M_n and w_n as follows.

$$M_n(\theta + 2\theta + \theta) = w_n L\left(\frac{1}{2} \times \theta \times \frac{L}{2}\right)$$

$$M_n = \frac{w_n L^2}{16}$$

$$w_n = \frac{16 M_n}{L^2}$$

For the 18-ft span used in Fig. 8-10 these values become

$$M_n = \frac{(w_n)(18)^2}{16} = 20.25\, w_n$$

$$w_n = \frac{M_n}{20.25}$$

Plastic analysis can be handled in a similar manner for the propped beam of Fig. 8-11. There the collapse mechanism is shown and the end rotations (which are equal to each other) are assumed to equal θ.

The work performed by the external load P_n as it moves through the distance $\theta \times L/2$ is equated to the internal work performed by the plastic moments at the

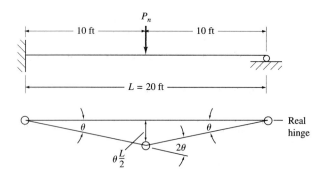

Figure 8-11

hinges, noting that there is no moment at the real hinge on the right end of the beam.

$$M_n(\theta + 2\theta) = P_n\left(\theta\frac{L}{2}\right)$$

$$M_n = \frac{P_n L}{6} \quad \text{(or 3.33 } P_n \text{ for the 20-ft beam shown)}$$

$$P_n = \frac{6M_n}{L} \quad \text{(or 0.3 } M_n \text{ for the 20-ft beam shown)}$$

The fixed-end beam of Fig. 8-12 together with its collapse mechanism and assumed angle rotations is next considered. From this figure the values of M_n and P_n can be determined by virtual work as follows.

$$M_n(2\theta + 3\theta + \theta) = P_n\left(2\theta \times \frac{L}{3}\right)$$

$$M_n = \frac{P_n L}{9} \quad \text{(or 3.33 } P_n \text{ for this beam)}$$

$$P_n = \frac{9M_n}{L} \quad \text{(or 0.3 } M_n \text{ for this beam)}$$

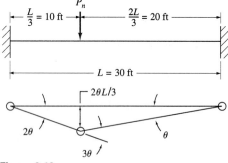

Figure 8-12

A person beginning the study of plastic analysis needs to learn to think of all the possible ways in which a particular structure might collapse. Such a habit is of the greatest importance when one begins to analyze more complex structures. In this light the plastic analysis of the propped beam of Fig. 8-13 is considered by the virtual-work method. The beam with its two concentrated loads is shown together with four possible collapse mechanisms and the necessary calculations. It is true that the mechanisms of parts (b), (d), and (e) of the figure do not control, but such a fact is not obvious to the average student until he or she makes the virtual-work calculations for each case. Actually, the mechanism of part (e) is based on the assumption that the plastic moment is reached at both the concentrated loads simultaneously (a situation that might very well occur).

The value for which the collapse load is the smallest in terms of M_n is the correct value (or the value where M_n is the greatest in terms of P_n). For this beam the second plastic hinge forms at the P_n concentrated load and P_n equals $0.154\ M_n$.

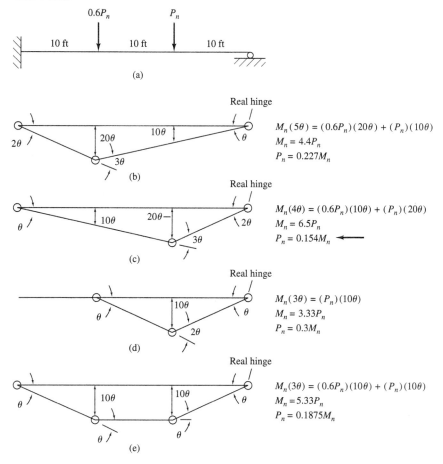

Figure 8-13

8-10 LOCATION OF PLASTIC HINGE FOR UNIFORM LOADINGS

There was no difficulty in locating the plastic hinge for the unformly loaded fixed-end beam, but for other beams with uniform loads, such as propped or continuous beams, the problem may be rather difficult. For this discussion the uniformly loaded propped beam of Fig. 8-14(a) is considered.

The elastic moment diagram for this beam is shown in part (b) of the figure. As the uniform load is increased in magnitude, a plastic hinge will first form at the fixed end. At this time the beam will, in effect, be a "simple" beam with a plastic hinge on one end and a real hinge on the other. Subsequent increases in the load will cause the moment to change as represented by the dashed line in part (b) of the figure. This process will continue until the moment at some other point (a distance x from the right support in the figure) reaches M_n and creates another plastic hinge.

The virtual-work expression for the collapse mechanism of this beam shown in part (c) of Fig. 8-14 is written as follows.

$$M_n\left(\theta + \theta + \frac{L-x}{x}\theta\right) = (w_n L)(\theta)(L - x)\left(\frac{1}{2}\right)$$

Solving this equation for M_n, taking $dM_n/dx = 0$, the value of x can be calculated to equal $0.414L$. This value is also applicable to uniformly loaded end spans of continuous beams with simple end supports as will be illustrated in the next section.

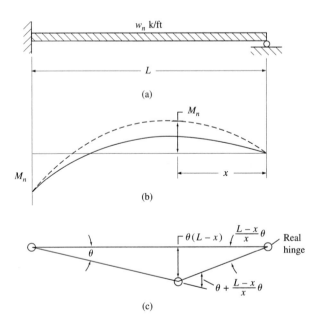

Figure 8-14

The beam and its collapse mechanism are redrawn in Fig. 8-15 and the following expression for the plastic moment is written using the virtual-work procedure.

$$M_n(\theta + 2.414\theta) = (w_n L)(0.586\theta L)\left(\frac{1}{2}\right)$$

$$M_n = 0.0858w_n L^2$$

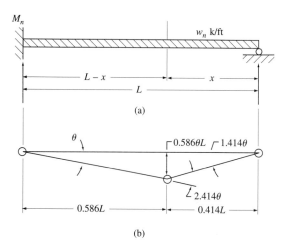

(a)

(b)

Figure 8-15

8-11 CONTINUOUS BEAMS

Continuous beams are very common in engineering structures. Their continuity causes analysis to be rather complicated in the elastic theory, and even though one of the complex "exact" methods is used for analysis, the resulting stress distribution is not nearly so accurate as is usually assumed.

Plastic analysis is applicable to continuous structures as it is to one-span structures. The resulting values definitely give a more realistic picture of the limiting strength of a structure than can be obtained by elastic analysis. Continuous statically indeterminate beams can be handled by the virtual-work procedure as they were for the single-span statically indeterminate beams. As an introduction to continuous beams Examples 8-2 and 8-3 are presented to illustrate two of the more elementary cases.

Here it is assumed that if any or all of a structure collapses failure has occurred. Thus, in the continuous beams to follow virtual-work expressions are written separately for each span. From the resulting expressions it is possible to determine the limiting or maximum loads that the beams can support.

■ EXAMPLE 8-2

A W18 × 55 (Z_x = 112 in.^{3}) has been selected for the beam shown in Fig. 8-16. Using A36 steel and assuming full lateral support determine the value of w_n.

Figure 8-16

Solution.

$$M_n = F_y Z = \frac{(36)(112)}{12} = 336 \text{ ft-k}$$

Drawing the (collapse) mechanisms for the two spans

Left-hand span

$$(M_n)(3.414\theta) = (24w_n)\left(\frac{1}{2}\right)(14.06\theta)$$

$$w_n = 0.0202 \, M_n = (0.0202)(336) = 6.8 \text{ klf}$$

Right-hand span

$$(M_n)(4\theta) = (30w_n)\left(\frac{1}{2}\right)(15\theta)$$

$$w_n = 0.0178 \, M_n = (0.0178)(336) = \underline{\underline{5.97}} \text{ klf} \leftarrow \quad ■$$

Additional spans have little effect on the amount of work involved in the plastic analysis procedure. The same cannot be said for elastic analysis. Example 8-3 illustrates the analysis of a three-span beam that is loaded with a concentrated load on each span. The student from his or her knowledge of elastic analysis can see that plastic hinges will initially form at the first interior supports and then at the center lines of the end spans, at which time each end span will have a collapse mechanism.

■ EXAMPLE 8-3

Using a W21 × 44 (Z_x = 95.4 in.³) consisting of A36 steel determine the value of P_n for the beam of Fig. 8-17.

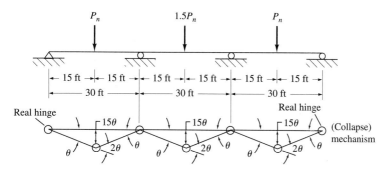

Figure 8-17

Solution.

$$M_n = F_y Z = \frac{(36)(95.4)}{12} = 286.2 \text{ ft-k}$$

For first and third spans

$$(P_n)(15\theta) = M_n(3\theta)$$
$$P_n = 0.2M_n = (0.2)(286.2) = 57.2 \text{ k}$$

For center span

$$(1.5P_n)(15\theta) = (M_n)(4\theta)$$
$$P_n = 0.178\,M_n = (0.178)(286.2) = 50.9 \text{ k} \qquad ■$$

8-12 BUILDING FRAMES

In this section plastic analysis is applied to a small building frame. It is not the author's purpose to consider frames at length in this chapter. Rather, he wants to show the reader that the virtual-work method is applicable to frames as well as to beams, and that there are other types of mechanisms besides the beam types.

For the frame considered it is assumed that the same W section is used for both the beam and the columns. If these members differed in size it would be necessary to take that into account in the analysis.

The pin-supported frame of Fig. 8-18 is statically indeterminate to the first degree. The development of one plastic hinge will cause it to become statically determinate, while the forming of a second hinge can create a mechanism. There are, however, several types of mechanisms that might feasibly occur in this frame. A possible beam mechanism is shown in part (b), a sidesway mechanism

(a) Frame and loads

(b) Beam mechanism

$$(P_n)(20\theta) = M_n(4\theta)$$

$$P_n = \frac{1}{5} M_n$$

(c) Sidesway mechanism

$$(0.6 P_n)(20\theta) = M_n(2\theta)$$

$$P_n = \frac{1}{6} M_n$$

(d) Combined beam and sidesway mechanism

$$(0.6 P_n)(20\theta) + (P_n)(20\theta) = M_n(4\theta)$$

$$P_n = \frac{1}{8} M_n \longleftarrow$$

Figure 8-18 Possible mechanisms for a frame.

is shown in part (c) and a combined beam and sidesway mechanism is shown in part (d). The critical condition is the one that will result in the smallest value of P_n.

Example 8-4 presents the plastic analysis of the frame of Fig. 8-18. The distances through which the loads tend to move in the various mechanisms should be carefully studied. The solution of this problem brings out a point of major significance: Superposition does not apply to plastic analysis. You can easily see this by studying the virtual work expressions for parts (b), (c), and (d) of the figure. The values of P_n obtained for the separate beam and sidesway mechanisms do not add up to the value obtained for the combined beam and sidesway mechanism.

For each mechanism we want to consider the situation in which we have the fewest possible number of plastic hinges that will cause collapse. If you look at one of the virtual-work expressions you will note that P_n becomes smaller as the number of plastic hinges decreases. With this in mind, look at part (d) of Fig. 8-18. The frame can feasibly sway to the right without the formation of a plastic hinge at the top of the left column. The two plastic hinges labeled A and B are sufficient for collapse to occur.

■ EXAMPLE 8-4

A W12 × 72 (Z_x = 108 in.) is used for the beam and columns of the frame shown in Fig. 8-18. If F_y = 50 ksi, determine the value of P_n.

Solution. The virtual-work expressions are written for parts (b), (c), and (d) of Fig. 8-18 and shown with the respective parts of the figure. The combined beam and sidesway case is found to be the critical case and from it the value of P_n is determined as follows:

$$P_n = \frac{1}{8}M_n = \left(\frac{1}{8}\right)(F_y Z) = \left(\frac{1}{8}\right)\left(\frac{50 \times 108}{12}\right) = \underline{\underline{56.25 \text{ k}}} \qquad ■$$

PROBLEMS

8–1 to 8–10. *Find the values of S, Z, and the shape factor about the x axes for the sections shown in the accompanying illustrations.*

Problem 8-1 (*Ans.* 229.3, 271, 1.18)

Problem 8-2

Problem 8-3 (*Ans.* 205.9, 252.5, 1.23)

Problem 8-4

Problem 8-5 (*Ans.* 100.3, 182, 1.81)

Problem 8-6

Problem 8-7 (*Ans.* 218.4, 254.4, 1.16)   **Problem 8-8**

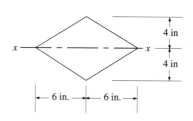

Problem 8-9 (*Ans.* 87.8, 121.3, 1.38)   **Problem 8-10**

8-11 to 8-20. *Determine the values of S , Z, and the shape factor about the x axes unless otherwise directed. Use the web and flange dimensions given in the LRFD Manual for making these calculations.*

8-11. A W27 × 235 (*Ans.* 671, 765, 1.14)

8-12. A W24 × 131 with one $\frac{3}{4}$ × 16 in. PL on each flange.

8-13. Two 8 × 6 × $\frac{5}{8}$-in. Ls long legs vertical and back to back. (*Ans.* 19.7, 35.8, 1.82)

8-14. Two C12 × 30s back-to-back.

8-15. Four 8 × 8 × $\frac{3}{4}$ in. Ls arranged as shown in the accompanying illustration. (*Ans.* 64.5, 104.2, 1.62)

Problem 8-15

8-16. The section of Prob. 5-9(b).

8-17. The section of Prob. 8-1 considering the *y* axis. (*Ans.* 33.6, 54.5, 1.62)

8-18. Rework Prob. 8-2 considering the *y* axis.

8-19. Rework Prob. 8-12 considering the *y* axis. (*Ans.* 106.5, 177.5, 1.67)

8-20. Rework Prob. 8-14 considering the *y* axis.

8-21 to 8-40. *Using the given sections all of A572 Grade 50 steel and the plastic theory determine the values of P_n and w_n as indicated.*

Problem 8-21 (*Ans. P_n 134.4 k*)

Problem 8-22

Problem 8-23 (*Ans. $w_n = 12.07$ k/ft*)

Problem 8-24

Problem 8-25 (*Ans. P_n = 155.3 k*)

Problem 8-26

Problem 8-27 (*Ans. P_n = 65 k*)

Problem 8-28

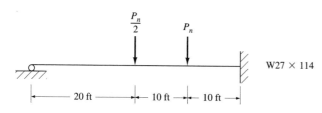

Problem 8-29 (*Ans. P_n = 214.4 k*)

Problem 8-30

Problem 8-31 (*Ans. w_n* = 12.26 k/ft)

Problem 8-32

Problem 8-33 (*Ans. w_n* = 2.69 k/ft)

Problem 8-34

Problem 8-35 (*Ans. w_n* = 23.62 k/ft)

Problem 8-36

Problem 8-37 (*Ans.* $P_n = 112.3$ k)

Problem 8-38

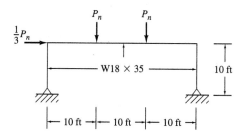

Problem 8-39 (*Ans.* $P_n = 45.5$ k)

8-40. Repeat Prob. 8-38 if the column bases are fixed.

Chapter 9

Design of Beams for Moments

9-1 INTRODUCTION

If gravity loads are applied to a fairly long, simply supported beam, the beam will bend downward, and its upper part will be placed in compression and will act as a compression member. The cross section of this "column" will consist of the portion of the beam cross section above the neutral axis. For the usual beam the "column" will have a much smaller moment of inertia about its y or vertical axis than about its x axis. If nothing is done to brace it perpendicular to the y axis, it will buckle laterally at a much smaller load than would otherwise have been required to produce a vertical failure. (You can verify these statements by trying to bend vertically a magazine held in a vertical position. The magazine will, just as a steel beam, always tend to buckle laterally unless it is braced in that direction.)

Lateral buckling will not occur if the compression flange of a member is braced laterally at frequent intervals. In this chapter the buckling moments of a series of compact ductile steel beams with different lateral bracing situations are considered. (As previously defined, a *compact section* is one that has a sufficiently stocky profile so that it is capable of developing a fully plastic stress distribution before buckling.)

In this chapter we will look at beams as follows:

1. First, the beams will be assumed to have continuous lateral bracing for their compression flanges.
2. Next, the beams will be assumed to be braced laterally at short intervals.
3. Finally, the beams will be assumed to be braced laterally at larger and larger intervals.

In Fig. 9-1 a typical curve showing the nominal resisting or buckling moments of one of these beams with varying unbraced lengths is presented.

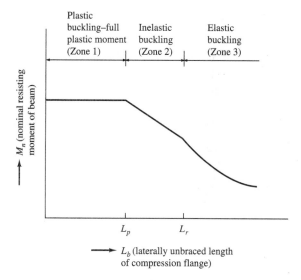

Figure 9.1 Nominal moment as function of unbraced length of compression flange.

An examination of Fig. 9-1 will show that beams have three distinct ranges or zones of buckling depending on their lateral bracing situation. If we have continuous or closely spaced lateral bracing, the beams will buckle plastically and fall into what is classified as Zone 1 buckling. As the distance between lateral bracing is increased further, the beams will begin to fail inelastically at smaller moments and fall into Zone 2. Finally, with even larger unbraced lengths, the beams will fail elastically and fall into Zone 3. A brief discussion of these three types of buckling is presented in this section, while the remainder of the chapter is devoted to a detailed discussion of each type, together with a series of numerical examples.

Plastic Buckling (Zone 1) If we were to take a compact beam whose compression flange is continuously braced laterally, we would find that we could load it until its full plastic moment M_p is reached; further loading then produces a redistribution of moments as was described in Chapter 8. In other words, the moments in these beams can reach M_p and then develop a rotation capacity sufficient for moment redistribution.

If we now take one of these compact beams and provide closely spaced intermittent lateral bracing for its compression flanges, we will find that we can still load it until the plastic moment plus moment redistribution is achieved if the spacing between the bracing does not exceed a certain value called L_p herein. (The value of L_p is dependent on the dimensions of the beam cross section and on its yield stress). *Most beams fall in Zone 1.*

Bridge over Allegheny River at Kittanning, PA. (Courtesy American Bridge Company.)

Inelastic Buckling (Zone 2)   If we now further increase the spacing between points of lateral bracing, the section may be loaded until some but not all of the compression fibers are stressed to F_y. The section will have insufficient rotation capacity to permit full moment redistribution and thus will not permit plastic analysis. In other words, in this zone we can bend the member until the yield strain is reached in some but not all of its compression elements before buckling occurs. This is referred to as *inelastic buckling.*

As we increase the unbraced length we will find that the moment the section resists will decrease until finally it will buckle before the yield stress is anywhere reached. The maximum unbraced length at which we can still reach F_y at one point is the end of the inelastic range. It's shown as L_r in Fig. 9-1; its value is dependent upon the properties of the beam cross section as well as on the yield and residual stresses of the beam. At this point, as soon as we have a moment which theoretically causes the yield stress to be reached anywhere (actually it's less than F_y because of residual stresses), the section will buckle.

Elastic Buckling (Zone 3)   If the unbraced length is greater than L_r the section will buckle elastically before the yield stress is reached anywhere. As the unbraced length is further increased, the buckling moment becomes smaller and

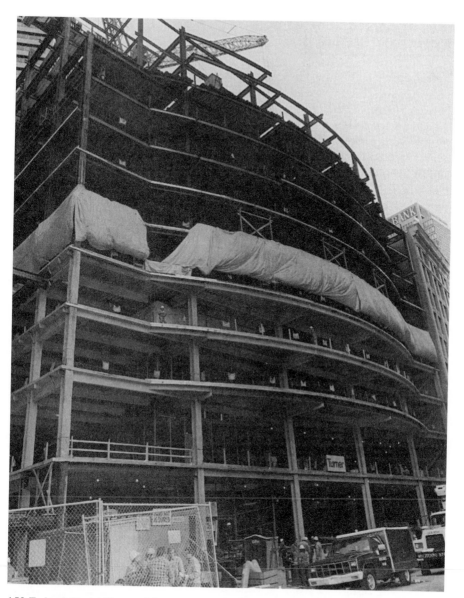

150 Federal Street, Boston, Mass. (Courtesy Owen Steel Company, Inc.)

smaller. As the moment is increased in such a beam, the beam will deflect more and more transversely until a critical moment value M_{cr} is reached. At this time the beam cross section will twist and the compression flange will move laterally. The moment M_{cr} is provided by the torsional resistance and the warping resistance of the beam as will be discussed in Section 9-7.

9-2 PLASTIC BUCKLING—FULL PLASTIC MOMENT, ZONE 1

In this section and the next two, beam formulas for plastic buckling (Zone 1) are presented, while in Sections 9-5 through 9-7 formulas are presented for inelastic buckling (Zone 2) and elastic buckling (Zone 3). After seeing some of these latter expressions the reader may become quite concerned that he or she is going to spend an enormous amount of time in formula substitution. This is not generally true, however, as the values needed are tabulated and graphed in simple form in the LRFD Manual.

When a steel section has a large shape factor appreciable inelastic deformations may occur under service loads if the section in designed so that M_p is reached at the factored load condition. As a result LRFD Specification F1.1 limits the amount of such deformation for sections with shape factors larger than 1.5. This is done by limiting M_p to a maximum value of 1.5 M_y.

If the unbraced length L_b of the compression flange of a compact I- or C-shaped section including hybrid members does not exceed L_p (if elastic analysis is being used) or L_{pd} (if plastic analysis is being used) then the member's bending strength about its major axis may be determined as follows:

$$M_n = M_p = F_y Z \leq 1.5\ M_y \qquad \text{(LRFD Equation F1-1)}$$

$$M_u = \phi_b M_n \quad \text{with} \quad \phi_b = 0.90$$

This part of the specification limiting M_n to 1.5 M_y for high shape factor sections such as WTs does not apply to hybrid sections with web yield stresses less than their flange yield stresses. Web yielding for such members does not result in significant inelastic deformations. For hybrid members the yield moment $M_y = F_{yf} S$

For elastic analysis L_b may not exceed the value L_p to follow if M_n is to equal $F_y Z$.

$$L_p = \frac{300 r_y}{\sqrt{F_{yf}}} \qquad \text{(LRFD Equation F1-4)}$$

For solid rectangular bars and box beams with A = cross-sectional area (in.^{2}) and J = torsional constant (in.^{4})

$$L_p = \frac{3,750\ r_y}{M_p} \sqrt{JA} \qquad \text{(LRFD Equation F1-5)}$$

For plastic analysis of doubly symmetric and singly symmetric I-shaped members with the compression flanges larger than their tension flanges (including hybrid members) loaded in the plane of the web L_b (which is defined as the laterally unbraced length of the compression flange at plastic hinge locations associated with failure mechanisms) may not exceed the value L_{pd} to follow if M_n is to equal $F_y Z$.

$$L_{pd} = \frac{3600 + 2200\ (M_1/M_2)}{F_y} r_y \qquad \text{(LRFD Equation F1-17)}$$

In this expression M_1 is the smaller moment at the end of the unbraced length of the beam and M_2 is the larger moment at the end of the unbraced length and the ratio M_1/M_2 is positive when the moments cause the member to be bent in double curvature $(\smile\frown)$ and negative if they bend it in single curvatures $(\frown\smile)$. Only steels with F_y values (F_y is the specified minimum yield stress of the compression flange) of 65 ksi or less may be considered. Higher-strength steels may not be ductile.

There is no limit of the unbraced length for circular or square cross sections or for I-shaped beams bent about their minor axes. (If I-shaped sections are bent about their minor or y axes they will not buckle before the full plastic moment M_p about the y axis is developed.) Equation F1-18 of the LRFD Specification also provides a value of L_{pd} for solid rectangular bars and symmetrical box beams.

For these sections to be compact the width-thickness ratios of the flanges and webs of I- and C-shaped sections are limited to the maximum values to follow, which are taken from Table B5.1 of the LRFD Specification. For flanges

$$\lambda_p = \frac{b}{t} \leq \frac{65}{\sqrt{F_y}}$$

For webs

$$\lambda_p = \frac{h}{t_w} \leq \frac{640}{\sqrt{F_y}}$$

In this last expression h is the distance from the web toe of the fillet in the top of the web to the web toe of the fillet in the bottom of the web (that is, twice the distance from the neutral axis to the inside face of the compression flange less the fillet or corner radius).

9-3 DESIGN OF BEAMS, ZONE 1

Included in the items that need to be considered in beam design are moments, shears, deflections, crippling, lateral bracing for the compression flanges, fatigue, and others. Beams will probably be selected that provide sufficient design moment capacities $(\phi_b M_n)$ and then checked to see if any of the other items are critical. The factored moment will be computed and a section having that much design moment capacity will be initially selected from the LRFD Manual.

A table is given in Section 4 of the LRFD Manual entitled "Load Factor Design Selection Table for Shapes Used as Beams." From this table steel shapes having sufficient plastic moduli to resist certain moments can quickly be selected. Two important items should be remembered in selecting shapes. These are:

1. These steel sections cost so many cents per pound and it is therefore desirable to select the lightest possible shape having the required plastic modulus (assuming that the resulting section is one that will reasonably fit into the structure). The table has the sections arranged in various groups having certain ranges of plastic moduli. The heavily typed section at the top of each group is the lightest section in that group and the others are arranged successively in the order of their plastic moduli. Normally the deeper sections will have the lightest weights giving the required plastic moduli, and they will be generally selected unless their depth causes a problem in obtaining the desired headroom, in which case a shallower but heavier section will be selected.

2. The plastic moduli values in the table are given about the horizontal axes for beams in their upright positions. If a beam is to be turned on its side the proper plastic modulus can be found in the tables giving dimensions and properties of shapes in Part 1 of the LRFD Manual. A W shape turned on its side may only be from 10 to 30 percent as strong as one in the upright position when subjected to gravity loads. In the same manner, the strength of a wood joist with the actual dimensions 2 × 10 in. turned on its side would be only 20 percent as strong as in the upright position.

The examples to follow illustrate the design of compact steel beams whose compression flanges have full lateral support or bracing, thus permitting plastic analysis. For the selection of such sections the designer may enter the tables either with the required plastic modulus or with the factored design moment (if $F_y = 36$ or 50 ksi).

Beam Weight Estimates In each of the examples to follow the weight of the beam is included in the calculation of the bending moment to be resisted, as the beam must support itself as well as the external loads. The estimates of beam weight are very close here because the author was able to perform a little preliminary paperwork before making his estimates. The beginning student is not expected to be able to glance at a problem and estimate exactly the weight of the beam required. A very simple method is available, however, with which the student can quickly and accurately estimate beam weights. He or she can calculate the maximum factored bending moment, not counting the effect of the beam weight, and pick a section from the LRFD table. Then the weight of that section or a little bit more (since the beam's weight will increase the moment somewhat) can be used as the estimated beam weight. The results will almost always be very close to the weight of the member selected in the final design.

■ EXAMPLE 9-1

Select a beam section for the span and loading shown in Fig. 9-2, assuming full lateral support is provided for the compression flange by the floor slab above (that is $L_b = 0$) and $F_y = 50$ ksi.

Figure 9.2

Solution.

Assume beam weight = 44 lb/ft

$$w_u = (1.2)(1.044) + (1.6)(3) = 6.05 \text{ klf}$$

$$M_u = \frac{(6.05)(21)^2}{8} = 333.5 \text{ ft-k}$$

$$Z_x \text{ required} = \frac{M_u}{\phi_b F_y} = \frac{(12)(333.5)}{(0.9)(50)} = 88.9 \text{ in.}^3$$

Use W21 × 44   ∎

■ EXAMPLE 9-2

The 5-in. reinforced-concrete slab shown in Fig. 9-3 is to be supported with steel W sections 8 ft 0 in. on centers. The beams, which will span 20 ft, are assumed to be simply supported. If the concrete slab is designed to support a live load of 100 psf, determine the lightest steel section required to support the slab. It is assumed that the compression flange of the beam will be fully supported laterally by the concrete slab. The concrete weighs 150 lb/ft³. $F_y = 50$ ksi.

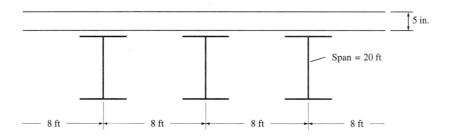

5 in.

Span = 20 ft

— 8 ft —|— 8 ft —|— 8 ft —|— 8 ft —

Figure 9.3

Solution.

$$\text{Dead loads: Slab} = (\tfrac{5}{12})(150)(8) = 500 \text{ lb/ft}$$

$$\text{Estimated beam wt} = \underline{22}$$

$$\text{Total} = 522 \text{ lb/ft}$$

$$w_u = (1.2)(522) + (1.6)(8 \times 100) = 1906 \text{ lb/ft} = 1.906 \text{ klf}$$

$$M_u = \frac{(1.906)(20)^2}{8} = 95.3 \text{ ft-k}$$

$$Z_x \text{ required} = \frac{(12)(95.3)}{(0.9)(50)} = 25.4 \text{ in.}^3$$

Use W10 × 22   ∎

Holes in Beams It is often necessary to have holes in steel beams. They are obviously required for the installation of bolts and sometimes for pipes, conduits, ducts, etc. If at all possible these latter types of holes should be completely avoided. When absolutely necessary they should be placed through the web if the shear is small and through the flange if the moment is small. Cutting a hole through the web of a beam does not reduce its section modulus greatly or its resisting moment; but, as will be described in Section 10-2, a large hole in the web tremendously reduces the shearing strength of a steel section. When large holes are put in beam webs, extra plates are sometimes connected to the webs around the holes to serve as reinforcing against possible web buckling.

When large holes are placed in beam webs the strength limit states of the beams such as local buckling of the compression flange, the web, or the tee-shaped compression zone above or below the opening—or the moment-shear interaction or serviceability limit states may control the size of the member. A general procedure for estimating these effects and the design of any required reinforcement is available for both steel and composite beams.[1,2]

The presence of holes of any type in a beam certainly does not make it stronger and in all probability weakens it somewhat. The effect of holes has been a subject which has been argued back and forth for many years. The questions, "Is the neutral axis affected by the presence of holes?" and "Is it necessary to subtract holes from the compression flange that are going to be plugged with bolts?" are frequently asked.

The theory that the neutral axis might move from its normal position to the theoretical position of its net section when bolt holes are present is rather questionable. Tests seem to show that flange holes for bolts do not appreciably change the location of the neutral axis. It is logical to assume that the location of the neutral axis will not follow the exact theoretical variation with its abrupt changes in position at bolt holes as shown in part (b) of Fig. 9-4. A more reasonable change in neutral axis location is shown in part (c) of this figure where it is assumed to have a more gradual variation in position.

It is interesting to note that flexure tests of steel beams seem to show that their failure is based on the strength of the compression flange even though there may be bolt holes in the tension flange. The presence of these holes does not seem to be as serious as might be thought, particularly as compared with holes in a pure tension member. These tests show little difference in the strengths of beams with no holes and in beams with an appreciable number of bolt holes in either flange.

---

[1] D. Darwin, *Steel and Composite Beams with Web Openings,"* AISC Design Guide Series No. 2 (Chicago: American Institute of Steel Construction, 1990).

[2] ASCE Task Committee on Design Criteria for Composite Structures in Steel and Concrete "Proposed Specification for Structural Steel Beams with Web Openings," D. Darwin, Chairman, *Journal of Structural Engineering,* ASCE, Vol. 118 (New York: ASCE December, 1992)

W beam with
bolt holes in
tension flange

(a)

Theoretical
variation of
N.A.

(b)

More probable
variation of
N.A.

(c)

Figure 9.4

The LRFD Specification does not require a reduction in flange area for bolts as long as the following expression is satisfied:

$$0.75F_u A_{fn} \geq 0.9F_y A_{fg}$$
(LRFD Equation B10-1)

In this equation A_{fn} is the net flange area and A_{fg} is the gross flange area. Substituting into this expression we find that no deduction is necessary if the net flange area is equal to or greater than 74 percent of the gross flange area for A36 steel or 92 percent for A572 Grade 50 steel. These values are shown in Table 9-1.

TABLE 9-1   WHEN TO IGNORE BOLT HOLES IN BEAM AND GIRDER FLANGES

Steel	No Reduction if $A_{fn}/A_{fg} \geq$
A36 ($F_u = 58$ ksi)	0.74
A572 Grade 50 ($F_u = 65$ ksi)	0.92
A588 ($F_y = 50$ ksi, $F_u = 70$ ksi)	0.86

Should $0.75F_u A_{fn}$ be less than $0.9F_y A_{fg}$, the LRFD Specification requires that the flexural properties of the section be based on an effective tension flange area A_{fe}, determined as follows:

$$A_{fe} = \frac{5}{6}\frac{F_u}{F_y}A_{fn}$$
(LRFD Equation B10-3)

Bolt holes in the webs of beams are generally considered to be insignificant as they have almost no effect on Z calculations.

Some specifications, notably the bridge ones, and some designers have not adopted the idea of ignoring the presence of all or part of the holes in tension flanges. As a result they follow the more conservative practice of deducting 100 percent of all holes. For such a case the reduction in Z_x will equal the statical moment of the holes (in both flanges) taken about the neutral axis.

The usual practice is to assume that we have the same holes in both flanges whether or not they are actually present in the compression flange. (If we have bolt holes filled with bolts in the compression flange only, we forget the whole thing. This is because it is felt that the fasteners can adequately transmit compression through the holes by means of the bolts.) For a section with holes in the tension flange only, where a reduction in the tension flange area is required by the specification or is considered necessary, we make the same deduction from both flanges.

Should the application of LRFD Equation B10-1 show that $0.75F_uA_{fn} < 0.9F_yA_{fg}$ we will need to reduce Z_x. Its value will equal Z_x from the Manual tables minus the statical moment of $A_{fg} - A_{fe}$ for each flange taken about the neutral axis. The review of a beam with flange bolt holes is considered in Example 9-3.

■ EXAMPLE 9-3

Determine M_u for the fully braced A36 W24 × 176 shown in Fig. 9-5 for the following situations:

(a) Using the LRFD Specification and assuming two lines of 1-in. bolts in each flange.
(b) Using the LRFD Specification and assuming four lines of 1-in. bolts in each flange.
(c) Same as (b) above, but conservatively assuming we must deduct all holes for computing beam properties.

1.340 in.

11.95 in.

25.24 in.

W24 × 176
($Z_x = 511$ in.³)

12.890 in.

Figure 9.5

Solution.

(a) With two holes in each flange (LRFD Specification)

$A_{fg} = (12.890)(1.340) = 17.27$ in.^{2}

$A_{fn} = 17.27 - (2)(1\frac{1}{8})(1.340) = 14.25$ in.^{2}

For no reduction $0.75F_u A_{fn} \geq 0.9F_y A_{fg}$

$(0.75)(58)(14.25) = 619.9$ k $> (0.9)(36)(17.27) = 559.5$ k

∴ No reduction in flange area is required.

$$M_u = \phi M_m = \frac{(0.9)(36)(511)}{12} = 1379.7 \text{ ft-k}$$

(b) With four holes in each flange (LRFD Specification)

$A_{fg} = 17.27$ in.^{2}

$A_{fm} = 17.27 - (4)(1\frac{1}{8})(1.340) = 11.24$ in.^{2}

For no reduction $0.75F_u A_{fn} \geq 0.9F_y A_{fg}$

$(0.75)(58)(11.24) = 488.9$ k $< (0.9)(36)(17.27) = 559.5$ k

∴ Flange area must be reduced

$$A_{fe} = \left(\frac{5}{6}\right)\left(\frac{58}{36}\right)(11.24) = 15.09 \text{ in.}^2$$

Reduced $Z_x = 511 - (17.27 - 15.09)(11.95)(2) = 458.9$ in.^{3}

$$M_u = \phi M_n = \frac{(0.9)(36)(458.9)}{12} = 1239.0 \text{ ft-k}$$

(c) Reducing each flange area by four holes (not LRFD Specification)

Area of four holes $= (4)(1\frac{1}{8})(1.340) = 6.03$ in.^{2}

Reduced $Z_x = 511 - (6.03)(11.95)(2) = 366.9$ in^3

$$M_u = \phi M_n = \frac{(0.9)(36)(366.9)}{12} = 990.6 \text{ ft-k}$$ ■

These LRFD requirements apply to the design of hybrid beams and girders whose flanges consist of a stronger grade of steel than their webs. This is true as long as these members are not required to resist an axial force larger than $\theta_b(0.15 \ F_{yf})$ of the flanges times their gross areas.

Should a hole be present in only one side of a flange of a W section, there will be no axis of symmetry for the net section of the shape. A correct theoretical solution of the problem would be very complex. Rather than going through such a lengthy process over a fairly minor point, it seems logical to assume holes in both sides of the flange. The results obtained will probably be just as satisfactory as those obtained by a laborious theoretical method.

A steel beam in place ready for installation of erection bolts. (Courtesy of the American Institute of Steel Construction, Inc.)

9-4 LATERAL SUPPORT OF BEAMS

Probably the large majority of steel beams are used in such a manner that their compression flanges are restrained against lateral buckling. (Unfortunately, however, the percentage has not been quite as high as the design profession has assumed.) The upper flanges of beams used to support concrete building and bridge floors are often incorporated in these concrete floors. For situations of this type, where the compression flanges are restrained against lateral buckling, the beams will fall into Zone 1.

Should the compression flange of a beam be without lateral support for some distance it will have a stress situation similar to that existing in columns. As is well known, the longer and slenderer a column becomes the greater becomes the danger of its buckling for the same loading condition. When the compression flange of a beam is long enough and slender enough, it may quite possibly buckle unless lateral support is provided.

There are many factors affecting the amount of stress that will cause buckling in the compression flange of a beam. Some of these factors are properties of the material, the spacing and types of lateral support provided, residual stresses in the sections, the types of end support or restraints, the loading conditions, etc.

The tension in the other flange of a beam tends to keep that flange straight and restrain the compression flange from buckling; but as the bending moment is increased, the tendency of the compression flange to buckle may become large enough to overcome the tensile restraint. When the compression flange does begin to buckle, twisting or torsion will occur, and the smaller the torsional strength of the beam the more rapid will be the failure. The W, S, and channel shapes so frequently used for beam sections do not have a great deal of resistance to lateral buckling and the resulting torsion. Some other shapes, notably the built-up box shapes, are tremendously stronger. These types of members have a great deal more torsional resistance than the W, S, and plate girder sections. Tests have shown that they will not buckle laterally until the strains developed are well in the plastic range.

Some judgment needs to be used in deciding what does and what does not constitute satisfactory lateral support for a steel beam. Perhaps the most common question asked by practicing steel designers is "What is lateral support?" A beam that is wholly encased in concrete or that has its compression flange incorporated in a concrete slab is certainly well supported laterally. When a concrete slab rests on the top flange of a beam, the engineer must study the situation carefully before he or she counts on friction to provide full lateral support. Perhaps if the loads on the slab are fairly well fixed in position, they will contribute to the friction and it may be reasonable to assume full lateral support. If, on the other hand, there is much movement of the loads and appreciable vibration, the friction may well be reduced and full lateral support cannot be assumed. Such situations occur in bridges due to traffic, and in buildings with vibrating machinery such as printing presses.

Should lateral support of the compression flange not be provided by a floor slab, it is possible that such support may be provided with connecting beams or with special members inserted for that purpose. Beams that frame into the sides of the beam or girder in question and are connected to the compression flange can usually be counted on to provide full lateral support at the connection. If the connection is made primarily to the tensile flange, little lateral support is provided to the compression flange. Before support is assumed from the these beams the designer should note whether they themselves are prevented from moving. The series of beams represented with horizontal dashed lines in Fig. 9-6 provide questionable lateral support for the main beams between columns. For a situation of this type some system of x-bracing may be desirable in one of the bays. Such a system is shown in Fig. 9-6. This one system will provide sufficient lateral support for the beams for several bays.

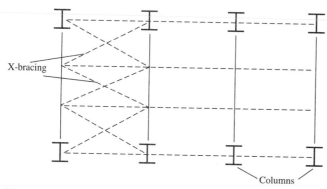

Figure 9.6

The intermittent welding of metal roof or floor decks to the compression flanges of beams will probably provide sufficient lateral bracing. The corrugated sheet-metal roofs that are usually connected to the purlins with metal straps probably furnish only partial lateral support. A similar situation exists when wood flooring is bolted to supporting steel beams. At this time the student quite naturally asks, "If only partial support is available, what am I to consider to be the distance between points of lateral support?" The answer to this question is for him or her to use his or her judgment. As an illustration, a wood floor is assumed to be bolted every 4 ft to the supporting steel beams in such a manner that it is thought only partial lateral support is provided at those points. After studying the situation the engineer might well decide that the equivalent of full lateral support at 8-ft intervals is provided. Such a decision seems to be within the spirit of the specification.

Should there be doubt in the designer's mind as to the degree of lateral support provided, he or she should probably be advised to assume there is none.

9-5 INTRODUCTION TO INELASTIC BUCKLING, ZONE 2

If intermittent lateral bracing is supplied for the compression flange of a beam section such that the member can be bent until the yield strain is reached in some but not all of its compression elements before lateral buckling occurs, we have inelastic buckling. In other words the bracing is insufficient to permit the member to reach a full plastic strain distribution before buckling occurs.

Because of the presence of residual stresses (discussed in Section 5-2) yielding will begin in a section at applied stresses equal to $F_y - F_r$ where F_y is the yield stress of the web and F_r equals the compressive residual stress assumed equal to 10 ksi for rolled shapes and 16.5 ksi for welded shapes. It should be noted that the definition of plastic moment $F_y Z$ in Zone 1 is not affected by residual stresses because the sum of the compressive residual stresses equals the sum of the tensile residual stresses in the section and the net effect is theoretically zero.

If the unbraced length, L_b, of a compact I- or C-shaped section is larger than L_p the beam will fail inelastically unless L_b is greater than a distance L_r (to be discussed) beyond which the beam will fail elastically before F_y is reached (thus falling into Zone 3).

Bending Coefficients   In the formulas presented in these next few sections for inelastic and elastic buckling we will use a term C_b. This is a moment coefficient that is included in the formulas to account for the effect of different moment gradients on lateral-torsional buckling. In other words, lateral buckling may be appreciably affected by the end restraint and loading conditions of the member.

As an illustration the reader can see that the moment in the unbraced beam of part (a) of Fig. 9-7 causes a worse compression flange situation than does the moment in the unbraced beam of part (b). For one reason the upper flange of the

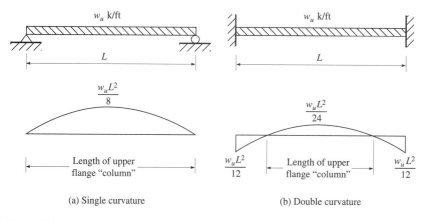

(a) Single curvature   (b) Double curvature

Figure 9.7

beam in part (a) is in compression for its entire length, while in (b) the length of the "column," that is, the length of the upper flange that is in compression, is much less (thus a much shorter "column").

For the simply supported beam of part (a) of the figure, C_b is taken as 1.0, while for the beam of part (b) it is taken as larger than 1.0. The basic moment capacity equations for Zones 2 and 3 were developed for laterally unbraced beams subject to single curvature with $C_b = 1.0$. Frequently, beams are not bent in single curvature with the result that they can resist more moment. We have seen this in Fig. 9-7. To handle this situation the LRFD Specification provides moment or C_b coefficients larger than 1.0 which are to be multiplied by the computed M_n values. The results are higher moment capacities. The designer who conservatively says "I'll always use $C_b = 1.0$" is missing out on the possibility of significant savings in steel weight for some situations. *When using C_b values the designer should clearly understand that the moment capacity obtained by multiplying M_n by C_b may not be larger than the plastic M_n of Zone I which is equal to $F_y Z$.* This situation is illustrated in Fig. 9-8.

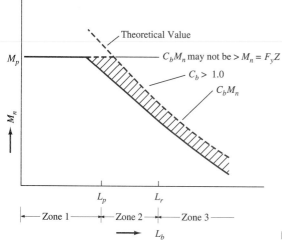

Figure 9.8

The value of C_b is determined from the expression to follow in which M_{max} is the largest moment in an unbraced segment of a beam while M_A, M_B, and M_c are, respectively, the moments at the 1/4 point, 1/2 point, and 3/4 point in the segment.

$$C_b = \frac{12.5 M_{max}}{2.5 M_{max} + 3M_A + 4M_B + 3M_c} \qquad \text{(LRFD Equation F1-3)}$$

C_b is equal to 1.0 for cantilevers or overhangs where the free end is unbraced. Some typical values of C_b calculated with the above equation are shown in Fig. 9-9 for various beam and moment situations.

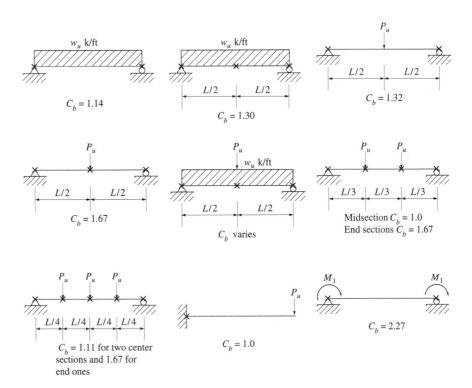

Figure 9.9 Sample C_b values.

9-6 MOMENT CAPACITIES, ZONE 2

As the unbraced length of the compression flange of a beam is increased beyond L_p the moment capacity of the section will become smaller and smaller. Finally, at an unbraced length L_r, the section will buckle elastically as soon as the yield stress is reached. Owing to the rolling operation, however, there is a residual stress in the section equal to F_r. Thus the elastically computed stress caused by bending can only reach $F_{yw} - F_r$. Assuming $C_b = 1.0$ the design moment capacity for a compact I- or C-shaped section bent about its x axis may be determined as follows if $L_b = L_r$.

$$M_u = \phi_b M_r = \phi_b S_x(F_{yw} - F_r)$$

L_r is a function of several of the section's properties such as its cross-sectional area, modulus of elasticity, yield stress, and warping and torsional properties. The very complex formulas needed for its computation are given in the LRFD Specification (F_1) and space is not taken to show them here. Fortunately, numerical values have been determined for sections normally used as beams and are given in the "Load Factor Design Selection Table."

Going backward from an unbraced length of L_r toward an unbraced length L_p we can see that buckling does not occur when the yield stress is first reached. We are in the inelastic range (Zone 2) where there is some penetration of the yield stress into the section from the extreme fibers. For these cases when the unbraced length falls between L_p and L_r the moment capacity will fall approximately on a straight line between $M_u = \phi_b F_y Z$ at L_p and $\phi_b S_x(F_{yw} - F_r)$ at L_r. For intermediate values of the unbraced length the moment capacity may be determined by proportions or by substituting into the expression at the end of this paragraph. If C_b is larger than 1.0 the section will resist additional moment but not more than $\phi_b F_y Z = \phi_b M_p$.

$$\phi_b M_n = C_b[\phi_b M_p - BF(L_b - L_p)] \leq \phi_b M_p$$

in which BF is a factor given in the "Load Factor Design Selection Table" for each section, which enables us to do the proportioning with a simple formula.

Alternately, the value of M_n can be determined from the equation to follow and multiplied by ϕ_b to obtain M_u.

$$M_n = C_b\left[M_p - (M_p - M_r)\left(\frac{L_b - L_p}{L_r - L_p}\right)\right] \leq M_p \qquad \text{(LRFD Equation F1-2)}$$

Example 9-4 illustrates the determination of the moment capacities of a section with L_b between L_p and L_r, while Example 9-5 demonstrates the design of a beam in the same range.

■ **EXAMPLE 9-4**

Determine the moment capacity of a W24 × 62 with $F_y = 36$ ksi and again with $F_y = 50$ ksi if $L_b = 8.0$ ft and $C_b = 1.0$.

Solution if F_y = 36 ksi.   From the Load Factor Design Selection Table for a W24 × 62

$$L_p = 5.8 \text{ ft}$$
$$L_r = 17.2 \text{ ft}$$
$$\phi_b M_r = 255 \text{ ft-k}$$
$$\phi_b M_p = 413 \text{ ft-k}$$
$$BF = 13.8 \text{ k}$$

Since $L_b > L_p < L_r$ the section is in Zone 2 for inelastic buckling and $\phi_b M_n$ can be determined as follows:

$$\phi_b M_n = C_b[\phi_b M_p - BF(L_b - L_p)]$$
$$= 1.0[413 - (13.8)(8.0 - 5.8)] = \underline{\underline{382.6 \text{ ft-k}}}$$

Or directly by proportions

$$\phi_b M_n = 255 + \left(\frac{17.2 - 8.0}{17.2 - 5.8}\right)(413 - 255) = 382.5 \text{ ft-k}$$

Solution if F_y = 50 ksi.   From the Load Factor Design Selection Table

$$L_p = 4.9 \text{ ft}$$
$$L_r = 13.3 \text{ ft}$$
$$\phi_b M_r = 393 \text{ ft-k}$$
$$\phi_b M_p = 574 \text{ ft-k}$$
$$BF = 21.4 \text{ k}$$

Since $L_b > L_p < L_r$

$$\phi_b M_n = 1.0[574 - (21.4)(8.0 - 4.9)] = 507.7 \text{ ft-k}$$ ■

■ **EXAMPLE 9-5**
Select the lightest available section for a factored moment of 290 ft-k if L_b = 10.0 ft. Use A36 steel and assume C_b = 1.0

Solution.   Enter the Load Factor Design Selection Table and notice that $\phi_b M_p$ for a W21 × 50 is 297 ft-k but L_p is 5.4 ft < L_b of 10.0 ft. Also $\phi_b M_r$ = 184 ft-k and BF = 10.5 k.

$$\therefore \ \phi_b M_n = 1.0[297 - (10.5)(10.0 - 5.4)]$$
$$= 248.7 \text{ ft-k} < 290 \text{ ft-k}$$ (NG)

Moving up in the Load Factor Design Selection Table try a W24 × 55 (with $L_p = 5.6$ ft, $L_r = 16.6$ ft, $\phi_b M_r = 222$ ft-k, $\phi_b M_p = 362$ ft-k, and $BF = 12.7$ k).

$$\phi_b M_n = 1.0[362 - (12.7)(10.0 - 5.6)]$$
$$= 306.1 \text{ ft-k} > 290 \text{ ft-k} \qquad \text{(OK)}$$
$$\underline{\underline{\text{Use W24} \times 55}}$$

Note: A much easier solution is presented in the next section. ∎

9-7 ELASTIC BUCKLING, ZONE 3

When a beam is not fully braced laterally, it may fail due to buckling laterally about the weaker axis between the points of lateral bracing. This will occur even though the beam is loaded so that it supposedly will bend about the stronger axis. The beam will bend initially about the stronger axis until a certain critical moment M_{cr} is reached. At that time it will buckle laterally about its weaker axis. As it bends laterally the tension in the other flange will try to keep the beam straight. As a result the buckling of the beam will be a combination of lateral bending and a twisting (or torsion) of the beam cross section. A sketch of this situation is shown in Fig. 9-10.

The critical moment or flexural-torsional moment M_{cr} in a beam will be made up of the torsional resistance (commonly called *St-Venant torsion*) and the warping resistance of the section. These are combined as follows:

$$M_{cr} = \sqrt{\left(\frac{\text{torsional}}{\text{resistance}}\right)^2 + \left(\frac{\text{warping}}{\text{resistance}}\right)^2}$$

Returning to the LRFD Specification, if the unbraced length of the compression flange of a beam section is greater than L_r the section will buckle elastically before the yield stress is reached anywhere in the section. In Section

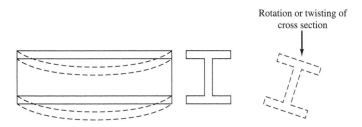

Figure 9.10 Lateral-torsional buckling of a simply supported beam.

F1.1.2b of the LRFD Specification the classic equation for determining this flexural-torsional buckling moment called M_{cr} is presented. This expression follows

$$M_{cr} = C_b \frac{\pi}{L_b} \sqrt{EI_y GJ + \left(\frac{\pi E}{L_b}\right)^2 I_y C_w} \qquad \text{(LRFD Equation F1-13)}$$

In this equation G is the shear modulus of elasticity of the steel = 11,200 ksi, J is a torsional constant (in.^{4}), while C_w is the warping constant (in.^{6}). The values of J and C_w are provided for rolled sections in the tables entitled "Torsion Properties" in Part 1 of the LRFD Manual.

This expression is applicable to compact doubly symmetric I-shaped members, channel sections loaded in the plane of their webs, and I-shaped singly symmetric sections with their compression flanges larger than their tension ones (⊥). Expressions also are given in Sections F1.1.2b and F1.1.2c of the LRFD Specification for M_{cr} in the elastic range for other sections such as solid rectangular bars, symmetric box sections, tees, and double angles.

It is not possible for lateral-torsional bucking to occur if the moment of inertia of the section about the bending axis is equal to or less than the moment of inertia out of plane. As a result the limit state of lateral-torsional buckling is not applicable for shapes bent about their minor axes, for shapes with $I_x \leq I_y$, or for circular or square shapes. Furthermore, yielding controls if the section is noncompact.

Example 9-6 illustrates the computation of $\phi_b M_{cr}$ for an elastic buckling situation.

■ EXAMPLE 9-6

Compute $M_u = \phi_b M_{cr}$ for a W18 × 97 consisting of A36 steel if the unbraced length L_b is 44 ft. Assume $C_b = 1.0$.

Solution. Noting L_b = 44 ft > L_r = 38.1 ft from Load Factor Design Selection Table (Part 4 LRFD Manual).

The following values for the W18 × 97 are also obtained from the Manual: I_y = 201 in.^{4}, J = 5.86 in.^{4}, and C_w = 15,800 in.^{6}

$$\phi_b M_n = \phi M_{cr} = M_u = (0.9)(1.0)\left(\frac{\pi}{12 \times 44}\right)$$

$$\times \sqrt{(29 \times 10^3)(201)(11,200)(5.86) + \left(\frac{\pi\, 29 \times 10^3}{44 \times 12}\right)^2 (201)(15,800)}$$

$$= 3698.9 \text{ in.k} = 308.2 \text{ ft-k}$$

(This value may be checked with the LRFD Manual charts described in the next few paragraphs. There we will be able to read $\phi M_n = M_u = 308$ ft-k.) ■

The LRFD Specification (F1.2b) also presents the elastic buckling equation in an alternate form as follows:

$$M_{cr} = \frac{C_b S_x X_1 \sqrt{2}}{L_b/r_y} \sqrt{1 + \frac{X_1^2 X_2}{2(L_b/r_y)^2}}$$

(LRFD Equation F1-13 Alternate Form)

In which

$$X_1 = \frac{\pi}{S_x} \sqrt{\frac{EGJA}{2}}$$

(LRFD Equation F1-8)

$$X_2 = 4\frac{C_w}{I_y}\left(\frac{S_x}{GJ}\right)^2$$

(LRFD Equation F1-9)

Values of X_1 and X_2 are shown for W shapes in the "Properties for W Shapes" section of Part 1 of the Manual.

9-8 DESIGN CHARTS

Fortunately the values of $\phi_b M_{cr} = \phi_b M_n$ for the sections normally used as beams are computed for different unbraced lengths and the results plotted as curves in Part 4 of the LRFD Manual. The values not only cover unbraced lengths in the elastic range but also the inelastic range, enabling us to very easily handle the problems presented in the last section as well as the ones in this section which fall in Zone 3. The moments are plotted for F_y values of 36 ksi and 50 ksi and for $C_b = 1.0$.

The curve for a typical W section is shown in Fig. 9-11. For each of the shapes L_p is indicated with a solid circle ($\bullet$) while L_r is shown with a hollow circle ($\circ$).

The charts were developed without regard to such things as shear, deflection, etc.—items that may occasionally control the design as described in Chapter 10. The curves extend to unbraced lengths equal to 30 times the section

Figure 9.11

Figure 9.12 Design moments for beams with different unbraced lengths.

depths. As such they cover almost all of the unbraced lengths encountered in practice. If C_b is greater than 1.0 the values given will be magnified somewhat, as illustrated in Fig. 9-8.

To select a member it is only necessary to enter the chart with the unbraced length L_b and the factored design moment M_u. For an illustration let's assume $F_y = 36$ ksi and assume that we wish to select a beam with $L_b = 20.0$ ft for a moment $M_u = 590$ ft-k. We enter the charts in Part 4 entitled "Beam Design Moments" and go through the pages where $F_y = 36$ ksi until we find in the left-hand column $\phi_b M_n$ equal to 590. We proceed up from the bottom of the chart for an unbraced length $= 20.0$ ft until we intersect a horizontal line from the 590. Any section to the right and above this intersection point ($\nearrow$) will have a greater unbraced length and a greater moment capacity. The appropriate page from the handbook is shown in Fig. 9-12 with the permission of the AISC.

Moving up and to the right we first encounter a W21 × 101. In this area of the charts this section is shown with a dashed line. This section will provide the necessary moment capacity but the dashed line indicates that it is in an uneconomical range. If we proceed farther upward and to the right the first solid line that we encounter will represent the lightest available section. In this case it's a W30 × 99. Other illustrations of the use of these charts are presented in Examples 9-7 and 9-8.

■ EXAMPLE 9-7

Using A36 steel select the lightest available section for the beam of Fig. 9-13 which has lateral bracing provided for its compression flange only at its ends.

Figure 9.13

Solution.

$$\text{Assume beam weight} = 60 \text{ lb/ft}$$

$$w_u = (1.2)(1.060) + (1.6)(2) = 4.472 \text{ klf}$$

$$M_u = \frac{(4.472)(20)^2}{8} = 223.6 \text{ ft-k}$$

$$\text{Noting } C_b = 1.0$$

Entering the Beam Design Moments charts with $L_b = 20$ ft and $M_u = 223.6$ ft-k.

Use W18 × 60

For the example problem that follows C_b is greater than 1.0. For such a situation the reader should look back to Fig. 9-8. There he or she will see that the design moment strength of a section can go to $\phi_b C_b M_n$ when $C_b > 1.0$ but may under no circumstances exceed $\phi_b M_p = \phi_b F_y Z$.

To handle such a problem we calculate an effective moment as follows (the numbers being taken from Example 9-8):

$$M_{\text{effective}} = \frac{M_u}{C_b} = \frac{868.7}{1.67} = 520.2 \text{ ft-k}$$

Then we enter the charts with our unbraced length of 17 ft and with $M_{\text{effective}} = 520.2$ ft-k and there select a W27 × 84. We must, however, check to see that our M_u (868.7) does not exceed $\phi_b F_y Z$ for the section. In this case it does and we must keep going until we find the lightest section that has a $\phi_b M_n \geq$ 520.2 ft-k at $L_b = 17$ ft and yet which has a $\phi_b F_y Z \geq 868.7$ ft-k. ∎

■ **EXAMPLE 9-8**

Using A36 steel select the lightest available section for the situation shown in Fig. 9-14. Bracing is provided only at the ends and center line of the member and thus $C_b = 1.67$.

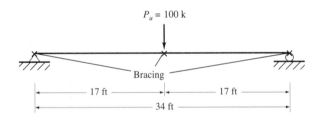

$P_u = 100$ k

Bracing

17 ft — 17 ft

34 ft

Figure 9.14

Solution.

$$\text{Assume beam weight} = 108 \text{ lb/ft}$$
$$w_u = (1.2)(0.108) = 0.1296 \text{ klf}$$
$$M_u = \frac{(100)(34)}{4} + \frac{(0.1296)(34)^2}{8} = 868.7 \text{ ft-k}$$

Entering the Beam Design Moments charts with $M_{u\,\text{effective}} = 868.7/1.67 = 520.2$ ft-k and $L_b = 17$ ft we find a W27 × 84 will do.

But $\phi_b M_n = 868.7$ ft-k may not exceed $\phi_b F_y Z$ of the section, and it does for a W27 × 84, as can be seen in the Load Factor Design Selection Table. Therefore, we continue looking in the table until we find the lightest section that has a $\phi_b F_y Z \geq 868.7$ ft-k.

Use W30 × 108

9-9 NONCOMPACT SECTIONS

You will remember from Section 5-7 of this textbook that a compact section is one that has a sufficiently stocky profile so that it is capable of developing a fully plastic stress distribution (assuming its compression flange has sufficient lateral bracing) before something buckles (web or flange). For a section to be compact the width thickness ratio of the flanges of W- or other I-shaped sections must not exceed a b/t value $\lambda_p = 65/\sqrt{F_y}$. Similarly, the webs in flexural compression must not exceed an h/t value $\lambda_p = 640/\sqrt{F_y}$. The values of b, t, h and t_f are shown in Fig. 9-15.

A noncompact section is one for which the yield stress can be reached in some but not all of its compression elements before buckling occurs. It is not capable of reaching a fully plastic stress distribution. The noncompact sections are those that have web thickness ratios greater than λ_p but not greater than λ_r. The λ_r values are provided in Table B5.1 of the LRFD Specification. For the noncompact range the width thickness ratios of the flanges must not exceed $\lambda_r = 141\sqrt{F_y - 10}$ while those for the webs must not exceed $\lambda_r = 970/\sqrt{F_y}$. Other values are provided in LRFD Table B5.1 for λ_p and λ_r for other shapes and for shapes subject to both axial load and bending.

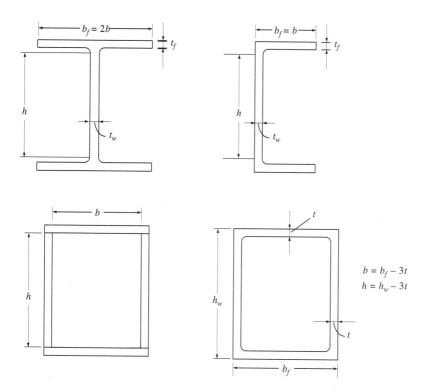

Figure 9-15 Values of h, b, and t to be used for computing λ = width-thickness ratios.

For noncompact beams the nominal flexural strength M_n is the lowest of the lateral-torsional buckling strength, the flange local buckling strength, or the web local buckling strength.

If we have a noncompact section, that is one with $\lambda_p < \lambda \leq \lambda_r$, the value of M_n may be obtained by linear interpolation between M_p and M_r as represented by the equations to follow:

For lateral-torsional buckling

$$M_n = C_b\left[M_p - (M_p - M_r)\left(\frac{\lambda - \lambda_p}{\lambda_2 - \lambda_p}\right)\right] \leq M_p$$

(LRFD Appendix Equation A-F1-2)

For flange and web local buckling

$$M_n = M_p - (M_p - M_r)\left(\frac{\lambda - \lambda_p}{\lambda_r - \lambda_p}\right)$$ (LRFD Appendix Equation A-F1-3)

Should $\lambda > \lambda_r$ the limit state of lateral-torsional buckling and flange local buckling is to be determined with the LRFD Appendix formula to follow in which S is the section modulus of the member and F_{cr} is the critical design stress for compression members as previously determined in Chapter 5 with LRFD Formulas E2-2 or E2-3 as applicable.

$$M_n = M_{cr} = SF_{cr} \leq M_p$$ (LRFD Appendix Equation A-F1-4)

There is one section listed in the Load Factor Design Selection Table that is noncompact when F_y is 36 ksi. It is the W6 × 15 and the LRFD Manual indicates that it is noncompact with a superscript c. There are 7 noncompact sections in the same table when $F_y = 50$ ksi. These sections, which are listed with a superscript b are the W40 × 174, W14 × 99, W14 × 90, W12 × 65, W10 × 12, W6 × 15, and the W8 × 10.

The equations listed here were used to obtain the values shown in the LRFD Manual "Load Factor Design Selection Table for shapes used as beams" for noncompact sections. The designer will have little trouble with noncompact sections when F_y is 36 or 50 ksi. He or she, however, will have to use the formulas presented in this section for shapes with F_y greater than 50 ksi.

9-10 COMPUTER EXAMPLE

In Example 9-9 the INSTEP program is used to obtain the design moment strength of a beam section for a certain unbraced length L_b. The strength for any other unbraced length can be quickly obtained by changing L_b. Notice the moment values are computed in in.-k.

■ EXAMPLE 9-9

With the enclosed computer program determine the design moment capacity of an A572 Grade 50 W36 × 300 with $L_b = 48$ ft. Assume $C_b = 1.0$.

Solution.

INPUT

STEEL & TRIAL MEMBER SIZE

| B | Type Steel: A572 Grade 50 |
| C | Trial Size: W36×300 |

FLEXURAL DESIGN PARAMETERS

| F | L_b = 576 |
| G | C_b = 1 |

FACTORED DESIGN LOADS

| J | M_u = 0 |
| K | V_u = 0 |

OUTPUT

SECTION CLASSIFICATION: COMPACT SECTION

L_r = 479.164

M_r = 1110(50−10) = 44400

M_p = 1260(50) = 63000

$$L_p = \frac{300(3.83)}{\sqrt{50}} = 162.493$$

M_n = 34530.1

$0.9 M_n$ = 31077.1   ∎

PROBLEMS

9-1 to 9-10. *Select the most economical sections using $F_y = 50$ ksi unless otherwise specified and assuming full lateral bracing for the compression flanges. Working or service loads are given for each case and beam weights are not included.*

Problem 9-1 (Ans. W30 × 90)

Problem 9-2

9-3. Repeat Prob. 9-2 using $F_y = 36$ ksi. (*Ans.* W30 × 108)

Problem 9-4

Problem 9-5 (Ans. W30 × 16)

Problem 9-6

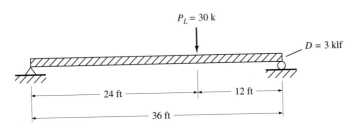

Problem 9-7 (Ans. W27 × 84)

9-8. The accompanying illustration shows the arrangement of beams and girders that are used to support a 6-in. reinforced-concrete floor for a small
industrial building. Design the beams and girders assuming they are simply supported. Assume full lateral support and a live load of 120 psf.
Concrete weight is 150 lb/ft³.

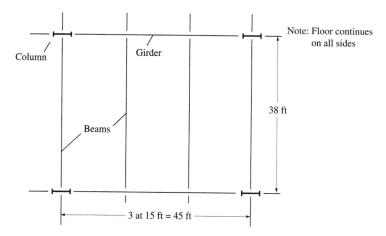

Problem 9-8

9-9. A beam consists of a W16 × 40 with a $\frac{1}{2}$ × 12-in. cover plate welded to each flange. Determine the design uniform load w_u the member can support in addition to its own weight for a 36-ft simple span. (*Ans.* 3.883 k/ft)

9-10. The member shown consists of A36 steel. Determine the maximum service live load that can be placed on the beam if in addition to its own weight it is supporting a service dead load of 1 klf. The member is used for a 30-ft simple span.

Problem 9-10

9-11 to 9-14. *Full lateral bracing is provided for these problems.*

9-11. Select a section for a 30-ft simple span to support a service dead uniform load of 2 klf and a live service uniform load of 3 klf if two holes for 1-in. bolts are assumed present in each flange at the section of maximum moment. Use LRFD Specification and A36 steel. (*Ans.* W30 × 99)

9-12. Rework Prob. 9-11 assuming two holes for 1-in. bolts pass through each flange at the point of maximum moment. Use LRFD Specification and A572 Grade 50 steel.

9-13. The section shown in the accompanying illustration has two 1-in. bolts passing through each flange. Find the design or factored load w_u which the section can support in addition to its own weight for a 30-ft simple span if it consists of a steel with $F_y = 50$ ksi. Deduct all holes for calculating section properties. (*Ans.* 12.171 k/ft)

Problem 9-13

9-14. A 40-ft simple span beam is to support two movable service 20-k loads a distance of 12 ft apart. Assuming a dead load of 1.2 klf in addition to the beam weight, select an A36 steel section to resist the largest possible moment.

9-15 to 9-28. *For these problems different values of L_b are given. Dead loads do not include beam weights.*

9-15. Determine the design moment strength $\phi_b M_n$ of a W24 × 68 for simple spans of 6, 12 and 22 ft if lateral bracing for the compression flanges is provided at the ends only. A572 Grade 50 steel. (*Ans.* 664, 563, 312)

9-16. Using $F_y = 50$ ksi select the lightest satisfactory section if bracing is provided at the ends only.

Problem 9-16

9-17. Select the lightest available section if $F_y = 50$ ksi. Lateral bracing is provided at the ends only. (*Ans.* W18 × 97)

$P_L = 36$ k $\quad$ $P_L = 36$ k

$D = 1$ klf

|← 9 ft →|← 9 ft →|← 9 ft →| **Problem 9-17**

9-18. Repeat Prob. 9-17 if lateral bracing is supplied at each of the concentrated loads as well as at the ends of the span.

9-19. A W33 × 141 consisting of A36 steel is used for a simple span of 34 ft and has lateral bracing provided at its ends only. If the only dead load present is the beam weight, what is the largest service concentrated live load that can be placed at the beam center line? (*Ans.* 67.3 k)

9-20. Repeat Prob. 9-19 is lateral bracing is supplied at beam ends and at the concentrated load.

9-21. If F_y is 36 ksi select the lightest section for the situation shown in the accompanying illustration if lateral bracing is supplied at the fixed end only. (*Ans.* W24 × 76)

$P_L = 20$ k

|← 14 ft →| **Problem 9-21**

9-22. Using a steel with $F_y = 50$ ksi, what uniformly distributed service live load can a simply supported W30 × 108 carry for a span of 40 ft when (a) the compression flange is braced laterally for its full length and when (b) lateral bracing is supplied at the ends only.

9-23. What service live uniform load can be placed on a W14 × 109 ($F_y = 50$ ksi) which has lateral bracing provided at its ends only? The beam is simply supported and has a span of 36 ft. The dead uniform service load is 1 klf plus the beam weight. (*Ans.* 2.41 k/ft)

9-24. A W33 × 130 is used to support the loads shown in the accompanying illustration. Using $F_y = 50$ ksi and neglecting the dead weight of the beam, find if the beam is overloaded.
(**a**) If lateral support is provided for the entire span.
(**b**) If lateral support is provided at the beam ends only.

$P_u = 55$ k $\quad$ $P_u = 55$ k

$M_u = 330$ ft-k

|← 12 ft →|← 12 ft →|← 12 ft →| **Problem 9-24**

9-25. Redesign the beam of Prob. 9-24 if lateral support is provided at the ends and at the concentrated loads. (*Ans.* W27 × 94)

9-26. The W27 × 94 shown has lateral support supplied at its center line as well as at its ends. If F_y is 50 ksi determine the maximum permissible service live load P_L. Neglect beam weight.

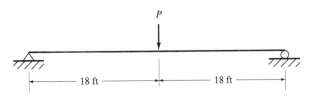

Problem 9-26

9-27. Repeat Prob. 9-26 using A36 steel if a dead uniform service load of 0.8 klf not including the beam weight is placed on the beam for its full length and if lateral bracing is provided at the ends and center line of the beam. (*Ans.* 40.06 k)

9-28. Using A36 steel select the lightest available section for the situation shown in the accompanying illustration if bracing is provided only at the ends and center line.

Problem 9-28

9-29. A W33 × 241 is required for a certain span but, because of a strike in the steel mills, cannot be obtained on time; however, an extra W33 × 118 of sufficient length is available along with a good supply of plates. Select $\frac{7}{8}$-in.-thick cover plates to be welded to the flanges of the W33 × 118 to provide a satisfactory substitute for the original beam. Use F_y = 50 ksi and assume full lateral bracing is to be provided for the compression flange. (*Ans.* PL$\frac{7}{8}$ × 18)

9-30. A W30 × 173 has been specified for use on a certain job. By mistake a W30 × 124 was shipped to the field. This beam must be erected today. Assuming that 1-in-thick plates are obtainable immediately, select cover plates to be welded to the flanges to obtain the necessary section. Use F_y = 50 ksi and assume full lateral bracing is supplied for the compression flange.

9-31. Repeat Prob. 9-29 if the original beam was to have been of a steel with $F_v = 50$ ksi and the substitute material is available only in A36 steel. Assume also that only $1\frac{1}{2}$-in. plates are available. (*Ans.* PL$1\frac{1}{2} \times 18$)

9-32. Design a beam of A36 steel with a depth no greater than 12.00 in. to support a uniform load of $w_u = 12.5$ klf (includes beam wt) for a 20-ft simple span. The member is to have full lateral bracing for its compression flange.

9-33 to 9-35. *Using the enclosed computer diskette repeat the following problems.*

9-33. Prob. 9-1 (*Ans.* W30 $\times$ 90)

9-34. Prob. 9-15

9-35. Prob. 9-28 (*Ans.* W30 $\times$ 116)

Chapter **10**

Design of Beams—Miscellaneous Topics

10-1 DESIGN OF CONTINUOUS BEAMS

Section A5 of the LRFD Specification permits the design of beams analyzed either on the basis of elastic analysis with factored loads or by plastic analysis with the same factored loads. Design based on plastic analysis is permitted only for sections with yield stresses no greater than 65 ksi and is subject to some special requirements in Sections B5.2, C2, E1.2, F1.2d, H1, and I1 of the LRFD Specification.

Both theory and tests show clearly that continuous ductile steel members meeting the requirements for compact sections with sufficient lateral bracing supplied for their compression flanges have the desirable ability of being able to redistribute moments caused by overloads. If plastic analysis is used this advantage is automatically included in the analysis.

If elastic analysis is used the LRFD handles the redistribution by a rule of thumb that approximates the real plastic behavior. The LRFD Specification (A5) states that for continuous compact sections the design *may* be made on the basis of nine-tenths of the maximum negative moments caused by gravity loads which are maximum at points of support if the positive moments are increased by one-tenth of the average negative moments at the adjacent supports. *(The 0.9 factor is applicable only to gravity loads and not to lateral loads such as those caused by wind and earthquake.* The factor can also be applied to columns that have axial stresses less than $0.15F_y$.) This moment reduction does not apply to members consisting of A514 steel, hybrid girders, or to moments produced by loading on cantilevers.

Example 10-1 illustrates the design of a three-span continuous beam analyzed by (a) plastic analysis and (b) elastic analysis.

■ **EXAMPLE 10-1**
The beam shown in Fig. 10-1 is assumed to consist of A36 steel. (a) Select the lightest W section available using plastic analysis and assuming full lateral support is provided for its compression flanges. (b) Design the beam using elastic analysis with the factored loads and the 0.9 rule and assuming full lateral support is provided for flanges. (c) Design the beam using elastic analysis with the factored loads and the 0.9 rule assuming full lateral support is provided for the upper flange *but* only for the lower flange at support points.

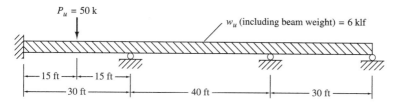

Figure 10-1

Solution.

    (a) Plastic analysis and design

Span 1 $\begin{cases} M_u 4\theta = (30w_u)\left(\dfrac{1}{2}\right)(15\theta) + (P_u)(15\theta) \\[4pt] M_u = 56.25w_u + 3.75P_u \\[4pt] M_u = (56.25)(6.0) + (3.75)(50) \\[4pt] M_u = 525 \text{ ft-k} \end{cases}$

Span 2 $\begin{cases} M_u(4\theta) = (40w_u)\left(\dfrac{1}{2}\right)(20\theta) \\[4pt] M_u = 100w_u = (100)(6.0) \\[4pt] M_u = 600 \text{ ft-k} \leftarrow \end{cases}$

Span 3 $\begin{cases} M_u(3.414\theta) = (30w_u)\left(\dfrac{1}{2}\right)(17.58\theta) \\[4pt] M_u = 77.24w_u = (77.24)(6.0) \\[4pt] = 463.4 \text{ ft-k} \end{cases}$

$$Z \text{ required} = \frac{M_u}{\phi_b F_y} = \frac{(12)(600)}{(0.9)(36)} = 222.2 \text{ in.}^3$$

Use W24 × 84

(b) Elastic analysis and design—full lateral bracing both flanges

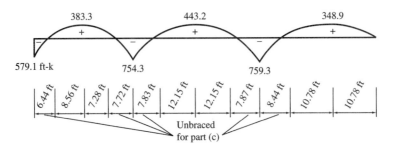

Figure 10-2

Maximum negative moment for design
$$= (0.9)(759.3) = -683.4 \text{ ft-k} \leftarrow$$
Maximum positive moment for design
$$= 443.2 + \frac{1}{10}\left(\frac{754.3 + 759.3}{2}\right) = +518.9 \text{ ft-k}$$
$$Z \text{ required} = \frac{(12)(683.4)}{(0.9)(36)} = 253.1 \text{ in.}^3$$

<u>Use W24 × 94</u>

(c) Elastic design—full lateral support supplied for top flange but only at support points for bottom flange

Referring to the moment diagram for the part (b) solution (Fig. 10-2) we see that the greatest unbraced length in the negative moment region is 8.44 ft and the reduced M_u is 683.4 ft-k.

From the Load Factor Design Selection Table with $M_u = 683.4$ ft-k we pick a $W30 \times 90$. Its $L_p = 8.7$ ft > 8.44 ft.

<u>Use W30 × 90</u> ∎

10-2 SHEAR

For this discussion the beam of Fig. 10-3(a) is considered. As the member bends, shear stresses occur because of the changes in length of its longitudinal fibers. For positive bending the lower fibers are stretched and the upper fibers are shortened while somewhere in between there is a neutral axis where the fibers do not change in length. Owing to these varying deformations a particular fiber has a tendency to slip on the fiber above or below.

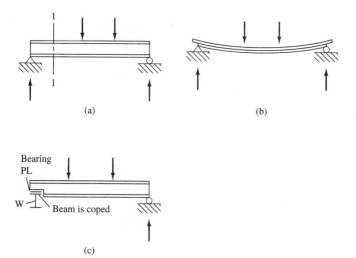

Figure 10-3

If a wooden beam was made by stacking boards on top of each other and not connecting them they would obviously tend to take the shape shown in part (b) of the figure. The student may have observed short heavily loaded timber beams with large transverse shears that split along horizontal planes.

This presentation may be entirely misleading in seeming to completely separate horizontal and vertical shears. In reality horizontal and vertical shear at any point are the same and may not be separated. Furthermore, one cannot occur without the other.

Generally, shear is not a problem in steel beams because the webs of rolled shapes are capable of resisting rather large shearing forces. Perhaps it is well, however, to list here the most common situations where shear might be excessive. These are as follows.

1. Should large concentrated loads be placed near beam supports they will cause large external shears without corresponding increases in bending moments. A fairly common example of this type occurs in tall buildings where on a particular floor the upper columns are offset with respect to the columns below. The loads from the upper columns applied to the beams on the floor level in question will be quite large if there are many stories above.

2. Probably the most common shear problem occurs where two members (as a beam and a column) are rigidly connected together so their webs lie in a common plane. This situation frequently occurs at the junction of columns and beams (or rafters) in rigid frame structures.

3. Where beams are notched or coped as shown in Fig. 10-3(c) shear can be a problem. For this case shear forces must be calculated for the remaining beam depth. A similar discussion can be made where holes are cut in beam webs for duct work or other items.

Combined welded and bolted joint, Transamerica Pyramid, San Francisco, Calif. (Courtesy of Kaiser Steel Corporation.)

4. Theoretically, very heavily loaded short beams can have excessive shears but practically this does not occur too often unless it is like Case 1.
5. Shear may very well be a problem even for ordinary loadings when very thin webs are used as in plate girders or in light gage cold formed steel members.

From his or her study of mechanics of materials the student is familiar with the shear stress formula $f_v = VQ/Ib$ where V is the external shear, Q is the statical moment of that portion of the section lying outside (either above or below) the line on which f_v is desired taken about the neutral axis, and b is the width of the section where the unit shearing stress is desired.

Figure 10.4(a) shows the variation in shear stresses across the cross section of an I-shaped member, while part (b) of the same figure shows the shear stress variation in a member with a rectangular cross section. It can be seen in part (a) of the figure that the shear in I-shaped sections is primarily resisted by the web.

If the load is increased on an I-shaped section until the bending yield stress is reached in the flange, the flange will be unable to resist shear stress and it will be carried in the web. If the moment is further increased the bending yield stress will penetrate farther down into the web and the area of web that can resist shear will be further reduced. Rather than assuming the nominal shear stress is resisted by part of the web, the LRFD Specification assumes that a reduced shear stress is resisted by the entire web area. This web area, A_w, is equal to the overall depth of the member, d, times the web thickness, t_w.

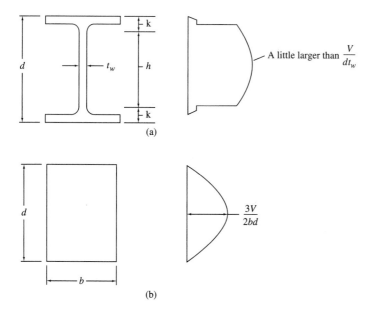

Figure 10-4

Shear strength expressions are given in LRFD Specification F2. In these expressions which follow, F_{yw} is the specified minimum yield stress of the web; h is the clear distance between the web toes of the fillets for rolled shapes, while for built-up welded sections it's the clear distance between flanges. For bolted built-up sections h is the distance between adjacent lines of bolts in the web. Different expressions are given for different h/t_w ratios depending on whether shear failures would be plastic, inelastic, or elastic.

1. **Web yielding.** Almost all rolled beam sections in the Manual fall into this classification.

   If $\dfrac{h}{t_w} \leq 418/\sqrt{F_{yw}} = 70$ for $F_y = 36$ ksi and 59 for $F_y = 50$ ksi

   $V_n = 0.6\, F_{yw}\, A_w$ (LRFD Equation F2-1)

2. **Inelastic buckling of web**

   If $418/\sqrt{F_{yw}} < \dfrac{h}{t_w} \leq 523/\sqrt{F_{yw}} = 87$ for $F_y = 36$ ksi and 74

   for $F_y = 50$ ksi

   $V_n = 0.6\, F_{yw}\, A_w\, (418/\sqrt{F_{yw}})/(h/t_w)$ (LRFD Equation F2-2)

3. **Elastic buckling of web**

   If $523/\sqrt{F_{yw}} < \dfrac{h}{t_w} \leq 260$

   $V_n = (132{,}000\, A_w)/(h/t_w)^2$ (LRFD Equation F2-3)

For each of the situations given $V_u = \phi_v V_n$ with $\phi_v = 0.90$. The LRFD Appendix F2.2 gives expressions for the general design shear strength of webs with or without stiffeners. Appendix G3 also provides an alternative method for plate girders that utilizes tension field action (see Chapter 18).

In Example 10-2 the shear strength of a beam is computed.

■ EXAMPLE 10-2

A W24 × 55 ($d = 23.57$ in., $t_w = 0.395$ in., $k = $ distance from extreme flange fiber to web toe of fillet $= 1\frac{5}{16}$ in.) consisting of 50 ksi steel is used for the beam and loading of Fig. 10-5. Check its adequacy in shear.

$w_u = 7.3$ klf (includes beam weight)

73 k |← ———————————— 20 ft ———————————— | 73 k

Figure 10-5

Solution.

$$h = 23.57 - (2)\left(1\frac{5}{16}\right) = 20.94 \text{ in.}$$

$$\frac{h}{t_w} = \frac{20.94}{0.395} = 53.01 < \frac{418}{\sqrt{50}} = 59$$

$$\phi_v V_n = (0.90)(0.6)(F_{yw})(A_w)$$
$$= (0.90)(0.6)(50)(23.57 \times 0.395) = 251.4 \text{ k} > 73\text{k} \qquad \text{(OK)} \quad ■$$

Note: In Part 4 of the LRFD Manual there is a large set of tables entitled "Beams W shapes maximum factored uniform load in kips for beams laterally supported." This title seems to the author to be quite inadequate because so much other information can be obtained from these tables, such as the design shear strengths or $\phi_v V_n$ values, and information concerning web yielding, web crippling, and other items to be discussed later.

Should V_u for a particular beam exceed the LRFD specified shear strength of the member, the usual procedure will be to select a little heavier section. If it is necessary, however, to use a much heavier section than required for moment, doubler plates (Fig. 10-6) may be welded to the beam web, or stiffeners may be connected to the webs in zones of high shear. Doubler plates must meet the width-thickness requirements for compact stiffened elements as prescribed in Section B.5 of the LRFD Specification. In addition, they must be welded sufficiently to the member webs to develop their proportionate share of the load.

The LRFD specified shear strength of a beam or girder is based on the entire area of the web. Somethimes, however, a connection is made to only a small

Figure 10-6 Increasing shear strength of beam by using doubler plates.

Figure 10-7 Block shear failure possible along dashed line.

portion or depth of the web. For such a case the designer may decide to assume that the shear is spread over only part of the web depth for purposes of computing shear strength. Thus he or she may compute A_w as being equal to t_w times the smaller depth for use in the shear strength expression.

When beams that have their top flanges at the same elevations (the usual situation) are connected to each other, it is frequently necessary to cope one of them, as shown in Fig. 10-7. For such cases there is a distinct possibility of a block shear failure along the broken lines shown. This subject was previously discussed in Section 3-7 and is continued in Chapter 15.

10-3 DEFLECTIONS

The deflections of steel beams are usually limited to certain maximum values. Among the several excellent reasons for deflection limitations are the following.

1. Excessive deflections may damage other materials attached to or supported by the beam in question. Plaster cracks caused by large ceiling joist deflections are one example.
2. The appearance of structures is often damaged by excessive deflections.
3. Extreme deflections do not inspire confidence in the persons using a structure, although it may be completely safe from a strength standpoint.
4. It may be necessary for several different beams supporting the same loads to deflect equal amounts.

Standard American practice for buildings has been to limit service live-load deflections to approximately $\frac{1}{360}$ of the span length. This deflection is supposedly the largest value that ceiling joists can deflect without causing cracks in underlying plaster. The $\frac{1}{360}$ deflection is only one of many maximum deflection values in use because of different loading situations, different engineers, and different specifications. For situations where precise and delicate machinery is supported, maximum deflections may be limited to 1/1500 or 1/2000 of the span lengths. The 1989 AASHTO Specifications limit deflections in steel beams and girders due to live load and impact to $\frac{1}{800}$ of the span. (For bridges in urban areas which are partly used by pedestrians, the AASHTO *recommends* a maximum value equal to 1/1000 of the span lengths.)

The LRFD Specification does not specify exact maximum permissible deflections. There are so many different materials, types of structures, and loadings that no one single set of deflection limitations is acceptable for all cases. Thus limitations must be set by the individual designer on the basis of his or her experience and judgment.

Before substituting blindly into a formula that will give the deflection of beam for a certain loading condition, the student should thoroughly understand the theoretical methods of calculating deflections. These methods include the moment area, conjugate beam, and virtual work procedures. From these methods various expressions can be determined such as the common one given at the end of this paragraph for the center line deflection of a uniformly loaded simple beam.

$$\Delta_E = \frac{5wL^4}{384EI}$$

To use deflection expressions such as this one the reader must be very careful to use consistent units. Example 10-3 illustrates the application of the preceding expression. The author has changed all units to pounds and inches. Thus the uniform load given in the problem as so many kips per foot is changed to so many lb/in.

■ **EXAMPLE 10-3**

A W24 × 55 (I_x = 1350 in.) has been selected for a 21-ft simple span to support a total service load of 3 klf (including beam weight). Is the center line deflection of this section satisfactory for the service live load if the maximum permissible value is 1/360 of the span?

Solution.

$$\Delta_E = \frac{5wL^4}{384EI} = \frac{(5)(3000/12)(12 \times 21)^4}{(384)(29 \times 10^6)(1350)} = 0.335 \text{ in.}$$

$$< \left(\frac{1}{360}\right)(12 \times 21) = 0.70 \text{ in.} \qquad\qquad \text{(OK)} \quad ■$$

In Part 4 of the LRFD Manual the following simple formula for determining maximum beam deflections for *W, M, HP, S, C,* and *MC* sections for several different loading conditions is presented.

$$\Delta = \frac{ML^2}{C_1 I_x}$$

In this expression M is the service load moment in ft-k based on a uniformly distributed load, C_1 is a constant whose value can be determined from Fig. 10-8, L is the span length (ft), and I_x is the moment of inertia (in.⁴).

If we want to use this expression for the beam of Example 10-3, C_1 from part (a) of Fig. 10-6 would be 161, the center line bending moment is $wL^2/8 = (3)(21)^2/8 = 165.375$ ft-k, and the center line deflection for the live service load would be

$$\Delta_L = \frac{(165.375)(21)^2}{(161)(1350)} = 0.336 \text{ in.}$$

Some specifications handle the deflection problem by requiring certain minimum depth-span ratios. For example, the AASHTO suggests the depth-span ratio be limited to a minimum value of 1/25. A shallower section is permitted, but it should have sufficient stiffness to prevent a deflection greater than would have occurred if the 1/25 ratio had been used. The LRFD Manual in their Table 4-2 presents recommended span/depth ratios for simply supported uniformly loaded beams with different service dead to live load (or to total load) ratios which, if followed, will effectively limit deflections.

A steel beam can be cold-bent or cambered an amount equal to the deflection caused by dead load or the deflection caused by dead load plus some percentage of the live load. Approximately 25 percent of the camber so produced is elastic and will disappear when the cambering operation is completed. Detailed information for particular shapes is given in Part 1 of the LRFD Manual in the section entitled "Standard Mill Practice." It should be remem-

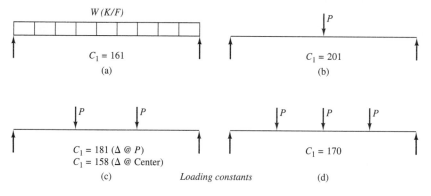

Figure 10-8 Values of constant C_1 for use in deflection expression.

bered that a beam which is bent upward looks much stronger and safer than one which sags downward (even a very small distance).

A camber requirement is quite common for longer steel beams. In fact, a rather large percentage of the beams used in composite construction today (see Chapter 16) are cambered. On many occasions, however, it is more economical to select heavier beams with their larger moments of inertia so as to reduce deflections and avoid the labor costs involved in cambering. One commonly used rule of thumb is that it takes about one extra man-work-hour to camber each beam.

Cambering is something of a nuisance to many fabricators and it may very well introduce some additional problems. For instance, when beams are cambered it may be necessary to adjust the connection details so that proper fitting of members is achieved. The end of a cambered beam will be rotated and thus it may be necessary to rotate the connection details by the same amount to insure proper fitting.

If we can move up one section in weight, thus reducing deflections so cambering is not needed, we may have a very desirable solution. Similarly, if a higher-strength steel is being used, it may be be desirable to switch the beams that need cambering to a lower yield stress steel. The results will be larger beams but smaller deflections and perhaps some economy if cambering can be eliminated.

Deflections may very well control the sizes of beams for longer spans or for short ones where deflection limitations are severe. To assist the designer in selecting sections where deflections may control, the LRFD Manual includes in Part 4 a table entitled "Moment of Inertia Selection Table," in which the I_x values are given in numerically descending order for all of the sections normally used as beams. In this table the sections are arranged in groups, with the lightest section in each group printed in bold type. Example 10-4 presents the design of a beam where deflections controls the design.

■ EXAMPLE 10-4

Select the lightest available section with $F_y = 36$ ksi to support a service dead load of 1.2 klf and a service live load of 3 klf for a 30-ft simple span. The section is to have full lateral bracing for its compression flange and the maximum total service load deflection is not to exceed 1/1500 the span length.

Solution.   After some scratch work assume beam weight = 167 lb/ft

$$w_u = (1.2)(1.367) + (1.6)(3) = 6.44 \text{ klf}$$

$$M_u = \frac{(6.44)(30)^2}{8} = 724.5 \text{ ft-k}$$

$$Z \text{ required} = \frac{(12)(724.5)}{(0.9)(36)} = 268.3 \text{ in.}^3$$

Try W30 × 90 ($I_x = 3620$ in.^{4})

Maximum permissible $\Delta = \left(\dfrac{1}{1500}\right)(12 \times 30) = 0.24$ in.

Actual $\Delta_L = \dfrac{ML^2}{C_1 I_x} = \dfrac{\dfrac{(4.367)(30)^2}{8}(30)^2}{(161)(3620)}$

$\qquad\qquad = 0.759$ in. > 0.24 in.   (NG)

Maximum I_x required to limit deflection to 0.24 in.

$= \left(\dfrac{0.759}{0.24}\right)(3620) = 11{,}448$ in.^{4}

From Moment of Inertia Selection Table

<u>Use W40 × 167</u>   ■

Vibrations

Though steel members may be selected that are satisfactory as to moment, shear, deflections, and so on, some very annoying floor vibrations may still occur. This is perhaps the most common serviceability problem faced by designers. Objectionable vibrations will frequently occur where long spans and large open floors without partitions or other items which might provide suitable damping are used. The reader may often have noticed this situation in the floors of large malls.

Damping of vibrations may be achieved by using framed-in-place partitions each attached to the floor system in at least 3 places or by installing "false" sheetrock partitions between ceilings and the underside of floor slabs. Further damping may be achieved by thickening the floor slabs, installing partitions, and by the office furniture and perhaps the equipment used in the building. A better procedure is to control the stiffness of the structural system.[1,2] (A rather common practice in the past has been to try to limit vibrations by selecting beams no shallower than 1/20th times span lengths.)

If the occupants of a building feel uneasy or are annoyed by vibrations the design is unsuccessful. It is rather difficult to correct a situation of this type in an existing structure. On the other hand, the situation can be easily predicted and

[1] T. M. Murray, "Controlling Floor Movement." *Modern Steel Construction* (AISC, Chicago, IL, June 1991) pp. 17–19.

[2] T. M. Murray, "Acceptability Criterion for Occupant-Induced Floor Vibrations," *Engineering Journal,* AISC, 18, 2 (2nd quarter, 1981), pp. 62–70.

corrected in the design stage. Several good procedures have been developed that enable the structural designer to estimate the acceptability of a given system by its users.

F. J. Hatfield[3] has prepared a useful chart that is used for estimating the perceptibility of vibrations of steel beams and concrete slabs for both office and residential buildings.

Ponding

If water on a flat roof accumulates faster than it runs off, the increased load causes the roof to deflect into a dish shape that can hold more water, which causes greater deflections and so on. This process of *ponding* continues until equilibrium is reached or until collapse occurs. Ponding is a serious matter, as illustrated by the large annual number of flat roof failures in the United States.

Ponding will occur on almost any flat roof to a certain degree, even though roof drains are present. Drains may be inadequate during severe storms, or they may become stopped up. Furthermore, they are often placed along the beam lines, which are actually the high points of the roof. The best method of preventing ponding is to have an appreciable slope on the roof ($\frac{1}{4}$ in./ft or more) together with good drainage facilities. It has been estimated that probably two thirds of the flat roofs in the United States have slopes less than this value, which is the minimum recommended by the National Roofing Contractors Association (NRCA). It costs approximately 3 to 6 percent more to construct a roof with this desired slope than building with no slope.[4]

When a very large flat roof (perhaps an acre or more) is being considered, the effect of wind on water depth may be quite important. A heavy rainstorm will frequently be accompanied by heavy winds. When a large quantity of water is present on the roof, a strong wind may very well push a great deal of water to one end, creating a dangerous depth of water as regards the load in pounds per square foot applied to the roof. For such situations *scuppers* are sometimes used. These are large holes or tubes in the walls or parapets which enable water above a certain depth to quickly drain off the roof.

Ponding failures will be prevented if the roof system (consisting of the roof deck and supporting beams and girders) has sufficient stiffness. The LRFD Specification (K2) describes a minimum stiffness to be achieved if ponding failures are to be prevented. If this minimum stiffness is not provided, it is necessary to make other investigations to be sure that a ponding failure is not possible.

---

[3] F. J. Hatfield, "Design Chart for Vibration of Office and Residential Floors," *Engineering Journal*, AISC, 29, 4 (4th quarter, 1992), pp. 141–144.

[4] Gary Van Ryzin, "Roof Design: Avoid Ponding by Sloping to Drain," *Civil Engineering* (New York: ASCE, January 1980), pp. 77–81.

Theoretical calculations for ponding are very complicated. The LRFD requirements are based on work by F. J. Marino[5] in which he considered the interaction of a two-way system of secondary or submembers supported by a primary system of main members or girders. A numerical ponding example is presented in Appendix F.

10-4 WEBS AND FLANGES WITH CONCENTRATED LOADS

When steel members have concentrated loads applied that are perpendicular to one flange and symmetric to the web, their flanges and webs must have sufficient flange and web design strength in the areas of flange bending, web yielding, web crippling, and sidesway web buckling. Should a member have concentrated loads applied to both flanges, it must have a sufficient web design strength in the areas of web yielding, web crippling, and column web buckling. In this section formulas for determining strengths in these areas are presented.

If flange and web strengths do not satisfy the requirements of LRFD Specification Section K1.2 to K1.6, it will be necessary to use transverse stiffeners at the concentrated loads. Should web design strengths not satisfy the requirements of LRFD Section K1.7, it will be necessary to use doubler plates or diagonal stiffeners. These situations are discussed in the paragraphs to follow.

Local Flange Bending   The flange must be sufficiently rigid so that it will not deform and cause a zone of high stress concentrated in the weld in line with the web. The nominal tensile load that may be applied through a plate welded to the flange of a W section is to be determined by the expression to follow in which F_{yf} is the specified minimum yield stress of the flange (ksi) and t_f is the flange thickness (in.)

$$\phi = 0.90$$
$$R_n = 6.25 \ t_f^2 F_{yf} \qquad \text{(LRFD Equation K1-1)}$$

It is not necessary to check this formula if the length of loading across the beam flange is less than 0.15 times the flange width b_f or if a pair of half-depth or deeper web stiffeners are provided. Fig. 10-10(a) shows a beam with local flange bending.

Local Web Yielding   The subject of local web yielding applies to all concentrated forces, tensile or compressive. Here we will try to limit the stress in the web of a member in which a force is being transmitted. Local web yielding is illustrated in part (b) of Fig. 10-10.

---

[5] F. J. Marino, "Ponding of Two-Way Roof System," *Engineering Journal* (New York: AISC, July 1966), pp. 93–100.

This beam is to be subjected to a heavy concentrated load from above. Partial web stiffeners are used to prevent web crippling and to keep the top flange from twisting or warping at the load. (Courtesy of American Institute of Steel Construction, Inc.).

The nominal strength of the web of a beam at the web toe of the fillet when a concentrated load or reaction is applied is to be determined by one of the following two expressions in which k is the distance from the outer edge of the flange to the web toe of the fillet, N is the bearing length (in.) of the force, F_{yw} is the specified minimum yield stress (ksi) of the web, and t_w is the thickness of the web.

If the force is a concentrated load or reaction that causes tension or compression and is applied at a distance greater than the member depth from the end of the member

$$\phi = 1.0$$
$$R_n = (5k + N)F_{yw}t_w \qquad \text{(LRFD Equation K1-2)}$$

If the force is a concentrated load or reaction applied at a distance d or less from the member end

$$\phi = 1.0$$
$$R_n = (2.5k + N)F_{yw}t_w \qquad \text{(LRFD Equation K1-3)}$$

Reference to Fig. 10-9 clearly shows where these expressions were obtained. The nominal strength R_n equals the length over which the force is assumed to be spread when it reaches the web toe of the fillet times the web thickness times the

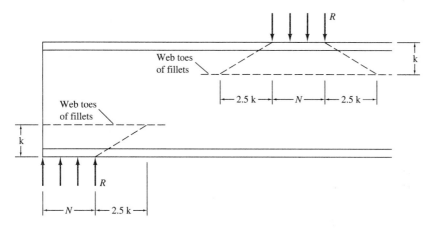

Figure 10-9 Local web yielding.

yield stress of the web. Should a stiffener extending for at least half the member depth be provided on each side of the web at the concentrated force, it is not necessary to check for web yielding.

Web Crippling   Should concentrated compressive loads be applied to a member with an unstiffened web (the load being applied in the plane of the web) the nominal web crippling strength of the web is to be determined by the appropriate equation of the two which follow (in which d is the overall depth of the member). If one or two web stiffeners or one or two doubler plates are provided and extend for at least half of the web depth, web crippling will not have to be checked. Research has shown that when web crippling occurs it is located in the part of the web adjacent to the loaded flange. Thus it is thought that stiffening the web in this area for half its depth will prevent the problem. Web crippling is illustrated in part (c) of Fig. 10-10.

If the concentrated load is applied at a distance not less than $d/2$ from the end of the member

$$\phi = 0.75$$

$$R_n = 135t_w^2\left[1 + 3\left(\frac{N}{d}\right)\left(\frac{t_w}{t_f}\right)^{1.5}\right]\sqrt{\frac{F_{yw}t_f}{t_w}} \qquad \text{(LRFD Equation K1-4)}$$

If the concentrated load is applied at a distance less than $d/2$ from the end of the member

For $\dfrac{N}{d} \le 0.2$

$$\phi = 0.75$$

$$R_n = 68t_w^2\left[1 + 3\left(\frac{N}{d}\right)\left(\frac{t_w}{t_f}\right)^{1.5}\right]\sqrt{\frac{F_{yw}t_f}{t_w}} \qquad \text{(LRFD Equation K1-5a)}$$

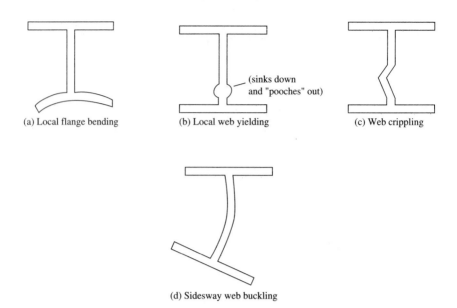

(a) Local flange bending   (b) Local web yielding   (c) Web crippling

(sinks down
and "pooches" out)

(d) Sidesway web buckling

Figure 10-10

For $\dfrac{N}{d} \leq 0.2$

$\phi = 0.75$

$$R_n = 68t_w^2 \left[1 + \left(4\frac{N}{d} - 0.2 \right)\left(\frac{t_w}{t_f}\right)^{1.5} \right] \sqrt{\frac{F_{yw}t_f}{t_w}}$$

(LRFD Equation K1-5b)

Sidesway Web Buckling   Should compressive loads be applied to laterally braced compression flanges the web will be put in compression and the tension flange may buckle as shown in Fig. 10-10(d).

It has been found that sidesway web buckling will not occur if the flanges are restrained against rotation with $(h/t_w)/(\ell/b_f) > 2.3$ or if $(h/t_w)/(\ell/b_f) > 1.7$ when flange rotation is not restrained. In these expressions h is the web depth between the web toes of the fillets, that is, $d - 2k$, and ℓ is the largest laterally unbraced length along either flange at the point of the load.

It is also possible to prevent sidesway web buckling with properly designed lateral bracing or stiffeners at the load point. The LRFD Commentary suggests that local bracing for *both* flanges be designed for 1 percent of the magnitude of the concentrated load applied at the point. If stiffeners are used they must extend from the point of load for at least one half of the member depth and should be designed to carry the full load. Flange rotation must be prevented if the stiffeners are to be effective.

Girders for all-welded Connecticut expressway bridge. (Courtesy of the Lincoln Electric Company.)

Should members not be restrained against relative movement by stiffeners or lateral bracing and be subject to concentrated compressive loads, their strength may be determined as follows:

When the loaded flange is braced against rotation and $(h/t_w)/(\ell/b_f)$ is ≤ 2.3

$$\phi = 0.85$$

$$R_n = \frac{C_r t_w^3 t_f}{h^2}\left[1 + 0.4\left(\frac{h/t_w}{\ell/b_f}\right)^3\right]$$ (LRFD Equation K1-6)

When the loaded flange is not restrained against rotation and $\dfrac{h/t_w}{\ell/b_f} \leq 1.7$

$$\phi = 0.85$$

$$R_n = \frac{C_r t_w^3 t_f}{h^2}\left[0.4\left(\frac{h/t_w}{\ell/b_f}\right)^3\right]$$ (LRFD Equation K1-7)

It is not necessary to check Equations K1-6 and K1-7 if $h/t_w/\ell/b_f$ is greater than 2.3 or 1.7, respectively, or if the webs are subject to distributed load. Furthermore, these equations were developed for bearing connections and *do not apply to moment connections*. In these expressions h is as defined for shear in Section 10-2 of this chapter and

$C_r = 960{,}000$ when $M_u < M_y$ at the location of the force, ksi.
$C_r = 480{,}000$ when $M_u \geq M_y$ at the location of the force, ksi.

Compression Buckling of the Web   This limit state applies to concentrated compression loads applied to both flanges of a member as, for instance, moment connections applied to both ends of a column. For such a situation it is necessary to limit the slenderness ratio of the web to avoid the possibility of buckling. Should the concentrated loads be larger than the value of ϕR_n given below, it will be necessary to provide either one stiffener or a pair of them extending for the full depth of the web and meeting the requirements of LRFD Specification K1.9. (The equation to follow is applicable to moment connections, but not to bearing ones.)

$$\phi = 0.90$$

$$R_n = \frac{4{,}100 \; t_w^3 \; \sqrt{F_{yw}}}{h} \qquad \text{(LRFD Formula K1-8)}$$

Example 10-5, which follows, presents the selection of a beam for moment and its review for the applicable items discussed in this section.

■ EXAMPLE 10-5

Select a W section for moment using A36 steel for the span and service loads shown in Fig. 10-11. Lateral bracing will be provided only at the ends and concentrated loads. If the end bearing length is 3.50 in. and the concentrated load bearing length is 3.00 in., check the beam for web yielding, web crippling, and sidesway web buckling.

Figure 10-11

Solution. Assume beam weight $= 44$ lb/ft

$$M_u = \frac{(1.2)(1.044)(15)^2}{8} + (1.6)(25)(5) = 235.2 \text{ ft-k}$$

$$Z \text{ required} = \frac{(12)(235.2)}{(0.9)(36)} = 87.1 \text{ in.}^3$$

Use W21 × 44 ($L_p = 5.3$ ft $> L_b$ of 5.0 ft, $d = 20.66$ in. $b_f = 6.500$ in., $t_w = 0.350$ in., $t_f = 0.450$ in., $k = 1\frac{3}{16}$ in.)

End reaction

$$R_u = (1.2)(1.044)\left(\frac{15}{2}\right) + (1.6)(25) = 49.4 \text{ k}$$

Concentrated load

$$P_u = (1.6)(25) = 40 \text{ k}$$

Local web yielding

End reaction

$$\phi R_n = \phi(2.5\,k + N)F_{yw}t_w = 1.0\left(2.5 \times 1\frac{3}{16} + 3.50\right)(36)(0.350)$$

$$= 81.5\,k > 49.4\,k \qquad\qquad\qquad\qquad \text{(OK)}$$

Concentrated load

$$\phi R_n = \phi(5k + N)F_{yw}t_w = (1.0)\left(5 \times 1\frac{3}{16} + 3.0\right)(36)(0.350)$$

$$= 112.6\,k > 40\,k \qquad\qquad\qquad\qquad \text{(OK)}$$

Web crippling

Concentrated loads

$$\phi R_n = \phi\,135t_w^2\left[1 + 3\left(\frac{N}{d}\right)\left(\frac{t_w}{t_f}\right)^{1.5}\right]\sqrt{\frac{F_{yw}t_f}{t_w}}$$

$$= (0.75)(135)(0.350)^2\left[1 + 3\left(\frac{3.00}{20.66}\right)\left(\frac{0.350}{0.450}\right)^{1.5}\right]\sqrt{\frac{(36)(0.450)}{0.350}}$$

$$= 109.6\,k > 40\,k \qquad\qquad\qquad\qquad \text{(OK)}$$

End reactions

$$\frac{N}{d} = \frac{3.5}{20.66} = 0.169 < 0.20$$

$$\phi R_n = \phi 68t_w^2\left[1 + 3\left(\frac{N}{d}\right)\left(\frac{t_w}{t_f}\right)^{1.5}\right]\sqrt{\frac{F_{yf}t_f}{t_w}}$$

$$= (0.75)(68)(0.350)^2\left[1 + 3\left(\frac{3.5}{20.66}\right)\left(\frac{0.350}{0.450}\right)^{1.5}\right]\sqrt{\frac{(36)(0.450)}{0.350}}$$

$$= 57.3\,k > 49.4\,k \qquad\qquad\qquad\qquad \text{(OK)}$$

Sidesway web buckling

$$\left(\frac{h}{t_w}\right)\left(\frac{\ell}{b_f}\right) = \frac{\left(20.66 - 2 \times 1\frac{3}{16}\right)}{0.350}\left(\frac{12 \times 5}{6.50}\right)$$

$$= 5.66 > 2.3$$

∴ sidesway web buckling does not have to be checked.   ■

Note that in Part 4 of the LRFD Manual sets of tables entitled "Beams W Shapes," "Beams S Shapes," and so on are presented with which the calculations for this beam and the other beams in this chapter can be greatly simplified. The definition of the terms in these tables is given in the Manual together with several useful example problems.

For instance, for the sections normally used as beams and with yield stresses of 36 and 50 ksi, the values $\phi_b M_p$, $\phi_v V_n$, L_p, and L_u are provided along with the quantities ϕR_1, ϕR_2, $\phi_r R_3$, and $\phi_r R_4$. The latter values are useful in making the calculations necessary to check web yielding and web crippling. If we have a W21 × 44 ($t_w = 0.350$ in., $k = 1\frac{3}{16}$ in; $f_{yw} = 36$ ksi) bearing on a length N of 3.5 in. we can compute ϕR_n for web yielding as follows:

$$\phi R_n = \phi(2.5k + N)F_{yw}t_w$$
$$= (1.0)(2.5 \times 1\frac{3}{16} + 3.5)(36)(0.350) = 81.5 \text{ k}$$

Or we can use the table values for ϕR_1 and ϕR_2

$$\phi R_n = \phi R_1 + \phi R_2 N = 37.4 + (12.6)(3.5) = 81.5 \text{ k}$$

If the requirements as to local flange bending, local web yielding, web crippling, or sidesway web buckling at concentrated loads are not met it is usually economical to take care of the situation by using a little larger section. If the problem cannot be handled by making a small increase in member size, we may decide to use web stiffeners as specified in detail in Section K1.9 of the LRFD Specification. Such stiffeners can be one-sided or two-sided. Should the load normal to the flange be tensile, the stiffeners need only be welded to the loaded flange. If, on the other hand, the load is compressive, the stiffeners must either bear against the compression flange or be welded to it.

10-5 UNSYMMETRICAL BENDING

From mechanics of materials it is remembered that each beam cross section has a pair of mutually perpendicular axes known as the principal axes for which the product of inertia is zero. Bending that occurs about any axis other than one of the principal axes is said to be unsymmetrical bending. When the external loads are not in a plane with either of the principal axes or when loads are simultaneously applied to the beam from two or more directions unsymmetrical bending is the result.

If a load is not perpendicular to one of the principal axes it may be broken into components which are perpendicular to those axes and the moments about each axis, M_{ux} and M_{uy}, determined as shown in Fig. 10-12.

When a section has one axis of symmetry, that axis is one of the principal axes and the calculations necessary for determining the moments are quite simple. For this reason unsymmetrical bending is not difficult to handle in the usual beam section, which is probably a W, S, M, or C. Each of these sections has at least one axis of symmetry and the calculations are appreciably reduced. A further simplifying factor is that the loads are usually gravity loads and probably perpendicular to the x axis.

Among the beams that must resist unsymmetrical bending are crane girders in industrial buildings and purlins for ordinary roof trusses. The x axes of purlins are parallel to the sloping roof surface while the large percentage of their loads

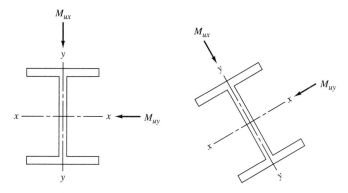

Figure 10-12

(roofing, snow, etc.) are gravity loads. These loads do not lie in a plane with either of the principal axes of the inclined purlins and the result is unsymmetrical bending. Wind loads are generally considered to act perpendicular to the roof surface and thus perpendicular to the x axes of the purlins with the result that they are not considered to cause unsymmetrical bending. The x axes of crane girders are usually horizontal, but the girders are subjected to lateral thrust loads from the moving cranes as well as to gravity loads.

To check the adequacy of members bent about both axes simultaneously the LRFD provides an equation in Section H1 of their Specification. The equation to follow is for combined bending and axial tension or compression if $P_u/\phi P_n < 0.2$.

$$\frac{P_u}{2\phi P_n} + \left(\frac{M_{ux}}{\phi_b M_{nx}} + \frac{M_{uy}}{\phi_b M_{ny}}\right) \le 1.0 \qquad \text{(LRFD Equation H1-1b)}$$

Since for the problem discussed here P_u is equal to zero the formula reduces to

$$\frac{M_{ux}}{\phi_b M_{nx}} + \frac{M_{uy}}{\phi_b M_{ny}} \le 1.0$$

This is an interaction or percentage equation. If M_{ux} is, say, 75 percent of the moment it could resist if bent about the x axis only ($\phi_b M_{nx}$), then M_{uy} can be no greater than 25 percent of what it could resist if bent about the y axis only ($\phi_b M_{ny}$).

Examples 10-6 and 10-7 illustrate the design of beams subjected to unsymmetrical bending. To illustrate the trial-and-error nature of the problem the author did not do quite as much scratchwork as in some of his earlier examples. The first design problems of this type which the student attempts may quite well take several trials. Consideration needs to be given to the question of lateral support for the compression flange. Should the lateral support be of questionable nature the engineer should reduce the design moment resistance by means of one of the expressions previously given for that purpose.

■ **EXAMPLE 10-6**

A certain beam in its upright position is estimated to have a vertical bending moment $M_{ux} = 160$ ft-k and a lateral bending moment $M_{uy} = 35$ ft-k. These moments include the effect of the estimated beam weight. The loads are assumed to pass through the centroid of the section. Select a W24 shape of A36 steel that can resist these moments assuming full lateral support for the compression flange.

Solution.

Try W24 × 62 ($\phi_b M_p = \phi_b M_{nx} = 413$ ft-k, $Z_y = 15.7$ in.^{3})

$$\phi_b M_{ny} = \frac{(0.9)(36)(15.7)}{12} = 42.39 \text{ ft-k}$$

$$\frac{M_{ux}}{\phi_b M_{nx}} + \frac{M_{uy}}{\phi_b M_{ny}} = \frac{160}{413} + \frac{35}{42.39} = 1.21 > 1.00 \qquad \text{(NG)}$$

Try W24 × 68 ($\phi_b M_p = \phi_b M_{nx} = 478$ ft-k, $Z_y = 24.5$ in.^{3})

$$\phi_b M_{ny} = \frac{(0.9)(36)(24.5)}{12} = 66.15 \text{ ft-k}$$

$$\frac{160}{478} + \frac{35}{66.15} = 0.864 < 1.00 \qquad \text{(OK)}$$

Use W24 × 68   ■

It will be noted in the solution for Example 10-6 that though the procedure used will yield a section that will adequately support the moments given the selection of the absolutely lightest section listed in the LRFD Manual could be quite lengthy because of the two variables Z_x and Z_y which affect the size. If we have a large M_{ux} and a small M_{uy} the most economical section will probably be quite deep and rather narrow, whereas if we have a large M_{uy} in proportion to M_{ux} the most economical section may be rather wide and shallow.

10-6 DESIGN OF PURLINS

To avoid bending in the top chords of roof trusses, it is theoretically desirable to place purlins only at panel points. For large trusses, however, it is more economical to space them at closer intervals. If this practice is not followed for large trusses the purlin sizes may become so large as to be impractical. When intermediate purlins are used the top chords of the truss should be designed for bending as well as axial stress as described in Chapter 11. Purlins are usually spaced from 2 to 6 ft apart depending on loading conditions, while their most desirable depth-to-span ratios are probably in the neighborhood of 1/24. Channels or S sections are the most frequently used sections, but on some occasions other shapes may be convenient.

As previously described in Chapter 4, the channel and S sections are very weak about their web axes and sag rods may be necessary to reduce the span lengths for bending about those axes. Sag rods, in effect, make the purlins continuous sections for their y axes and the moments about these axes are greatly reduced, as shown in Fig. 10-13. These moment diagrams are developed on the assumption that the changes in length of the sag rods are negligible. It is further assumed that the purlins are simply supported at the trusses. This as-

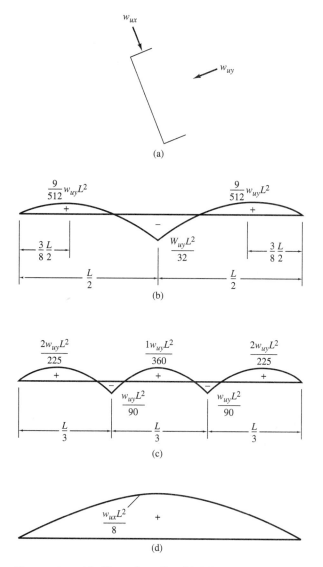

Figure 10-13 (a) Channel purlin, (b) Moment about web axis of purlins—sag rods at midspan. (c) Moment about web axis of purlins—sag rods at one-third points. (d) Moment about X axis of purlin.

sumption is on the conservative side since they are often continuous over two or more trusses and appreciable continuity may be achieved at their splices. The student can easily reproduce these diagrams from his or her knowledge of the moment distribution or other analysis methods. In the diagrams L is the distance between trusses, w_{uy} is the load component *perpendicular* to the web axis of the purlin, and w_{ux} is the load component *parallel* to the web axis.

If sag rods were not used the maximum moment about the web axis of a purlin would be $w_{uy}L^2/8$. When sag rods are used at midspan this moment is reduced to a maximum of $w_{uy}L^2/32$ (a 75 percent reduction), and when used at one-third points is reduced to a maximum of $w_{uy}L^2/90$ (a 91 percent reduction). In Example 10-7, sag rods are used at the midpoints and the purlins are designed for a moment of $w_{ux}L^2/8$ parallel to the web axis and $w_{uy}L^2/32$ perpendicular to the web axis.

In addition to being of advantage in reducing moments about the web axes of purlins, sag rods can serve other useful purposes. First, they can provide lateral support for the purlins; second, they are useful in keeping the purlins in proper alignment during erection until the roof deck is installed and connected to the purlins.

■ EXAMPLE 10-7

Select a W6 purlin for the roof shown in Fig. 10-14. The trusses are 15 ft 0 in. on center and sag rods are used at the midpoints between trusses. Full lateral support is assumed to be supplied from the roof above. Use A36 steel and the LRFD Specification and assume there is no torsion. Thus the full value of Z_y is to be used. Loads are as follows in terms of pounds per square foot of roof surface.

$$\text{Snow} = 30 \text{ psf}$$
$$\text{Roofing} = 6 \text{ psf}$$
$$\text{Estimated purlin weight} = 3 \text{ psf}$$
$$\text{Wind pressure} = 15 \text{ psf} \perp \text{to roof surface}$$

Figure 10-14

Solution.   Computing the load w_{ux} to be supported per foot by the purlins (which are 4.42 ft on center).

$$w_{ux} = 1.2D + 1.6S + 0.8W$$

$$= (1.2)(6 + 3)(4.42)\left(\frac{2}{\sqrt{5}}\right) + (1.6)(30)(4.42)\left(\frac{2}{\sqrt{5}}\right)$$

$$+ (0.8)(15)(4.42) = 285.6 \text{ lb/ft} \leftarrow$$

$$w_{ux} = 1.2D + 1.3W + 0.5S$$

$$= (1.2)(6 + 3)(4.42)\left(\frac{2}{\sqrt{5}}\right) + (1.3)(15)(4.42)$$

$$+ (0.5)(30)(4.42)\left(\frac{2}{\sqrt{5}}\right) = 188.2 \text{ lb/ft} \leftarrow$$

$$w_{uy} = (1.2D + 1.6S)\frac{1}{\sqrt{5}}$$

$$= [(1.2)(6 + 3)(4.42) + (1.6)(30)(4.42)]\frac{1}{\sqrt{5}}$$

$$= 116.5 \text{ lb/ft}$$

$$M_{ux} = \frac{(0.2856)(15)^2}{8} = 8.03 \text{ ft-k}$$

$$M_{uy} = \frac{(0.1165)(15)^2}{32} = 0.819 \text{ ft-k}$$

Try W6 × 9 ($\phi_b M_p = 16.8$ ft-k, $Z_y = 1.73$ in.^{3})

$$\frac{M_{ux}}{\phi_b M_{nx}} + \frac{M_{uy}}{\phi_b M_{ny}} \leq 1.0 \quad \text{since} \quad \frac{P_u}{\phi P_n} < 0.2$$

$$\frac{8.03}{16.8} + \frac{0.819}{(0.9)(36)(1.73)/12} = 0.653 < 1.0 \qquad \text{(OK)}$$

Use W6 × 9   ■

10-7 THE SHEAR CENTER

The shear center is defined as the point on the cross section of a beam through which the resultant of the transverse loads must pass so that the stresses in the beam may be calculated only from the theories of pure bending and transverse shear. Should the resultant pass through this point, it is unnecessary to analyze the beam for torsional moments. For a beam with two axes of symmetry the shear center will fall at the intersection of the two axes, thus coinciding with the centroid of the section. For a beam with one axis of symmetry the shear center will fall somewhere on that axis but not necessarily at the centroid of the section. This surprising statement means that to avoid torsion in some beams the lines of action of the loads should not pass through the centroids of the sections.

The shear center is of particular importance for beams whose cross sections are composed of thin parts which provide considerable bending resistance but little resistance to torsion. Many common structural members such as the W, S, and C sections, angles, and various beams made up of thin plates (as in aircraft construction) fall into this class and the problem has wide application.

The location of shear centers for several open sections are shown in Fig. 10-15. Sections such as these are relatively weak in torsion and for them and similar shapes the location of the resultant of the external loads can be a very serious matter. A previous discussion has indicated that the addition of one or more webs to these sections so they are changed into box shapes greatly increases their torsional resistance.

The average designer probably does not take the time to go through the sometimes tedious computations involved in locating the shear center and calculating the effect of twisting. He or she may instead simply ignore the situation, or may make a rough estimate of the effect of torsion on the bending stresses. One very poor estimate used occasionally is that of reducing $\phi_b M_{ny}$ by 50 percent for use in the interaction equation.

Shear centers can be located quickly for beams with open cross sections and relatively thin webs. For other beams the shear centers can probably be found, but only with considerable difficulty. The term *shear flow* is often used when

Figure 10-15

Figure 10-16

speaking of thin-wall members, although there is really no flowing involved. It refers to the shear per inch of the cross section and equals the unit shearing stress times the thickness of the member. (The unit shearing stress has been determined by the expression VQ/bI and the shear flow can be determined by VQ/I if the shearing stress is assumed to be constant across the thickness of the section.) The shear flow acts parallel to the sides of each element of a member.

The channel section of Fig. 10-16(a) will be considered for this discussion. In this figure the shear flow is shown with the small arrows and in part (b) the values are totaled for each component of the shape and labeled H and V. The two H values are in equilibrium horizontally and the internal V value balances the external shear at the section. Although the horizontal and vertical forces are in equilibrium, the same cannot be said for the moment forces unless the lines of action of the resultant of the external forces passes through a certain point called the *shear center*. The horizontal H forces in part (b) of the figure can be seen to form a couple. The moment produced by this couple must be opposed by an equal and opposite moment which can only be produced by the two V values. The location of the shear center is a problem in equilibrium; therefore, moments should be taken about a point that eliminates the largest number of forces possible.

From this information the following equation can be written from which the shear center can be located.

$$Ve = Hh \quad \text{(moments taken about c.g. of web)}$$

Examples 10-8 and 10-9 illustrate the calculations involved in locating the shear center for two shapes. Note that the location of the shear center is independent of the value of the external shear. (*Shear center locations are provided for channels in the Properties Table of Part I of the LRFD Manual.*)

■ EXAMPLE 10-8

The channel section shown in Fig. 10-17(a) is subjected to an external shear of V. Locate the shear center.

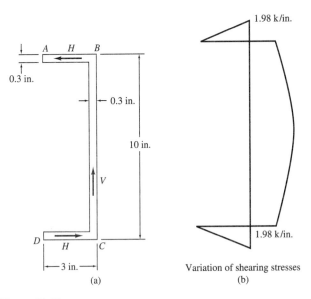

(a)

Variation of shearing stresses
(b)

Figure 10-17

Solution.   Properties of section:

$$I_x = \left(\frac{1}{12}\right)(3)(10)^3 - \left(\frac{1}{12}\right)(2.7)(9.4)^3 = 63 \text{ in.}^4$$

$$f_v \text{ at } B = \frac{(V)(2.85 \times 0.3 \times 4.85)}{63} = 0.06582 \, V/\text{in.}$$

$$\text{Total } H = \left(\frac{1}{2}\right)(2.85)(0.06582V) = 0.0938V$$

Location of shear center:

$$Ve = Hh$$
$$Ve = (0.0938V)(9.7)$$
$$e = 0.91 \text{ in. from ⊄ of web} \qquad ■$$

The student should clearly understand that shear stress variation across the corners, where the webs and flanges join, cannot be determined correctly with the mechanics of materials expression (VQ/bI or VQ/I for shear flow) and cannot be determined too well even after a complicated study with the theory of elasticity. The author as an approximation in the two examples presented here has assumed that shear flow continues up to the middle of the corners on the same pattern (straight-line variation for horizontal members and parabolic for others). The values of Q are computed for the corresponding dimensions. Other assumptions could have been made such as assuming a shear flow variation continuing vertically for the full depth of webs and only for the protruding parts of flanges horizontally, or vice versa. It is rather disturbing to find that whichever assumption is used the values do not check out perfectly. For instance, Example 10-9, which follows the sum of the vertical shear flow values, does not check out very well with the external shear.

■ EXAMPLE 10-9

The open section of Fig. 10-18 is subjected to an external shear of V. Locate the shear center.

Figure 10-18

Solution.   Properties of section:

$$I_x = \left(\frac{1}{12}\right)(0.25)(16)^3 + (2)(3.75 \times 0.25)(4.87)^2 = 129.8 \text{ in.}^4$$

Values of shear flow (labeled q):

$$q_A = 0$$

$$q_B = \frac{(V)(3.12 \times 0.25 \times 6.44)}{129.8} = 0.0387V/\text{in.}$$

$$q_C = q_B + \frac{(V)(3.75 \times 0.25 \times 4.87)}{129.8} = 0.0739V/\text{in.}$$

$$q_E = q_C + \frac{(V)(4.87 \times 0.25 \times 2.44)}{129.8} = 0.0968V/\text{in.}$$

Figure 10-19

These shear flow values are shown in Fig. 10-19(a) and the summation for each part of the member is given in part (b) of the figure.

Taking moments about the center line of *CD*

$$-(0.211V)(9.75) + (2)(0.0807V)(3.75) + Ve = 0$$
$$e = 1.45 \text{ in.} \qquad \blacksquare$$

The theory of the shear center is a very useful one in design, but it has certain limitations that should be clearly understood. For instance, the approximate analysis given in this section is valid only for thin sections. In addition, steel beams often have variable cross sections along their spans with the result that the loci of the shear centers are not straight lines along those spans. Thus, if the resultant of the loads passes through the shear center at one cross section, it might not do so at other cross sections.

When designers are faced with the application of loads to thin-walled sections such that twisting of those sections will be a problem, they will usually provide some means by which the twisting can be constrained. They may specify special bracing at close intervals, or attachments to flooring or roofing, or other similar devices. Should such solutions not be feasible they will probably consider selecting sections with greater torsional stiffnesses. Two references on this topic are given here.[6,7]

[6] C. G. Salmon and J. E. Johnson, *Steel Structures, Design and Behavior,* 2d ed. (New York: Harper & Row, 1980), pp. 672–675.

[7] AISC Engineering Staff, *Torsional Analysis of Steel Members* (Chicago: AISC, 1983).

10-8 BEAM-BEARING PLATES

When the ends of beams are supported by direct bearing on concrete or other masonry construction it is frequently necessary to distribute the beam reactions over the masonry by means of beam-bearing plates. The reaction is assumed to be spread uniformly through the bearing plate to the masonry and the masonry is assumed to push up against the plate with a uniform pressure equal to the factored reaction R_u over the area of the plate A_1. This pressure tends to curl up the plate and the bottom flange of the beam. The LRFD Manual recommends that the bearing plate be considered to take the entire bending moment produced and that the critical section for moment be assumed to be a distance k from the center line of the beam (see Fig. 10-20). The distance k is the same as the distance from the outer face of the flange to the web toe of the fillet given in the tables for each section (or it equals the flange thickness plus the fillet radius).

The determination of the true pressure distribution in a beam-bearing plate is a very formidable task, and the uniform pressure distribution assumption is usually made. This assumption is probably on the conservative side as the pressure is usually larger at the center of the beam than at the edges. The outer edges of the plate and flange tend to bend upward and the center of the beam tends to go down, concentrating the pressure there.

The required thickness of a 1-in.-wide strip of plate can be determined as follows with reference being made to Fig. 10-20:

$$M_u = \frac{R_u}{A_1} n \frac{n}{2} = \frac{R_u n^2}{2A_1}$$

Z of a 1 in.-wide piece of plate of t thickness $= (1)\left(\frac{t}{2}\right)\left(\frac{t}{4}\right)(2) = \frac{t^2}{4}$

Figure 10-20

$$M_u = \phi_b F_y Z \quad \text{with} \quad \phi_b = 0.9$$

$$\frac{R_u n^2}{2A_1} = (0.9)(F_y)\left(\frac{t^2}{4}\right)$$

$$t = \sqrt{\frac{2.22 R_u n^2}{A_1 F_y}}$$

In the absence of code regulations specifying different values the design strength for bearing on concrete is to be taken equal to $\phi_c P_p$ according to LRFD Specification J9. This specification states that when a bearing plate extends for the full area of a concrete support, the bearing strength of the concrete can be determined as follows

$$\phi_c = 0.60$$

$$P_p = 0.85 f_c' A_1 \qquad \text{(LRFD Equation J9-1)}$$

In this expression f_c' is the compression strength of the concrete in psi and A_1 is the area of the plate (in.^{2}) bearing concentrically on the concrete. For the design of such a plate, its required area A_1 can be determined by dividing the factored reaction R_u by $\phi_c P_p$:

$$A_1 = \frac{R_u}{\phi_c P_p} = \frac{R_u}{\phi_c 0.85 f_c'}$$

(The values provided are the same as those given in Section 10.15 of the 1989 ACI *Building Code for Reinforced Concrete*.)

Should the bearing load be applied to an area less than the full area of the concrete support $\phi_c P_p$ is to be determined with the following equation in which A_2 is the maximum area of the supporting surface that is geometrically similar to and concentric with the loaded area with $\sqrt{A_2/A_1}$ having a maximum value of 2.

$$\phi_c = 0.60$$

$$P_p = 0.85 f_c' A_1 \sqrt{\frac{A_2}{A_1}} \qquad \text{(LRFD Equation J9-2)}$$

For this case in which the full area of the concrete support is not loaded, A_1 can be computed as follows:

$$A_1 = \frac{R_u}{\phi_c P_p} = \frac{R_u}{\phi_c f_c' A_1 \sqrt{\dfrac{A_2}{A_1}}}$$

$$= \frac{1}{A_2}\left(\frac{R_u}{\phi_c 0.85 f_c'}\right)^2 \quad \text{when} \quad A_2 < 4A_1$$

or

$$A_1 = \frac{R_u}{1.7 \, \phi_c f_c'} \quad \text{when} \quad A_2 \geq 4A_1.$$

After A_1 is determined, its length (parallel to the beam) and its width are selected. The length may not be less than the N required to prevent web yielding or web crippling of the beam, nor may it be less than about $3\frac{1}{2}$ or 4 in. for practical construction reasons. It may not be greater than the thickness of the wall or other support, and actually it may have to be less than that thickness, particularly at exterior walls, to prevent the steel from being exposed.

Example 10-10 illustrates the calculations involved in designing a beam bearing plate. Notice that the width and the length of the plate are desirably taken to the nearest full inch.

■ EXAMPLE 10-10

A W18 × 71 beam ($d = 18.47$ in., $t_w = 0.495$ in., $b_f = 7.635$ in., $t_f = 0.810$ in., $k = 1\frac{1}{2}$ in.) has one of its ends supported by a reinforced-concrete wall with $f'_c = 3$ ksi. Design a bearing plate for the beam with A36 steel. The factored end reaction is 120 k and the maximum length of end bearing $\perp$ to the wall is the full wall thickness 8.0 in.

Solution.

$$\text{Area required} = A_1 = \frac{R_u}{\phi_c 0.85 f'_c} = \frac{120}{(0.60)(0.85)(3)} = 78.43 \text{ in.}^2$$

Minimum length of bearing $\perp$ to wall required for local web yielding

$$R_u = \phi(2.5 \, k + N)F_{yw}t_w$$
$$120 = 1.0(2.5 \times 1\tfrac{1}{2} + N)(36)(0.495)$$
$$N = 2.98 \text{ in.}$$

Minimum length of bearing $\perp$ to wall for web crippling

$$\text{Assume } \frac{N}{d} = \frac{8}{18.47} = 0.433 > 0.2$$

$$R_u = \phi 68 t_w^2 \left[1 + \left(4\frac{N}{d} - 0.2 \right)\left(\frac{t_w}{t_f}\right)^{1.5} \right] \sqrt{\frac{F_{yw}t_f}{t_w}}$$

$$120 = (0.75)(68)(0.495)^2 \left[1 + \left(4\frac{N}{18.47} - 0.2 \right)\left(\frac{0.495}{0.810}\right)^{1.5} \right] \sqrt{\frac{(36)(0.810)}{0.495}}$$

$$N = 3.36 \text{ in.}$$

Try 8 × 10 in. PL

$$n = 5 - 1\tfrac{1}{2} = 3.5 \text{ in.}$$
$$t = \sqrt{\frac{2.22 \, R_u n^2}{A_1 F_y}} = \sqrt{\frac{(2.22)(120)(3.5)^2}{(80.0)(36)}} = 1.06 \text{ in.}$$

Use $1\frac{1}{8}$ × 8 × 10 in. PL   ■

On some occasions the beam flanges alone probably provide sufficient bearing area, but bearing plates are nevertheless recommended as they are useful in erection and ensure an even bearing surface for the beam. They can be placed separately from the beams and carefully leveled to the proper elevations. When the ends of steel beams are enclosed by the concrete or masonry walls, it is considered desirable to use some type of wall anchor to prevent the beam from moving longitudinally with respect to the wall. The usual anchor consists of a bent steel bar called a government anchor passing through the web of the beam and running parallel to the wall. Details of these anchors and their sizes are given in Fig. 12-13 of the LRFD Manual. Occasionally, clip angles attached to the web are used instead of government anchors. Should longitudinal loads of considerable size be anticipated, regular vertical anchor bolts may be used at the beam ends.

If we were to check to see if the flange thickness alone is sufficient, we would have $\left(\text{with } n = \dfrac{b_f}{2} - k \right)$.

$$t = \sqrt{\frac{2.22\,R_u n^2}{A_1 F_y}} = \sqrt{\frac{(2.22)(120)(2.94)^2}{(8 \times 7.635)(36)}} = 1.02 \text{ in.} > t_f \text{ of W18} \times 71$$

$$\therefore \text{ Flanges not sufficient alone.}$$

10-9 COMPUTER EXAMPLE

A beam base plate is designed in Example 10-11 using the enclosed computer program. Here it is necessary as it was for column base plates to assume an initial plate size. The computer will print out the area and thickness required. You can then revise the initial dimensions and run the program again.

■ EXAMPLE 10-11
Repeat Example 10-10 using the computer program INSTEP.

Solution.

INPUT

STEEL & BEAM PARAMETERS

B	Type Steel: A36
C	Beam: W18×71
D	Design Load: 120

TRIAL BEARING PLATE SIZE

G	B = 10
H	N = 8
I	t = 1.125
J	Location: End

CONCRETE STRENGTH

M	$f_c' = 3$
N	$b_1 = 0$

OUTPUT
COMPUTE ACTUAL AREAS

$A_1 = 10(8) = 80$
$A_{1_{min}} = 120/0.51(3) = 78.4314$ OK
$B_1 = 2(0)+10 = 10$
$N_1 = 2(0)+8 = 8$
$A_2 = 10(8) = 80$

ALLOWABLE BEARING FORCE

$P_p = 0.85(3)80 = 204$

$P_p = 0.85(3)80 \sqrt{\dfrac{80}{80}} = 204$

$0.6P_p = 122.4 > 120$ OK

CHECK WEB YIELDING

$R_n = [2.5(1.5)+8](36).495 = 209.385$
$1.0R_n = 209.385 > 120$ OK

CHECK WEB CRIPPLING

$R_n = 207.267$
$0.75R_n = 155.45 > 120$ OK

COMPUTE MINIMUM THICKNESS

$t = \sqrt{\dfrac{2.22(120)(3.5)^2}{80(36)}} = 1.06448$

Specified Thickness is Adequate   ■

PROBLEMS

10-1 to 10-5. *Considering moment only and assuming full lateral support for the compression flanges, select the lightest sections available using 50 ksi steel. The loads shown include the effect of the beam weights. Use elastic analysis, factored loads, and the 0.9 rule.*

$w_u = 3.6$ klf

|← 40 ft →|← 40 ft →|← 30 ft →|

Problem 10-1   (*Ans.* W24 × 55)

Problem 10-2

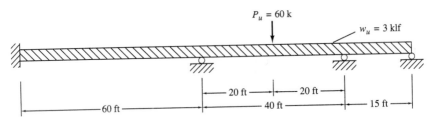

Problem 10-3   (*Ans.* W24 × 84)

Problem 10-4

Problem 10-5   (*Ans.* W24 × 62)

10-6. Repeat Prob. 10-1 using plastic analysis.

10-7. Repeat Prob. 10-3 using plastic analysis. (*Ans.* W24 × 76, W24 × 68 slightly overstressed)

10-8. Repeat Prob. 10-5 using plastic analysis.

10-9. Three methods of supporting a roof are shown in the accompanying illustration. Using an elastic analysis with factored loads, $F_y = 50$ ksi, and assuming full lateral support in each case, select the lightest section if a dead uniform service load (not including beam weight) of 1 klf and a live uniform service load of 1.2 klf is to be supported. Consider moment only.

(a)   (*Ans.* W21 × 62)

(b)   (*Ans.* W18 × 50)

(c)   (*Ans.* W18 × 50)

Problem 10-9

10-10. The welded section shown, made from A36 steel, has full lateral support for its compression flange and is bent about its major axis. If C_b is 1.0, determine its design moment and shear strengths.

Problem 10-10

10-11 to 10-13. *Using $F_y = 50$ ksi select the lightest available section for the span and loading shown. Consider moment and shear only and neglect beam weight in all calculations. The member is assumed to have full lateral bracing.*

Problem 10-11 (*Ans.* W30 × 90)

Problem 10-12

10-13. Repeat Prob. 10-11 using $F_y = 36$ ksi. (*Ans.* W33 × 118)

10-14. If a fully braced W14 × 34 section consisting of 50 ksi steel is used for a simple span of 6 ft 6 in., determine the maximum uniform load w_u that it can support in addition to its own factored weight. Use elastic analysis and consider shear and moment only.

10-15. A W16 × 40 consisting of 50 ksi steel is used as a simple beam for a span of 8 ft. If it has full lateral support determine the maximum uniform load w_u which it can support in addition to its own factored weight. Use an elastic analysis and consider shear and moment only. (*Ans.* 32.95 klf)

10-16. A fully braced W36 × 245 consisting of A36 steel is used as a simple beam for a span of 16 ft. Considering moment and shear only determine the maximum uniform load w_u it can support in addition to its own factored weight using an elastic analysis.

10-17. A 30-ft simply supported beam is to support a moving concentrated load $P_u = 80$ k. Using 50 ksi steel, select the most economical section considering moment and shear only. Use an elastic analysis and neglect beam weight. (*Ans.* W24 × 68)

10-18. A 40-ft simple beam that supports a service concentrated load $P_L = 30$ k at midspan is laterally unbraced except at its ends and center line. If the maximum permissible center line deflection under service loads equals 1/1000 of the span, select the most economical W section of 50 ksi steel considering moment, shear, and deflection. Neglect beam weight.

10-19. Design a beam for a 24-ft simple span to support the working uniform loads $w_D = 1.2$ klf (includes beam weight) and $w_L = 2.8$ klf. The maximum permissible deflection under working loads is $1/1200$ of the span. Use 50 ksi steel and consider moment, shear, and deflection. The beam is to be braced laterally for its full length. (*Ans.* W30 × 108)

10-20. Select the lightest available W section of A36 steel for the span and service loads shown. The beam will have full lateral support for its compression flange. Its maximum service load center-line deflection may not exceed $1/1500$ of the span under working loads. Consider moment, shear, and deflection only.

$w_D = 2$ klf (not including beam weight)
$w_L = 4$ klf

28 ft

Problem 10-20

10-21. Select the lightest section available if $F_y = 50$ ksi for the span and working loads shown if the section is to be fully braced laterally and have a maximum factored load deflection of $1/800$ of its span length. Neglect beam weight in all calculations. Consider moment, shear and deflection. (*Ans.* W44 × 230)

$P_L = 12$ k $P_L = 25$ k

10 ft   10 ft

Problem 10-21

10-22. If the maximum permissible service load deflection for the fully braced beam shown is $1/1200$ of the span, select the lightest available section using 50 ksi steel. Consider moment, shear, and deflections. Working loads are shown.

$P_L = 20$ k $P_L = 10$ k

$w_D = 1$ klf
(not including beam weight)

5 ft   10 ft

Problem 10-22

10-23. Select the lightest available W section (50 ksi steel) for the working live loads and span shown in the accompanying illustration. Lateral support is provided only at the 12-ft points and a maximum deflection (under working loads) equal to 1/1500 of the 24-ft span is permitted. Neglect beam weight. Consider moment, shear, and deflections. (*Ans.* W30 × 108)

Problem 10-23

10-24. A simply supported W24 × 146 consisting of A36 steel is supporting a 300-k service live load as shown in the figure. If the length of bearing at the left support is 8 in. and at the concentrated load is 12 in., check the beam for shear, web yielding, and web crippling.

Problem 10-24

10-25. A 32-ft beam with full lateral support for its compression flange is supporting a moving service concentrated load of 60 k. Using 50 ksi steel, select the lightest section available for moment. Then check to see if the section is satisfactory in shear and compute the minimum length of bearing required at the supports from the standpoint of web yielding and web crippling. (*Ans.* W24 × 84)

10-26. Select the lightest available W30 section consisting of 50 ksi steel to resist a gravity moment $M_{ux} = 500$ ft-k and a lateral bending moment of $M_{uy} = 200$ ft-k. The section is assumed to have full lateral support.

10-27. The 20-ft simple beam shown has full lateral support for its compression flange and consists of A36 steel. The beam supports a gravity service

20°

Problem 10-27

dead load of 1.2 klf (includes beam weight) and a gravity live load of 3.4 klf. The loads are assumed to act through the c.g. of the section. Select the lightest available W33 section. (*Ans.* W33 × 130)

10-28. Repeat Prob. 10-27 if F_y = 50 ksi and if the gravity live load is 4.8 klf.

10-29. Design a steel bearing plate from A36 steel for a W21 × 68 beam with an end reaction R_u = 100 k. The beam will bear on a reinforced-concrete wall with f'_c = 4 ksi. In a direction perpendicular to the wall the bearing plate may not be longer than 8 in. (*One ans.* PL 1 × 8 × 0 ft 9 in.)

10-30. Design a steel bearing plate of 50 ksi steel for a W30 × 116 beam supported by a reinforced-concrete wall with f'_c = 3 ksi. The maximum beam reaction R_u is 170 k. Assume the width of the plate perpendicular to the wall is 8 in.

10-31. Repeat Prob. 10-30 using a steel with F_y = 36 ksi. (*One ans.* $1\frac{3}{4}$ × 8 × 1 ft 2 in.)

10-32 to 10-33. *Solve the problems shown using the computer program INSTEP.*

10-32. Repeat Prob. 10-29.

10-33. Repeat Prob. 10-31. (*One ans.* $1\frac{3}{4}$ × 8 × 1 ft 2 in.)

Chapter 11

Bending and Axial Force

11-1 OCCURRENCE

Structural members that are subjected to a combination of bending and axial force are far more common than the student may realize. This section is devoted to listing a few of the more obvious cases. Columns that are part of a steel building frame must nearly always resist sizable bending moments in addition to the usual compressive loads. It is almost impossible to erect and center loads exactly on columns, even in a testing lab, and in an actual building the student can see that it is even more impossible. Even if building loads could be perfectly centered at one time they would not stay in one place. Furthermore, columns may be initially crooked or have other flaws with the result that lateral bending is produced. The beams framing into columns are quite commonly supported with framing angles or brackets on the sides of the columns. These eccentrically applied loads produce moments. Wind and other lateral loads cause columns to bend laterally, and the columns in rigid frame buildings are subjected to moments even when the frame is supporting gravity loads alone. The members of bridge portals must resist combined forces as do building columns. Among the causes of the combined forces are heavy lateral wind loads, vertical traffic loads whether symmetrical or not, and the centrifugal effect of traffic on curved bridges.

The previous experience of the student has probably been to assume truss members axially loaded only. Purlins for roof trusses, however, are frequently placed in between truss joints, causing the top chords to bend. Similarly, the bottom chords may be bent by the hanging of light fixtures, ductwork, and other items between the truss joints. All horizontal and inclined truss members have moments caused by their own weights, while all truss members—whether vertical or not—are subjected to secondary bending forces. Secondary forces are

Inryco Building, Chicago, Ill. (Courtesy of I. R. Construction Products Co.)

developed because the members are not connected with frictionless pins, as assumed in the usual analysis, the member centers of gravities or those of their connectors do not exactly coincide at the joints, etc.

Moments in tension members are not as serious as those in compression members because tension tends to reduce lateral deflections while compression increases them. Increased lateral deflection in turn results in larger moments, which result in larger lateral deflections, etc. It is hoped that members in such situations are quite stiff so as to keep the additional lateral deflections from becoming excessive.

11-2 MEMBERS SUBJECT TO BENDING AND AXIAL TENSION

A few types of members subject to both bending and axial tension are shown in Fig. 11-1.

In Section H1 of the LRFD Specification the following interaction equations are given for symmetric shapes subjected simultaneously to bending and axial tensile forces. These equations are also applicable to members subjected to bending and axial compression forces as will be described in Sections 11-7 to 11-9.

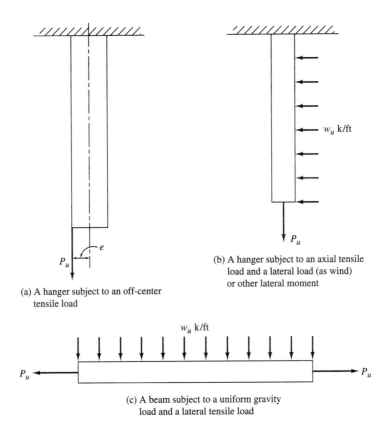

(a) A hanger subject to an off-center
tensile load

(b) A hanger subject to an axial tensile
load and a lateral load (as wind)
or other lateral moment

(c) A beam subject to a uniform gravity
load and a lateral tensile load

Figure 11-1 Some members subject to bending and axial tension.

If $\dfrac{P_u}{\phi_t P_n} \geq 0.2$

$$\frac{P_u}{\phi_t P_n} + \frac{8}{9}\left(\frac{M_{ux}}{\phi_b M_{nx}} + \frac{M_{uy}}{\phi_b M_{ny}}\right) \leq 1.0 \qquad \text{(LRFD Equation H1-1a)}$$

If $\dfrac{P_u}{\phi_t P_n} < 0.2$

$$\frac{P_u}{2\phi_t P_n} + \left(\frac{M_{ux}}{\phi_b M_{nx}} + \frac{M_{uy}}{\phi_b M_{ny}}\right) \leq 1.0 \qquad \text{(LRFD Equation H1-1b)}$$

The terms in these equations have previously been defined where P_u and M_u are the required tensile and flexural strengths, P_n and M_n are the nominal tensile and flexural strengths, and the resistance factors ϕ_t and ϕ_b are determined as in previous chapters. Usually only a first-order analysis (that is, not including any secondary forces as described in the next section) is made for members subject to bending and axial tension. However, the analyst, may—in fact, is encouraged to—make second-order analyses for these members and use the results in his or her designs. Examples 11-1 and 11-2 illustrate the use of the interaction equations to review members subjected simultaneously to bending and axial tension, while Example 11-3 illustrates the design of such a member.

■ EXAMPLE 11-1

A W12 × 35 tension member with no holes consisting of 50 ksi steel is subjected to a factored tensile force P_u of 80 k and a factored bending moment M_{uy} of 35 ft-k. Is the member satisfactory if $L_b < L_p$?

Solution.   Using a W12 × 35 ($A = 10.3$ in.^{3}, $Z_y = 11.5$ in.^{3})

$$\phi_t P_n = \phi_t F_y A_g = (0.9)(50)(10.3) = 463.5 \text{ k}$$

$$\frac{P_u}{\phi_t P_n} = \frac{80}{463.5} = 0.173 < 0.2$$

$$\therefore \text{ Use LRFD Equation H1-1b}$$

$$\phi_b M_{ny} = \phi_b F_y Z_y = \frac{(0.9)(50)(11.5)}{12} = 43.12 \text{ ft-k}$$

$$\frac{P_u}{2\phi_t P_n} + \left(\frac{M_{ux}}{\phi_b M_{nx}} + \frac{M_{uy}}{\phi_b M_{ny}} \right) = \frac{80}{(2)(463.5)} + \left(0 + \frac{35}{43.12} \right)$$

$$= 0.898 < 1.0 \qquad \text{(OK)} \quad ■$$

■ EXAMPLE 11-2

A W10 × 30 tensile member with no holes consisting of 50 ksi steel and with $L_b = 12.0$ ft is subjected to a factored tensile force $P_u = 120$ k and to the factored moments $M_{ux} = 90$ ft-k and $M_{uy} = 0$. If $C_b = 1.0$, is the member satisfactory?

Solution.   Using a W10 × 30 ($A = 8.84$ in.^{2}, $L_p = 4.8$ ft, $L_r = 14.5$ ft)

$$\phi_t P_n = (0.9)(50)(8.84) = 397.8 \text{ k}$$

$$\frac{P_u}{\phi_t P_n} = \frac{120}{397.8} = 0.302 > 0.2$$

$$\therefore \text{ Use LRFD Equation H1-1a}$$

Noting $L_b > L_p$, from Load Factor Design Selection Table

$$\phi_b M_p = 137 \text{ ft-k}$$

$$\phi_b M_r = 97.2 \text{ ft-k}$$

$$\text{BF} = 4.13$$

$$\phi_b M_n = C_b[\phi_b M_p - \text{BF}(L_b - L_p)]$$

$$= 1.0[137 - 4.13(12.0 - 4.8)] = 107.26 \text{ ft-k}$$

$$\frac{P_u}{\phi_t P_n} + \frac{8}{9} \left(\frac{M_{ux}}{\phi_b M_{nx}} + \frac{M_{uy}}{\phi_b M_{ny}} \right) = \frac{120}{397.8} + \frac{8}{9} \left(\frac{90}{107.26} + 0 \right)$$

$$= 1.048 > 1.0 \qquad \text{(NG)} \quad ■$$

■ EXAMPLE 11-3

Select an 8-ft long W10 to support a factored tensile load of 110 k applied with an eccentricity of 5 in. with respect to the x axis. The A36 member is to be welded and braced laterally only at its supports. Assume C_b is equal to 1.0.

Solution. Try a W10 × 26 ($A = 7.61$ in.^{2}, $L_p = 5.7$ ft, $L_r = 18.5$ ft, $M_{ux} = (5)(110) = 550$ in.-k $= 45.8$ ft-k, $\phi_t P_n = (0.9)(36)(7.61) = 246.6$ k.

$$\frac{P_u}{\phi_t P_n} = \frac{110}{246.6} = 0.446 > 0.2$$

∴ Use LRFD Equation H1-1a

Noting that $L_b > L_p < L_r$, we determine $\phi_b M_n$ as follows, with reference to the Load Factor Design Selection Table in the Manual.

$$\phi_b M_p = 84.5 \text{ ft-k}$$
$$\phi_b M_r = 54.4 \text{ ft-k}$$
$$\text{BF} = 2.34$$
$$\phi_b M_{nx} = C_b[\phi_b M_p - \text{BF}(L_b - L_p)]$$
$$= 1.0[84.5 - (2.34)(8.0 - 5.7)] = 79.1 \text{ ft-k}$$
$$\frac{P_u}{\phi_t P_n} + \frac{8}{9}\left(\frac{M_{ux}}{\phi_b M_{nx}} + \frac{M_{uy}}{\phi_b M_{ny}}\right) = \frac{110}{246.6} + \frac{8}{9}\left(\frac{45.8}{79.1} + 0\right)$$
$$= 0.961 < 1.0 \qquad \text{(OK)}$$

Subsequent check of W10 × 22 shows it will not do.
$$\underline{\underline{\text{Use W10 × 26}}}$$ ■

11-3 COMPUTER EXAMPLES FOR MEMBERS SUBJECT TO BENDING AND AXIAL TENSION

The computer program INSTEP is used in Examples 11-4 and 11-5 to repeat Examples 11-2 and 11-3. For convenience the solutions of the problems use a few different but obvious abbreviations such as P_{na} for the design axial load strength $\phi_t P_n$ of a member, and M_{na} which is the design bending strength of the member $\phi_b M_n$.

■ EXAMPLE 11-4

Repeat Example 11-2 using INSTEP.

Solution.

INPUT

A Type Steel: A572 Grade 50
B Trial Size: W10×30

BEAM/COLUMN DESIGN PARAMETERS

D L = 144 inches
E L_b = 144 inches
F C_b = 1
G C_M = 1
H B_2 = 0
I A_e = 8.84 sq.in.
J K_b = 1

FACTORED DESIGN LOADS

L P_u = 120 kips
M $M_{nt/u}$ = 1080 kip-inches
N M_{1t} = 0 kip-inches

OUTPUT

From Axial Tension Considerations
  P_{na} = 397.8 kips

From Flexural Considerations
  M_{na} = 1292.61 kip-inches

Check the Interaction Equation
$$UC = \frac{120}{397.8} + \left[\frac{8}{9}\right]\left[\frac{1080}{1292.61}\right]$$
  UC = 1.04434

This Member is NOT Acceptable

■ EXAMPLE 11-5
Repeat Example 11-3 using INSTEP.

Solution.

INPUT

A Type Steel: A36
B Trial Size: W10

BEAM/COLUMN DESIGN PARAMETERS

D L = 96 inches
E L_b = 96 inches
F C_b = 1
G C_M = 1
H B_2 = 0
I A_e = 0 sq.in.
J K_b = 1

FACTORED DESIGN LOADS

- L P_u = 110 kips
- M $M_{nt/u}$ = 550 kip-inches
- N M_{1t} = 0 kip-inches

OUTPUT

The lightest acceptable W10 shape
is being determined. The effective
net area is assumed to be 90% of
the gross area when finding the
lighest section. The adequacy of
the suggested section must be
verified by the design engineer!

Suggested Trial Size: W10×26

P_{na} = 246.564 kips
M_{na} = 948.524 kip-inches
Unity Check = .961552

11-4 FIRST-ORDER AND SECOND-ORDER MOMENTS FOR MEMBERS SUBJECT TO AXIAL COMPRESSION AND BENDING

When a beam-column is subjected to moment along its unbraced length it will be displaced laterally in the plane of bending. The result will be an increased or secondary moment equal to the axial compression load times the lateral displacement or eccentricity. In Fig. 11-2 we can see that the member moment is increased by an amount $P_u\delta$. This moment will cause additional lateral deflection, which will in turn cause a larger column moment, which will cause a larger lateral deflection, and so on until equilibrium is reached.

If a frame is subject to sidesway where the ends of the columns can move laterally with respect to each other, additional secondary moments will result. In Fig. 11-3 the secondary moment produced due to sidesway is equal to $P_u\Delta$.

The moment M_2 is assumed by the LRFD Specification to equal M_{lt} (which is the moment due to the lateral loads) plus the moment due to $P_u\Delta$.

The required total flexural strength of a member must equal at least the sum of the first-order and second-order moments. Several methods are available for determining this required strength, ranging from very simple approximations to very rigorous procedures.

The LRFD Specification C.1 states that we can either (1) make a second-order analysis to determine the maximum factored load strength or (2) use a first-order elastic analysis and amplify the moments obtained with some amplification factors called B_1 and B_2 and described in the paragraphs to follow.

Should the designer make a second-order analysis, he or she should realize that it must account for the interaction of the factored load effects. That is, we must consider combinations of factored loads acting at the same time. We cannot correctly make separate analyses and superimpose the results.

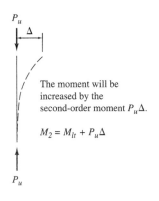

Figure 11-2 Moment amplification of a column that is braced against sidesway.

Figure 11-3 Column in an unbraced frame.

In this chapter the author presents the approximate method of analysis given in the LRFD Manual. We will make two first-order analyses—one an analysis where the frame is assumed to be braced so that it cannot sway. We will call these moments M_{nt} and will multiply them by a magnification factor called B_1 to account for the P-δ effect (see Fig. 11-2). Then we will analyze the frame again, allowing it to sway. We will call these moments M_{lt} and will multiply them by a magnification factor called B_2 to account for the P-Δ effect (see Fig. 11-3). The final moment in a particular member will equal

$$M_u = B_1 M_{nt} + B_2 M_{lt} \qquad \text{(LRFD Equation C1-1)}$$

Instead of using the LRFD empirical procedure described here, the designer may and is encouraged to use a theoretical second-order analysis, provided he or she meets certain requirements of Sections C1 and C2 of the Specification. These requirements pertain to axial deformations, maximum axial forces permitted in members, bracing, K factors, and so on.

11-5 MAGNIFICATION FACTORS

The magnification factors are B_1 and B_2. With B_1 the analyst attempts to estimate the $P_u\delta$ effect for a column whether the frame is braced or unbraced against sidesway. With B_2 he or she attempts to estimate the $P_u\Delta$ effect in unbraced frames.

These factors are theoretically applicable when the connections are fully restrained or when they are completely unrestrained. The LRFD Manual indicates that the determination of secondary moments in between these two classifications for partially restrained moment connections is beyond the scope of their specification. The terms *fully restrained* and *partially restrained* are discussed at length in Chapter 15.

In the expression for B_1 that follows, C_m is a term that will be defined in the next section, P_u is the required axial strength of the member, and P_{e1} is the member's Euler buckling strength equal to $A_g F_y/\lambda_c^2$ for a braced frame. In this expression $\lambda_c = (KL/r\pi)\sqrt{F_y/E}$ as given by LRFD Equation E2-4. I and KL are both taken in the plane of bending determined in accordance with LRFD Specification C2.1 for a *braced frame*.

Substituting this value of λ_c into the expression for P_{e1} and replacing A_g with I/r^2, it becomes

$$P_{e1} = \frac{A_g F_y}{\lambda_c^2} = \frac{\dfrac{I}{r^2}F_y}{\left(\dfrac{kL}{2\pi}\sqrt{\dfrac{F_y}{E}}\right)^2}$$

$$P_{e1} = \frac{\pi^2 EI}{(KL)^2}$$

In a similar fashion P_{e2} is the Euler buckling strength $\pi^2 EI/(KL)^2$ with K determined in the plane of bending for the *unbraced frame*.

In Table 8 of the LRFD Specification Appendix the value of P_{e2}/A_g can be obtained directly. Multiplying it by A_g will provide the value of P_{e2}. Should we compute a B_1 or B_2 value that is less than 1.0 we must use 1.0. (If we are supposedly magnifying moments, we don't want to multiply them by a number less than 1.0.)

The expression to follow for B_1 was derived for a member braced against sidesway. It will be used only to magnify the M_{nt} moments (those moments computed assuming there is no lateral translation of the frame).

$$B_1 = \frac{C_m}{1 - \dfrac{P_u}{P_{e1}}} \geq 1.0 \qquad\qquad \text{(LRFD Equation C1-5)}$$

The horizontal deflection of a multistory building due to wind or seismic load is called *drift*. It is represented by Δ in Figs. 11-3 and 11-4. In the formula

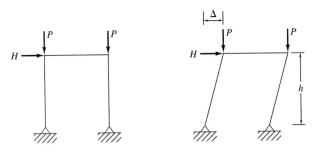

Figure 11-4 Drift in a building frame.

to follow the term Δ_{oh} is used. It represents the lateral deflection of the story in question calculated with service loads with respect to the story below. Drift is measured with the so-called *drift index* Δ_{oh}/h where Δ_{oh} is the lateral deflection and h is the height or distance to the lower level. For the comfort of occupants of a building the index usually is limited at working or service loads to a value between 0.0015 and 0.0030, and at factored loads to about 0.0040.

The designer may use either of the following expressions for B_2. In the first of these expressions ΣP_u represents the total required axial strength of all the columns on the floor in question, Δ_{oh}/L represents the story drift index, and ΣH is the sum of all the story horizontal forces producing Δ_{oh}. The value of P_{e2} is defined as before for P_{e1}, except that the effective length factor K is determined in the plane of bending for *an unbraced frame.*

The values shown for ΣP_u and ΣP_{e2} are for all of the columns on the floor in question. This is considered to be necessary because the B_2 term is used to magnify column moments for sidesway. For sidesway to occur in a particular column, it is necessary for all of the columns on the floor to sway simultaneously. The ΣH value used in the first of the B_2 expressions represents the sum of the lateral loads acting above the floor being considered. To compute the ratio $\Delta_{oh}/\Sigma H$ we may use either factored or unfactored loads.

$$B_2 = \frac{1}{1 - \Sigma P_u(\Delta_{oh}/\Sigma HL)} \qquad \text{(LRFD Equation C1-4)}$$

or

$$B_2 = \frac{1}{1 - \Sigma P_u/\Sigma P_{e2}} \qquad \text{(LRFD Equation C1-5)}$$

We must remember that the amplification factor B_2 is only applicable to moments caused by forces that cause sidesway and is to be computed for an entire story. (Of course, if you want to be conservative you can multiply B_2 times the sum of the no-sway and the sway moments—i.e., M_{nt} and M_{lt}, but that's probably overdoing it.) To use the B_2 value given by LRFD Equation C1-5 we must select initial member sizes (that is so we can compute a value for P_{e2}). Should you be designing a building frame where you are limiting the drift index Δ_{oh}/L to a certain maximum value, you can compute B_2 with LRFD Equation C1-4 before you design the member. In this way you can set a drift limit in advance so that secondary bending is insignificant.

To calculate the values of ΣP_u and ΣP_e some designers will calculate the values for the columns in the one frame under consideration (or for that single line of columns perpendicular to the wind). This, however, is a rather bad practice unless all the other frames are exactly the same as the one under study.

Of the two expressions given for B_2, the first one (C1-4), which involves a drift index, is better suited for design office practice. If we assume drifts as large as these values, we are probably being quite conservative. That is, the real structures may not drift this much.

11-6 MOMENT MODIFICATION OR C_m FACTORS

In Sections 11-4 and 11-5 the subject of moment magnification due to lateral deflections was introduced, and the factors B_1 and B_2 were presented with which the moment increases could be estimated. In the expression for B_1 a term C_m, called the *modification factor,* was included. The magnification factor B_1 was developed for the largest possible lateral displacement. On many occasions the displacement is not that large and B_1 overmagnifies the column moment. As a result, the moment needs to be reduced or modified with the C_m factor. You can see that this is the case in Fig. 11-5, where we have a column bent in single curvature with equal end moments such that the column bends laterally by an amount δ at mid-depth. The maximum total moment occuring in the column will clearly equal M plus the increased moment $P_u\delta$. As a result no modification is required and $C_m = 1.0$.

An entirely different situation is considered in Fig. 11-6 where the end moments tend to bend the member in reverse curvature. The initial maximum moment occurs at one of the ends, and we shouldn't increase it by a value $P_u\delta$ that occurs some distance out in the column because we will be overdoing the moment magnification. The purpose of the modification factor is to modify or reduce the magnified moment when the variation of the moments in the column is such that B_1 is made too large. If we didn't use a modification factor we would end up with the same total moments in the columns of both Figs. 11-5 and 11-6, assuming the same dimensions and initial moments and load.

Modification factors are based on the rotational restraint at the member ends and on the moment gradients in the members. The LRFD Specification (C1) includes two categories of C_m as described in the next few paragraphs.

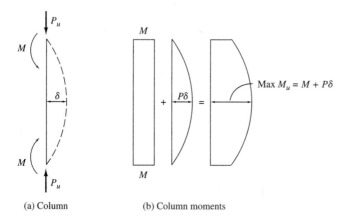

(a) Column   (b) Column moments

Figure 11-5 Moment magnification for column bent in single curvature.

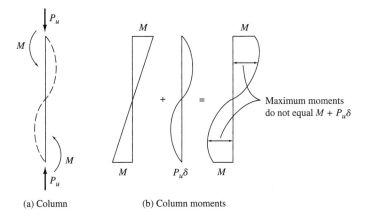

(a) Column   (b) Column moments

Figure 11-6 Moment magnification for column bent in double curvature.

In Category 1 the members are prevented from joint translation or sidesway and they are not subject to transverse loading between their ends. For such members the modification factor is based on an elastic first-order analysis.

$$C_m = 0.6 - 0.4 \frac{M_1}{M_2} \qquad \text{(LRFD Equation C1-3)}$$

In this expression M_1/M_2 is the ratio of the smaller moment to the larger moment at the ends of the unbraced length in the plane of bending under consideration. The ratio is negative if the moments cause the member to bend in single curvature and positive if they bend the members in reverse or double curvature. As previously described, a member in single curvature has larger lateral deflections than a member bent in reverse curvature. With larger lateral deflections the moments due to the axial loads will be larger.

Category 2 applies to members that are subjected to transverse loading between the joints in the plane of loading. The compression chord of a truss with a purlin load between its joints is a typical example of this category. The LRFD Specification states that the value of C_m is to be determined either by rational analysis or by using one of the values to follow.

(a) For members with restrained ends $C_m = 0.85$.
(b) For members with unrestrained ends $C_m = 1.0$.

If either one of these two values is used, or if we interpolate between them where we feel our conditions warrant, the results will probably be quite reasonable.

Instead of using these values for transversely loaded members, the values of C_m for Category 2 may be determined for various end conditions and loads by the values given in Table 11-1, which is a reproduction of Table C-C1.1 of the Commentary on the LRFD Specification. In the expressions given in the table P_u is the factored column axial load and P_{el} is the elastic buckling load for a braced column for the axis about which bending is being considered.

$$P_{el} = \frac{\pi^2 EI}{(KL)^2}$$

In Table 11-1 note that some members have restrained ends and some do not.

Sample values of C_m are calculated for four beam columns and shown in Fig. 11-7.

TABLE 11-1   MODIFICATION FACTORS FOR BEAM
COLUMNS SUBJECT TO TRANSVERSE
LOADS BETWEEN JOINTS.

Case	ψ	C_m
	0	1.0
	−0.4	$1 - 0.4\dfrac{P_u}{P_{el}}$
	−0.4	$1 - 0.4\dfrac{P_u}{P_{el}}$
	−0.2	$1 - 0.2\dfrac{P_u}{P_{el}}$
	−0.3	$1 - 0.3\dfrac{P_u}{P_{el}}$
	−0.2	$1 - 0.2\dfrac{P_u}{P_{el}}$

(a) No sidesway and no transverse loading.
Moments bend member in single curvature

$$C_m = 0.6 - (0.4)\left(-\frac{40}{50}\right) = 0.92$$

(b) No sidesway and no transverse loading.
Moments bend member in reverse curvature

$$C_m = 0.6 - 0.4\left(+\frac{60}{80}\right) = 0.30$$

(c) Member has restrained ends and transverse loading
and is bent about x axis.
$C_m = 0.85$ or can be determined from Table 11-1
(LRFD Table C-C1.1) as below

$$P_{e1} = \frac{\pi^2 EI}{(KL)_x^2} = \frac{(\pi^2)(29 \times 10^3)(272)}{(12 \times 20)^2}$$
$$= 1351 \ k$$
$$C_m = 1 - 0.4\left(\frac{280}{1351}\right) = 0.92$$

(d) Member has unrestrained ends and transverse loading
and is bent about x axis.
$C_m = 1.0$ or can be determined from Table 11-1
(LFRD Table C-C1.1)

$$P_{e1} = \frac{(\pi^2)(29 \times 10^3)(475)}{(12 \times 20)^2} = 2360 \ k$$
$$C_m = 1 - 0.2\left(\frac{200}{2360}\right) = 0.98$$

Figure 11-7 Example modification or C_m factors.

11-7 REVIEW OF BEAM-COLUMNS IN BRACED FRAMES

The same interaction equations are used for members subject to axial compression and bending as were used for members subject to axial tension and bending. However, some of the terms involved in the equations are defined somewhat differently. For instance, P_u and P_n refer to compressive forces rather than tensile forces, ϕ_c is 0.85 for axial compression, and ϕ_b is 0.9 for bending.

To analyze a particular beam column or a member subject to both bending and axial compression we need to make both a first-order and a second-order analysis to obtain the bending moments. The first-order moment is usually obtained by making an elastic analysis and consists of the moments M_{nt} (these are the moments in beam columns caused by gravity loads) and the moments M_{lt} (these are the moments in beam columns due to the lateral loads, that is, due to lateral translation).

Theoretically, if both the loads and frame are symmetrical, M_{lt} will be zero. Similarly, if the frame is braced, M_{lt} will be zero. For practical purposes you can have lateral deflections in taller buildings with symmetrical dimensions and loads.

Examples 11-6 to 11-8 illustrate the application of the interaction equations to beam-columns that are members of braced frames. Thus, only B_1 will be computed, as B_2 is not applicable. *It is to be remembered that C_m was developed for braced frames and thus must be used in these three examples for calculating B_1.*

■ **EXAMPLE 11-6**
A 12-ft W12 × 96 (A36 steel) is used as a beam-column in a braced frame. It is bent in single curvature with equal and opposite end moments and is not subjected to intermediate transverse loads. Is the section satisfactory if $P_u = 500$ k and first-order moment $M_{ntx} = 125$ ft-k?

Solution. Using a W12 × 96 ($A = 28.2$ in.^{2}, $I_x = 833$ in.^{4}, $L_p = 12.9$ ft, $\phi_b M_b = 397$ ft-k)

For a braced frame $K = 1.0$

$\therefore \underline{K_x L_x = K_y L_y = (1.0)(12) = 12 \text{ ft}}$

$\phi_c P_n = 770$ k from LRFD column tables

$\dfrac{P_u}{\phi_c P_n} = \dfrac{500}{770} = 0.649 > 0.2$

$\therefore$ Must use LRFD Equation H1-1a

As the only moment is M_{ntx} there is no lateral translation of the frame, that is, $M_{lt} = 0$.

$$\therefore M_{ux} = B_1 M_{ntx}$$

$$C_m = 0.6 - 0.4\left(\frac{M_1}{M_2}\right) = 0.6 - 0.4\left(-\frac{125}{125}\right) = 1.0$$

$$P_{e1} = \frac{\pi^2 E I_x}{(K_x L_x)^2} = \frac{(\pi^2)(29 \times 10^3)(833)}{(12 \times 12)^2} = 11,498 \text{ k}$$

Notice this value can be determined from the bottom of the column tables where $P_e(KL^2)/10^4$ is given for both the x and y axes. (*Note:* the value of P_e can also be determined from Table 8 in Part 6 of the Manual. First $K_x L_x/r_x$ is computed and the P_e/A_g is picked from the table. Finally, $P_e = A_g$ times the table value.)

$$B_1 = \frac{C_m}{1 - \dfrac{P_u}{P_{e1}}} = \frac{1.0}{1 - \dfrac{500}{11,498}} = 1.045$$

$$M_{ux} = (1.045)(125) = 130.6 \text{ ft-k}$$

Applying LRFD Equation H1-1a

Since $L_b = 12 \text{ ft} < L_p = 12.9 \text{ ft}$, $\phi_b M_{nx} = \phi_b M_p = 397 \text{ ft-k}$

$$\frac{P_u}{\phi_c P_n} + \frac{8}{9}\left(\frac{M_{ux}}{\phi_b M_{nx}} + \frac{M_{uy}}{\phi_b M_{ny}}\right)$$

$$= \frac{500}{770} + \frac{8}{9}\left(\frac{130.6}{397} + 0\right) = 0.942 < 1.0$$

$\therefore$ Section is satisfactory.   ■

■ EXAMPLE 11-7

A 14-ft W14 × 120 (A36 steel) is used as a beam-column in a braced frame. It is bent in single curvature with equal and opposite moments. Its ends are restrained and it is not subjected to intermediate transverse loads. Is the section satisfactory if $P_u = 180 \text{ k}$ and if it has the first-order moments $M_{ntx} = 150 \text{ ft-k}$ and $M_{nty} = 100 \text{ ft-k}$?

Solution.   Using a W14 × 120 ($A = 35.3 \text{ in.}^2$, $L_p = 15.6 \text{ ft}$, $I_x = 1380 \text{ in.}^4$, $I_y = 495 \text{ in.}^4$, $Z_x = 212 \text{ in.}^3$, $Z_y = 102 \text{ in.}^3$)

For a braced frame $K_x L_x = K_y L_y = (1.0)(14) = 14 \text{ ft}$

$\phi_c P_n = 971 \text{ k}$ from LRFD column tables

$$\frac{P_u}{\phi_c P_n} = \frac{180}{971} = 0.185 < 0.2$$

$\therefore$ Must use LRFD Equation H1-1b

As we have only the moments M_{ntx} and M_{nty} there is no lateral translation of the frame and thus $M_{ltx} = M_{lty} = 0$.

$$M_{ux} = B_1 M_{ntx} \quad \text{and} \quad M_{uy} = B_1 M_{nty}$$

$$C_m = 0.6 - 0.4\left(-\frac{150}{150}\right) = 1.0$$

$$P_{elx} \text{ from column table} = \frac{(10)^4(39,300)}{(12 \times 14)^2} = 13,924 \text{ k}$$

$$B_{1x} = \frac{C_m}{1 - \dfrac{P_u}{P_{elx}}} = \frac{1.0}{1 - \dfrac{180}{13,924}} = 1.013$$

$$M_{ux} = (1.013)(150) = 152 \text{ ft-k}$$

$$P_{ely} \text{ from column table} = \frac{(10)^4(14,100)}{(12 \times 14)^2} = 4996 \text{ k}$$

$$B_{1y} = \frac{C_m}{1 - \dfrac{P_u}{P_{ely}}} = \frac{1.0}{1 - \dfrac{180}{4996}} = 1.037$$

$$M_{uy} = (1.037)(100) = 103.7 \text{ ft-k}$$

Since $L_b = 14 \text{ ft} < L_p = 15.6 \text{ ft}$

$$\phi_b M_{nx} = \phi_b F_y Z = \frac{(0.9)(36)(212)}{12} = 572.4 \text{ ft-k}$$

$$\phi_b M_{ny} = \phi_b F_y Z = \frac{(0.9)(36)(102)}{12} = 275.4 \text{ ft-k}$$

Applying LRFD Equation H1-1b

$$\frac{P_u}{2\phi P_n} + \left(\frac{M_{ux}}{\phi_b M_{nx}} + \frac{M_{uy}}{\phi_b M_{ny}}\right)$$

$$= \frac{180}{(2)(971)} + \left(\frac{152}{572.4} + \frac{103.7}{275.4}\right)$$

$$= 0.735 < 1.0 \qquad\qquad \text{(OK but overdesigned)} \quad\blacksquare$$

■ EXAMPLE 11-8

For the truss shown in Fig. 11-8(a) a W8 × 31 is used as a continuous top chord member from joint L_o to joint U_3. If the member consists of A36 steel, does it have sufficient strength to resist the factored loads shown in part (b) of the figure. Part (b) shows the portion of the chord from L_0 to U_1 and the 16-k load represents the effect of a purlin. It is assumed that lateral support is provided for this member at its ends and center line.

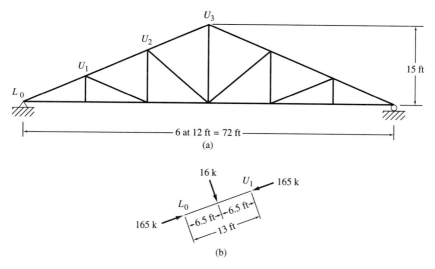

Figure 11-8 A truss whose top chord is subject to intermediate loads.

Solution. Using a W8 × 31 (A = 9.13 in.², I_x = 110 in.⁴, r_x = 3.47 in., r_y = 2.02 in., L_p = 8.4 ft, Z_x = 30.4 in.³)

For a frame braced against sidesway K = 1.0

$$\left(\frac{KL}{r}\right)_x = \frac{(1.0)(12 \times 13)}{3.47} = 44.96$$

$$\left(\frac{KL}{r}\right)_y = \frac{(1.0)(12 \times 6.5)}{2.02} = 38.61$$

$\phi_c F_{cr}$ = 27.50 ksi (from LFRD Table 3-36)

$\phi_c P_n$ = (27.50)(9.13) = 251 k

$$\frac{P_u}{\phi_c P_n} = \frac{165}{251} = > 0.2$$

∴ Must use LRFD Equation H1-1a

For a braced frame M_{lt} (the moment due to lateral loads) = 0

$$P_{e1x} = \frac{(10)^4(3150)}{(12 \times 13)^2} \text{ from column table} = 1294 \text{ k}$$

From Table 11-1

For $C_m = 1 - 0.2\dfrac{P_u}{P_{e1}} = 1 - 0.2\left(\dfrac{165}{1294}\right) = 0.974$

For $C_m = 1 - 0.3\dfrac{P_u}{P_{e1}} = 1 - 0.3\left(\dfrac{165}{1294}\right) = 0.962$

Average C_m = 0.97

Computing moment

For $\quad$ $M_{nt} = \dfrac{(16)(13)}{4} = 52$ ft-k

For $\quad$ $M_{nt} = \dfrac{(3)(16)(13)}{16} = 39$ ft-k

$$\text{Avg } M_{nt} = \frac{52 + 39}{2} = 45.5 \text{ ft-k}$$

Computing P_{e1x} and B_{1x}

$$B_{1x} = \frac{C_{mx}}{1 - \dfrac{P_u}{P_{elx}}} = \frac{0.97}{1 - \dfrac{165}{1294}} = 1.112$$

$M_{ux} = (1.112)(45.5) = 50.6$ ft-k

Since $L_b = 6.5$ ft $< L_p = 8.4$ ft

$$\phi_b M_{nx} = \phi_b F_y Z_x = \frac{(0.9)(36)(30.4)}{12} = 82.1 \text{ ft-k}$$

Applying LRFD Equation H1-1a

$$\frac{P_u}{\phi P_n} + \frac{8}{9}\left(\frac{M_{ux}}{\phi_b M_{nx}} + \frac{M_{uy}}{\phi_b M_{ny}}\right)$$

$$= \frac{165}{251} + \frac{8}{9}\left(\frac{50.6}{82.1} + 0\right) = 1.21 > 1.0 \qquad \text{(NG)} \quad \blacksquare$$

11-8 REVIEW OF BEAM COLUMNS IN UNBRACED FRAMES

The maximum primary moments in unbraced frames almost always occur at the column ends. As you can see in Fig. 11-3, the maximum sidesway moments always occur at the member ends and the total moment for a particular column is determined by adding its primary end moment to its sidesway moment. Therefore it is not necessary to use a modification factor and C_m is not used in the B_2 expressions.

■ **EXAMPLE 11-9**

A 10-ft W12 × 106 ($F_y = 50$ ksi) is used as a beam-column in an unbraced frame. It is bent in reverse curvature with equal and opposite moments and is not subject to intermediate transverse loads. Is the section satisfactory if $P_u = 250$ k, $M_{ntx} = 60$ ft-k, $M_{nty} = 40$ ft-k, $M_{ltx} = 100$ ft-k and $M_{lty} = 80$ ft-k?

The total factored gravity load ΣP_u above this level has been calculated to equal 5000 k. Assume that $\Sigma P_{ex} = 40{,}000$ k and $\Sigma P_{ey} = 20{,}000$ k. $K_x = K_y = 1.2$.

Solution.   Using a W12 × 106 ($A = 31.2$ in.², $I_x = 933$ in.⁴, $I_y = 301$ in.⁴, $L_p = 11.0$ ft, $Z_x = 164$ in.³, $Z_y = 75.1$ in.³)

$$K_x L_x = K_y L_y = (1.2)(10) = 12 \text{ ft}$$

$$\phi_c P_n = 1130 \text{ k from LRFD column tables}$$

$$\frac{P_u}{\phi_c P_n} = \frac{250}{1130} = 0.221 > 0.2$$

∴ Must use LRFD Equation H1-1a

As we have the moments M_{ntx} and M_{nty} as well as M_{ltx} and M_{lty}

$$M_{ux} = B_1 M_{ntx} + B_2 M_{ltx}$$

$$M_{uy} = B_1 M_{nty} + B_2 M_{lty}$$

For reverse curvature and equal end moments

$$C_{mx} = C_{my} = 0.6 - 0.4(+1.0) = 0.2$$

$$P_{elx} = \frac{\pi^2 EI_x}{(K_x L_x)^2} = \frac{(\pi^2)(29 \times 10^3)(933)}{(12 \times 12)^2} = 12{,}877 \text{ k}$$

$$B_{1x} = \frac{C_m}{1 - \dfrac{P_u}{P_{elx}}} = \frac{0.2}{1 - \dfrac{250}{12{,}877}} = 0.204 < 1.0 \quad \underline{\underline{\text{Use } 1.0}}$$

$$P_{ely} = \frac{\pi^2 EI_y}{(K_y L_y)^2} = \frac{(\pi^2)(29 \times 10^3)(301)}{(12 \times 12)^2} = 4154 \text{ k}$$

$$B_{1y} = \frac{C_m}{1 - \dfrac{P_u}{P_{ely}}} = \frac{0.2}{1 - \dfrac{250}{4154}} = 0.213 < 1.0 \quad \underline{\underline{\text{Use } 1.0}}$$

$$B_{2x} = \frac{1}{1 - \dfrac{\Sigma P_u}{\Sigma P_{e2x}}} = \frac{1}{1 - \dfrac{5000}{40{,}000}} = 1.143$$

$$B_{2y} = \frac{1}{1 - \dfrac{\Sigma P_u}{\Sigma P_{e2y}}} = \frac{1}{1 - \dfrac{5000}{20{,}000}} = 1.333$$

$$M_{ux} = (1.0)(60) + (1.143)(100) = 174.3 \text{ ft-k}$$

$$M_{uy} = (1.0)(40) + (1.333)(80) = 146.6 \text{ ft-k}$$

Since $L_b = 10 \text{ ft} < L_p = 11.0 \text{ ft}$

$$\phi_b M_{nx} = \phi_b F_y Z_x = \frac{(0.9)(50)(164)}{12} = 615 \text{ ft-k}$$

$$\phi_b M_{ny} = \phi_b F_y Z_y = \frac{(0.9)(50)(75.1)}{12} = 281.6 \text{ ft-k}$$

Applying LRFD Equation Hl-la

$$\frac{P_u}{\phi P_n} + \frac{8}{9}\left(\frac{M_{ux}}{\phi_b M_{nx}} + \frac{M_{uy}}{\phi_b M_{ny}}\right) = \frac{250}{1130} + \frac{8}{9}\left(\frac{174.3}{615} + \frac{146.6}{281.6}\right)$$

$$= 0.936 < 1.0 \hspace{3cm} \text{(OK)} \hspace{0.3cm} \blacksquare$$

11-9 DESIGN OF BEAM-COLUMNS—BRACED OR UNBRACED

The design of beam-columns involves a trial-and-error procedure. A trial section is selected by some process and is then checked with the appropriate interaction formula. If the section does not satisfy the equation or if it's too much on the safe side (that is, if it's overdesigned) another section is selected and the interaction equation is applied again. Probably the first thought of the reader is "I sure hope that we can select a good section the first time and not have to go through all of that rigamarole more than once." We certainly can make a good initial estimate, and that's the topic of the remainder of this section.

A common method used for selecting sections to resist both moments and axial loads is the *equivalent axial load or effective axial load method*. In this method the axial load (P_u) and the bending moment (M_{ux} and/or M_{uy}) are replaced with a fictitious concentric load P_{ueq} equivalent to the actual design axial load plus the design moment.

It is assumed for this discussion that it is desired to select the most economical section to resist both a moment and an axial load. By a trial-and-error procedure it is possible eventually to select the lightest section. Somewhere, however, there is a fictitious axial load that will require the same section as the one required for the actual moment and the actual axial load. This fictitious load is called the equivalent axial load or the effective axial load P_{ueq}.

Equations are used to convert the bending moment into an estimated equivalent axial load P_u' which is added to the actual design axial load P_u. The total of $P_u + P_u'$ is the equivalent or effective axial load P_{ueq}, and it is used to enter the concentric column tables of the LRFD Manual for choice of a trial section. In the approximate formula for P_{ueq} which follows, m is a factor given in Table 11-2 of this chapter (its Table 3-2 in Part 3 of the LRFD Manual).

$$P_{ueq} = P_u + M_{ux}m + M_{uy}mu$$

To apply this expression a value of m is taken from the first approximation section of Table 11-2 and u is assumed equal to 2. *In applying the equation the moments M_{ux} and M_{uy} must be used in ft-k.* The equation is solved for P_{ueq} and a column is selected from the concentrically loaded column tables for that load. Then the equation for P_{ueq} is solved again using a revised value of m from the subsequent approximations part of the table and the value of u is taken from the

TABLE 11-2 PRELIMINARY BEAM-COLUMN DESIGN
F_y = 36 ksi, F_y = 50 ksi

Values of m

F_y	36 ksi							50 ksi						
KL(ft)	10	12	14	16	18	20	22 and over	10	12	14	16	18	20	22 and over
							1st Approximation							
All Shapes	2.0	1.9	1.8	1.7	1.6	1.5	1.3	1.9	1.8	1.7	1.6	1.4	1.3	1.2
							Subsequent Approximation							
W4	3.1	2.3	1.7	1.4	1.1	1.0	0.8	2.4	1.8	1.4	1.1	1.0	0.9	0.8
W5	3.2	2.7	2.1	1.7	1.4	1.2	1.0	2.8	2.2	1.7	1.4	1.1	1.0	0.9
W6	2.8	2.5	2.1	1.8	1.5	1.3	1.1	2.5	2.2	1.8	1.5	1.3	1.2	1.1
W8	2.5	2.3	2.2	2.0	1.8	1.6	1.4	2.4	2.2	2.0	1.7	1.5	1.3	1.2
W10	2.1	2.0	1.9	1.8	1.7	1.6	1.4	2.0	1.9	1.8	1.7	1.5	1.4	1.3
W12	1.7	1.7	1.6	1.5	1.5	1.4	1.3	1.7	1.6	1.5	1.5	1.4	1.3	1.2
W14	1.5	1.5	1.4	1.4	1.3	1.3	1.2	1.5	1.4	1.4	1.3	1.3	1.2	1.2

Source: This table is from a paper in AISC *Engineering Journal* by Uang, Wattar, and Leet (1990).

column tables for the column initially selected. Another shape is selected and the process is continued until m and u stabilize (that is, until the column size selected does not change).

Finally, it is necessary to check the trial column size with the appropriate LRFD interaction equation (H1-1a) or (H1-1b). The equivalent axial load equation yields sections that are generally on the conservative side. For this reason the designer may select a section by the equivalent axial load method and then try the interaction equation on a section one or two sizes smaller. This procedure may very well provide significant savings in steel weight.

Limitations of P_{ueq} Formula

The application of the equivalent axial load formula and Table 11-2 results in economical beam-column designs unless the moment becomes quite large in comparison with the axial load. For such cases the members selected will be capable of supporting the loads and moments but may very well be rather uneconomical. The tables for concentrically loaded columns of Part 3 of the Manual are limited to the W14s, W12s, and shallower sections, but when the moment is large in proportion to the axial load there will often be a much deeper and appreciably lighter section such as a W27 or W30 which may be a good bit lighter and which will satisfy the appropriate interaction equations.

As there are so many different loading situations that can occur, the author presents quite a few example problems (11-10 to 11-15) in the pages to follow. After a section is selected by the approximate P_{ueq} formula, it is necessary to check it with the appropriate interaction equation. As we have been doing just this in the last several examples (11-6 to 11-9), the checking part of the examples to follow is somewhat abbreviated in some cases.

■ EXAMPLE 11-10

Select a 14-ft 50 ksi beam-column to support the following: $P_u = 600$ k, first-order moment $M_{ux} = 200$ ft-k and $M_{uy} = 0$. The member is to be bent in single curvature with no transverse loading and the frame is to be braced. Use $K = 1.0$ and $C_m = 0.85$.

Solution.   For first approximation $m = 1.7$ (from Table 11-2) and $u = 2.0$.

$$P_{ueq} = P_u + M_{ux}m + M_{uy}mu$$
$$= 600 + (200)(1.7) + 0 = 940 \text{ k}$$

Entering column tables with $P_u = 940$ k and $K_x L_x = K_y L_y = (1.0)(14) = 14$ ft

Try W14 × 90 ($m = 1.4$ from Table 11-2 and $u = 1.94$ from column tables)

$$P_{ueq} = 600 + (200)(1.4) + 0 = 880 \text{ k}$$

Returning to the column tables we find that the W14 × 90($\phi_c P_n = 969$ k) still is required. Checking it with the appropriate interaction equation:

$$\left(\frac{KL}{r}\right)_x = \frac{12 \times 14}{6.14} = 27.36$$

$$P_{e1} \text{ from column table} = \frac{(10)^4(28,600)}{(12 \times 14)^2} = 10,133 \text{ k}$$

$$B_1 = \frac{C_m}{1 - \dfrac{P_u}{P_{e1}}} = \frac{0.85}{1 - \dfrac{600}{10,133}} < 1.0 \quad \therefore \text{ Use } 1.0$$

$$\therefore M_{ux} = (1.0)(200) = 200 \text{ ft-k}$$

$\phi_b M_{nx} = 587$ ft-k from Load Factor Design Selection Table as $L_b < L_p$

$$\frac{P_u}{\phi_c P_n} = \frac{600}{969} > 0.2$$

$\therefore$ Must use LRFD Equation H1-1a

$$\frac{P_u}{\phi_c P_n} + \frac{8}{9}\left(\frac{M_{ux}}{\phi_b M_{nx}} + \frac{M_{uy}}{\phi_b M_{ny}}\right)$$

$$= \frac{600}{969} + \frac{8}{9}\left(\frac{200}{587}\right) = 0.922 < 1.0$$

Use W14 × 90   (OK) ■

■ EXAMPLE 11-11

A 12-ft beam-column of A36 steel is to be used in a braced frame for the following: $P_u = 200$ k; first-order moment $M_{ux} = 150$ ft-k and $M_{uy} = 100$ ft-k. Assuming it is to have restrained ends and to be subject to intermediate transverse loads, select a W12 section. Assume $K = 1.0$.

Solution.   1st approximation making use of Table 11-2 ($m = 1.9$, $u = 2.0$)

$$P_{ueq} = P_u + M_{ux}m + M_{uy}m$$
$$= 200 + (150)(1.9) + (100)(1.9)(2.0) = 865 \text{ k}$$

Entering column tables with $P_u = 865$ k and $K_x L_x = K_y L_y = (1)(12) = 12$ ft.

Try W12 × 120 ($m = 1.7$ from Table 11-2 and $u = 2.12$ from column tables)

$$P_{ueff} = 200 + (150)(1.7) + (100)(1.7)(2.12) = 815 \text{ k}$$

Try W12 × 106 ($m = 1.7$, $u = 2.12$)

$$P_{ueq} = 200 + (150)(1.7) + (100)(1.7)(2.12) = 815 \text{ k}$$

The column load tables still call for a W12 × 106.

Checking a W12 × 106 $(A = 31.2 \text{ in.}^2, \ L_p = 13.0 > L_b, \ \phi_b M_{nx} = 443 \text{ ft-k}, Z_y = 75.1 \text{ in.}^3, \ \phi_b M_{ny} = \dfrac{(0.9)(36)(75.1)}{12} = 202.8 \text{ ft-k}, r_x = 5.47 \text{ in.}, r_y = 3.11 \text{ in.}, \phi_c P_n = 853 \text{ k})$

$C_{mx} = C_{my}$ for members with restrained ends subject to transverse loads between ends = 0.85.

$$P_{ex1} = \frac{(10)^4(26,700)}{(12 \times 12)^2} = 12,876 \text{ k}$$

$$B_{1x} = \frac{C_{mx}}{1 - \dfrac{P_u}{P_{e1x}}} = \frac{0.85}{1 - \dfrac{200}{12,876}} = 0.86 \quad \text{Use 1.0}$$

$$M_{ux} = (1.0)(150) = 150 \text{ ft-k}$$

$$P_{ey1} = \frac{(10)^4(8640)}{(12 \times 12)^2} = 4167 \text{ k}$$

$$B_{1y} = \frac{C_{my}}{1 - \dfrac{P_u}{P_{e1y}}} = \frac{0.85}{1 - \dfrac{200}{4167}} = 0.89 \quad \text{Use 1.0}$$

$$M_{uy} = (1.0)(100) = 100 \text{ ft-k}$$

$$\frac{P_u}{\phi_c P_n} = \frac{200}{853} = 0.234 > 0.2$$

∴ Must use LRFD Equation H1-1a

$$\frac{P_u}{2\phi P_n} + \frac{8}{9}\left(\frac{M_{ux}}{\phi_b M_{nx}} + \frac{M_{uy}}{\phi_b M_{ny}}\right)$$

$$= \frac{200}{853} + \frac{8}{9}\left(\frac{150}{443} + \frac{100}{202.8}\right)$$

$$= 0.974 < 1.00$$

Use W12 × 106

■ EXAMPLE 11-12

Select a 10-ft W12 beam-column (A36 steel) for a braced symmetrical frame with $P_u = 200$ k; first-order moments $M_{ltx} = 120$ ft-k, $M_{lty} = 80$ ft-k, $M_{ntx} = M_{nty} = 0$. The computed ΣP_u for all of the columns on this level is 3600 k while ΣP_{ex2} has been calculated to equal 50,000 k and ΣP_{ey2} to equal 25,000 k. Assume $K_x = K_y = 1.2$. The column is to be bent in single curvature with no intermediate loads.

Solution.   Determining M_{ux} and M_{uy} from LRFD Equation C1-2

$$M_u = B_1 M_{nt} + B_2 M_{lt}$$

$$B_{2x} = \frac{1.0}{1 - \dfrac{\Sigma P_u}{\Sigma P_{ex2}}} = \frac{1.0}{1 - \dfrac{3600}{50{,}000}} = 1.078$$

$$B_{2y} = \frac{1.0}{1 - \dfrac{\Sigma P_u}{\Sigma P_{ey2}}} = \frac{1.0}{1 - \dfrac{3600}{25{,}000}} = 1.168$$

The second order moment strengths are

$$M_{ux} = (1.078)(120) = 129.4 \text{ ft-k}$$
$$M_{uy} = (1.168)(80) = 93.4 \text{ ft-k}$$
$$KL = (1.2)(10) = 12 \text{ ft}$$

For first approximation $m = 1.9$ and $u = 2.0$

$$P_{ueq} = 200 + (129.4)(1.9) + (93.4)(1.9)(2.0) = 800.8 \text{ k}$$

Try W12 × 106 ($m = 1.7$, $u = 2.12$)

$$P_{ueq} = 200 + (129.4)(1.7) + (93.4)(1.7)(2.12) = 756.6 \text{ k}$$

Try W12 × 96 ($m = 1.7$, $u = 2.10$)

$$P_{ueq} = 200 + (129.4)(1.7) + (93.4)(1.7)(2.10) = 753.4 \text{ k}$$

The column tables still call for a W12 × 96. Checking a W12 × 96 ($\phi_c P_n = 770$ k, $\phi_b M_{nx} = 397$ ft-k since $L_b < L_p$, $Z_y = 67.5$ in.^{3}, $\phi_b M_{ny} = \dfrac{(0.9)(36)(67.5)}{12} = 182.2$ ft-k)

$$\frac{P_u}{\phi_c P_n} = \frac{200}{770} = 0.260 > 0.2$$

$\therefore$ Must use LRFD Equation H1-1a

$$\frac{P_u}{\phi P_n} + \frac{8}{9}\left(\frac{M_{ux}}{\phi_b M_{nx}} + \frac{M_{uy}}{\phi_b M_{ny}}\right)$$

$$= \frac{200}{770} + \frac{8}{9}\left(\frac{129.4}{397} + \frac{93.4}{182.8}\right)$$

$$= 1.006 \approx 1.00 \qquad\qquad\qquad (\text{OK})$$

$$\underline{\underline{\text{Use W12 × 96}}}\qquad\qquad\qquad\qquad\blacksquare$$

■ EXAMPLE 11-13

Select a W14 beam-column of A36 steel with $K_x = 1.2 = K_y$ for the following: $P_u = 350$ k, first-order moment $M_{lty} = 100$ ft-k due to wind, with all other moments equal to 0. The 14-ft member is to be used in a symmetrical unbraced frame with an allowable story drift index $\Delta_{oh}/L = 0.0020$ as a result of a total service or unfactored load of 100 k. The total factored gravity load above has been calculated to equal 5000 k.

Solution.

$$K_x L_x = K_y L_y = (1.2)(14) = 16.8 \text{ ft}$$

Having set a drift index $= 0.002$ we can backfigure B_2 with LRFD Formula C1-4.

$$B_2 = \cfrac{1}{1 - \Sigma P_u\left(\cfrac{\Delta_{oh}}{\Sigma HL}\right)} = \cfrac{1}{1 - 5000\left(\cfrac{0.0020}{100}\right)} = 1.111$$

$$\therefore B_2 M_{lty} = (1.111)(100) = 111.1 \text{ ft-k}$$

Trying a W14 with $m = 1.36$ and $u = 2.0$

$$P_{ueq} = P_u + M_{ux}m + M_{uy}mu$$
$$= 350 + 0 + (111.1)(1.36)(2.0) = 652.2 \text{ k}$$

Try W14 × 90 ($m = 1.36$ and $u = 2.02$)

$$P_{ueq} = 350 + 0 + (111.1)(1.36)(2.02) = 655.2 \text{ k}$$

Column tables still call for a W14 × 90. Checking a W14 × 90 ($Z_y = 75.6 \text{ in.}^3$)

$$\phi_c P_n = 693.6 \text{ k from column tables by interpolation}$$

$$\frac{P_u}{\phi_c P_n} = \frac{350}{693.6} > 0.2$$

$$\therefore \text{ Must use LRFD Equation H1-1a}$$

$$\phi_b M_{ny} = \frac{(0.9)(36)(75.6)}{12} = 204.1 \text{ ft-k}$$

$$\frac{P_u}{\phi P_n} + \frac{8}{9}\left(\frac{M_{ux}}{\phi_b M_{nx}} + \frac{M_{uy}}{\phi_b M_{ny}}\right)$$

$$= \frac{350}{693.6} + \frac{8}{9}\left(0 + \frac{111.1}{204.1}\right) = 0.989 < 1.0 \qquad \text{(OK)}$$

Use W14 × 90   ∎

■ EXAMPLE 11-14

Design a 12-ft W12 (A36) for a symmetrical unbraced frame to support the following: $P_u = 280$ k, $M_{ntx} = 60$ ft-k, $M_{nty} = 40$ ft-k, $M_{ltx} = 100$ ft-k, and $M_{lty} = 70$ ft-k. The member is to be subjected to intermediate transverse loads and its ends are restrained. It is to have an allowable drift index $= 0.0020$ due to the total horizontal service or unfactored loads of 120 k perpendicular to the x axis and 80 k perpendicular to the y axis. Assume $K_x = K_y = 1.2$ and $\Sigma P_u = 5000$ k.

Solution.   We do not know the value of P_{e2x}, P_{e2y}, B_{1x}, or B_{1y}. Assume $B_{1x} = B_{1y} = 1.0$ and compute the values of B_{2x} and B_{2y} as follows using LRFD Equation C1-4.

$$B_{2x} = \frac{1}{1 - \Sigma P_u \left(\dfrac{\Delta_{oh}}{\Sigma HL}\right)} = \frac{1}{1 - 5000\left(\dfrac{0.002}{120}\right)} = 1.09$$

$$B_{2y} = \frac{1}{1 - 5000\left(\dfrac{0.002}{80}\right)} = 1.14$$

$M_{ux} = (1.00)(60) + (1.09)(100) = 169$ ft-k
$M_{uy} = (1.00)(40) + (1.14)(70) = 119.8$ ft-k

Selecting a trial section

$KL = (1.2)(12) = 14.4$ ft
$m = 1.58$ and $u = 2.0$
$P_{ueq} = 280 + (169)(1.58) + (119.8)(1.58)(2.0) = 925.6$ k

Try W12 × 120 ($m = 1.58$, $u = 2.12$)

$P_{ueq} = 280 + (169)(1.58) + (119.8)(1.58)(2.12) = 948.3$ k

Tables call for W12 × 136, but try a W12 × 120 ($\phi_c P_n = 920$ k, $L_p = 13.0$ ft, $\phi_b M_{nx} = 502$ ft-k, $Z_y = 85.4$ in³.)

$$\frac{P_u}{\phi_c P_n} = \frac{280}{920} = 0.304 > 0.2$$

$\therefore$ Must use LRFD Equation H1-1a

$$\phi_b M_{ny} = \frac{(0.9)(36)(85.4)}{12} = 230.6 \text{ ft-k}$$

$$\frac{P_u}{\phi_c P_n} + \frac{8}{9}\left(\frac{M_{ux}}{\phi_b M_{nx}} + \frac{M_{uy}}{\phi_b M_{ny}}\right)$$

$$= \frac{280}{920} + \frac{8}{9}\left(\frac{169}{502} + \frac{119.8}{230.6}\right)$$

$$= 1.07 > 1.00 \qquad\qquad\qquad \text{(NG)}$$

Use W12 × 136

■

■ EXAMPLE 11-15

Select the lightest satisfactory W12 section (A36 steel) for a 12-ft column in a building frame for the situation to be described. It is to be unbraced in the plane of the frame but is to be braced at each story out of plane so that $K_y = 1.0$. In the plane of the frame K_x has been estimated to equal 1.50.

A first-order analysis has been made with the factored loads and the results are given in Fig. 11-9. It is assumed that the 1.2D + 0.5L + 1.3W load combination controls. Assume that ΣP_u for all of the columns on this floor has been calculated to be 7200 k, while ΣP_{el} has been estimated to equal 114,000 k.

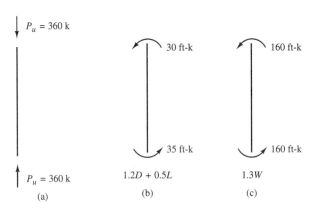

Figure 11-9 Column factored load and moments considered in Example 11-13.

Solution.

$$K_x L_x = (1.5)(12) = 18 \text{ ft}$$
$$m = 1.5$$
$$P_{ueq} = 360 + (35 + 160)(1.5) = 652.5 \text{ k}$$

Try W12 × 96 ($A = 28.2$ in.², $r_x = 5.44$ in., $r_y = 3.09$ in., $L_p = 12.9$ ft)

$$\left(\frac{KL}{r}\right)x = \left(\frac{1.5 \times 12 \times 12}{5.44}\right) = 39.71$$

$$\left(\frac{KL}{r}\right)y = \left(\frac{1.0 \times 12 \times 12}{3.09}\right) = 46.60 \leftarrow$$

$$\phi_c F_{cr} = 27.3 \text{ ksi}$$
$$\phi_c P_n = (28.2)(27.3) = 770 \text{ k}$$
$$\frac{P_u}{\phi_c P_n} = \frac{360}{770} = 0.468 > 0.2$$

∴ Must use LRFD Equation H1-1a

$$P_{ex1} \text{ from column table} = \frac{(10)^4(23,900)}{(1.5 \times 12 \times 12)^2} = 5123 \text{ k}$$

$$C_{mx} = 0.6 - 0.4\left(+\frac{30}{35}\right) = 0.257$$

$$B_1 = \frac{C_m}{1 - \dfrac{P_u}{P_{ex1}}} = \frac{0.257}{1 - \dfrac{360}{5123}} = < 1.0 \quad \underline{\underline{\text{Use } 1.0}}$$

$$B_2 = \frac{1}{1 - \left(\dfrac{\Sigma P_u}{\Sigma P_{e1}}\right)} = \frac{1}{1 - \dfrac{7200}{114{,}000}} = 1.07$$

$$M_{ux} = (1.0)(35) + (1.07)(160) = 206.2 \text{ ft-k}$$

$$\phi_b M_{nx} = 397 \text{ ft-k}$$

$$\frac{P_u}{\phi_c P_n} + \frac{8}{9}\left(\frac{M_{ux}}{\phi_b M_{nx}} + \frac{M_{uy}}{\phi_b M_{ny}}\right)$$

$$= \frac{360}{770} + \frac{8}{9}\left(\frac{206.2}{397} + 0\right)$$

$$= 0.929 < 1.0 \qquad\qquad \text{(OK)}$$

$$\underline{\underline{\text{Use W12} \times 96}} \qquad\qquad \blacksquare$$

11-10 COMPUTER EXAMPLES FOR MEMBERS SUBJECT TO BENDING AND AXIAL COMPRESSION

Example 11-16 illustrates the use of INSTEP to check a member subject to bending and axial compression. Then in Example 11-17 INSTEP is used to select the lightest W14 shape for given moment and axial compression values. You can quickly repeat this design for the lightest W10, W12, etc. These two examples are repeats of Examples 11-6 and 11-10.

■ **EXAMPLE 11-16**
Repeat Example 11-6 using INSTEP.

Solution.

```
                    INPUT
        A   Type Steel: A36
        B   Trial Size: W12×96

BEAM/COLUMN DESIGN PARAMETERS
        D   L = 144 inches
        E   Lb = 144 inches
        F   Cb = 1
        G   CM = 1
        H   B2 = 1
        I   Ae = 0 sq.in.
        J   Kb = 1
```

FACTORED DESIGN LOADS

L $P_u = -500$ kips
M $M_{nt/u} = 1500$ kip-inches
N $M_{1t} = 0$ kip-inches

OUTPUT

EVALUATE BEAM–COLUMN INTERACTION

From Axial Compression Considerations

$P_{na} = 769.693$ kips

From Flexural Considerations

$M_{na} = 4762.8$ kip-inches

Calculate Design Bending Moment

$P_e = 11519.1$ kips
$B_1 = 1.04538$
$M_u = 1.04538(1500) + 1(0)$
$M_u = 1568.06$ kip-inches

Check the Interaction Equation

$$UC = \frac{500}{769.693} + \left[\frac{8}{9}\right]\left[\frac{1568.06}{4762.8}\right]$$
$$UC = .94226$$

This Member is Acceptable

■ EXAMPLE 11-17

Repeat Example 11-10 using INSTEP.

Solution.   **INPUT**

A Type Steel: A572 Grade 50
B Trial Size: W14

BEAM/COLUMN DESIGN PARAMETERS

D $L = 168$ inches
E $L_b = 168$ inches
F $C_b = 1$
G $C_M = .85$
H $B_2 = 0$
I $A_e = 0$ sq.in.
J $K_b = 1$

FACTORED DESIGN LOADS

L $P_u = -600$ kips
M $M_{nt/u} = 2400$ kip-inches
N $M_{1t} = 0$ kip-inches

OUTPUT

Suggested Trial Size: W14×90

$P_{na} = 968.65$ kips
$M_{na} = 7044$ kip-inches
Unity Check = 0.922277

PROBLEMS

Bending and Axial Tension

11-1. A W14 × 26 tension member with no holes consisting of A36 steel is subjected to a factored tensile load $P_u = 160$ k and a factored bending moment M_{ux} of 20 ft-k. Is the member satisfactory if $L_b = 4.0$ ft and if $C_b = 1.0$? (*Ans.* 0.805 < 1.00 OK)

11-2. A W12 × 58 tension member with no holes and $F_y = 50$ ksi is subjected to $P_u = 90$ k and $M_{ux} = 100$ ft-k. Is the member satisfactory if $L_b = 8.0$ ft and $C_b = 1.0$?

11-3. Repeat Prob. 11-1 if the member is also subject to a factored bending moment $M_{uy} = 10$ ft-k. (*Ans.* 1.40 > 1.00 NG)

11-4. Select the lightest W10 available to support a factored tensile load $P_u = 50$ k and a factored bending moment $M_{uy} = 40$ ft-k. Assume the member consists of A36 steel, has no holes, and is braced laterally at its supports.

11-5. Select the lightest W12 section available to support a factored tensile load $P_u = 150$ k using A36 steel to support a factored bending moment $M_{ux} = 50$ ft-k, $L_b = 8.0$ ft, and $C_b = 1.0$. The member is assumed to have no holes. (*Ans.* W12 × 30)

11-6. Select the lightest available W12 section (A36 steel) to support an axial factored tensile load equal to 150 k placed with an eccentricity of 5 in. with respect to the x axis. The member which is to be welded is to be 8 ft long and is to be braced laterally only at its supports.

11-7. Repeat Prob. 11-6 if a W14 section with $F_y = 50$ ksi is to be used. (*Ans.* W14 × 26)

11-8. A W12 × 58 member consisting of A36 steel has continuous lateral bracing. If M_{ux} is 70 ft-k and M_{uy} is 0 determine the maximum axial tensile force P_u that it can support. It is assumed to have no holes.

11-9. Repeat Prob. 11-8 if $F_y = 50$ ksi (*Ans.* 618.1 k)

Bending and Axial Compression

11-10. A W10 × 49 pin-connected beam-column that is not subjected to sidesway is 15 ft long. A load $P_u = 230$ k is applied to the column at its upper end with an eccentricity of 2 in. so as to cause bending about the major axis of the section. Check the adequacy of the member if it consists of A36 steel and is used in a braced frame so that $M_{ltx} = M_{lty} = 0$. $K_y = 1.0$ while $K_x = 1.00$. Assume $C_b = 1.0$ and $C_{mx} = C_{my} = 0.85$.

11-11. Sidesway is prevented for the beam-column shown in the accompanying illustration. If the first-order moments shown are about the major axis, is the member satisfactory if it consists of A36 steel? Assume $K_x = K_y = 1.0$. (*Ans.* 0.888 < 1.0 OK)

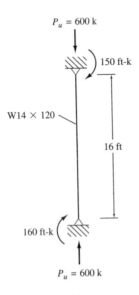

$P_u = 600$ k

150 ft-k

W14 × 120

16 ft

160 ft-k

$P_u = 600$ k

Problem 11-11

11-12. A pin-connected W10 × 49 consisting of A36 steel is used as a beam-column. If sidesway is prevented and a load $P_u = 300$ k is applied 1.50 in. off center at both ends so as to cause single curvature bending about the x axis, is the member satisfactory? Length = 12 ft. $K_x = 1.4$, $K_y = 1.0$.

11-13. The W10 × 30 truss member shown in the accompanying illustration consists of A36 steel and is fixed at its ends in the x and y directions. The purlin shown is assumed to provide pinned support in the y direction at the midpoint of the member. Is the member satisfactory to support the factored loads shown? (*Ans.* 0.702 < 1.0 OK)

110 k

12 k

Purlin

8 ft

110 k

8 ft

Problem 11-13

11-14. Repeat Prob. 11-13 if the ends of the member are assumed to be pinned and braced against lateral movement.

11-15. Select a W14 section consisting of A36 steel to support a factored load $P_u = 700$ k and the factored moments $M_{ntx} = 100$ ft-k and $M_{nty} = 30$ ft-k. The 12-ft member is to be used in a braced frame with M_{ltx} and $M_{lty} = 0$. Assume $K_x = K_y = 1.0$ and $C_m = 0.85$. (*Ans.* W14 × 120)

11-16. Repeat Prob. 11-15 if a W14 section consisting of 50-ksi steel is to be used.

11-17. A W14 × 109 section (A36 steel) is used for a 15-ft column in a building frame. It is unbraced in the plane of the frame but is braced at each story out of plane so that $K_y = 1.00$. K_x has been estimated to equal 1.40. A first-order analysis has been made with factored loads, and the results are shown in the accompanying illustration. Is the member satisfactory? The total factored load above this level has been calculated to equal 6000 k and ΣP_{ex} is estimated to equal 36,000 k. (*Ans.* 0.773 < 1.00 OK)

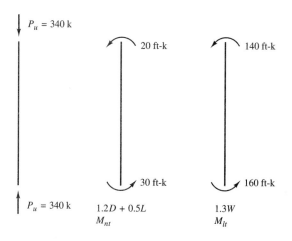

Problem 11-17

11-18. Repeat Prob. 11-17 if the moments bend the member in single curvature.

11-19. Select a W12 with $F_y = 50$ ksi for the load and moments of Prob. 11-17. (*Ans.* W12 × 72)

11-20. Select a 14-ft A36 W14 beam-column for $P_u = 750$ k and the first-order moments $M_{ntx} = 200$ ft-k and $M_{nty} = 0$. This pinned-ended member is to be located in a braced frame and is to be bent in single curvature with no intermediate transverse loads.

11-21. Repeat Prob. 11-20 if a W14 is to be selected and $F_y = 50$ ksi. (*Ans.* W14 × 99)

11-22. Select a 12-ft A572 Grade 50 W12 beam-column; for $P_u = 300$ k and the first-order moments $M_{ntx} = 200$ ft-k and $M_{nty} = 100$ ft-k. The member is to be pinned at its ends and is subjected to transverse loads. Assume the frame is braced $C_b = 1.0$ and $C_m = 0.85$.

11-23. Repeat Prob. 11-22 if M_{ntx} is 300 ft-k and $F_y = 50$ ksi. (*Ans.* W12 $\times$ 120)

11-24. Select a 14-ft W14 (A36 steel) with $K = 1.2$, $P_u = 300$ ft-k, and a first-order moment $M_{ltx} = 120$ ft-k due to wind. There are no other calculated moments and the maximum drift index is 0.0020 due to an unfactored wind load of 75 k. Assume the total factored gravity load above this story is 5000 k. The frame is symmetrical and unbraced.

11-25. Repeat Prob. 11-24 using a W12 section with $F_y = 36$ ksi. (*Ans.* W12 $\times$ 79)

11-26. Select a 15-ft W14 (A36 steel) with $K_x = K_y = 1.2$, $P_u = 200$ k; the first-order moments $M_{ltx} = 120$ ft-k, $M_{lty} = 80$ ft-k; and $M_{ntx} = M_{nty} = 0$. The frame is symmetrical and unbraced and ΣP_u for all of the loads above this story is 2500 k, while $\Sigma P_{e2x} = 50,000$ k for bending about the x axis and $\Sigma P_{e2y} = 30,000$ k for bending about the y axis. Column is bent in single curvature with no intermediate loads.

11-27. Repeat Prob. 11-26 if a 14-ft W12 with $F_y = 50$ ksi is to be used. (*Ans.* W12 $\times$ 79)

11-28 to 11-35. *Repeat the problems listed using the enclosed diskette.*

11-28. Prob. 11-1

11-29. Prob. 11-10 (*Ans.* 0.901 $<$ 1.0 OK)

11-30. Prob. 11-12

11-31. Prob. 11-13 (*Ans.* 0.702 $<$ 1.00 OK)

11-32. Prob. 11-17

11-33. Prob. 11-20 (*Ans.* W14 $\times$ 132)

11-34. Prob. 11-21

11-35. Prob. 11-24 (*Ans.* W14 $\times$ 82)

Chapter **12**

Bolted Connections

12-1 INTRODUCTION

For quite a few decades riveting was the accepted method used for connecting the members of steel structures. For the last few decades, however, bolting and welding have been the methods used for making structural steel connections, and riveting is almost extinct. This chapter and the next are almost entirely devoted to bolted connections, although some brief remarks are presented at the end of Chapter 13 concerning rivets.

Bolting of steel structures is a very rapid field erection process that requires less skilled labor than does riveting or welding. This gives bolting a distinct economic advantage over the other connection methods in the United States where labor costs are so very high. Even though the purchase price of a high-strength bolt is several times that of a rivet, the overall cost of bolted construction is cheaper than that for riveted construction because of reduced labor and equipment costs and the smaller number of bolts required to resist the same loads.

12-2 TYPES OF BOLTS

There are several types of bolts that can be used for connecting steel members. They are described in the following paragraphs.

Unfinished bolts are also called *ordinary* or *common bolts*. They are classified by the ASTM as A307 bolts and are made from carbon steels with stress-strain characteristics very similar to those of A36 steel. They are available in diameters from $\frac{5}{8}$ to $1\frac{1}{2}$ in. in $\frac{1}{8}$-in. increments.

A307 bolts generally have square heads and nuts to reduce costs, but hexagonal heads are sometimes used because they have a little more attractive appearance, are easier to turn and easier to hold with the wrenches, and require less turning space. As they have relatively large tolerances in shank and thread dimensions, their design strengths are appreciably smaller than those for high-strength bolts. They are primarily used in light structures subjected to static loads and for secondary members (such as purlins, girts, bracing, platforms, small trusses, and so forth).

Designers often are guilty of specifying high-strength bolts for connections when common bolts would be satisfactory. *The strength and advantages of common bolts have usually been greatly underrated in the past.* The analysis and design of A307 bolted connections are handled exactly as are riveted connections in every way except that the allowable stresses are slightly different.

High-strength bolts are made from medium carbon heat-treated steel and from alloy steel and have tensile strengths two or more times those of ordinary bolts. There are two basic types, the A325 bolts (made from a heat-treated medium carbon steel) and the higher strength A490 bolts (also heat-treated but made from an alloy steel). High-strength bolts are used for all types of structures, from small buildings to "skyscrapers" and monumental bridges. These bolts were developed to overcome the weaknesses of rivets—primarily insufficient tension in their shanks after cooling. The resulting rivet tensions may not be large enough to hold them in place during the application of severe impactive and vibrating loads. The result is that they may become loose and vibrate and may eventually have to be replaced. High-strength bolts may be tightened until they have very high tensile stresses so that the connected parts are clamped tightly together between the bolt and nut heads, permitting loads to be transferred primarily by friction.

Sometimes high-strength bolts are made from A449 steel in sizes larger than the $1\frac{1}{2}$-in. maximum diameter A325 and A490 bolts. These larger bolts may also be used as high-strength anchor bolts and for threaded rods of many different diameters.

12-3 HISTORY OF HIGH-STRENGTH BOLTS

The joints obtained using high-strength bolts are superior to riveted joints in performance and economy and they are the leading field method of fastening structural steel members. C. Batho and E. H. Bateman first claimed in 1934 that high-strength bolts could satisfactorily be used for the assembly of steel structures,[1] but it was not until 1947 that the Research Council on Riveted and Bolted

---

[1] C. Batho and E. H. Bateman, "Investigations on Bolts and Bolted Joints," H. M. Stationery Office (London, 1934).

High-strength bolt. (Courtesy of Bethlehem Steel Corporation.)

Structural Joints of the Engineering Foundation was established. This group issued their first specifications in 1951, and high-strength bolts were adopted by both building and bridge engineers for both static and dynamic loadings with amazing speed. They not only quickly became the leading method of making field connections, they also were found to have many applications for shop connections. The construction of the Mackinac Bridge in Michigan involved the use of more than one million high-strength bolts.

Connections that were formerly made with ordinary bolts and nuts were not too satisfactory when they were subjected to vibratory loads because the nuts frequently became loose. For many years this problem was dealt with by using some type of locknut, but the modern high-strength bolts furnish a far superior solution.

12-4 ADVANTAGES OF HIGH-STRENGTH BOLTS

Among the many advantages of high-strength bolts, partly explaining their great success, are the following.

1. Smaller crews are involved as compared with riveting. Two two-person bolting crews can easily turn out over twice as many bolts in a day as the number of rivets driven by the standard four-person riveting crew. The result is quicker steel erection.
2. In comparison with rivets, fewer bolts are needed to provide the same strength.
3. Good bolted joints can be made by people with a great deal less training and experience than is necessary to produce welded and riveted connections of equal quality. The proper installation of high-strength bolts can be learned in a matter of hours.
4. No erection bolts are required that may have to be later removed (depending on specifications) as in welded joints.
5. Though quite noisy, bolting is not nearly as bad as riveting.
6. Cheaper equipment is used to make bolted connections.
7. No fire hazard is present, nor danger from the tossing of hot rivets.

A spud wrench used by ironworkers for erecting structural steel and tightening bolts. One end of the wrench is sized for the hexagonal ends of bolts and nuts, while the other end is tapered to a rounded point and is used to align bolt holes between different connection pieces. (Courtesy of the AISC.)

8. Tests on riveted joints and fully tensioned bolted joints under identical conditions definitely show that bolted joints have a higher fatigue strength. Their fatigue strength is also equal to or greater than that obtained with equivalent welded joints.

9. Where structures are to be later altered or disassembled, changes in connections are quite simple because of the ease of bolt removal.

12-5 SNUG-TIGHT AND FULLY TENSIONED BOLTS

High-strength bolts are said to be either *snug-tight* or *fully tensioned.* These terms refer to the degree to which bolts are tightened.

For most connections bolts are tightened only to the snug-tight condition. Snug-tight is the situation existing when all the plies of a connection are in firm contact with each other. It usually means the tightness obtained by the full effort of a person using a spud wrench, or the tightness obtained after a few impacts of an impact wrench. Obviously, there is some variation in the degree of tightness obtained under these conditions. Snug-tight bolts must be clearly identified on both design and erection drawings.

When loads are applied to snug-tight bolts there may be a little slippage, as the holes are a little larger than the shanks of the bolts. As a result the parts of the connection may bear against the bolts. You can see that if we have a fatigue situation with constantly changing loads, this is not a desirable situation.

For fatigue situations, and for connections subject to direct tension, it is desirable to use connections that will not slip. These are referred to as *slip-critical connections.* To achieve this situation the bolts must be tightened until they reach a fully tensioned condition in which they are subject to extremely large tensile forces.

Fully tensioning bolts is an expensive process, as is the inspection necessary to see that they are fully tensioned. Thus, they should be used only where absolutely necessary, as where the working loads cause large numbers of stress changes resulting in fatigue problems. Section J1.11 of the LRFD Specification gives a detailed list of connections that must be made with fully tensioned bolts. Included are connections for supports of running machinery or for live loads producing impact and stress reversal; column splices in all tier structures 200 ft or more in height; connections of all beams and girders to columns and other beams or girders on which the bracing of the columns is dependent for structures over 125 ft in height; and so on. In all other cases we may use either snug-tight (bearing type) high-strength bolts or A307 common bolts.

Snug-tight bolts have several advantages over fully tensioned ones. One worker can properly tighten bolts to a snug-tight condition with an ordinary spud wrench or with only a few impacts of an impact wrench. The installation is quick and only a visual inspection of the work is needed. (Such is not the case for fully tensioned bolts.) Furthermore, snug-tight bolts may be installed with electric wrenches, thus eliminating the need for air compression on the site. As a result the use of snug-tight bolts saves time and money and is safer than the procedure needed for fully tensioned bolts. Therefore, for most situations snug-tight bolts should be used.

Table 12-1 on page 358, which is a reproduction of Table J3.1 of the LRFD Specification, presents the fastener tensions required for slip-resistant connections and for connections subject to direct tension. To be fully tensioned both A325 and A490 bolts are tightened to at least 70 percent of their specified minimum tensile strengths.

The quality-control provisions specified in the manufacture of the A325 and A490 bolts are more stringent than those for the A449 bolts. As a result, despite the method of tightening, the A449 bolts may not be used in slip-resistant connections.

Although many engineers felt that there would be some slippage as compared with rivets (because of the fact that the hot driven rivets more nearly filled the holes), the results show that there is less slippage in fully tensioned high-strength bolted joints than in riveted joints under similar conditions.

It is interesting to note that the nuts used for fully tensioned high-strength bolts need no special provisions for locking. Once these bolts are installed and

TABLE 12-1   FASTENER TENSION (KIPS) REQUIRED FOR SLIP-CRITICAL CONNECTIONS AND CONNECTIONS SUBJECT TO DIRECT TENSION*

Bolt size (in.)	A325 Bolts	A490 Bolts
$\frac{1}{2}$	12	15
$\frac{5}{8}$	19	24
$\frac{3}{4}$	28	35
$\frac{7}{8}$	39	49
1	51	64
$1\frac{1}{8}$	56	80
$1\frac{1}{4}$	71	102
$1\frac{3}{8}$	85	121
$1\frac{1}{2}$	103	148

*Equal to 0.70 of minimum tensile strength of bolts, rounded off to nearest kip, as specified in ASTM specifications for A325 and A490 bolts with UNC threads.

sufficiently tightened to produce the tension required, there is almost no tendency for the nuts to come loose. There are, however, a few situations where they will work loose under heavy vibrating loads. What do we do then? Some steel erectors have replaced the offending bolts with longer ones with two fully tightened nuts. Others have welded the nuts onto the bolts. Apparently the results have been somewhat successful.

12-6 METHODS FOR FULLY TENSIONING HIGH-STRENGTH BOLTS

We have already commented on the tightening required for snug-tight bolts. For fully tensioned bolts several methods of tightening are available. These methods, including the turn-of-the-nut method, the calibrated wrench method, and the use of alternate design bolts and direct tension indicators, are permitted without preference by the LRFD Specification.

Turn-of-the-Nut Method   The bolts are brought to a snug-tight condition and then, with an impact wrench, they are given from one-third to one full turn, depending on their length and the slope of the surfaces under their heads and nuts. (The amount of turn given can easily be controlled by marking the snug-tight position with paint or crayon.)

Calibrated Wrench Method   With this method the bolts are tightened with an impact wrench that is adjusted to stall at that certain torque which is theoretically necessary to tension a bolt of that diameter and ASTM classification to the desired tension. It is necessary that wrenches be calibrated daily and that hardened washers be used. Particular care needs to be given to protecting the bolts from dirt and moisture at the job site. The reader should refer to the "Specification for Structural Joints Using ASTM A325 or A490 Bolts" in Part 6 of the Manual for additional tightening requirements.

Direct Tension Indicator   The direct tension indicator (which was originally a British device) consists of a hardened washer which has protrusions on one face in the form of small arches. The arches will be flattened as a bolt is tightened. The amount of gap at any one time is a measure of the bolt tension. For fully tensioned bolts the gaps should measure about 0.015 in. or less.

Alternate Design Fasteners   In addition to the preceding methods there are some alternate design fasteners that can be tensioned quite satisfactorily. Bolts with splined ends that extend beyond the threaded portion of the bolts, called "twist-off bolts," are one example. Special wrench chucks are used to tighten the nuts until the splined ends shear off. This method of tightening bolts is quite satisfactory and will result in smaller labor costs.

A maximum bolt tension is not specified in any of the preceding tightening methods. This means that the bolt can be tightened to the highest load that will not break it and the bolt still will do the job. Should the bolt break, another one is put in with no damage done. It might be noted that the nuts are stronger than

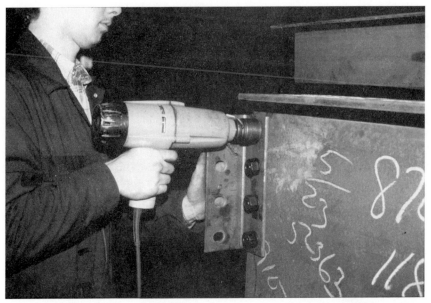

An impact wrench used to tighten bolts to either a snug-tight or a fully tensioned condition. It may be electric, such as this one, or it may be pneumatic. (Courtesy of the AISC.)

Torquing the nut for a high-strength bolt with an air-driven impact wrench. (Courtesy of Bethlehem Steel Corporation.)

the bolt and the bolt will break before the nut strips. (The bolt specification mentioned above requires that a tension measuring device be available at the job site to ensure that specified tensions are achieved.)

For fatigue situations where members are subjected to constantly fluctuating loads, the slip-resistant connection is very desirable. If, however, the force to be carried is less than the frictional resistance and thus no forces are applied to the bolts, how could we ever have a fatigue failure of the bolts? The slip-resistant connection is a serviceability limit state in that it is based on working loads. For a slip-resistant joint the working loads are not permitted to exceed the permissible frictional resistance.

Other situations where slip-resistant connections are desirable include joints where bolts are used in oversized holes, joints where bolts are used in slotted holes where the loads are applied parallel or nearly so to the slots, joints that are subjected to significant force reversals, and joints where bolts and welds resist shear together on a common *faying surface*. (The faying surface is the contact or shear area between the members.)

A "twist-off bolt," also called a "load-indicator bolt" or a "tension control bolt." Notice the spline at the end of the bolt shank. (Courtesy of AISC.)

12-7 SLIP-RESISTANT CONNECTIONS AND BEARING-TYPE CONNECTIONS

When high-strength bolts are fully tensioned they clamp the parts being connected tightly together. The result is a considerable resistance to slipping on the faying surface. This resistance is equal to the clamping force times the coefficient of friction.

If the shearing load is less than the permissible frictional resistance the connection is referred to as slip-resistant. If the load exceeds the frictional resistance the members will slip on each other and will tend to shear off the bolts, and at the same time the connected parts will push or bear against the bolts as shown in Fig. 12-1.

The surfaces of joints, including the area adjacent to washers, need to be free of loose scale, dirt, burrs, and other defects that might prevent the parts from solid seating. It is necessary for the surface of the parts to be connected to have slopes of not more than 1 to 20 with respect to the bolt heads and nuts, unless beveled washers are used. For slip-resistant joints the faying surfaces must also be free from oil, paint, and lacquer. (Actually, paint may be used if it is proved to be satisfactory by test.)

If the faying surfaces are galvanized, the slip factor will be reduced to almost half of its value for clean mill scale surfaces. The slip factor, however, may be significantly improved if the surfaces are subjected to hand wire brushing or to "brush off" grit blasting. However, such treatments do not seem to provide increased slip resistance for sustained loadings where there seems to be a creep-like behavior.[2]

The 1989 AASHTO Specifications permit hot-dip galvanization if the coated surfaces are scored with wire brushes or sandblasted after galvanization and before steel erection.

The ASTM Specification permits the galvanization of the A325 bolts themselves, but not the A490 bolts. There is a danger of embrittlement of this higher-strength steel during galvanization due to the possibility that hydrogen may be introduced into the steel in the pickling operation of the galvanization process.

If special faying surface conditions (such as blast-cleaned surfaces or blast-cleaned surfaces with special slip-resistant coatings applied) are used to increase the slip resistance, the designer may increase the values used here to the ones given by the Research Council on Structural Joints in Part 6 of the LRFD Manual.

---

[2] J. W. Fisher and J. H. A. Struik, *Guide to Design Criteria for Bolted and Riveted Joints* (New York: John Wiley & Sons, 1974) pp. 205–206.

12-8 MIXED JOINTS

Bolts may on occasion be used in combination with welds and on other occasions with rivets (as where they are added to old riveted connections to enable them to carry increased loads). The LRFD Specification contains some specific rules for these situations.

Bolts in Combination with Welds   For new work neither A307 common bolts nor high-strength bolts designed for bearing or snug-tight connections may be considered to share the load with welds. (Before the connection's ultimate strength is reached the bolts will slip, with the result that the welds will carry a larger proportion of the load—the actual proportion being difficult to determine.) For such circumstances welds will have to be proportioned to resist the entire loads.

If high-strength bolts are designed for slip-critical conditions, they may be allowed to share the load with welds. For such situations the LRFD Commentary J1.9 states that it is necessary to fully tighten the bolts before the welds are made. If the weld is made first, the heat from the weld may very well distort the connection so that we will not get the slip-critical resistance desired from the bolts. If the bolts are placed and fully tightened before the welds are made, the heat of the welding will not change the mechanical properties of the bolts. Thus, for such a case the slip-critical bolts and the welds may be assumed to share the load.

If we are making alterations for an existing structure that is connected with bearing or snug-tight bolts or with rivets, we can assume that any slipping that is going to occur has already taken place. Thus, if we are using welds in the alteration, we will design those welds neglecting the forces that would be produced by the existing dead load.

High-Strength Bolts in Combination with Rivets   High-strength bolts may be considered to share loads with rivets for new work or for alterations of existing connections that were designed as slip-critical. (The ductility of the rivets allows the capacity of both sets of fasteners to act together.)

12-9 SIZES OF BOLT HOLES

In addition to the standard size bolt holes (STD) which are $\frac{1}{16}$ in. larger in diameter than the bolts, there are three types of enlarged holes: oversized, short-slotted, and long-slotted. Oversized holes will on occasion be very useful in speeding up steel erection. In addition, they give some latitude for adjustments in plumbing frames during erection. The use of nonstandard holes requires the approval of the designer and is subject to the requirements of Section J3 of the LRFD Specification. Table 12-2, which is Table J3.3 of the LRFD

TABLE 12-2   NOMINAL HOLE DIMENSIONS

			Hole dimensions	
Bolt dia.	Stan- dard (dia.)	Over- size (dia.)	Short-slot (width × length)	Long-slot (width × length)
$\frac{1}{2}$	$\frac{9}{16}$	$\frac{5}{8}$	$\frac{9}{16} \times \frac{11}{16}$	$\frac{9}{16} \times 1\frac{1}{4}$
$\frac{5}{8}$	$\frac{11}{16}$	$\frac{13}{16}$	$\frac{11}{16} \times \frac{7}{8}$	$\frac{11}{16} \times 1\frac{9}{16}$
$\frac{3}{4}$	$\frac{13}{16}$	$\frac{15}{16}$	$\frac{13}{16} \times 1$	$\frac{13}{16} \times 1\frac{7}{8}$
$\frac{7}{8}$	$\frac{15}{16}$	$1\frac{1}{16}$	$\frac{15}{16} \times 1\frac{1}{8}$	$\frac{15}{16} \times 2\frac{3}{16}$
1	$1\frac{1}{16}$	$1\frac{1}{4}$	$1\frac{1}{16} \times 1\frac{5}{16}$	$1\frac{1}{16} \times 2\frac{1}{2}$
$\geq 1\frac{1}{8}$	$d + \frac{1}{16}$	$d + \frac{5}{16}$	$(d + \frac{1}{16}) \times (d + \frac{3}{8})$	$(d + \frac{1}{16}) \times (2.5 \times d)$

Source: American Institute of Steel Construction, *Manual of Steel Construction Load & Resistance Factor Design,* 2nd ed. (Chicago: AISC, 1994). Table J3.3, p. 6-82. Reprinted with the permission of the AISC.

Manual, provides the nominal dimensions of the various kinds of enlarged holes permitted for the different bolt sizes.

The situations in which we may use the various types of enlarged holes are now described.

Oversized holes (OVS) may be used in all plies of connections as long as the applied load does not exceed the permissible slip resistance. They may not be used in bearing-type connections. It is necessary for hardened washers to be used over oversized holes that are located in outer plies. The use of oversized holes permits the use of larger construction tolerances.

Short-slotted holes (SSL) may be used regardless of the direction of the applied load for slip-critical or bearing-type connections. Should the load be applied in a direction approximately normal (between 80 and 100 degrees) to the slot, these holes may be used in any or all plies of connections for bearing-type connections. It is necessary to use washers (hardened if high-strength bolts are being used) over short-slotted holes in an outer ply. The use of short-slotted holes provides for some mill and fabrication tolerances, but does not result in the necessity for slip-critical procedures.

Long-slotted holes (LSL) may be used in *only one* of the connected parts of slip-critical or bearing-type connections at any one faying surface. For slip-critical joints these holes may be used in any direction, but for bearing-type connections the loads must be normal (between 80 and 100 degrees) to the axes of the slotted holes. If long-slotted holes are used in an outer ply they will need to be covered with plate washers or a continuous bar. For high-strength bolted connections the washers or bar do not have to be hardened, but they must be made of structural-grade material and may not be less than 5/16 in. thick. Long-slotted holes are usually used when connections are being made to existing structures where the exact positions of the members being connected are not known.

Generally, washers are used to prevent scoring or galling of members when bolts are tightened. Most persons think that they also serve the purpose of spreading out the clamping forces more uniformly to the connected members. Tests have shown, however, that standard size washers don't affect the pressure very much except when oversized or short-slotted holes are used.

12-10 LOAD TRANSFER AND TYPES OF JOINTS

The following paragraphs present a few of the elementary types of bolted joints subjected to axial forces (that is, the loads are assumed to pass through the centers of gravities of the groups of connectors). For each of these joint types some comments are made about the methods of load transfer. Eccentrically loaded connections are discussed in Chapter 13.

For this initial discussion the reader is referred to part (a) of Figure 12-1. It is assumed that the plates shown are connected with a group of snug-tight bolts. In other words, the bolts are not tightened sufficiently so as to significantly squeeze the plates together. If there is assumed to be little friction between the plates they will slip a little due to the applied loads shown. As a result the loads in the plates will tend to shear the connectors off on the plane between the plates and press or bear against the sides of the bolts as shown in the figure. These connectors are said to be in *single shear and bearing* (also called *unenclosed bearing*). They must have sufficient strength to satisfactorily resist these forces, and the members forming the joint must be sufficiently strong to prevent the connectors from tearing through.

Should rivets be used instead of the snug-tight bolts, the situation is somewhat different because hot-driven rivets cool and shrink and squeeze or clamp

Figure 12-1 (a) Lap joint. (b) Butt joint.

the connected pieces together with sizable forces that greatly increase the friction between the pieces. As a result a large portion of the loads being transferred between the members is transferred by friction. The clamping forces produced in riveted joints, however, are generally not considered to be dependable, and as a result specifications normally consider such connections to be snug-tight with no frictional resistance. The same assumption is made for A307 common bolts as they are not tightened to large dependable tensions.

Fully tensioned high-strength bolts are in a different class altogether. Using the tightening methods previously described, a very dependable tension is obtained in the bolts resulting in large clamping forces and large dependable amounts of frictional resistance to slipping. Unless the loads to be transferred are larger than the frictional resistance, the entire forces are resisted by friction and the bolts are not really placed in shear or bearing. If the load exceeds the frictional resistance there will be slippage, with the result that the bolts will be placed in shear and bearing.

The Lap Joint

The joint shown in part (a) of Figure 12-1 is referred to as a *lap joint*. This type of joint has a disadvantage in that the center of gravity of the force in one member is not in line with the center of gravity of the force in the other member. A couple is present that causes an undesirable bending in the connection, as shown in the figure. For this reason the lap joint, which is desirably used only for minor connections, should be designed with at least two fasteners in each line parallel to the length of the member to minimize the possibility of a bending failure.

The Butt Joint

A *butt joint* is formed when three members are connected as shown in Figure 12-1(b). If the slip resistance between the members is negligible the members will slip a little and tend to shear off the bolts simultaneously on the two planes of contact between the members. Again, the members are bearing against the bolts and the bolts are said to be in *double shear and bearing* (also called *enclosed bearing*). The butt joint is more desirable than the lap joint for two main reasons. These are:

1. The members are arranged so that the total shearing force P is split into two parts, causing the force on each plane to be only about one-half of what is would be on a single plane if a lap joint were used. From a shear standpoint, therefore, the load-carrying ability of a group of bolts in double shear is theoretically twice as great as the same number of bolts in single shear.

2. A more symmetrical loading condition is provided. (In fact, the butt joint does provide a symmetrical situation if the outside members are the same thickness and resist the same forces. The result is a reduction or elimination of the bending described for a lap joint.)

Figure 12-2 (a) Hanger connection. (b) Bolts in multiple shear.

Double-Plane Connections

The double-plane connection is one in which the bolts are subjected to single shear and bearing, but in which bending moment is prevented. This type of connection, which is shown for a hanger in Fig. 12-2(a), subjects the bolts to single shear on two different planes.

Miscellaneous

Bolted connections generally consist of lap or butt joints or some combination of them, but there are other cases. For instance, there are occasionally joints in which more than three members are being connected and the bolts are in multiple shear, as shown in Fig. 12-2(b). In this figure you can see how the loads are tending to shear this bolt on four separate planes (quadruple shear). Although the bolts in this connection are being sheared on more than two planes, the usual practice is to consider no more than double shear for strength calculations. It seems rather unlikely that shear failures can occur simultaneously on three or more planes. Several other types of bolted connections are discussed in this chapter and the next. These include bolts in tension, bolts in shear and tension, etc.

12-11 FAILURE OF BOLTED JOINTS

Figure 12-3 shows several ways in which failure of bolted joints can occur. To design bolted joints satisfactorily it is necessary to understand these possibilities. These are described as follows.

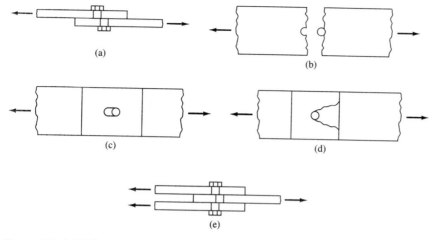

Figure 12-3 (a) Failure by single shearing of bolt. (b) Tension failure of plate. (c) Crushing failure of plate. (d) Shear failure of plate behind bolt. (e) Double shear failure of a butt joint.

1. The possibility of failure in a lap joint by shearing of the bolt on the plane between the members (single shear) is shown in part (a).
2. The possibility of a tension failure of one of the plates through a bolt hole is shown in part (b).
3. A possible failure of the bolts and/or plates by bearing between the two is given in (c).
4. The possibility of failure due to the shearing out of part of the member as shown in part (d).
5. The possibility of a shear failure of the bolts along two planes (double shear) as shown in part (e).

12-12 SPACING AND EDGE DISTANCES OF BOLTS

Before minimum spacings and edge distances can be discussed it is necessary for a few terms to be explained. The following definitions are given for a group of bolts in a connection and are shown in Fig. 12-4.

Pitch is the center-to-center distance of bolts in a direction parallel to the axis of the member.

Gage is the center-to-center distance of bolt lines perpendicular to the axis of the member.

The *edge distance* is the distance from the center of a bolt to the adjacent edge of a member.

The *distance between bolts* is the shortest distance between fasteners on the same or different gage lines.

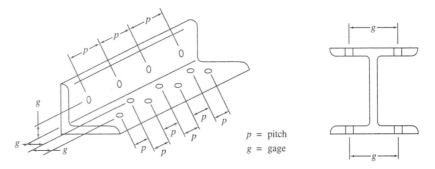

Figure 12-4

Minimum Spacings

Bolts should be placed a sufficient distance apart to permit efficient installation and to prevent bearing failures of the members between fasteners. The LRFD Specification (J3.3) provides a minimum center-to-center distance for standard, oversized, or slotted fastener holes equal to not less than $2\frac{2}{3}$ diameters (with 3 diameters being preferred). Test results have clearly shown that bearing strengths are directly proportional to the 3d center-to-center value up to a maximum of 3d.

As we will learn in the pages to follow, bearing strengths will have to be reduced when holes are lengthened in the direction parallel to the line of force. Table 12-3 (which is Table J3.7 of the LRFD Specification) shows the values of increments that have to be added to the 3d value to account for increases in hole dimensions (that is, for oversized and slotted holes) parallel to the line of force.

TABLE 12-3   VALUES OF SPACING INCREMENT C_1 FOR DETERMINING MINIMUM SPACINGS OF ENLARGED HOLES

Nominal dia. of fastener	Oversize holes	Slotted holes		
		Perpendicular to line of force	Parallel to line of force	
			Short slots	Long slots*
$\leq \frac{7}{8}$	$\frac{1}{8}$	0	$\frac{3}{16}$	$1\frac{1}{2}d - \frac{1}{16}$
1	$\frac{3}{16}$	0	$\frac{1}{4}$	$1\frac{7}{16}$
$\geq 1\frac{1}{8}$	$\frac{1}{4}$	0	$\frac{5}{16}$	$1\frac{1}{2}d - \frac{1}{16}$

* When length of slot is less than maximum allowed in Table 12.2, C_1 may be reduced by the difference between the maximum and actual slot lengths.

Source: American Institute of Steel Construction, *Manual of Steel Construction Load & Resistance Factor Design,* 2nd ed. (Chicago: AISC, 1994), Table J3.7, p. 6-86. Reprinted with the permission of the AISC.

Minimum Edge Distances

Bolts should not be placed too near the edges of a member for two major reasons. First, the punching of holes too close to the edges may cause the steel opposite the hole to bulge out or even crack. The second reason applies to the ends of members where there is danger of the fastener tearing through the metal. The usual practice is to place the fastener a minimum distance from the edge of the plates equal to about 1.5 to 2.0 times the fastener diameter so the metal there will have a shearing strength at least equal to that of the fasteners. For more exact information it is necessary to refer to the specification being used. The LRFD Specification (J3.4) states that the distance from the center of a standard hole to the edge of a connected part may not be less than the applicable value given in Table 12-4 (which is Table J3.4 of the LRFD Manual).

A reduced minimum edge distance is permitted (it's $1\frac{1}{4}$ in.) by the LRFD Specification for end connections bolted to beam webs and designed for beam shear reactions only. This information is shown as a footnote in Table 12-4.

The minimum edge distance from the center of an oversized hole or a slotted hole to the edge of a connected part must equal the minimum distance required for a standard hole plus an increment C_2, the values of which are provided in

TABLE 12-4 MINIMUM EDGE DISTANCE [a], in. (CENTER OF STANDARD HOLE [b] TO EDGE OF CONNECTED PART)

Nominal rivet or bolt diameter (In.)	At sheared edges	At rolled edges of plates, shapes or bars, or gas cut edges [c]
$\frac{1}{2}$	$\frac{7}{8}$	$\frac{3}{4}$
$\frac{5}{8}$	$1\frac{1}{8}$	$\frac{7}{8}$
$\frac{3}{4}$	$1\frac{1}{4}$	1
$\frac{7}{8}$	$1\frac{1}{2}$ [d]	$1\frac{1}{8}$
1	$1\frac{3}{4}$ [d]	$1\frac{1}{4}$
$1\frac{1}{8}$	2	$1\frac{1}{2}$
$1\frac{1}{4}$	$2\frac{1}{4}$	$1\frac{5}{8}$
Over $1\frac{1}{4}$	$1\frac{3}{4} \times$ Diameter	$1\frac{1}{4} \times$ Diameter

[a] Lesser edge distances are permitted to be used provided equations from LRFD Specification J3.10, as appropriate, are satisfied.

[b] For oversized or slotted holes, see Table 12-5.

[c] All edge distances in this column may be reduced $\frac{1}{8}$-in. when the hole is at a point where stress does not exceed 25 percent of the maximum design strength in the element.

[d] These may be $1\frac{1}{4}$-in. at the ends of beam connection angles and shear end plates.

Source: American Institute of Steel Construction, *Manual of Steel Construction Load & Resistance Factor Design,* 2nd ed. (Chicago: AISC, 1994) Table J3.4, p. 6-82. Reprinted with permission of AISC.

TABLE 12-5   VALUES OF EDGE DISTANCE INCREMENT C_2, in.

Nominal diameter of fastener (In.)	Oversized holes	Slotted holes		Long axis parallel to edge
		Long axis perpendicular to edge		
		Short slots	Long slots [a]	
$\leq \frac{7}{8}$	$\frac{1}{16}$	$\frac{1}{8}$		
1	$\frac{1}{8}$	$\frac{1}{8}$	$\frac{3}{4}d$	0
$\leq 1\frac{1}{8}$	$\frac{1}{8}$	$\frac{3}{16}$		

[a] When length of slot is less than maximum allowable (see Table 12-2), C_2 may be reduced by one-half the difference between the maximum and actual slot lengths.

Source: American Institute of Steel Construction, *Manual of Steel Construction Load & Resistance Factor Design,* 2nd ed. (Chicago: AISC, 1994) Table J3.8, p. 6-86. Reprinted with permission of AISC.

Table 12-5 (Table J3.8 of the LRFD Specification). As will be shown in the pages to follow, the bearing strengths of connections will have to be reduced if these requirements are not met.

Maximum Spacing and Edge Distances

Structural steel specifications provide maximum edge distances for bolted connections. The purpose of such requirements is to reduce the chances of moisture getting between the parts. When fasteners are too far from the edges of parts being connected, the edges may sometimes separate, thus permitting the entrance of moisture. When this happens and there is a failure of the paint, corrosion will develop and accumulate, causing increased separations between the parts. The LRFD maximum permissible edge distance (J3.5) is 12 times the thickness of the connected part, but not more than 6 in.

The maximum edge distances and spacings of bolts used for weathering steel are smaller than they are for regular painted steel subject to corrosion, or for regular unpainted steel not subject to corrosion. One of the requirements for using weathering steel is that it must not be allowed to be constantly in contact with water. As a result the LRFD Specification tries to insure that the parts of a built-up weathering steel member are connected tightly together at frequent intervals to prevent forming of pockets that might catch water. The LRFD Specification (J3.5) states that the maximum spacing of bolts center-to-center for painted members or for unpainted members not subject to corrosion is 24 times the thickness of the thinner plate, not to exceed 12 in. For unpainted members consisting of weathering steel subject to atmospheric corrosion, the maximum is 14 times the thickness of the thinner plate, not to exceed 7 in.

Holes cannot be punched very close to the web-flange junction of a beam or the junction of the legs of an angle. They can be drilled, but this rather expensive practice should not be followed unless there is an unusual situation. Even if the holes are drilled in these locations, there may be considerable difficulty in placing and tightening the bolts in the limited available space.

12-13 BEARING-TYPE CONNECTIONS—LOADS PASSING THROUGH CENTER OF GRAVITY OF CONNECTIONS

Shearing Strength   In bearing-type connections it is assumed that the loads to be transferred are larger than the frictional resistance caused by tightening the bolts, with the result that the members slip a little on each other, putting the bolts in shear and bearing. The design strength of a bolt in single shear equal ϕ times the nominal shearing strength of the bolt in ksi times its cross-sectional area. The LRFD ϕ values for shear are 0.75 for high-strength bolts, rivets, and A307 common bolts.

The nominal shearing strengths of bolts and rivets are given in Table 12.6 (Table J3.2 in the LRFD Specification). For A325 bolts the values are 48 ksi if threads are not excluded from shear planes and 60 ksi if threads are excluded. (The values are 60 ksi and 75 ksi for A490 bolts.) Should a bolt be in double shear, its shearing strength is considered to be twice its single shear value.

The student may very well wonder what is done in design practice concerning threads excluded or not excluded from the shear planes. If normal bolt and member sizes are used, the threads will almost always be excluded from the shear plane. It is true, however, that some extremely conservative individuals always assume the threads are not excluded from the shear plane.

Sometimes the designer needs to use high-strength bolts with diameters larger than those available with A325 and A490 bolts. One example is the use of very large bolts for fastening machine bases. For such situations LRFD Specification A3.3 permits the use of the quenched and tempered A449 bolts. The LRFD Specification A3.3 says that such bolts can be used only when diameters larger than $1\frac{1}{2}$ in. are required, and then only for bearing type connections.

Bearing Strength   The bearing strength of a bolted connection is not, as you might expect, determined from the strength of the bolts themselves; rather, it is based upon the strength of the parts being connected and the arrangement of the bolts. In detail, its computed strength is dependent upon the spacing of the bolts and their edge distances, the specified tensile strength F_u of the connected parts, and their thicknesses.

The bearing design strength of a bolt equals ϕ (which is 0.75) times the nominal bearing strength of the connected part (R_n). Expressions for R_n are provided in Section J3.10 of the LRFD Specification. The various expressions listed there include bolt diameters (d) and the thicknesses of members bearing against the bolts (t). Some of the expressions also contain a spacing term (s)

TABLE 12-6   DESIGN STRENGTH OF FASTENERS

Description of fasteners	Tensile strength		Shear strength in bearing-type connections	
	Resistance factor ϕ	Nominal strength, ksi	Resistance factor ϕ	Nominal strength, ksi
A307 bolts	0.75	45.0[a]	0.75	24[b, e]
A325 bolts, when threads are not excluded from shear planes		90[d]	0.75	48[e]
A325 bolts, when threads are excluded from shear planes		90[d]		60[e]
A490 bolts, when threads are not excluded from shear planes		113[d]		60[e]
A490 bolts, when threads are excluded from shear planes		113[d]		75[e]
Threaded parts meeting the requirements of LRFD Sect. A3, when threads are excluded from shear planes		$0.75F_u$[a, c]	0.75	$0.40F_u$
Threaded parts meeting the requirements of LRFD Sect. A3, when threads are excluded from shear planes		$0.75F_u$[a, c]		$0.50F_u$[a, c]
A502, Gr. 1, hot-driven rivets		45.0[a]		25[e]
A502, Gr. 2 & 3, hot-driven rivets		60.0[a]		33[e]

[a] Static loading only.

[b] Threads permitted in shear planes.

[c] The nominal tensile strength of the threaded portion of an upset rod, based upon the cross-sectional area at its major thread diameter, A_D shall be larger than the nominal body area of the rod before upsetting times F_y.

[d] For A325 and A490 bolts subject to tensile fatigue loading, see Appendix K3.

[e] When bearing-type connections used to splice tension members have a fastener pattern whose length, measured parallel to the line of force, exceeds 50 in., tabulated values shall be reduced by 20 percent.

Source: American Institute of Steel Construction, *Manual of Steel Construction Load & Resistance Factor Design,* 2nd ed. (Chicago: AISC, 1994) Table J3.2, p. 6-81. Reprinted with permission of AISC.

which represents the distance between the centers of standard size holes in the direction of the line of force. Should we have short- or long-slotted holes with their slots perpendicular to the line of force, s is the distance between the centers of those holes. For oversized holes and for slotted holes parallel to the line of force, s is increased by the appropriate spacing increment C_1 of Table 12-3 (LRFD Table J3.7).

(a) When $L_e \geq 1.5d$ and $s \geq 3d$, and when there are two or more bolts in the line of force.

If deformation around bolt holes is a design consideration (that is, if we want deformations to be ≤ 0.25 in.)

$$R_n = 2.4 \, dtF_u \qquad\qquad \text{(LRFD Equation J3-1a)}$$

For the problems considered in this text we will normally assume that deformations around the bolt holes are important. Thus, unless specifically stated otherwise, the value $2.4 \, dtF_u$ will be used for bearing calculations.

If deformation around bolt holes is not a design consideration (that is, if deformations > 0.25 in. are acceptable)

For the bolt hole nearest the edge

$$R_n = L_e tF_u \leq 3.0 \, dtF_u \qquad\qquad \text{(LRFD Equation J3-1b)}$$

For the remaining bolts

$$R_n = \left(s - \frac{d}{2}\right)tF_u \leq 3.0 dtF_u \qquad\qquad \text{(LRFD Equation J3-1c)}$$

For long-slotted bolt holes perpendicular to the line of force

$$R_n = 2.0 \, dtF_u \qquad\qquad \text{(LRFD Equation J3-1d)}$$

(b) When $L_e < 1.5d$ or $s < 3d$, or where there is only a single bolt in the line of force

For a single bolt or for the bolt nearest the edge when there are two or more bolts in the line of force

$$R_n = L_e tF_u \leq 2.4 \, dtF_u \qquad\qquad \text{(LRFD Equation J3-2a)}$$

For the remaining bolts

$$R_n = \left(s - \frac{d}{2}\right)tF_u \leq 2.4 \, dtF_u \qquad\qquad \text{(LRFD Equation J3-2b)}$$

Values for R_n also are provided in Section J3.10 of the specification for long-slotted holes perpendicular to the line of force. It is to be remembered as described in Section 12.9 of this chapter that oversized holes and short- and long-slotted holes can be used only for slip-critical connections unless the loads act perpendicular to the slots.

Tests of bolted joints have shown that neither the bolts nor the metal in contact with the bolts actually fail in bearing. However, these tests also have shown that the efficiency of the connected parts in tension and compression is affected by the magnitude of the bearing stress. Therefore, the nominal bearing strengths given by the LRFD Specification are values above which they feel the strength of the connected parts is impaired. In other words, these apparently very high design bearing stresses are not really bearing stresses at all, but, rather, indexes of the efficiencies of the connected parts. If bearing stresses larger than the values given are permitted, the holes seem to elongate more than about $\frac{1}{4}$ in. and impair the strength of the connections.

From the preceding we can see that the bearing strengths given are not specified to protect fasteners from bearing failures because they do not need such protection. Thus the same bearing values will be used for a particular joint regardless of the grades of bolts used and regardless of the presence or absence of bolt threads in the bearing area.

Minimum Connection Strength   The LRFD Specification (Section J1.7) states that except for lacing, sag rods, and girts, connections must be provided with design strengths sufficient to support factored loads of at least 10 kips.

Example 12-1 illustrates the calculations involved in determining the strength of a bearing-type connection. Using a similar procedure the number of bolts required for a certain loading condition is calculated in Example 12-2. In each case the bearing thickness to be used equals the smaller total thickness on one side or the other. For instance, in Fig. 12-6 it equals the smaller of $2 \times \frac{1}{2}$ in. on the left or $\frac{3}{4}$ in. on the right.

The values given for the strength of bolts in this chapter, whether bearing-type or slip-critical, can be obtained from the tables entitled "Design Shear Strength of One Bolt" and "Design Bearing Strength at Bolt Holes" in Section 8 of the Manual.

In the connection tables of the LRFD Manual and in various bolt literature we constantly see abbreviations used when referring to various types of bolts. For instance, we may see A325-SC, A325-N, A325-X, A490-SC, and so on. These are used to represent the following

A325-SC—slip-critical or fully tensioned A325 bolts

A325-N—snug-tight or bearing A325 bolts with threads *included* in the shear planes

A325-X—snug-tight or bearing A325 bolts with threads *excluded* from the shear planes

■ **EXAMPLE 12-1**

Determine the design strength P_u of the bearing-type connection shown in Fig. 12-5. The steel is A572 Grade 50, the bolts are $\frac{7}{8}$-in. A325, the holes are standard sizes, the threads are excluded from the shear plane, edge distances are $> 1\frac{1}{2}d$, and the distance center-to-center of the holes is $> 3d$.

Figure 12-5

Solution. Design strength of plates:

$A_g = (\frac{1}{2})(12) = 6.0$ in.^{2}
$A_n = 6.00 - (2)(1.0)(\frac{1}{2}) = 5.0$ in.$^2 = A_e$
$P_u = \phi_t F_y A_g = (0.9)(50)(6.0) = 270$ k
$P_u = \phi_t F_u A_e = (0.75)(65)(5.0) = 243.7$ k

Bolts in single shear and bearing on $\frac{1}{2}$ in.:

$P_u = \phi(0.6)(60)(4) = (0.75)(0.6)(60)(4) = 108$ k $\leftarrow$
$P_u = \phi 2.4 dt F_u = (0.75)(2.4)(\frac{7}{8})(\frac{1}{2})(65)(4) = 204.7$ k

$$\text{Design } P_u = 108 \text{ k} \qquad \text{(LRFD Equation J3-1a)} \qquad \blacksquare$$

■ EXAMPLE 12-2

How many $\frac{3}{4}$-in. A325 bolts in standard-size holes with threads excluded from the shear plane are required for the bearing-type connection shown in Fig. 12-6? Use A36 steel and assume edge distances $>1\frac{1}{2}d$ and the distance center-to-center of holes to be $>3d$.

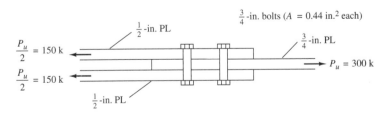

Figure 12-6

Solution. Bolts in double shear and bearing on $\frac{3}{4}$ in.:

$$\text{Design shear strength per bolt} = (0.75)(2 \times 0.44)(60) = 39.6 \text{ k} \leftarrow$$
$$\text{Design bearing strength per bolt} = (0.75)(2.4)(\tfrac{3}{4})(\tfrac{3}{4})(58) = 58.7 \text{ k}$$
$$\text{No. of bolts required} = \frac{300}{39.6} = 7^+$$

Use 8 or 9 bolts (depending on arrangement) ■

Where cover plates are bolted to the flanges of W sections the bolts must carry the longitudinal shear on the plane between the plates and the flanges. With reference to the cover-plated beam of Fig. 12-7 the unit longitudinal shearing stress to be resisted between a cover plate and the W flange can be determined with the expression $f_v = VQ/Ib$. The total shear force across the flange for a 1-in. length of the beam equals $(b)(1.0)(VQ/Ib) = VQ/I$.

The LRFD Specification (E4) provides a maximum permissible spacing for bolts used in the outside plates of built-up members. It equals the thinner outside plate thickness times $127/\sqrt{F_y}$ or 12 in., whichever is smaller.

The spacing of pairs of bolts in Fig. 12-7 can be determined by dividing the design strength of two bolts by the shear force to be taken per inch at a particular section. The theoretical spacings will vary as the external shear varies along the span. Example 12-3 illustrates the calculations involved in determining bolt spacing for a cover-plated beam.

■ EXAMPLE 12-3

At a certain section in the cover-plated beam of Fig. 12-7 the external factored shear V_u is 275 k. Determine the spacing required for $\frac{7}{8}$-in. A325 bolts used in a bearing-type connection. Assume the $1\frac{1}{2}d$ and $3d$ edge and center-to-center distance requirements are met and that the bolt threads are excluded from the shear plane. The steel is A36.

PL $\frac{3}{4}$ in. × 16

0.75 in.

W21 × 147

(I_x = 3630, t_f = 1.150)

23.56 in.

22.06 in.

0.75 in.

PL $\frac{3}{4}$ in. × 16

P

Figure 12-7

Solution.

$$I_x = 3630 + (2)(16 \times \tfrac{3}{4})(11.405)^2 = 6752 \text{ in.}^4$$

Factored shear to be taken per inch $= \dfrac{V_u Q}{I}$

$$= \frac{(275)(16 \times \tfrac{3}{4} \times 11.405)}{6752} = 5.574 \text{ k/in}$$

Bolts in single shear and bearing on 0.75 in.

Design shear strength for 2 bolts $= (2)(0.75)(0.6)(60) = 54 \text{ k} \leftarrow$

Design bearing strength for 2 bolts $= (2)(0.75)(2.4)(\tfrac{7}{8})(\tfrac{3}{4})(58) = 137 \text{ k}$

$$p = \frac{54}{5.574} = 9.69 \text{ in.} \quad \text{(Say 9 in. on center)}$$

$$\text{Maximum } p = \frac{3}{4} \frac{127}{\sqrt{36}} = 15.88 \text{ in. or 12 in.}$$

$$\text{Place bolts 9 in. o.c.} \qquad \blacksquare$$

Example 12-4 illustrates the calculation of the design strength of a connection where the edge distance L_e is $< 1.5d$ and the spacing of the bolts is $< 3d$.

■ EXAMPLE 12-4

Compute the design strength of the bearing-type connection shown in Fig. 12-8. The bolts are 1-in. A325 with threads excluded from the shear plane and the steel is A36. It is assumed that the plate strength is satisfactory.

Figure 12-8

Solution. Bolts in single shear and bearing on $\tfrac{3}{8}$ in.

Design shear strength of 4 bolts $= (0.75)(0.785)(60)(4) = 141.3 \text{ k}$

Noting that $L_e = 1.25 \text{ in.} < 1.5d = 1.50 \text{ in.}$

and $s = 2.00 \text{ in.} < 3d = 3.00 \text{ in.}$

Design bearing strength of each end bolt

$$= \phi L_e t F_u = (0.75)(1.25)(\tfrac{3}{8})(58) = 20.39 \text{ k}$$

Design bearing strength of each interior bolt

$$= \phi(s - \tfrac{d}{2})t F_u = 0.75(2 - \tfrac{1.00}{2})(\tfrac{3}{8})(58) = 24.47 \text{ k}$$

Total design bearing strength of 4 bolts

$$= (2)(20.39) + (2)(24.47) = 89.7 \text{ k} \leftarrow$$

$$P_u = 89.7 \text{ k}$$

The assumption has been made that the loads applied to a bearing-type connection are equally divided between the bolts if edge distances and spacings are satisfactory. For this distribution to be correct the plates must be perfectly rigid and the bolts perfectly elastic, but actually the plates being connected are elastic too and have deformations that decidedly affect the bolt stresses. The effect of these deformations is to cause a very complex distribution of load in the elastic range.

Should the plates be assumed to be completely rigid and nondeforming, all bolts would be deformed equally and have equal stresses. This situation is shown in part (a) of Fig. 12-9. Actually, the loads resisted by the bolts of a group are probably never equal (in the elastic range) when there are more than two bolts in a line. Should the plates be deformable, the plate stresses and thus the deformations will decrease from the ends of the connection to the middle, as

Figure 12-9 (a) Assuming nondeforming plates. (b) Assuming deformable plates. (c) Stepped joint (impractical).

Three high-strength bolted structures in Constitution Plaza Complex, Hartford, Conn., using approximately 195,000 bolts. (Courtesy of Bethlehem Steel Corporation.)

shown in part (b) of Fig. 12-9. The result is that the highest stressed elements of the top plate will be over the lowest stressed elements of the lower plate, and vice versa. The slip will be greatest at the end bolts and smallest at the middle bolts. The bolts at the ends will then have stresses much greater than those in the inside bolts.

The greater the spacing of bolts in a connection the greater will be the variation in bolt stresses due to plate deformation; therefore, the use of compact joints is very desirable as they will tend to reduce the variation in bolt stresses. It might be interesting to consider a theoretical (although not practical) method of roughly equalizing bolt stresses. The theory would involve the reduction of the thickness of the plate toward its end in proportion to the reduced stresses by stepping. This procedure, which is shown in Fig. 12-9 (c), would tend to equalize the deformations of the plate and thus the bolt stresses. A similar procedure would be to scarf the overlapping plates.

The calculation of the theoretically correct elastic stresses in a bolted group based on plate deformations is a tedious problem and is rarely handled in the design office. On the other hand the analysis of a bolted joint based on the plastic theory is a very simple problem. In this theory the end bolts are assumed to be stressed to their yield point. Should the total load on the connection be increased, the end bolts will deform without resisting additional load, the next bolts in the line will have their stresses increased until they too are at the yield point, and so forth. Plastic analysis seems to justify to a certain extent the assumption of rigid plates and equal bolt stresses that is usually made in design practice. This assumption is used in the example problems of this chapter.

When there are only a few bolts in a line the plastic theory of equal stresses seems to be borne out very well, but when there are a large number of bolts in a line the situation changes. Tests have clearly shown that the end bolts will fail before the full redistribution takes place.[3]

The LRFD Specification requires that the design shearing strength of a bearing-type connection used to splice a tension member be reduced if it has a fastener pattern longer than 50 in. parallel to the line of force. As a footnote to their Table J3.2 (it's Table 12.6 herein) it is stated that tabulated values for such situations must be reduced by 20 percent.

For load-carrying bolted joints it is common for specifications to require a minimum of two or three fasteners. The feeling is that a single connector may fail to live up to its specified strength because of improper installation, material weakness, etc., but if several fasteners are used the effects of one bad fastener in the group will be overcome.

12-14 SLIP-CRITICAL CONNECTIONS—LOADS PASSING THROUGH CENTER OF GRAVITY OF CONNECTIONS

Slip-critical connections may be designed either for service load conditions as described in LRFD Specification J3.8a, or they may be designed for factored loads as described in LRFD Appendix J3.8b. Though the number of bolts required by the two methods will be approximately the same, there can be a little variation with different live-to-dead load ratios. The author uses the service load approach herein. Regardless of the method used, it is necessary, when the direction of loading is toward the edge of a connected part, to provide adequate bearing capacity at factored loads.

If bolts are tightened to their required tensions for slip-critical connections (see Table 12-1) there is very little chance of their bearing against the plates that they are connecting. In fact, tests show that there is very little chance of slip occurring unless there is a calculated shear of at least 50 percent of the total bolt tension. This means, as we have said all along, that slip-critical bolts are not stressed in shear; however, the LRFD Specification J3.8a provides allowable shear strengths (they are really permissible friction values on the faying surfaces) so that the designer can handle the connections in just about the same manner he or she uses for bearing-type connections. The LRFD assumes the bolts are in "shear" and no bearing. The nominal shear strengths for such bolts are given in Table 12-7 (which is Table J3.6 of the Manual). $\phi = 1.0$, except it's 0.85 for long-slotted holes with load parallel to slot.

---

[3] *Trans. ASCE* 105 (1940), p. 1193.

TABLE 12-7 SLIP-CRITICAL NOMINAL RESISTANCE TO SHEAR,
ksi, OF HIGH-STRENGTH BOLTS[a]

	Nominal Resistance to Shear		
Type of bolt	Standard size holes	Oversized and short-slotted holes	Long-slotted holes
A325	17	15	12
A490	21	18	15

[a] For each shear plane.

It is permissible to introduce finger shims up to $\frac{1}{4}$ in. thick into slip-critical connections with standard holes without the necessity of reducing the bolt design strength values to those specified for slotted holes (LRFD J3.8a).

The values given in Table 12.7 for nominal slip-critical shear strengths are based on Class A, clean mill-scale and blast-cleaned surfaces with Class A coatings with a slip coefficient of 0.33. The designer may specify connections with special surface conditions and increase the nominal slip resistance to the applicable values given in the RCSC Load and Resistance Factor Design Specification.[4]

The preceding discussion concerning slip-critical joints does not present the whole story because during erection the joints may be assembled with bolts, and as the members are erected their weights will often push the bolts against the side of the holes before they are tightened and put them in some bearing and shear.

Example 12-5 presents the design of a slip-critical connection for a lap joint. First, the number of bolts required for the service load limit state of no slippage is determined. Then the number of bolts required for the factored load limit state is computed assuming the slip resistance is overcome and the bolts subjected to bearing.

■ EXAMPLE 12-5

It is desired to design a slip-critical connection for the plates shown in Fig. 12-10 to resist the axial service loads $P_D = 30$ k and $P_L = 50$ k using 1-in. A325 high-strength bolts with threads excluded from the shear plane and with standard-size holes, $L_e > 1\frac{1}{2}d$ and distance center-to-center of bolts $> 3d$. A36 steel.

Figure 12-10

[4] Research Council on Structural Connections, *Load and Resistance Factor Design Specifications for Structural Joints Using ASTM A325 or A490 bolts* (Chicago: AISC, June 8, 1988).

Solution.   Slip-critical design (service loads)
Bolts in "single shear" and no bearing

"ss" strength of one bolt $= (\phi)(0.785)(17) = (1.0)(0.785)(17) = 13.35$ k

$$\text{No. of bolts required} = \frac{80}{13.35} = 5.99 \quad \underline{\text{say 6}}$$

Design as bearing-type connection (factored loads)
$P_u = (1.2)(30) + (1.6)(50) = 116$ k

Bolts in single shear and bearing on $\frac{5}{8}$ in. ss strength of one
bolt $= (0.75)(0.785)(60) = 35.32$ k $\leftarrow$

Bearing strength of one bolt $= (0.75)(2.4)(1.0)(\frac{5}{8})(58) = 65.25$ k

$$\text{No. of bolts required} = \frac{116}{35.32} = 3^+$$

$$\underline{\underline{\text{Use 6 bolts}}} \qquad\qquad \blacksquare$$

In Example 12-6 bearing-type bolts are used to connect a beam to a pair of gusset plates. The tensile strength of the W section and the plates along with the strength of the bolts are determined. Finally, the block shear strength of the W section is calculated.

Bridge over Allegheny River at Kittaning, PA. (Courtesy American Bridge Company.)

■ EXAMPLE 12-6

The connection shown in Fig. 12-11 is made with $\frac{7}{8}$-in. A325 bearing-type bolts in standard-size holes with the threads excluded from the shear planes. The beam and gusset plates consist of A36 steel. Check the following items: (a) the tensile strengths of the W section and the gusset plates, (b) the strength of the bolts in single shear and bearing, and (c) the block shear strength of the W section crosshatched areas shown in part (b) of the figure.

Figure 12-11

Solution.

(a) Tensile design strength of W section

$$P_u = \phi_t F_y A_g = (0.9)(36)(11.2) = 362.9 \text{ k} > 330 \text{ k} \qquad \text{(OK)}$$
$$A_n = 11.2 - (4)(1)(0.515) = 9.14 \text{ in.}^2$$

$$U \text{ from Chapter 3} = 1 - \frac{\bar{x}}{L} = 1 - \frac{1.54}{L} = 0.74$$

$$A_e = (0.74)(9.14) = 6.76 \text{ in.}^2$$
$$P_u = \phi_t F_u A_e = (0.75)(58)(6.76) = 294 \text{ k} < 330 \text{ k} \qquad \text{(NG)}$$

Tensile design strength of gusset plates

$$P_u = \phi_t F_y A_g = (0.9)(36)(\tfrac{1}{2} \times 12)(2) = 388.8 \text{ k} > 330 \text{ k} \quad \text{(OK)}$$
$$A_n \text{ of 2 plates} = [(\tfrac{1}{2})(12) - (2)(1)(\tfrac{1}{2})]2 = 10 \text{ in.}^2 \leftarrow$$
$$0.85 \, A_g = (0.85)(\tfrac{1}{2})(12)(2) = 10.2 \text{ in.}^2$$
$$P_u = \phi_t F_u A_n = (0.75)(58)(10) = 435 \text{ k} > 330 \text{ k} \quad \text{(OK)}$$

(b) Bolts in ss and bearing on $\tfrac{1}{2}$ in.

$$\text{ss design strength of bolts} = (0.75)(0.6)(60)(12)$$
$$= 324 \text{ k} < 330 \text{ k} \quad \text{(NG)}$$

Bearing design strength of bolts

$$= (0.75)(2.4)(\tfrac{7}{8})(\tfrac{1}{2})(58)(12) = 548.1 \text{ k} > 330 \text{ k} \quad \text{(OK)}$$

(c) Block shearing strength for W section

Total areas for top and bottom

$$A_{gv} = \text{gross area subject to shear}$$
$$= (4)(0.515)(7\tfrac{1}{2}) = 15.45 \text{ in.}^2$$
$$A_{gt} = \text{gross area subject to tension}$$
$$= (4)(0.515)(1.635) = 3.37 \text{ in.}^2$$
$$A_{nv} = \text{net area subject to shear}$$
$$= (4)(7.50 - 2.5 \times 1.0)(0.515) = 10.30 \text{ in.}^2$$
$$A_{nt} = \text{net area subject to tension}$$
$$= (4)(1.635 - \tfrac{1}{2} \times 1)(0.515) = 2.34 \text{ in.}^2$$

Checking to see which equation is applicable

$$F_u A_{nt} = (58)(2.34) = 135.7 \text{ k}$$
$$< 0.6 \, F_u A_{nv} = (0.6)(58)(10.30) = 358.4 \text{ k}$$
$$\therefore \text{ Must use LRFD Equation J4-3b}$$
$$\phi R_n = \phi[0.6 \, F_u A_{nv} + F_y A_{gt}]$$
$$= 0.75 \, [(0.6)(58)(10.30) + (36)(3.37)]$$
$$= 359.8 \text{ k} > 330 \text{ k} \quad \text{(OK)} \quad \blacksquare$$

The calculations for riveted connections and for connections made with A307 common bolts are made almost exactly as are the ones for high-strength bolts in bearing-type connections. The only difference is that the shear strength values for these connections are much smaller. The LRFD Specification does not permit the design of slip-critical joints using rivets or common bolts. Rivet and common bolt examples are presented in Chapter 13.

12-15 COMPUTER EXAMPLE

In Example 12-7 INSTEP is used to determine the strength of a bolted joint. A few remarks are made here to help you input the correct data.

To select the type bolts to be used press the letter B as shown on the screen repeatedly until the type desired appears on the screen and then press the return or enter key. The same situation occurs for keys D, E and F where the bolt diameter, stress and loading conditions are specified. To specify the bolt pattern press the letter I. When the picture appears on the screen each bolt and its position are specified by pressing the appropriate letter and number. The rest of the work should be self-explanatory.

■ EXAMPLE 12-7

Repeat Example 12-1 using INSTEP.

Solution.

INPUT

|A| Type Steel: A572 Grade 50
|B| Type Bolt: A325 Bolts

Connection Details

|D| Bolt Size: .875 inches
|E| Threads: SS Bearing, Excluded
|F| Joint Type: Shear and Bearing
|G| t_1 = .5 inches
|H| t_2 = .5 inches
|I| Specify Bolt Pattern

Joint Forces Acting

|K| P = 0 kips
|L| V = 0 kips
|M| e = 0 inches

Set Horizontal Spacing

|H| S_h = 3 inches

OUTPUT

Design Loads

V_u = 108.2 kips (Factored Shear)
P_u = 0 kips (Factored Tension) ■

PROBLEMS

For each of the problems listed, the following information is to be used unless otherwise indicated: (a) LRFD Specification; (b) standard-size holes; (c) members have clean mill-scale surfaces (class A); (d) edge distances are $\geq 1.5\,d$ and distances center-to-center of holes are $\geq 3d$.

12-1 to 12-5. *Determine the design tensile strength P_u for the member and the connection shown assuming a bearing-type connection.*

Problems 12-1 to 12-5

12-1. A36 steel, $\frac{7}{8}$-in. A325 bolts, threads excluded from shear plane. (*Ans.* 243 k)

12-2. A36 steel, $\frac{3}{4}$-in. A325 bolts, threads excluded from shear plane.

12-3. A36 steel, 1-in. A325 bolts, threads not excluded from shear plane. (*Ans.* 254.3 k)

12-4. A572, grade 50 steel, 1-in. A490 bolts, threads excluded from shear plane.

12-5. A572, grade 50 steel, $\frac{7}{8}$-in. A490 bolts, threads not excluded from shear plane. (*Ans.* 243 k)

12-6 to 12-10. *Determine the design tensile strength P_u for the member and the bearing-type connection shown.*

Problems 12-6 to 12-10

12-6. A36 steel, $\frac{7}{8}$-in. A325 bolts, threads excluded from shear planes.

12-7. A36 steel, 1-in. A325 bolts, threads not excluded from shear planes. (*Ans.* 328.4 k)

12-8. A36 steel, $\frac{3}{4}$-in. A325 bolts, threads not excluded from shear plane.

12-9. A steel with $F_y = 50$ ksi, $F_u = 70$ ksi, 1-in. A490 bolts, threads excluded from shear plane. (*Ans.* 396.4 k)

12-10. A steel with $F_y = 50$ ksi, $F_u = 70$ ksi, $\frac{3}{4}$-in. A490 bolts, threads excluded from shear planes.

12-11 to 12-13. *How many bolts are required for the bearing-type connection shown?*

Problems 12-11 to 12-13

12-11. A36 steel, $\frac{3}{4}$-in. A325 bolts, threads excluded from shear plane. (*Ans.* 11.11, use 12)

12-12. $F_y = 50$ ksi, $F_u = 70$ ksi, $\frac{7}{8}$-in. A325 bolts, threads excluded from shear plane.

12-13. A36 steel, 1-in. A325 bolts, threads not excluded from shear plane. (*Ans.* 7.78, use 8 or 9)

12-14 to 12-16. *How many bolts are required for the bearing type connection shown?*

Problems 12-14 to 12-16

12-14. A36 steel, $\frac{7}{8}$-in. bolts, threads excluded from shear planes.

12-15. A36 steel, $\frac{7}{8}$-in. bolts, threads not excluded from shear planes. (*Ans.* 8.33, use 9 or 10)

12-16. $F_y = 50$ ksi, $F_u = 65$ ksi, $\frac{7}{8}$-in. A325 bolts, threads excluded from shear planes.

12-17. The truss member shown in the accompanying illustration consists of two C12 × 25s (A36 steel) connected to a 1-in. gusset plate. How many $\frac{7}{8}$-in. A325 bolts (threads excluded) from shear plane are required to develop the full design tensile capacity of the member if it is used as a bearing-type connection? Assume $U = 0.85$. (*Ans.* 8.82, use 9 or 10)

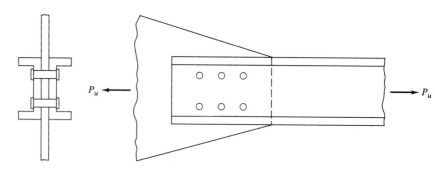

Problem 12-17

12-18. Repeat Prob. 12-17 using A242 Grade 46 steel and A490 bolts.

12-19. Rework Prob. 12-3 if 1-in. A490 bolts and A572 steel ($F_y = 50$ ksi, $F_u = 65$ ksi) are used. (*Ans.* 315.4 k)

12-20. For the connection shown in the accompanying illustration P_u equals 750 k. Determine the number of 1-in. A325 bolts required for a bearing-type connection using A36 steel. Threads excluded from shear plane.

Problem 12-20

12-21. Rework Prob. 12-20 using $\frac{7}{8}$-in. A490 bolts. (*Ans.* 11.11, use 12)

12-22. How many $\frac{3}{4}$-in. A325 bolts (threads excluded from shear planes) in a bearing-type connection are required to develop the design tensile strength of the member shown? Assume A36 steel is used and that there are two lines of bolts in each flange (at least three in a line 4 in. o.c.).

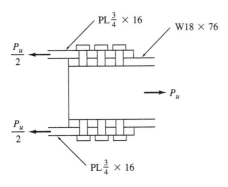

PL$\frac{3}{4}$ × 16   W18 × 76

$\frac{P_u}{2}$

P_u

$\frac{P_u}{2}$

PL$\frac{3}{4}$ × 16

Problem 12-22

12-23. For the beam shown in the accompanying illustration, what is the required spacing of $\frac{7}{8}$-in. A325 bolts (thread excluded from shear plane) in a bearing-type connection at a section where the external shear V_u is 300 k? A36 steel. (*Ans.* 7.76 in, use $7\frac{3}{4}$ in.)

1 × 12 PL

W24 × 94

1 × 12 PL

Problem 12-23

12-24. The cover-plated section shown in the accompanying illustration is used to support a uniform load $w_D = 12$ klf (includes beam weight effect and $w_L = 15$ klf for an 18-ft simple span). If $\frac{7}{8}$-in. A325 bolts (threads excluded) are used in a bearing-type connection, work out a spacing diagram for the entire span. A36 steel.

$\frac{5}{16} \times 16$ PL

W27 × 178

$\frac{5}{16} \times 16$ PL

Problem 12-24

12-25. For the section shown in the accompanying illustration determine the required spacing of $\frac{3}{4}$-in. A490 bolts (threads excluded) for a bearing-type connection if the member consists of A572 grade 60 steel ($F_u = 75$ ksi). $V_D = 100$ k and $V_L = 140$ k. (*Ans.* 9.21 in, use 9 in.)

$2\frac{1}{2}$ in.

PL$\frac{1}{2} \times 48$

$48\frac{1}{2}$ in.

$43\frac{1}{2}$ in.

$2\frac{1}{2}$ in.

Ls 8 × 4 × $\frac{1}{2}$

(short legs back-to-back)

Problem 12-25

12-26. For an external shear V_u of 900 k determine the spacing required for $\frac{3}{4}$-in. A325 web bolts (threads excluded) in a bearing-type connection for the built-up section shown in the accompanying illustration. A36 steel.

PLs $\frac{1}{2}$ × 20

Ls 8 × 8 × $\frac{1}{2}$

PL $\frac{5}{8}$ × 54

Note: bolts are not staggered

3 in.

3 in.

42$\frac{1}{2}$ in.

54$\frac{1}{2}$ in.

3 in.

3 in.

Problem 12-26

12-27. Determine the design strength P_u of the bearing-type connection shown if $\frac{7}{8}$ in. A325 bolts (threads excluded) and A36 steel are used. Notice that $L_e < 1\frac{1}{2}d$ and $s < 3d$. (*Ans.* 154.9 k)

PL $\frac{1}{2}$ × 12

P_u

PL $\frac{1}{2}$ × 12

P_u

P_u

3 in.

6 in.

3 in.

P_u

1 in.   2 in.   2 in.   1 in.

Problem 12-27

12-28 to 12-33. *Repeat these problems using slip-critical connections, Class A surfaces and the service loads given below.*

12-28. Prob. 12-11. $P_D = 60$ k, $P_L = 90$ k

12-29. Prob. 12-7. $P_D = 50$ k, $P_L = 100$ k (*Ans.* 14.71, say 15)

12-30. Prob. 12-9. $P_D = 75$ k, $P_L = 160$ k

12-31. Prob. 12-14. $P_D = 60$ k, $P_L = 80$ k (*Ans.* 6.86, say 8 or 9)

12-32. Prob. 12-13. $P_D = 40$ k, $P_L = 100$ k

12-33. Prob. 12-21. $P_D = 120$ k, $P_L = 360$ k (*Ans.* 23.53, say 24)

12-34 to 35. *Repeat these problems using slip-critical connections.*

12-34. Prob. 12-24

12-35. Prob. 12-25 (*Ans.* 4.50 in.)

12-36. Determine the design tensile strength P_u of the connection shown if A36 steel and eight $\frac{7}{8}$-in. A325 bearing-type bolts (threads excluded from shear plane) are used in each flange. Include block shear in your calculations.

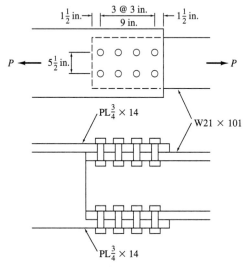

Problem 12-36

12-37. Repeat Prob. 12-36 using A572 Grade 50 steel and 1-in. A325 bearing-type bolts. (*Ans.* 565.2 k)

12-38 to 12-43. *Repeat the problems listed using the enclosed computer diskette.*

12-38. Prob. 12-1

12-39. Prob. 12-6 (*Ans.* bolt strength 486 k)

12-40. Prob. 12-11

12-41. Prob. 12-14 (*Ans.* 6.67)

12-42. Prob. 12-28

12-43. Prob. 12-31 (*Ans.* 6.86)

Chapter **13**

Eccentrically Loaded Bolted Connections and Historical Notes on Rivets

13-1 BOLTS SUBJECTED TO ECCENTRIC SHEAR

Eccentrically loaded bolt groups are subjected to shears and bending moments. The student may feel such situations are rare but the truth is that they are much more common than he or she suspects. For instance, in a truss it is desirable to have the center of gravity of a member lined up exactly with the center of gravity of the bolts at its end connections. This feat is not quite as easy to accomplish as it may seem, and connections are often subjected to moments.

Eccentricity is quite obvious in Fig. 13-1(a), where a beam is connected to a column with a plate. In part (b) of the figure another beam is connected to a column with a pair of web angles. It is obvious that this connection must resist some moment because the center of gravity of the load from the beam does not coincide with the reaction from the column.

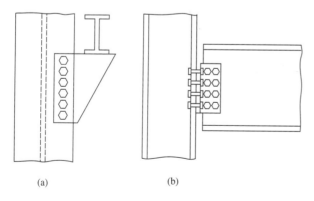

(a)   (b)

Figure 13-1 Eccentrically loaded bolt groups

In general, specifications for bolts and welds clearly state that the center of gravity of the connection should coincide with the center of gravity of the member unless the eccentricity is accounted for in the calculations. However, Section J1.8 of the LRFD Specification provides some exceptions to this rule. It states that the rule is not applicable to the end connections of statically loaded single angles, double angles, and similar members. In other words, the eccentricities between the centers of gravity of these members and the centers of gravity of the connections may be ignored unless fatigue loadings are involved. *Furthermore, the eccentricity between the gravity axes and the gage lines of bolted members may be neglected for statically loaded members.*

The LRFD Specification presents values for computing the design strengths of individual bolts, but does not specify a method for computing the forces on these fasteners when they are eccentrically loaded. As a result the method of analysis to be used is left up to the designer.

Three general approaches for the analysis of eccentrically loaded connections have been developed through the years. The first of the methods is the very conservative *elastic method* in which friction or slip resistance between the connected parts is neglected. In addition these connected parts are assumed to be perfectly rigid. This type of analysis has been commonly used since at least 1870.[1,2]

Tests have shown that the elastic method usually provides very conservative results. As a consequence various *reduced* or *effective eccentricity methods* have been proposed.[3] The analysis is handled just as it is in the elastic method except that smaller eccentricities and thus smaller moments are used in the calculations.

The third method, called the *ultimate strength method,* provides the most realistic values as compared with test results, but is extremely tedious to apply, at least with hand-held calculators. The tables in Part 8 of the LRFD Manual for eccentrically loaded connections are based on the ultimate strength method and enable us to solve most of these types of problems quite easily, as long as the bolt patterns are symmetrical. The remainder of this section is devoted to these three analysis methods.

Elastic Analysis   For this discussion the bolts of Fig. 13-2(a) are assumed to be subjected to a load P that has an eccentricity of e from the c.g. (center of gravity) of the bolt group. To consider the force situation in the bolts, an upward and downward force—each equal to P—is assumed to act at the c.g. of the bolt group. This situation, shown in part (b) of the figure, in no way changes the bolt forces. The force in a particular bolt will, therefore, equal P divided by the

---

[1] W. McGuire, *Steel Structures* (Englewood Cliffs, N.J.: Prentice-Hall, 1968), p. 813.

[2] C. Reilly, "Studies of Iron Girder Bridges," *Proc. Inst. Civil Engrs.* 29 (London, 1870).

[3] T. R. Higgins, "New Formulas for Fasteners Loaded Off Center," *Engr. News Record* (May 21, 1964).

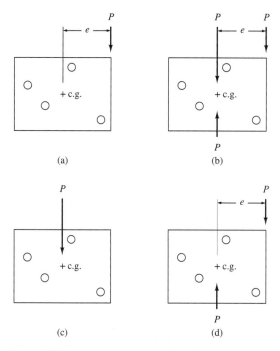

(a)

(b)

(c)

(d)

Figure 13-2

number of bolts in the group as seen in part (c), plus the force due to the moment caused by the couple shown in part (d) of the figure.

The magnitude of the forces in the bolts due to the moment Pe will now be considered. The distances of each bolt from the c.g. of the group are represented by the values d_1, d_2, etc., in Fig. 13-3. The moment produced by the couple is assumed to cause the plate to rotate about the c.g. of the bolt connection with

Figure 13-3

Bridge over New River Gorge in Fayette County, West VA, near Charleston. (Courtesy American Bridge Company.)

the amount of rotation or strain at a particular bolt being proportional to its distance from the c.g. (For this derivation the gusset plates are again assumed to be perfectly rigid and the bolts are assumed to be perfectly elastic.) Rotation is greatest at the bolt that is the greatest distance from the c.g., as will be the stress since stress is proportional to strain in the elastic range.

The rotation is assumed to produce reactions of r_1, r_2, r_3, and r_4 respectively from the bolts in the figure. The moment transferred to the bolts must be balanced by resisting moments of the bolts as follows.

$$M_{c.g.} = Pe = r_1 d_1 + r_2 d_2 + r_3 d_3 + r_4 d_4 \tag{1}$$

As the force caused on each bolt is assumed to be directly proportional to the distance from the c.g., the following expression can be written.

$$\frac{r_1}{d_1} = \frac{r_2}{d_2} = \frac{r_3}{d_3} = \frac{r_4}{d_4}$$

and writing each r value in terms of r_1 and d_1

$$r_1 = \frac{r_1 d_1}{d_1} \qquad r_2 = \frac{r_1 d_2}{d_1} \qquad r_3 = \frac{r_1 d_3}{d_1} \qquad r_4 = \frac{r_1 d_4}{d_1}$$

Substituting these values into equation (1) and simplifying,

$$M = \frac{r_1 d_1^2}{d_1} + \frac{r_1 d_2^2}{d_1} + \frac{r_1 d_3^2}{d_1} + \frac{r_1 d_4^2}{d_1} = \frac{r_1}{d_1}(d_1^2 + d_2^2 + d_3^2 + d_4^2)$$

Therefore,

$$M = \frac{r_1 \Sigma d^2}{d_1}$$

The force on each bolt can now be written as follows.

$$r_1 = \frac{M d_1}{\Sigma d^2} \qquad r_2 = \frac{d_2}{d_1} r_1 = \frac{M d_2}{\Sigma d^2} \qquad r_3 = \frac{M d_3}{\Sigma d^2} \qquad r_4 = \frac{M d_4}{\Sigma d^2}$$

Each value of r is perpendicular to the line drawn from the c.g. to the particular bolt. It is usually more convenient to break these down into vertical and horizontal components. In accomplishing this purpose reference is made to Fig. 13-4.

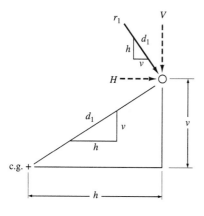

Figure 13-4

The horizontal and vertical components of the distance d_1 are represented by h and v respectively, and the horizontal and vertical components of force r_1 are represented by H and V respectively in this figure. It is now possible to write the following ratio from which H can be obtained.

$$\frac{r_1}{d_1} = \frac{H}{v}$$

$$H = \frac{r_1 v}{d_1} = \left(\frac{Md_1}{\Sigma d^2}\right)\left(\frac{v}{d_1}\right)$$

Therefore,

$$H = \frac{Mv}{\Sigma d^2}$$

By a similar procedure V is found to equal

$$V = \frac{Mh}{\Sigma d^2}$$

■ EXAMPLE 13-1

Determine the force in the most stressed bolt of the group shown in Fig. 13-5 using the elastic analysis method.

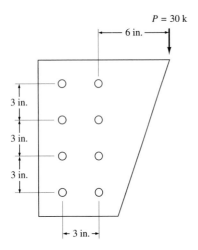

Figure 13-5

Solution. A sketch of each bolt and the forces applied to it by the direct load and the clockwise moment are shown in Fig. 13-6. From this sketch the student will see that the upper right-hand bolt and the lower right-hand bolt are the most stressed and have equal stresses.

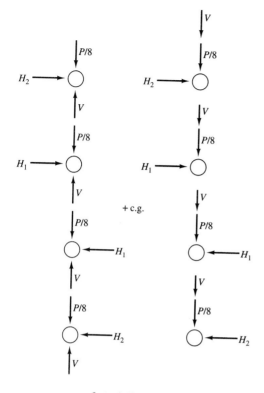

Figure 13-6

$$e = 6 + 1.5 = 7.5 \text{ in.}$$
$$M = Pe = (30)(7.5) = 225 \text{ in.-k}$$
$$\Sigma d^2 = \Sigma h^2 + \Sigma v^2$$
$$\Sigma d^2 = (8)(1.5)^2 + (4)(1.5^2 + 4.5^2) = 108$$
$$H = \frac{Mv}{\Sigma d^2} = \frac{(225)(4.5)}{108} = 9.38 \text{ k}$$
$$V = \frac{Mh}{\Sigma d^2} = \frac{(225)(1.5)}{108} = 3.13 \text{ k}$$
$$\frac{P}{8} = \frac{30}{8} = 3.75 \text{ k}$$

These components are sketched below for the lower right-hand bolt.

The resultant force applied to this bolt is

$$R = \sqrt{(3.13 + 3.75)^2 + (9.38)^2} = 11.63 \text{ k}$$

■

Should the eccentric load be inclined it can be broken down into vertical and horizontal components and the moment of each about the c.g. of the bolt group determined. Various design formulas can be developed that will enable the engineer to directly design eccentric connections, but in all probability the process of assuming a certain number and arrangement of bolts, checking stresses, and redesigning is probably just as satisfactory.

The trouble with this inaccurate but very conservative method of analysis is that we in effect are assuming that there is a linear relation between loads and deformations in the fasteners and, further, that their yield stress is not exceeded when the ultimate load on the connection is reached. Various experiments have shown that these assumptions are incorrect.

Reduced Eccentricity Method   The elastic analysis method just described appreciably overestimates the moment forces applied to the connectors. As a result quite a few proposals have been made through the years that make use of an effective eccentricity, thus in effect taking into account the slip resistance on the faying or contact surfaces. One set of reduced eccentricity values that were fairly common at one time follow:

1. With one gage line of fasteners and where n is the number of fasteners in the line,

$$e_{\text{effective}} = e_{\text{actual}} - \frac{1 + 2n}{4}$$

2. With two or more gage lines of fasteners symmetrically placed and where n is the number of fasteners in each line,

$$e_{\text{effective}} = e_{\text{actual}} - \frac{1 + n}{2}$$

The reduced eccentricity values for two fastener arrangements are shown in Fig. 13-7.

To analyze a particular connection with the reduced eccentricity method, the value of $e_{\text{effective}}$ is computed as described above and used to compute the

Figure 13-7

eccentric moment. Then the elastic procedure is used for the remainder of the calculations.

Ultimate Strength Method   The elastic and reduced eccentricity methods for analyzing eccentrically loaded fastener groups both are based upon the assumption that the behavior of the fasteners is elastic. A much more realistic method of analysis is the ultimate strength method, which is described in the next few paragraphs. The values given in the tables of Part 8 of the LRFD Manual for eccentrically loaded fastener groups were computed using this method.

If one of the outermost bolts in an eccentrically loaded connection begins to slip or yield, the connection will not fail. Instead, the magnitude of the eccentric load may be increased, the inner bolts will resist more load, and failure will not occur until all of the bolts slip or yield.

The eccentric load tends to cause both a relative rotation and translation of the connected material. In effect this is equivalent to pure rotation of the connection about a single point called the *instantaneous center of rotation*. An eccentrically loaded bolted connection is shown in Fig. 13-8, and the instantaneous center is represented by point 0. It is located a distance e' from the center of gravity of the bolt group.

The deformations of these bolts are assumed to vary in proportion to their distances from the instantaneous center. The ultimate shear force that one of them can resist is not equal to the pure shear force that a bolt can resist. Rather, it is dependent upon the load-deformation relationship in the bolt. Studies by Crawford and Kulak[4] have shown that this force may be closely estimated with the following expression:

$$R = R_{ult}(1 - e^{-10\Delta})^{0.55}$$

Figure 13-8

---

[4] S. F. Crawford and G. L. Kulak, "Eccentrically Loaded Bolt Connections," *Journal of Structural Division,* ASCE 97, ST3 (March 1971), pp. 765–783.

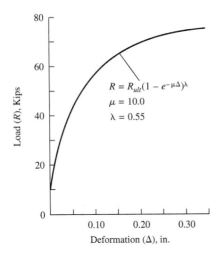

Figure 13-9 Ultimate shear force R in a single bolt at any given deformation.

In this formula R_{ult} is the ultimate shear load for a single fastener equaling 74 k for an A325 bolt, e is the base of the natural logarithm (2.718), and Δ is the total deformation of a bolt. Its maximum value experimentally determined is 0.34 in. The Δ values for the other bolts are assumed to be in proportion to R as their d distances are to d for the bolt with the largest d. The coefficients 10.0 and 0.55 also were experimentally obtained. Figure 13-9 illustrates this load-deformation relationship.

This expression clearly shows that the ultimate shear load taken by a particular bolt in an eccentrically loaded connection is affected by its deformation. Thus, the load applied to a particular bolt is dependent upon its position in the connection with respect to the instantaneous center of rotation.

The resisting forces of the bolts of the connection of Fig. 13-8 are represented with the letters R_1, R_2, R_3, and so on. Each of these forces is assumed to act in a direction perpendicular to a line drawn from point 0 to the center of the bolt in question. For this symmetrical connection the instantaneous center of rotation will fall somewhere on a horizontal line through the center of gravity of the bolt group. This is the case because the sum of the horizontal components of the R forces must be zero as also must be the sum of the moments of the horizontal components about point 0. The position of point 0 on the horizontal line may be found by a tedious trial-and-error procedure to be described here.

With reference to Fig. 13-8, the moment of the eccentric load about point 0 must be equal to the summation of the moments of the R resisting forces about the same point. If we knew the location of the instantaneous center we could compute R values for the bolts with the Crawford-Kulak formula and determine

P_u from the expression to follow in which e and e' are distances shown in Figs. 13-8 and 13-11.

$$P_u(e' + e) = \Sigma Rd$$

$$P_u = \frac{\Sigma Rd}{e' + e}$$

To determine the design strength of such a connection according to the LRFD Specification, we can replace $R_{u\ell t}$ in the Crawford-Kulak formula with the design shearing strength of one bolt in a connection where the load is not eccentric. For instance, if we have $\frac{7}{8}$-in. A325 bolts (threads excluded from shear plane) in single shear bearing on a sufficient thickness so bearing does not control, $R_{u\ell t}$ will equal $(0.75)(0.60)(60) = 27.0$ k.

The location of the instantaneous center is not known, however. Its position is estimated, the R values determined, and P_u calculated as described. It will be noted that P_u must be equal to the summation of the vertical components of the R resisting forces (ΣR_v). If the value is computed and equals the P_u computed by the formula above we have the correct location for the instantaneous center. If not, we try another location, and so on.

In Example 13-2 the author demonstrates the very tedious trial-and-error calculations necessary to locate the instantaneous center of rotation for a symmetrical connection consisting of four bolts. In addition the design strength P_u of the connection is determined.

To solve such a problem it is very convenient to set the calculations up in a table similar to the one used in the solution to follow. In the table shown the h and v values given are the horizontal and vertical components of the d distances from point 0 to the center of gravities of the bolts. The bolt that is located at the greatest distance from point 0 is assumed to have a Δ value of 0.34 in. The Δ values for the other bolts are assumed to be proportionate to their distances from point 0. The Δ values so determined are used in the R formula.

A set of tables entitled "Coefficients C for Eccentrically Loaded Bolt Groups" is presented in Part 8 of the LRFD Manual. The values in these tables were determined by the procedure described here. A large percentage of the practical cases that the designer will encounter are included in the tables. Should some other situation not covered be faced, the designer may very well decide to use the more conservative elastic procedure previously described.

■ EXAMPLE 13-2

The bearing-type $\frac{7}{8}$-in. A325 bolts of the connection of Fig. 13-10 have a design shear strength $R_u = (0.75)(0.60)(60) = 27.0$ k. Locate the instantaneous center of rotation of the connection using the trial-and-error procedure and determine the value of P_u.

Figure 13-10

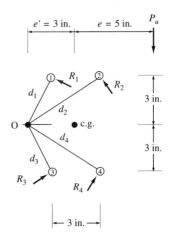

Figure 13-11

Solution. **By trial and error:** Try a value of $e' = 3$ in., reference being made to Fig. 13-11. In the following table Δ for bolt 1 equals $(3.3541/5.4083)(0.34) = 0.211$ in. and R for the same bolt equals $74(1 - e^{-(10)(0.211)})^{0.55}$.

Bolt No.	h(in.)	v(in.)	d(in.)	Δ(in.)	R(kips)	R_v(kips)	Rd(in. -kips)
1	1.5	3	3.3541	0.211	25.15	11.25	84.36
2	4.5	3	5.4083	0.34	26.50	22.05	143.32
3	1.5	3	3.3541	0.211	25.15	11.25	84.36
4	4.5	3	5.4083	0.34	26.50	22.05	143.32
						$\Sigma = 66.60$	$\Sigma = 455.36$

$$P_u = \frac{\Sigma Rd}{e' + e} = \frac{455.36}{3 + 5} = 56.92 \text{ k not} = 66.60 \text{ k} \qquad \text{(NG)}$$

After several trials assume $e' = 2.40$ in.

Bolt No.	h(in.)	v(in.)	d(in.)	Δ(in.)	R(kips)	R_v(kips)	Rd(in. -kips)
1	0.90	3	3.1321	0.216	25.24	7.23	79.06
2	3.90	3	4.9204	0.34	26.50	21.00	130.39
3	0.90	3	3.1321	0.216	25.24	7.23	79.06
4	3.90	3	4.9204	0.34	26.50	21.00	130.39
						$\Sigma = 56.46$	$\Sigma = 418.90$

Then

$$P_u = \frac{\Sigma Rd}{e' + e} = \frac{418.90}{2.4 + 5} = 56.61 \text{ k almost} = 56.46 \text{ k} \qquad \text{(OK)}$$

$$\underline{\underline{P_u = 56.6 \text{ k}}} \qquad\qquad\qquad\qquad\qquad\qquad\qquad \blacksquare$$

Although the development of this method of analysis was actually based on bearing-type connections where slip may occur, both theory and load tests have shown that it may conservatively be applied to slip-critical connections.[5]

The ultimate-strength method may be expanded to include inclined loads and unsymmetrical bolt arrangements, but the trial-and-error calculations are extraordinarily long for such situations.

Examples 13-3 and 13-4 provide illustrations of the use of the ultimate strength tables in Part 8 of the LRFD Manual for both analysis and design.

■ **EXAMPLE 13-3**
Repeat Example 13-2 using ultimate strength Table 8-19 in Part 8 of the Manual. This table is entitled "Coefficients C for Eccentrically Loaded Bolt Groups Angle $= 0°$."

Solution.

$$P_u = C \times \phi r$$
$$C = 2.10 \text{ from Table 8-19}$$
$$\underline{\underline{P_u = (2.10)(27.0) = 56.7 \text{ k}}} \qquad\qquad\qquad \text{(OK)} \quad \blacksquare$$

■ **EXAMPLE 13-4**
Determine the number of $\frac{7}{8}$-in. A325 bolts in standard size holes required for the connection shown in Fig. 13-12. Use A36 steel and assume the connection is to be a bearing type with threads excluded from the shear plane. Further assume that the bolts are in single shear and bearing on $\frac{1}{2}$ in. Use the ultimate strength analysis method presented in Part 8 of the LRFD Manual.

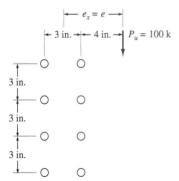

Figure 13-12

---

[5] G. L. Kulak, "Eccentrically Loaded Slip-Resistant Connections," *Engineering Journal,* AISC, 12, No. 2 (second quarter, 1975), pp. 52–55.

Solution.   Using Table 8-19 in Part 8 of the LRFD Manual

$$e_x = e = 5\tfrac{1}{2} \text{ in.} = 5.5 \text{ in.}$$

Bolts in single shear and bearing on $\tfrac{1}{2}$ in.

ϕr_v = shear design strength per fastener
$\qquad = (0.75)(0.6)(60) = 27.0 \text{ k} \leftarrow$
ϕr_v = bearing design strength per fastener
$\qquad = (0.75)(2.4)(\tfrac{7}{8})(\tfrac{1}{2})(58) = 45.7 \text{ k}$

With reference to Table 8-19 in Part 8 of the Manual the necessary value of C required to provide a sufficient number of bolts can be determined as follows:

$$C_{\text{reqd}} = \frac{P_u}{\phi r_n} = \frac{100}{27.0} = 3.70$$

Then with $e_x = 5\tfrac{1}{2}$ in. and a vertical spacing s of 3 in. we move horizontally in the table until we find the number of bolts in each vertical row so as to provide a C of 3.70 or more. It's 4.

<u>Use 4 bolts in each vertical row.</u>   ■

Note: If the situation faced by the designer does not fit the ultimate strength tables given in Part 8 of the LRFD Manual, it is recommended that he or she use the conservative elastic procedure to handle the problem, whether analysis or design.

13-2 BOLTS SUBJECTED TO SHEAR AND TENSION

The bolts used for a good many structural steel connections are subjected to a combination of shear and tension. One obvious case is shown in Fig. 13-13 where a diagonal brace is attached to a column. The 111.8-k component shown in the figure is trying to shear the bolts off at the face of the column while the 223.6 k component is tending to pull off their heads.

Tests on bearing-type bolts subject to combined shear and tension show that their ultimate strengths can be represented with an elliptical interaction curve as shown in Fig. 13-14. Particularly note the values F_t and F_v. F_t is the limiting tensile stress if there is no shear, and F_v is the limiting shearing stress if there is no externally applied tension.

The three straight dashed lines shown in Fig. 13-14 very closely represent the test result interaction curve. Equations for these lines are given in Table 13.1

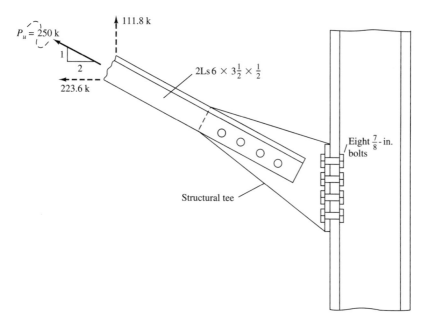

Figure 13-13 Combined shear and tension connection.

Figure 13-14 Bolts in a bearing-type connection subject to combined shear and tension.

(Table J3.5 of the LRFD Specification). In these expressions f_v and f_t are respectively the computed shear and tension stresses in the bolts due to the factored loads.

The design tensile stress for a bolt subjected to combined shear and tension equals ϕF_t, while its total design strength is $\phi F_t A_b$ with $\phi = 0.75$. The nominal tensile stress F_t is to be computed with the appropriate equation given in Table 13-1.

TABLE 13-1   TENSION STRESS LIMIT (F_t), ksi, FOR FASTENERS IN BEARING-TYPE CONNECTIONS

Description of fasteners	Threads included in the shear plane	Threads excluded from the shear plane
A307 bolts	$59 - 1.9f_v \leq 45$	
A325 bolts	$117 - 1.9f_v \leq 90$	$117 - 1.5f_v \leq 90$
A490 bolts	$147 - 1.9f_v \leq 113$	$147 - 1.5f_v \leq 113$
Threaded parts A449 bolts over $1\frac{1}{2}$-in. diameter	$0.98F_u - 1.9f_v \leq 0.75F_u$	$0.98F_u - 1.5f_v \leq 0.75F_u$
A502 grade 1 rivets	$59 - 1.8f_v \leq 45$	
A502 grade 2 rivets	$78 - 1.8f_v \leq 60$	

Example 13-5 illustrates the calculations involved in checking a high-strength bearing-type bolted connection for shear and tension. (There is another stress situation which may affect this connection. It is called *prying action,* but as it is not discussed until Section 13-4 it is not considered in this example.)

■ EXAMPLE 13-5

The tension member shown in Fig. 13-13 is connected to the column shown with eight $\frac{7}{8}$-in. A325 high-strength bolts in a bearing-type connection with the threads excluded from the shear plane and standard-size holes. Is this a sufficient number of bolts to resist the applied load according to the LRFD Specification?

Solution.

$$\text{Shearing stress } f_v = \frac{111.8}{(8)(0.6)} = 23.29 \text{ ksi}$$

$$\text{Tension stress } f_t = \frac{223.6}{(8)(0.6)} = 46.58 \text{ ksi}$$

$$\text{Limiting tension stress } \phi F_t = 0.75[117 - (1.5)(23.29)]$$
$$= 61.54 \text{ ksi} > 46.58 \text{ ksi} \qquad \text{(OK)}$$

$$\underline{\text{Connection is satisfactory}} \qquad ■$$

When an axial tension force is applied to a slip-critical connection, the clamping force will be reduced and the design shear strength must be decreased in some proportion to the loss in clamping or prestress. This is accomplished in the LRFD Specification (Section J3.9) by requiring that the nominal slip-critical shear strengths given in Table 12.7 (LRFD Table J3.6) be multiplied by the reduction factor $(1 - T/T_b)$. T is the service tensile force applied to one bolt and T_b is the minimum pretension load for one bolt in a slip-critical connection as given in Table 12.1 (LRFD Table J3.1). For such a situation ϕ is equal to 1.0 unless we have long-slotted holes with the load applied in the direction of the slot. In the latter case ϕ is to be taken as 0.85, says the LRFD Specification.

For bolts in slip-critical connections with standard holes the nominal shear strengths at service load conditions are to be determined as follows where T is the tension force per bolt due to service loads.

$$\text{For A325 bolts} \le \left(1 - \frac{T}{T_b}\right)(17)$$

$$\text{For A490 bolts} \le \left(1 - \frac{T}{T_b}\right)(21)$$

■ EXAMPLE 13-6

Rework Example 13-5 assuming that a slip-critical connection is to be used. The service loads are $P_D = 75$ k and $P_L = 90$ k.

Solution.

Service load $P = 75 + 90 = 165$ k

Components of P

$$P_v = \left(\frac{1}{\sqrt{5}}\right)(165) = 73.8 \text{ k}$$

$$P_h = \left(\frac{2}{\sqrt{5}}\right)(165) = 147.6 \text{ k}$$

Then, calculating shearing and tension stresses

$$f_v = \frac{73.8}{(8)(0.6)} = 15.38 \text{ ksi}$$

$$f_t = \frac{147.6}{(8)(0.6)} = 30.75 \text{ ksi}$$

$$\text{Limiting shear strength} = \left(1 - \frac{T}{T_b}\right)(17)$$

$$= \left(1 - \frac{147.6/8}{39}\right)(17) = 8.95 \text{ ksi} < 15.38 \text{ ksi} \qquad \text{(NG)}$$

Connection is unsatisfactory if slip-critical   ■

13-3 TENSION LOADS ON BOLTED JOINTS

Bolted and riveted connections subjected to pure tensile loads have been avoided as much as possible in the past by designers. The use of tensile connections was probably "forced on" them more often for wind-bracing systems in tall buildings than for any other situation. There are some other locations where they have

Hanger connection

Figure 13-15 Hanger connection.

been used, however, such as hanger connections for bridges, flange connections for piping systems, etc. Figure 13-15 shows a hanger-type connection with an applied tensile load.

Hot-driven rivets and fully tensioned high-strength bolts are not free to shorten, with the result that large tensile forces are produced in them during their installation. These initial tensions are actually close to the yield points. There has always been considerable reluctance among designers to apply tensile loads to connectors of this type for fear that the external loads might easily increase their already present tensile stresses and cause them to fail. The truth of the matter, however, is that when external tensile loads are applied to connections of this type the connectors will probably experience little if any change in stress.

Hot-driven rivets that have cooled and shrunk or fully tensioned high-strength bolts actually prestress the joints in which they are used against tensile loads. (To follow this discussion the student may like to think of a prestressed concrete beam that has external compressive loads applied at each end.) The tensile stresses in the connectors squeeze together the members being connected. If a tensile load is applied to this connection at the contact surface, it cannot exert any additional load on the bolts or rivets until the members are pulled apart and additional strains put on the bolts or rivets. The members cannot be pulled apart until a load is applied that is larger than the total tension in the connectors of the connection. This statement means that the joint is prestressed against tensile forces by the amount of stress initially put in the shanks of the connectors.

Another way of saying this is that if a tensile load P is applied at the contact surface it tends to reduce the thickness of the plates somewhat, but at the same time the contact pressure between the plates will be correspondingly reduced and the plates will tend to expand by the same amount. The theoretical result, then, is no change in plate thickness and no change in connector tension. This situation continues until P equals the connector tension. At this time an increase in P will result in separation of the plates and thereafter the tension in the connector will equal P.

Should the load be applied to the outer surfaces there will be some immediate strain increase in the connector. This increase will be accompanied by an expansion of the plates even though the load does not exceed the prestress, but the increase will be very slight because the load will go to the plate and connectors roughly in proportion to their stiffness. As the plate is stiffer it will receive most of the load. An expression can be developed for the elongation of the bolt based on the bolt area and the assumed contact area between the plates. Depending on the contact area assumed, it will be found that unless P is greater than the bolt tension, its stress increase will be in the range of 10 percent. Should the load exceed the prestress, the bolt stress will rise appreciably.

The preceding rather lengthy discussion is approximate, but should explain why an ordinary tensile load applied to a riveted or bolted joint will not change the stress situation very much.

The LRFD tensile design strength of bolts, rivets, and threaded parts is given by the expression to follow, which is independent of any initial tightening force.

$$P_u = \phi 0.75 \, A_b F_u$$

When fasteners are loaded in tension there is usually some bending due to the deformation of the connected parts. As a result the value of ϕ in this expression is a rather small 0.75. Table 12-6 of this text (Table J3.2 of the LRFD Specification) gives values of F_t the nominal tensile strength (ksi) for the different kinds of connectors with the values of rivets and threaded parts being quite conservative.

In this expression A_b is the nominal body area of a rivet, or the unthreaded portion of a bolt, or its threaded part not including upset rods. An upset rod has its ends made larger than the regular rod and the threads are placed in the enlarged section so that the area at the root of the thread is larger than that of the regular rod. An upset rod was shown in Fig. 4-3. The use of upset rods is not usually economical and should be avoided unless a large order is being made.

If an upset rod is used the nominal tensile strength of the threaded portion is set equal to $0.75 \, F_u$ times the cross-sectional area at its major thread diameter. This value must be larger than F_y times the nominal body area of the rod before upsetting.

Example 13.7 illustrates the determination of the strength of a tension connection.

■ **EXAMPLE 13-7**

Determine the design tensile strength of the bolts of the hanger connection of Fig. 13-15 if eight $\frac{7}{8}$-in. A490 high-strength bolts are used.

Solution.

$$P_u = (0.75)(8)(0.6)(113) = \underline{\underline{406.8 \text{ k}}}$$

The load applied to a tensile connection shall be the sum of the factored external loads plus any tension forces that result from prying action as described in the next section.

13-4 PRYING ACTION

A further consideration that should often be given to tensile connections is the possibility of prying action. A tensile connection is shown in Fig. 13-16(a) that is subjected to prying action as illustrated in part (b) of the same figure. Should the flanges of the connection be quite thick and stiff or have stiffener plates as shown in Fig. 13-16(c), the prying action will probably be negligible, but this is not the case if they are thin and flexible and have no stiffeners.

It is usually desirable to limit the number of rows of bolts in a tensile connection because a large percentage of the load is carried by the inner rows of multirow connections even at ultimate load. The tensile connection shown in Fig. 13-17 illustrates this point as the prying action will throw a large part of the load to the inner connectors, particularly if the plates are thin and flexible. For connections subjected to pure tensile loads, estimates should be made of possible prying action and its magnitude.

The additional force in the bolts resulting from prying action should be added to the tensile force resulting directly from the applied forces. The actual determination of prying forces is quite complex, and research on the subject is still being conducted. Several empirical formulas have been developed that approximate test results. Among these are the LRFD expressions included in this section. The reader should realize that we don't know very much about prying action and our formulas keep changing almost yearly.

Only fully tensioned bolts should be used for connections for which the applied loads subject the bolts to axial tension. This is true whether the connections are classified as slip-critical and whether they are subject to fatigue loads and whether there is prying action. If snug-tight bolts are used for any of these situations, the tensile loads will immediately start increasing bolt tensions.

Figure 13-16

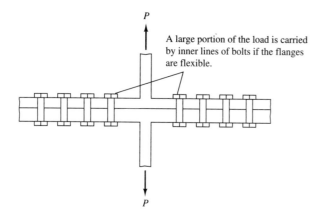

A large portion of the load is carried by inner lines of bolts if the flanges are flexible.

Figure 13-17

Hanger and other tension connections should be so designed as to prevent significant deformations. The most important item in such designs is the use of rigid flanges. Rigidity is more important than bending resistance. To achieve this goal the distance b shown in Fig. 13-18 should be made as small as possible with a minimum value equal to the space required to use a wrench for tightening the bolts. Information concerning wrench clearance dimensions is presented in a table entitled "Entering and Tightening Clearances" in Table 8-4 in Part 8 of the LRFD Manual.

In Part 11 of the LRFD Manual a detailed procedure is presented for designing hanger connections and computing prying forces. This empirical method is for factored forces only. If it is desired to consider service loads for the purposes of investigating fatigue, deflections, or drift limitations, the reader may refer to the service load formulas for prying that were presented in the ninth edition of the *AISC Manual of Steel Construction Allowable Stress Design*.

For reasons of space limitations only one numerical prying action example is presented here. In this example the author checks to see if the flange thickness of a WT hanger connection is satisfactory according to the AISC procedure. For a much more detailed example the reader can refer to Section 11 of the Manual.

Figure 13-18

Quite a few abbreviations and equations are used in the LRFD prying analysis procedure. These are listed below with reference made to Fig. 13-18.

ϕr_n = design tensile strength of each bolt = $\phi F_t A_b$

$b = \dfrac{g - t_w}{2}$ where g is the gage. It must be sufficient for wrench clearance as given in Table 8-4 in Part 8 of the Manual.

a = distance from bolt center line to the edge of tee flange or angle leg but not less than $1.25b = \dfrac{b_f - g}{2}$

$b' = b - \dfrac{d}{2}$ where d = bolt diameter

$a' = a + \dfrac{d}{2}$

$p = \dfrac{b'}{a'}$

$r_{ut} = \dfrac{P_u}{\text{number of bolts}}$

$\beta = \dfrac{1}{p}\left(\dfrac{\phi r_n}{r_{ut}} - 1\right)$

If $\beta \geq 1$, set $\alpha' = 1.0$

If $\beta < 1$, set α' = the lesser of 1.0 or $\dfrac{1}{\delta}\left(\dfrac{\beta}{1 - \beta}\right)$ where δ is the ratio of the net area at the bolt line to the gross area at the face of the stem or angle leg and d' = width of bolt hole parallel to tee stem

$\delta = 1 - \dfrac{d'}{p}$ where p is the length of connection tributary to each bolt

$\alpha' = \dfrac{1}{\delta}\left(\dfrac{\beta}{1 - \beta} \leq 1.0\right)$

Finally the required flange thickness t is $t_{req} = \sqrt{\dfrac{4.44\, r_{ut} b'}{p F_y (1 + \delta \alpha')}}$

If the flange thickness is $\geq t_{req}$, the bolt tension will be $\leq \phi r_n$ and no other calculations are necessary. In Part 11 of the LRFD Manual, however, formulas are presented for the calculation of the total bolt tension.

■ **EXAMPLE 13-8**
A 10-in.-long WT8 × 22.5 (t_f = 0.565 in., t_w = 0.345 in., and b_f = 7.035 in.) is connected to a W36 × 150 as shown in Fig. 13-19 with four $\frac{7}{8}$-in. A325 high-strength bolts. If A36 steel is used is the flange sufficiently thick if prying action is considered?

W36 × 150

b_f = 7.035 in.

g = gage = 4 in.

0.565 in. = t

b' — a'

b — a

2 in.

P_u = 100 k

Figure 13-19

Solution.

$$\phi r_n = \phi F_t A_b = (0.75)(90)(0.6) = 40.5 \text{ k}$$

$$b = \frac{g - t_w}{2} = \frac{4 - 0.345}{2} = 1.827 \text{ in.} > 1\tfrac{3}{8} \text{ in.}$$

wrench clearance required (OK)

$$a = \frac{b_f - g}{2} = \frac{7.035 - 4}{2} = 1.518 \text{ in.}$$

$$b' = b - \frac{d}{2} = 1.827 - \frac{0.875}{2} = 1.389 \text{ in.}$$

$$a' = a + \frac{d}{2} = 1.518 + \frac{0.875}{2} = 1.955 \text{ in.}$$

$$p = \frac{b'}{a} = \frac{1.389}{1.518} = 0.915$$

$$r_{ut} = \frac{P_u}{\text{number of bolts}} = \frac{100}{4} = 25 \text{ k}$$

$$\beta = \frac{1}{p}\left(\frac{\phi r_n}{r_{ut}} - 1\right) = \frac{1}{0.915}\left(\frac{40.5}{25} - 1\right) = 0.678$$

Since $\beta < 1.0$

$$\delta = 1 - \frac{d'}{p}$$ where with four bolts the maximum effective length is $2g = (2)(4) = 8$ in. Therefore, there are 4 in. of tee length tributary to each pair of bolts and $p = 4$ in.

$$\delta = 1 - \frac{15/16}{4} = 0.766$$

$$\alpha' = \frac{1}{\delta}\left(\frac{\beta}{1 - \beta}\right) = \frac{1}{0.766}\left(\frac{0.678}{1 - 0.678}\right) = 2.75 \quad \text{Use 1.0}$$

Then the required flange thickness is

$$t_{req} = \sqrt{\frac{4.44 \, r_{ut} b'}{p F_y (1 + \delta \alpha')}} = \sqrt{\frac{(4.44)(25)(1.389)}{(4)(36)(1 + 0.766 \times 1.0)}}$$

$$= 0.779 \text{ in.} > 0.565 \text{ in.} \qquad \text{(NG)} \quad \blacksquare$$

13-5 HISTORICAL NOTES ON RIVETS

For many years rivets were the accepted method for connecting the members of steel structures. Today, however, they no longer provide the most economical connections and are obsolete. It is doubtful if you could find a steel fabricator who can do riveting. It is, however, desirable for the designer to be familiar with rivets even though he or she will seldom if ever design riveted structures. He or she may have to analyze an existing riveted structure for new loads or for an expansion of the structure. The purpose of these sections is to present only a very brief introduction to the analysis and design of rivets.

The rivets used in construction work were usually made of a soft grade of steel that would not become brittle when heated and hammered with a riveting gun to form the head. The usual rivet consisted of a cylindrical shank of steel with a rounded head on one end. It was heated in the field to a cherry-red color (approximately 1800°F), inserted in the hole, and a head formed on the other end probably with a portable rivet gun operated by compressed air. The rivet gun, which had a depression in its head to give the rivet head the proper shape, applied a rapid succession of blows to the rivet.

For riveting done in the shop the rivets were probably heated to a light cherry-red color and driven with a pressure-type riveter. This type of riveter, usually called a "bull" riveter, squeezed the rivet with a pressure of perhaps as high as 50 to 80 tons (445 to 712 kN) and drove the rivet with one stroke. Because of this great pressure the rivet in its soft state was forced to fill the hole very satisfactorily. This type of riveting was much to be preferred over that done

Handling hot rivets, Chicago, Ill. (Courtesy of IR Construction Products Co.)

with the pneumatic hammer, but no greater nominal strengths were allowed by riveting specifications. The bull riveters were built for much faster operation than were the portable hand riveters, but the latter riveters were needed for places that were not easily accessible (i.e., field erection).

As the rivet cooled it shrank, or contracted and squeezed together the parts being connected. The squeezing effect actually caused considerable transfer of stress between the parts being connected to take place by friction. The amount of friction was not dependable, however, and the specifications did not permit its inclusion in the strength of a connection. Rivets shrink diametrically as well as lengthwise and actually become somewhat smaller than the holes that they are assumed to fill. (Permissible strengths for rivets are actually given in terms of the nominal cross-sectional areas of the rivets before driving.)

Some shop rivets were driven cold with tremendous pressures. Obviously, the cold-driving process worked better for the smaller size rivets (probably $\frac{3}{4}$ in. in diameter or less) although larger ones have been successfully used. Cold-driven rivets fill the holes better, eliminate the cost of heating, and are stronger because the steel is cold worked. There is, however, a reduction of clamping force since the rivets do not shrink after driving.

13-6 TYPES OF RIVETS

The sizes of rivets used in ordinary construction work were $\frac{3}{4}$ in. and $\frac{7}{8}$ in. in diameter, but they could be obtained in standard sizes from $\frac{1}{2}$ in. to $1\frac{1}{2}$ in. in $\frac{1}{8}$-in. increments. (The smaller sizes were used for small roof trusses, signs, small towers, etc., while the larger sizes were used for very large bridges or towers and very tall buildings.) The use of more than one or two sizes of rivets or bolts on a single job is usually undesirable because it is expensive and inconvenient to punch different-size holes in a member in the shop, and the installation of different-size rivets or bolts in the field may be confusing. Some cases arise where it is absolutely necessary to have different sizes, as where smaller rivets or bolts are needed for keeping the proper edge distance in certain sections, but these situations should be avoided if possible.

Rivet heads were usually round in shape, called button heads; but if clearance requirements dictated, the head was flattened or even countersunk and chipped flush. These situations are shown in Fig. 13-20.

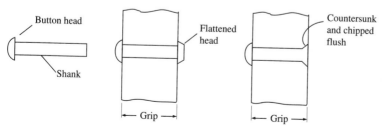

Figure 13-20 Types of rivets.

The countersunk and chipped-flush rivets did not have sufficient bearing areas to develop full strength and the designer usually discounted their computed strengths by 50 percent. A rivet with a flattened head was to be preferred to a countersunk rivet, but if a smooth surface was required the countersunk and chipped-flush rivet was necessary. This latter type of rivet was appreciably more expensive than the button head type in addition to being weaker, and they were not used unless absolutely necessary.

There are three ASTM classifications for rivets for structural steel applications, as described in the following paragraphs.

ASTM Specification A502, Grade 1

These rivets were used for most structural work. They had a low carbon content of about 0.80 percent, were weaker than the ordinary structural carbon steel, and had a higher ductility. The fact that these rivets were easier to drive than the higher-strength rivets was the main reason that when rivets were used they probably were A502, grade 1, regardless of the strength of the steel used in the structural members.

ASTM Specification A502, Grade 2

These carbon-manganese rivets have higher strengths than the grade 1 rivets and were developed for the higher-strength steels. Their higher strength permits the designer to use fewer rivets in a connection and thus smaller gusset plates.

ASTM Specification A502, Grade 3

These rivets have the same nominal strengths as the grade 2 rivets, but they have much higher resistance to atmospheric corrosion equal to approximately four times that of carbon steel without copper.

13-7 STRENGTH OF RIVETED CONNECTIONS—RIVETS IN SHEAR

The factors determining the strength of a rivet are its grade, its diameter, and the thickness and arrangement of the pieces being connected. The actual distribution of stress around a rivet hole is difficult to determine, if it can be determined at all; and to simplify the calculations it is assumed to vary uniformly over a rectangular area equal to the diameter of the rivet times the thickness of the plate.

The strength of a rivet in bearing equals the design bearing unit stress of the rivet times the diameter of the shank of the rivet times the thickness of the member that bears on the rivet. The strength of a rivet in single shear is the nominal shearing strength times the cross-sectional area of the shank of the

TABLE 13-2 LRFD NOMINAL TENSILE AND SHEARING STRENGTHS FOR RIVETS AND A307 BOLTS

Fastener type	Tensile strength (ksi)	Shearing strength in bearing-type connections (ksi)
A502, grade 1, hot-driven rivets	45.0, $\phi = 0.75$	25.0, $\phi = 0.75$
A502, grade 2 or 3, hot-driven rivets	60.0, $\phi = 0.75$	33.0, $\phi = 0.75$
A307 bolts	45.0, $\phi = 0.75$	24.0, $\phi = 0.75$

rivet. Should a rivet be in double shear, its shearing strength is considered to be twice its single-shear value.

The nominal tension and shearing strengths for rivets and A307 bolts were given in Table 12-6 (LRFD Table J3.2). These values are for static loads only and are repeated in Table 13-2. Notice that the nominal shear strength of the A307 bolts is not affected if the bolt threads are in the shear plane.

Examples 13-9 and 13-10 illustrate the calculations necessary to determine the design strengths of existing connections or to design riveted connections. Little comment is made here concerning A307 bolts. The reason is that all the calculations for these fasteners are made exactly as they are for rivets, except that the shearing strengths given by the LRFD Specification are different. Only one brief example with these common bolts (13-11) is included.

■ **EXAMPLE 13-9**

Determine the design strength P_u of the bearing-type connection shown in Fig. 13-21. A36 steel and A502, grade 1 rivets are used in the connection and it is assumed that standard-size holes are used and that edge distances and center-to-center distances are $> 1\frac{1}{2}d$ and $3d$ respectively.

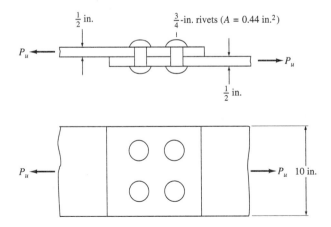

Figure 13-21

Solution.   Design tensile force applied to plates

$$A_g = (\tfrac{1}{2})(10) = 5.00 \text{ in.}^2$$
$$A_n = [(\tfrac{1}{2})(10) - (2)(\tfrac{7}{8})(\tfrac{1}{2})] = 4.125 \text{ in}^2 = A_e$$
$$P_u = \phi_t F_y A_g = (0.90)(36)(5.00) = 162 \text{ k}$$
$$P_u = \phi_t F_u A_e = (0.75)(58)(4.125) = 179.4 \text{ k}$$

Rivets in single shear and bearing on $\tfrac{1}{2}$ in.

Design shearing strength of rivets = $(0.75)(0.44)(25)(4) = 33.0 \text{ k} \leftarrow$
Design bearing strength of rivets = $(0.75)(2.4)(\tfrac{3}{4})(\tfrac{1}{2})(58)(4) = 156.6$ k
$$P_u = 33.0 \text{ k} \qquad \blacksquare$$

■ EXAMPLE 13-10
How many $\tfrac{7}{8}$-in. A502, grade 1 rivets are required for the connection shown in Fig. 13-22 if the plates are A36, if standard-size holes are used, and the edge distances and center-to-center distances are $>1\tfrac{1}{2}d$ and $3d$ respectively?

Figure 13-22

Solution.   Bolts in double shear and bearing on $\tfrac{1}{2}$ in.:

Design shear strength of 1 rivet = $(0.75)(2 \times 0.6)(25) = 22.5 \text{ k} \leftarrow$
Design bearing strength of 1 rivet = $(0.75)(2.4)(\tfrac{7}{8})(\tfrac{1}{2})(58) = 45.7 \text{ k}$
$$\text{No. of rivets required} = \frac{170}{22.5} = 7.56$$
Use eight $\tfrac{7}{8}$-in. rivets   ■

■ EXAMPLE 13-11
Repeat Example 13-10 using $\tfrac{7}{8}$-in. A307 bolts.

Solution.   Bolts in double shear and bearing on $\tfrac{1}{2}$ in.:

Design shear strength of 1 bolt = $(0.75)(2 \times 0.6)(24) = 21.6 \text{ k} \leftarrow$
Design bearing strength of 1 bolt = $(0.75)(2.4)(\tfrac{7}{8})(\tfrac{1}{2})(58) = 45.7 \text{ k}$
$$\text{No. of bolts required} = \frac{170}{21.6} = 7.87$$
Use eight $\tfrac{7}{8}$-in. A307 bolts   ■

13-8 COMPUTER EXAMPLE

In Example 13-12 the computer program INSTEP is used to determine the ultimate load which an eccentrically loaded connection can support.

■ EXAMPLE 13-12

Using the enclosed computer disk determine the value of P_u for the bearing type connection of Fig. 13-23 if $\frac{7}{8}$-in. A325 bolts threads excluded from shear plane and A572 Grade 50 steel are used.

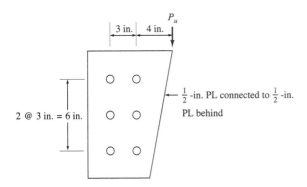

Figure 13-23

Solution.

                    INPUT
        A   Type Steel: A572 Grade 50
        B   Type Bolt: A325 Bolts

            Connection Details
        D   Bolt Size: .875 inches
        E   Threads: SS Bearing, Excluded
        F   Joint Type: Eccentric Shear Only
        G t_1 = .5 inches
        H t_2 = .5 inches
        I   Specify Bolt Pattern

            Joint Forces Acting
        K   P = 0 kips
        L   V = 100 kips
        M   e = 5.5 inches

        Set Vertical Spacing
        V S_v = 3 inches

        Set Horizontal Spacing
        H S_h = 3 inches

OUTPUT

SUMMARY OF FINAL RESULTS

CG_x = 1.5 inches
CG_y = 3 inches

ULTIMATE LOAD CONDITIONS

V_u = 65.2152 kips
e' = 1.50781 inches from CG
R_{max} = 26.558 kips

Iterations: 14

PROBLEMS

For each of the problems listed the following information is to be used unless otherwise indicated: (a) A36 steel; (b) standard-size holes; (c) edge distances and center-to-center distances of bolts $\geq 1\frac{1}{2}d$ and $3d$ respectively; (d) threads of bolts excluded from shear plane.

13-1 to 13-7. *Determine the resultant load on the most stressed bolt in the eccentrically loaded connections shown using the elastic method.*

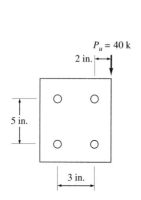

Problem 13-1 *(Ans.* 19.17 k)

Problem 13-2

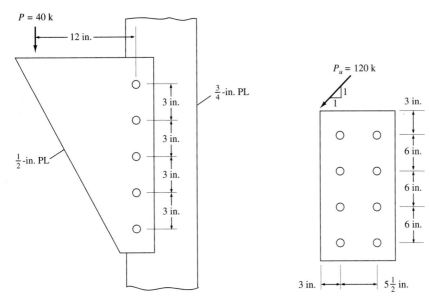

$P = 40$ k

12 in.

3 in.

3 in.

3 in.

3 in.

$\frac{3}{4}$-in. PL

$\frac{1}{2}$-in. PL

Problem 13-3 (*Ans.* 32.98 k)

$P_u = 120$ k

1
1

3 in.

6 in.

6 in.

6 in.

3 in. $5\frac{1}{2}$ in.

Problem 13-4

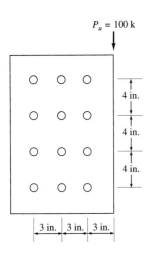

$P_u = 100$ k

4 in.

4 in.

4 in.

3 in. 3 in. 3 in.

Problem 13-5 (*Ans.* 18.22 k)

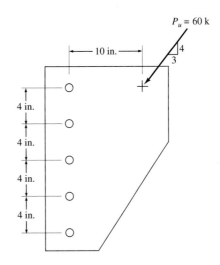

$P_u = 60$ k

10 in.

4
3

4 in.

4 in.

4 in.

4 in.

4 in.

4 in.

4 in.

Problem 13-6

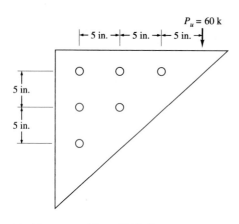

Problem 13-7 (*Ans.* 40.51 k)

13-8. Repeat Prob. 13-2 using the reduced eccentricity method.

13-9. Using the elastic method determine the design strength of the bearing-type connection shown. The bolts are $\frac{7}{8}$-in. A325 and are in single shear and bearing on $\frac{1}{2}$ in. (*Ans.* 78.7 k)

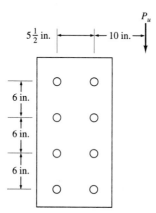

Problem 13-9

13-10. Using the elastic method determine the design strength P_u for the slip-critical connection shown. The $\frac{3}{4}$-in. A325 bolts are in "double shear."

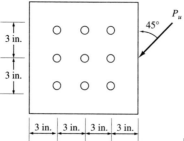

Problem 13-10

13-11. Repeat Prob. 13-9 using the ultimate strength tables entitled "Coefficients C for Eccentrically Loaded Bolt Groups" in Part 8 of the LRFD Manual. (*Ans.* 99.1 k)

13-12. Repeat Prob. 13-10 using the ultimate strength tables entitled "Coefficients C for Eccentrically Loaded Bolt Groups" in Part 8 of the LRFD Manual.

13-13. Is the bearing-type connection shown in the accompanying illustration sufficient to resist the 180-k load which passes through the center of gravity of the bolt group? (*Ans.* $\phi F_t = 53.37$ ksi > 40.74 ksi (OK))

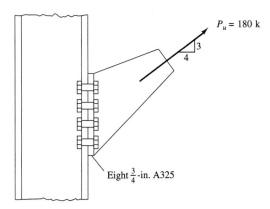

Eight $\frac{3}{4}$-in. A325

Problem 13-13

13-14. Repeat Prob. 13-13 if fully-tensioned bolts are used and if $P_D = 60$ k and $P_L = 50$ k.

13-15. If the load shown in the accompanying bearing-type illustration passes through the center of gravity of the bolt group, how large can it be? (*Ans.* 473.6 k)

Ten $\frac{7}{8}$-in. A490

Problem 13-15

13-16. Repeat Prob. 13-15 if bolts are A325.

13-17. Determine the number of $\frac{3}{4}$-in. A325 bolts required in the angles and in the flange of the W shape shown in the accompanying illustration if a bearing-type (snug-tight) connection is used. Use A36 steel. (*Ans.* 5 in angles and 8 in W flange)

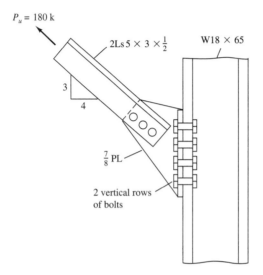

Problem 13-17

13-18. Are the bolts shown in this 16 in.-long hanger satisfactory to resist direct tension and prying? There are eight $\frac{7}{8}$-in. A325 bolts and they are spaced 4 in. on center longitudinally.

Problem 13-18

13-19. Determine the design strength P_u of the connection shown if $\frac{7}{8}$-in. A502, grade 1 rivets and A36 steel are used. (*Ans.* 135 k)

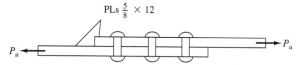

PLs $\frac{5}{8}$ × 12

Problem 13-19

13-20. The truss tension member shown consists of a single-angle 5 × 3 × $\frac{5}{16}$ angle and is connected to a $\frac{1}{2}$-in. gusset plate with five $\frac{7}{8}$-in. A502, grade 1 rivets. Determine P_u if U assumed to equal 0.9. Neglect block shear.

Problem 13-20

13-21. How many $\frac{3}{4}$-in. A502, grade 1 rivets are needed to carry the load shown in the accompanying illustration? (*Ans.* 18.18, use 18–21 rivets)

$\frac{3}{8}$ PL

P_u = 150 k

P_u = 150 k

$\frac{3}{8}$ PL

Problem 13-21

13-22. Repeat Prob. 13-21 if A307 bolts are used.

13-23. How many A502, grade 1 rivets with 1-in. diameters need to be used for the butt joint shown? (*Ans.* 6.11, use 6–9 rivets)

Problem 13-23

13-24. How many $\frac{7}{8}$-in. A307 bolts are required for the connection shown in the accompanying illustrations? Factored loads are shown.

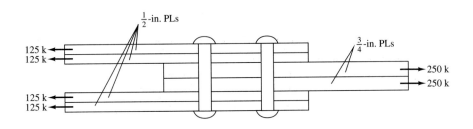

Problem 13-24

13-25. For the connection shown in the accompanying illustration $P_u = 650$ k, determine the number of $\frac{7}{8}$-in. A502, grade 2 rivets required. (*Ans.* 21.88, use 22–25 rivets)

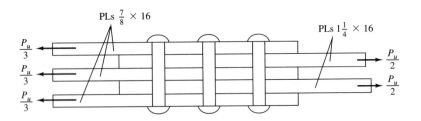

Problem 13-25

13-26. Determine the design strength P_u for the connection shown if $\frac{3}{4}$-in. A502, grade 2 rivets are used.

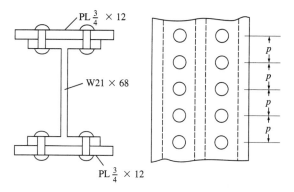

Problem 13-26

13-27. For the beam shown in the accompanying illustration, what is the required spacing of $\frac{7}{8}$-in. A307 bolts if $V_u = 150$ k? (*Ans.* 5.32 in., use $5\frac{1}{4}$ in.)

Problem 13-27

13-28. Is the connection shown sufficient to resist the 90-k load that passes through the center of gravity of the rivet group?

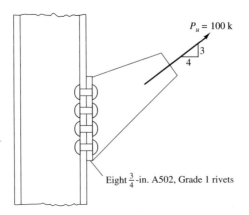

$P_u = 100$ k

3

4

Eight $\frac{3}{4}$-in. A502, Grade 1 rivets

Problem 13-28

13-29. If the load shown in the accompanying illustration passes through the center of gravity of the rivet group, how large can it be? (*Ans.* 152.8 k)

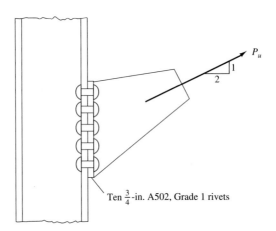

P_u

1

2

Ten $\frac{3}{4}$-in. A502, Grade 1 rivets

Problem 13-29

13-30 to 13-31. *Repeat the problems listed using the enclosed computer diskette.*

13-30. Prob. 13-9

13-31. Prob. 13-10 (*Ans.* 54 k)

Chapter **14**

Welded Connections

14-1 GENERAL

Welding is a process in which metallic parts are connected by heating their surfaces to a plastic or fluid state and allowing the parts to flow together and join (with or without the addition of other molten metal). It is impossible to determine when welding originated, but it was several thousand years ago. Metalworking, including welding, was quite an art in ancient Greece at least three thousand years ago, but welding had undoubtedly been performed for many centuries before those days. Ancient welding was probably a forging process in which the metals were heated to a certain temperature (not to the melting stage) and hammered together.

Although modern welding has been available for a good many years, it has only come into its own in the last few decades for the building and bridge phases of structural engineering. The adoption of structural welding was quite slow for several decades because many engineers thought that welding had two great disadvantages—(1) that welds had reduced fatigue strength as compared with riveted and bolted connections and (2) that it was impossible to ensure a high quality of welding without unreasonably extensive and costly inspection.

These negative feelings persisted for many years, although tests seemed to indicate that neither reason was valid. Regardless of the validity of these fears, they were widely held and undoubtedly slowed down the use of welding—particularly for highway bridges and, to an even greater extent, for railroad bridges. Today most engineers agree that welded joints have considerable fatigue strength. They will also admit that the rules governing the qualification of welders, the better techniques applied, and the excellent workmanship requirements of the AWS (American Welding Society) specifications make the inspection of welding a much less difficult problem. Furthermore, the chemistry of

Fabrication of plate girders for Connecticut expressway bridge. (Courtesy of the Lincoln Electric Company.)

steels manufactured today is especially formulated to improve their weldability. Consequently, today welding is permitted for almost all structural work other than for some bridges.

On the subject of fear of welding, it is interesting to consider welded ships. Ships are subjected to severe impactive loadings that are difficult to predict, yet naval architects use all-welded ships with great success. A similar discussion can be made for airplanes and aeronautical engineers. The slowest adoption of structural welding was for railroad bridges. These bridges are undoubtedly subjected to heavier live loads than highway bridges, larger vibrations, and more stress reversals; but are their stress situations as serious and as difficult to predict as those for ships and planes?

14-2 ADVANTAGES OF WELDING

Today it is possible to make use of the many advantages that welding offers since the fatigue and inspection fears have been largely eliminated. Several of the many welding advantages are discussed in the following paragraphs.

1. To most persons the first advantage is in the area of economy, because the use of welding permits large savings in pounds of steel used. Welded structures allow the elimination of a large percentage of the gusset and splice plates necessary for bolted structures as well as the elimination of bolt heads. In some bridge trusses it may be possible to save up to 15 percent or more of the steel weight by using welding.

2. Welding has a much wider range of application than bolting. Consider a steel pipe column and the difficulties of connecting it to other steel members by bolting. A bolted connection may be virtually impossible, but a welded connection will present few difficulties. The student can visualize many other similar situations in which welding has a decided advantage.

3. Welded structures are more rigid because the members are often welded directly to each other. Frequently, the connections for bolted structures are made through intermediate connection angles or plates that deform due to load transfer, making the entire structure more flexible. On the other hand, greater rigidity can be a disadvantage where simple end connections with little moment resistance are desired. For such cases designers must be careful as to the type of joint they specify.

4. The process of fusing pieces together gives the most truly continuous structures. It results in one-piece construction, and because welded joints are as strong as or stronger than the base metal, no restrictions have to be placed on the joints. This continuity advantage has permitted the erection of countless slender and graceful steel statically indeterminate frames throughout the world. Some of the more outspoken proponents of welding have referred to bolted structures, with their heavy plates and abundance of bolts, as looking like tanks or armored cars when compared with the clean, smooth lines of welded structures. For a graphic illustration of this advantage the student should compare the moment-resisting connections of Fig. 15-5.

5. It is easier to make changes in design and to correct errors during erection (and at less expense) if welding is used. A closely related advantage has certainly been illustrated in military engagements during the past few wars by the quick welding repairs made to military equipment under battle conditions.

6. Another item that is often important is the relative silence of welding. Imagine the importance of this fact when working near hospitals or schools or when making additions to existing buildings. Anyone with close-to-normal hearing who has attempted to work in an office within several hundred feet of a bolted job will attest to this advantage.

7. Fewer pieces are used and, as a result, time is saved in detailing, fabrication, and field erection.

14-3 AMERICAN WELDING SOCIETY

The American Welding Society's *Structural Welding Code*[1] is the generally recognized standard for welding in the United States. The LRFD Specification clearly states that the provisions of the AWS Code apply under the LRFD Specification with a very few minor exceptions, and these are listed in LRFD

---

[1] American Welding Society, *Structural Welding Code-Steel,* AWS D.1.1-92 (Miami: AWS: 1992), 447 pages.

Specification J2. Both the AWS and the AASHTO Specifications cover dynamically loaded structures. Normally, however, the AWS Specification is used for designing buildings subject to dynamic loads unless the contract documents state otherwise.

14-4 TYPES OF WELDING

Although both gas and arc welding are available, almost all structural welding is arc welding. Sir Humphry Davy discovered in 1801 how to create an electric arc by bringing close together two terminals of an electric circuit of relatively high voltage. Although he is generally given credit for the development of modern welding, a good many years elapsed after his discovery before welding was actually performed with the electric arc. (His work was of the greatest importance to the modern structural world, but it is interesting to note that many people say his greatest discovery was not the electric arc, but rather a laboratory assistant whose name was Michael Faraday.) Several Europeans formed welds of one type or another in the 1880s with the electric arc, while in the United States the first patent for arc welding was given to Charles Coffin of Detroit in 1889.[2]

The figures to follow in this chapter show the necessity of supplying additional metal to the joints being welded to give satisfactory connections. In electric-arc welding the metallic rod, which is used as the electrode, melts off into the joint as it is being made. When gas welding is used it is necessary to introduce a metal rod known as a *filler* or *welding rod.*

In gas welding a mixture of oxygen and some suitable type of gas is burned at the tip of a torch or blowpipe held in the welder's hand or by an automatic machine. The gas used in structural welding is probably acetylene, and the process is called oxyacetylene welding. The flame produced can be used for flame cutting of metals as well as for welding. Gas welding is rather easy to learn and the equipment used is rather inexpensive. It is, however, a somewhat slow process as compared with other means of welding, and normally is used for repair and maintenance work and not for the fabrication and erection of large steel structures.

In arc welding, an electric arc is formed between the pieces being welded and an electrode held in the operator's hand with some type of holder, or by an automatic machine. The arc is a continuous spark that upon contact brings the electrode and the pieces being welded to the melting point. The resistance of the air or gas between the electrode and the pieces being welded changes the electrical energy into heat. A temperature of somewhere between 6000 and 10,000°F is produced in the arc. As the end of the electrode melts, small droplets or globules of the molten metal are formed and are actually forced by the arc

---

[2] Lincoln Electric Company, *Procedure Handbook of Arc Welding Design and Practice,* 11th ed., 1957, Part I.

across to the pieces being connected, penetrating the molten metal to become a part of the weld. The amount of penetration can be controlled by the amount of current consumed. Since the molten droplets of the electrodes are actually propelled to the weld, arc welding can be successfully used for overhead work.

A pool of molten steel can hold a fairly large amount of gases in solution, and if not protected from the surrounding air will chemically combine with oxygen and nitrogen. After cooling the welds will be relatively porous due to the little pockets formed by the gases. Such welds are relatively brittle and have much less resistance to corrosion. A weld can be shielded by using an electrode coated with certain mineral compounds. The electric arc causes the coating to melt and creates an inert gas or vapor around the area being welded. The vapor acts as a shield around the molten metal and keeps it from coming freely in contact with the surrounding air. It also deposits a slag in the molten metal, which has less density than the base metal and comes to the surface to protect the weld from the air while the weld cools. After cooling, the slag can easily be removed by peening and wire brushing (such removal being absolutely necessary before painting or application of another weld layer). A picture showing the elements of the shielded arc welding process is shown in Fig. 14-1. This figure is taken from the *Procedure Handbook of Arc Welding Design and Practice* published by the Lincoln Electric Company. Shielded metal arc welding is frequently abbreviated here with the letters SMAW.

The type of welding electrode used is very important as it decidedly affects the weld properties such as strength, ductility, and corrosion resistance. Quite a number of different types of electrodes are manufactured, the type to be used for a certain job being dependent upon the type of metal being welded, the amount of material that needs to be added, the position of the work, etc. The electrodes fall into two general classes—the *lightly coated electrodes* and the *heavily coated electrodes*.

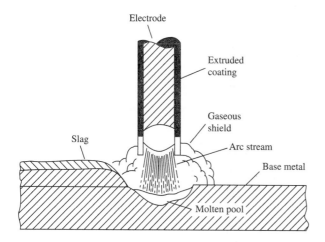

Figure 14-1 Elements of the shielded metal arc welding process (SMAW).

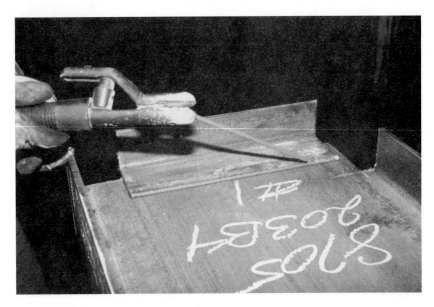

Shielded metal arc welding (SMAW) and electrode just before starting an arc to fillet weld the clip angle to the beam web (Courtesy of the American Institute of Steel Construction, Inc.)

The heavily coated electrodes are normally used in structural welding because the melting of their coatings produces very satisfactory vapor shields around the work as well as slag in the weld. The resulting welds are stronger, more resistant to corrosion, and more ductile than are those produced with lightly coated electrodes. When the lightly coated electrodes are used, no attempt is made to prevent oxidation and no slag is formed. The electrodes are lightly coated with some arc-stabilizing chemical such as lime.

Submerged (or hidden) arc welding (SAW) is an automatic process in which the arc is covered with a mound of granular fusible material and thus hidden from view. A bare metal electrode is fed from a reel and melted and deposited as filler material. The electrode, power source, and a hopper of flux are attached to a frame that is placed on rollers and that moves at a certain rate as the weld is formed. SAW welds are quickly and efficiently made and are of high quality, exhibiting high impact strength and corrosion resistance and good ductility. Furthermore, they provide deeper penetration with the result that the area effective in resisting loads is larger. A large percentage of the welding done for bridge structures is SAW. If a single electrode is used, the size of the weld obtained with a single pass is limited. Multiple electrodes may be used, however, permitting much larger welds.

Submerged arc welding (SAW) (Courtesy of the American Institute of Steel Construction, Inc.)

Welds made by the SAW process (automatic or semiautomatic) are consistently of high quality and are very suitable for long welds. One disadvantage is that the work must be positioned for near flat or horizontal welding.

Another type of welding is *flux-cored arc welding* (FCAW). In this process a flux-filled steel tube electrode is continuously fed from a reel. Gas shielding and slag are formed from the flux. The AWS Specification (4.14) provides limiting sizes for welding electrode diameters and weld sizes, as well as other requirements pertaining to welding procedures.

14-5 PREQUALIFIED WELDING

The AWS accepts four welding processes as being prequalified. The word *prequalified* is used here to mean that processes are acceptable without the necessity of further proof of their suitability by procedure qualification tests. We are saying that, based on many years of experience, sound weld metal with the desired properties can be deposited if the work is performed in accordance with the requirements of the Structural Welding Code of the AWS. The processes that are listed in AWS Specification 1.3.1 are (1) shielded metal arc welding (SMAW), (2) submerged arc welding (SAW), (3) gas metal arc welding (GMAW), and (4) flux cored arc welding (FCAW). The SMAW process is the usual process used for hand welding, while the other three are usually automatic or semiautomatic.

14-6 WELDING INSPECTION

Three steps must be taken to ensure good welding for a particular job. These are: (1) establishment of good welding procedures, (2) use of prequalified welders, and (3) employment of competent inspectors in shop and field.

When the procedures established by the AWS and AISC for good welding are followed and when welders are used who have previously been required to prove their ability, good results will probably be obtained. However, to make absolutely sure, well-qualified inspectors are needed.

Good welding procedure involves the selection of proper electrodes, current, and voltage; the properties of base metal and filler; and the position of welding, to name only a few factors. The usual practice for large jobs is to employ welders who have certificates showing their qualifications. In addition, it is not a bad practice to have each person make an identifying mark on each weld so that those persons frequently doing poor work can be located. This practice probably improves the general quality of work performed.

Lincoln ML-3 Squirtwelder mounted on a self-propelled trackless trailer deposits this $\frac{1}{4}$-in. web-to-flange weld at 28 in./min. (Courtesy of the Lincoln Electric Company.)

Visual Inspection

Another factor that will cause welders to perform better work is simply the presence of an inspector who they feel knows good welding when he or she sees it. A good inspector should have done welding and spent much time observing the work of good welders. From this experience he or she should be able to know if a welder is obtaining satisfactory fusion and penetration. He or she should be able to recognize good welds as to their shape, size, and general appearance. For instance, the metal in a good weld should approximate its original color after it has cooled. If it has been overheated it may have a rusty and reddish-looking color. An inspector can use various scales and gages to check the sizes and shapes of welds.

Visual inspection by a competent person will probably give a good indication of the quality of welds, but is not a perfect source of information as to the subsurface condition of the weld. It is surely the most economical inspection method and is particularly useful for single-pass welds. This method, however, is only good for picking up surface imperfections. There are several methods for determining the internal soundness of a weld. These include the use of penetrating dyes and magnetic particles, ultrasonic testing, and radiographic procedures. These methods can be used to detect internal defects such as porosity, weld penetration, and presence of slag.

Liquid Penetrants

Various types of dyes can be spread over weld surfaces. These will penetrate into surface cracks of the weld. After the dye has penetrated into the crack the excess surface material is wiped off and a powdery developer is used to draw the dye out of the cracks. The outlines of the cracks can then be seen with the eye. Several variations of this method are used to improve the visibility of the defects such as the use of fluorescent dyes. After the dye is drawn from the cracks the cracks are made to stand out brightly by examination under black light.[3] This method enables us to detect cracks that are open to the surface just as does visual inspection.

Magnetic Particles

In this method the weld being inspected is magnetized electrically. Cracks that are at or near the surface of the weld cause north and south poles to form on each side of the cracks. Dry iron powdered filings or a liquid suspension of particles are placed on the weld. The patterns of these particles formed when many of them cling to the cracks show the locations of cracks and indicate their size and

---

[3] James Hughes, "It's Superinspector," *Steelways* 25, no. 4 (New York: American Iron and Steel Institute, September/October, 1969), pp. 19–21.

shape. Only cracks, seams, inclusions, etc., within about $\frac{1}{10}$ in. of the surface can be located by this method. A disadvantage of this method is that if multilayer welds are used the method has to be applied to each layer.

Ultrasonic Testing

In recent years the steel industry has applied ultrasonics to the manufacture of steel. Although the equipment is expensive the method is quite useful in welding inspection as well. Sound waves are sent through the material being tested and are reflected from the opposite side of the material. These reflections are shown on a cathode ray tube. Defects in the weld will affect the time of the sound transmission. The operator can read the picture on the tube and then locate flaws and learn how severe they are. Ultrasonic testing can sucessfully be used to locate discontinuities in carbon and low-alloy steels, but it doesn't work too well for some stainless steels and for extremely coarse-grained steels.

Radiographic Procedures

The more expensive radiographic methods can be used to check occasional welds in important structures. From these tests it is possible to make good estimates of the percentage of bad welds in a structure. Portable x-ray machines where access is not a problem and radium or radioactive cobalt for making pictures are excellent but expensive methods of testing welds. These methods are satisfactory for butt welds (such as for the welding of important stainless steel piping at chemical and nuclear projects), but are not satisfactory for fillet welds because the pictures are difficult to interpret. A further disadvantage of these methods is the radioactive danger. Careful procedures have to be used to protect the technicians as well as nearby workers. On a construction job this danger probably requires night inspection of welds when only a few workers are near the inspection area. (Normally a very large job would be required before the use of the extremely expensive radioactive materials could be justified.)

A properly welded connection can always be made much stronger (perhaps $1\frac{1}{2}$ or 2 times) than the plates being connected. As a result, the actual strength is much higher than is required by the specifications. The reasons for this extra strength are as follows: the electrode wire is made from premium steel, the metal is melted electrically (as is done in the manufacture of high-quality steels), and the cooling rate is quite rapid. As a result, it is probably a rare occasion for a welder to make a weld of less strength than required by the design.

14-7 CLASSIFICATION OF WELDS

There are three separate classifications of welds described in the following paragraphs. These classifications are based upon the types of welds made, the positions of welds, and the types of joints.

Type of Weld

The two main types of welds are the *fillet welds* and the *groove welds*. In addition, there are plug and slot welds, which are not as common in structural work. These four types of welds are shown in Fig. 14-2.

The fillet welds will be shown to be weaker than groove welds; however, most structural connections (about 80 percent) are made with fillet welds. Any person who has had experience in steel structures will understand why fillet welds are more common than are groove welds. Groove welds (which are welds made in grooves between the members to be joined) are used when the members to be connected are lined up in the same plane. To use them in every situation would mean that the members would have to fit almost perfectly, and, unfortunately, the average steel structure does not fit together in that manner. Many students have seen steel workers pulling and ramming steel members to get them in position. When members are allowed to lap over each other larger tolerances are allowable in erection, and fillet welds are the welds used. Nevertheless, groove welds are quite common for many connections such as column splices, butting of beam flanges to columns, etc., and they comprise about 15 percent of

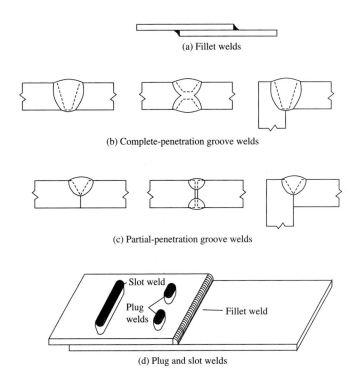

(a) Fillet welds

(b) Complete-penetration groove welds

(c) Partial-penetration groove welds

(d) Plug and slot welds

Figure 14-2 Four types of structural welds.

structural welding. Groove welds may be complete-penetration welds, which extend for the full thickness of the part being connected, or partial penetration welds, which extend for only part of the member thickness.

Groove welds are generally more expensive than fillet welds because of the costs of preparation. In fact, groove welds can cost up to 50–100 percent more than fillet welds.

A plug weld is a circular weld passing through one member to another and joining the two together. A slot weld is a weld formed in a slot or elongated hole which joins one member to the other member through the slot. The slot may be partly or fully filled with weld material. These two expensive types of welds may occasionally be used when members lap over each other and the desired length of fillet welds cannot be obtained. They may also be used to stitch together parts of a member, such as the fastening of cover plates to a built-up member.

A plug or slot weld is not generally considered to be suitable for transferring tensile forces perpendicular to the faying surface. The reason is that there is not usually much penetration of the weld into the member behind the plug or slot—and yet resistance to tension is provided primarily by penetration.

Structural designers accept plug and slot welds as being satisfactory for stitching the different parts of a member together, but many designers are not happy to use these welds for the transmission of shear forces. The penetration of the welds from the slots or plugs into the other members is questionable and, in addition, there can be critical voids down in the welds that cannot be detected with the usual inspection procedures.

Position

Welds are referred to as being flat, horizontal, vertical, and overhead. They are listed in the preceding sentence in order of their economy, with the flat welds being the most economical and the overhead welds being the most expensive. A fairly good welder can do a very satisfactory job with a flat weld, but it takes the very best welder to do a good job with an overhead weld. Although the flat welds can often be made with an automatic machine, much structural welding is done by hand. It has previously been indicated that the assistance of gravity is not necessary for the forming of good welds, but it does speed up the process. The globules of the molten electrodes can be forced into the overhead welds against gravity, and good welds will result; however, they are slow and expensive to make, so it is desirable to avoid them whenever possible. These types of welds are shown in Fig. 14-3.

Type of Joint

Welds can be further classified according to the type of joint used: butt, lap, tee, edge, corner, etc. These joint types are shown in Fig. 14-4.

Figure 14-3 Weld positions.

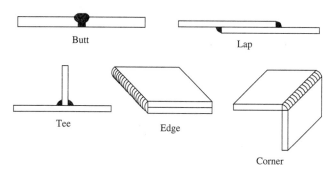

Figure 14-4 Types of weld joints.

14-8 WELDING SYMBOLS

Figure 14-5 presents the method of making welding symbols, developed by the American Welding Society. With this excellent shorthand system a great deal of information can be presented in a small space on engineering plans and drawings. Requiring only a few lines and numbers, these symbols remove the necessity of drawing in the welds and making long descriptive notes. It is certainly desirable for steel designers and draftsmen to use this standard system. Should most of the welds on a drawing be of the same size, a note to that effect can be given and the symbols omitted, except for the off-size welds.

The purpose of this section is not to show every possible type of symbol, but rather to give a general idea of the appearance of welding symbols and the information they can show. The reader can refer to the detailed information published by the AWS and reprinted in many handbooks (including the LRFD Manual). At first glance the information presented in Fig. 14-5 is probably quite confusing. For this reason a few very common symbols for fillet welds are presented in Fig. 14-6, together with an explanation of each.

Prequalified Welded Joints

Figure 14-5 Source: American Institute of Steel Construction, *Manual of Steel Construction Load & Resistance Factor Design,* 2nd ed. (Chicago: AISC, 1994), Table 8-36, p. 8-135. Reprint with the permission of the AISC.

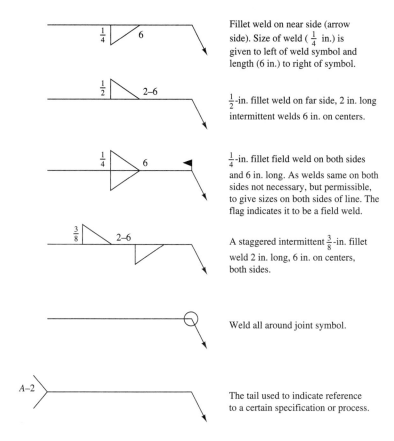

Fillet weld on near side (arrow side). Size of weld ($\frac{1}{4}$ in.) is given to left of weld symbol and length (6 in.) to right of symbol.

$\frac{1}{2}$ -in. fillet weld on far side, 2 in. long intermittent welds 6 in. on centers.

$\frac{1}{4}$ -in. fillet field weld on both sides and 6 in. long. As welds same on both sides not necessary, but permissible, to give sizes on both sides of line. The flag indicates it to be a field weld.

A staggered intermittent $\frac{3}{8}$ -in. fillet weld 2 in. long, 6 in. on centers, both sides.

Weld all around joint symbol.

The tail used to indicate reference to a certain specification or process.

Figure 14-6 Sample weld symbols.

14-9 GROOVE WELDS

When complete penetration groove welds are subjected to axial tension or axial compression the weld stress is assumed to equal the load divided by the net area of the weld. Three types of groove welds are shown in Fig. 14-7. The square groove joint, shown in part (a) of the figure, is used to connect relatively thin

$\frac{1}{4}$ –in. maximum

(a) Square groove joint

Reinforcement

(b) Single vee

(c) Double vee

Figure 14-7 Groove welds.

material up to roughly $\frac{5}{16}$ in. (8 mm) thickness. As the material becomes thicker it is necessary to use the single-vee groove welds and the double-vee groove welds illustrated in parts (b) and (c) respectively of Fig. 14-7. For these two welds the members are bevelled before welding to permit full penetration of the weld.

The groove welds shown in Fig. 14-7 are said to have *reinforcement*. Reinforcement is added weld metal that causes the throat dimension to be greater than the thickness of the welded material. Because of reinforcement, groove welds may be referred to as 100, 125, 150 percent, etc., according to the amount of extra thickness at the weld. There are two major reasons for having reinforcement. These are: (1) reinforcement gives a little extra strength because the extra metal takes care of pits and other irregularities and (2) the welder can easily make the weld a little thicker than the welded material. It would be a difficult, if not impossible, task to make a perfectly smooth weld with no places that were thinner or thicker than the material welded.

Reinforcement undoubtedly makes groove welds stronger and better when they are to be subjected to static loads. When the connection is to be subjected to repeated and vibrating loads, however, reinforcement is not as satisfactory because stress concentrations seem to develop in the reinforcement and contribute to earlier failure. For such cases as these a common practice is to provide reinforcement and grind it off flush with the material being connected (AASHTO Section 10.34.2.1).

In Fig. 14-8 some of the edge preparations that may be necessary for groove welds are shown. In part (a) a bevel with a feathered edge is shown. When feathered edges are used there is a problem with burn-through. This may be lessened if a *land* is used such as shown in part (b) of the figure, or a backup strip or backing bar as shown in part (c). The backup strip is often a $\frac{1}{4}$- in. copper plate. Weld metal does not stick to copper, and copper also has a very high conductivity that is useful in carrying away excess heat and reducing distortion. Sometimes steel backup strips are used, but they will become a part of the weld and are left in place. A land should not be used together with a backup strip because there is a high possibility that a gas pocket might be formed, preventing full penetration. When double bevels are used as shown in part (d) of the figure, spacers are sometimes provided to prevent burn-through. The spacers are removed after one side is welded.

From the standpoints of strength, resistance to impact and stress repetition, and amount of filler metal required, groove welds are much to be preferred to fillet welds. From other standpoints, however, they are not so attractive, and the vast majority of structural welding is fillet welding. Groove welds have higher residual stresses, and the preparations (such as the scarfing and veeing) of the edges of members for groove welds are expensive, but the major disadvantages probably lie in the problems involved in getting the pieces to fit together in the field. The advantages of fillet welds in this respect were described in Section 14-7. For these reasons field groove joints are not used very often, except on small jobs and where members may be fabricated a little long and cut in the field to the lengths necessary for precise fitting.

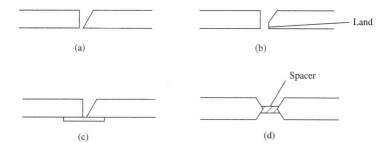

Figure 14-8 Edge preparation for groove welds. (a) Bevel with feathered edge. (b) Bevel with a land. (c) Bevel with a backup plate. (d) Double bevel with a spacer.

14-10 FILLET WELDS

Tests have shown that fillet welds are stronger in tension and compression than they are in shear, so the controlling fillet weld stresses given by the various specifications are shearing stresses. When practical, it is desirable to try to arrange welded connections so they will be subjected to shearing stresses only, and not to a combination of shear and tension or shear and compression.

Fillet welds when tested to failure with loads parallel to the weld axes seem to fail by shear at angles of about 45° through the throat. Their strength is therefore assumed to equal the design shearing stress times the theoretical throat area of the weld. The theoretical throats of several fillet welds are shown in Fig. 14-9. The throat area equals the theoretical throat distance times the length of the weld. In this figure the root of the weld is the point where the faces of the original metal pieces intersect, and the theoretical throat of the weld is the shortest distance from the root of the weld to its diagrammatic face.

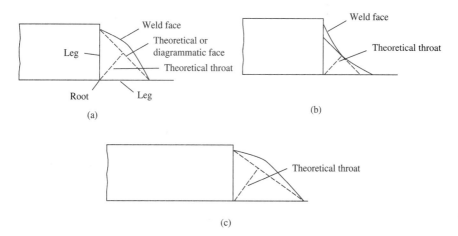

Figure 14-9 (a) Convex surface. (b) Concave surface. (c) Unequal leg fillet weld.

For the 45° or equal leg fillet the throat dimension is 0.707 times the leg of the weld, but it has a different value for fillet welds with unequal legs. The desirable fillet weld has a flat or slightly convex surface, although the convexity of the weld does not add to its calculated strength. At first glance the concave surface would appear to give the ideal fillet weld shape because stresses could apparently flow smoothly and evenly around the corner with little stress concentration. Years of experience, however, have shown that single-pass fillet welds of a concave shape have a greater tendency to crack upon cooling and this factor has proved to be of greater importance than the smoother stress distribution of convex types.

When a concave weld shrinks the surface is placed in tension tending to cause cracks, whereas when the surface of a convex weld shrinks it does not place the outer surface in tension. Rather, as the face shortens it is placed in compression.

Another item of importance pertaining to the shape of fillet welds is the angle of the weld with the pieces being welded. The desirable value of this angle is in the vicinity of 45°. For 45° fillet welds the leg sizes are equal, and such welds are referred to by the leg sizes (as a $\frac{1}{4}$-in. fillet weld). Should the leg sizes be different (not a 45° weld), both leg sizes are given in describing the weld (as a $\frac{3}{8}$-by-$\frac{1}{2}$ in. fillet weld).

The automatic submerged arc welding method (SAW) provides a deeper penetration than does the usual shielded arc welding process, As a result the LRFD permits the designer to use a larger throat area for welds made by this process. In Section J2.2a the LRFD Specification states that the effective throat thickness for SAW fillet welds with legs $\frac{3}{8}$ in. or less may equal the leg sizes. For larger leg sizes the effective throat thicknesses are considered to equal the theoretical throat thicknesses plus 0.11 in.

14-11 STRENGTH OF WELDS

For this discussion reference is made to Fig.14-10. The stress in a fillet weld is usually said to equal the load divided by the effective throat area of the weld with no consideration given to the direction of the load. Tests have shown, however, that transversely loaded fillet welds are appreciably stronger than ones loaded parallel to the weld's axis.

Figure 14-10 (a) Longitudinal fillet weld. (b) Transverse fillet weld.

Transverse fillet welds are stronger for two reasons. First, they are more uniformly stressed over their entire lengths, while longitudinal fillet welds are stressed unevenly due to varying deformations along their lengths. Second, tests show that failure occurs at angles other than 45°, giving them larger effective throat areas.

The method of determining the strength of fillet welds along their longitudinal axes regardless of the load directions is usually used to simplify computations. For the first time, in its Appendix J2.4, the December 1, 1993 AISC Specification on which the LRFD Manual is based permits a higher strength for fillet welds depending on the angle of the load with respect to the axis of the weld. The author takes advantage of this specification for the problems presented in this book.

14-12 LRFD REQUIREMENTS

When welds are made, the electrode material should have properties of the base metal. If the properties are comparable the weld metal is referred to as the matching base metal (that is, their nominal strengths are similar).

Table 14-1 (which is Table J2.5 of the LRFD Specification) provides nominal strengths for various types of welds, including fillet welds, plug and slot welds, and complete-penetration and partial-penetration groove welds.

The design strength of a particular weld is taken as the lower value of ϕF_W (where F_W is the nominal strength of the weld) and ϕF_{BM} (Where F_{BM} is the nominal strength of the base material).

For fillet welds, the nominal strength for shear on the effective area of the weld is $0.60 F_{EXX}$ (where F_{EXX} is the classification strength of the weld metal) and ϕ is 0.75. If we have tension or compression parallel to the axis of the weld the nominal strength of the base metal F_{BM} is F_y and ϕ is 0.90. The design shear strength of members being connected is taken as $\phi F_n A_{ns}$ where ϕ is 0.75, F_n is $0.6 F_u$, and A_{ns} is the net area subject to shear.

TABLE 14-1   DESIGN STRENGTH OF WELDS

Types of weld and stress[a]	Material	Resistance factor ϕ	Nominal strength F_{BM} or F_w	Required weld strength level [b,c]
Complete-joint-penetration groove weld				
Tension normal to effective area	Base	0.90	F_y	"Matching" weld must be used.
Compression normal to effective area	Base	0.90	F_y	Weld metal with a strength level equal to or less than matching weld metal is permitted to be used.
Tension or compression parallel to axis of weld				
Shear on effective area	Base Weld electrode	0.9 0.80	$0.60F_y$ $0.60F_{EXX}$	
Partial-joint-penetration groove weld				
Compression normal to effective area	Base	0.90	F_y	Weld metal with a strength level equal to or less than matching weld metal is permitted to be used.
Tension or compression parallel to axis of weld [d]				
Shear parallel to axis of weld	Base Weld electrode	0.75	[e] $0.60F_{EXX}$	
Tension normal to effective area	Base Weld electrode	0.90 0.80	F_y $0.60F_{EXX}$	
Fillet welds				
Shear on effective area	Base Weld electrode	0.75	[e] $0.60F_{EXX}$	Weld metal with a strength level equal to or less than matching weld metal is permitted to be used.
Tension or compression parallel to axis of weld [d]	Base	0.90	F_y	
Plug or slot welds				
Shear parallel to faying surfaces (on effective area)	Base Weld electrode	0.75	[e] $0.60F_{EXX}$	Weld metal with a strength level equal to or less than matching weld metal is permitted to be used.

Source: American Institute of Steel Construction, *Manual of Steel Construction Load & Resistance Factor Design*, 2nd ed. (Chicago: AISC, 1994), Table J2.5, p. 6-78. Reprinted with the permission of the AISC.

[a] For definition of effective area, see Section J2.
[b] For matching weld metal, see Table 4.1, AWS D1.1.
[c] Weld metal one strength level stronger than matching weld metal is permitted.
[d] Fillet welds and partial-joint-penetration groove welds joining component elements of built-up members, such as flange-to-web connections, are not required to be designed with the tensile or compressive stress in these elements parallel to the axis of the welds.
[e] The design of connected material is governed by Sections J4 and J5.

The all-welded 56-story Toronto Dominion Bank Tower. (Courtesy of the Lincoln Electric Company.)

The filler metal electrodes for shielded arc welding are listed as E60XX, E70XX, etc. In this classification the letter E represents an electrode, while the first set of digits (as 60, 70, 80, 90, 100, or 110) indicates the minimum tensile strength of the weld in ksi. The remaining digits specify the type of coating. As strength is the most important factor to the structural designer, we usually specify electrodes as E70XX, E80XX, or simply E70, E80, and so on. For the usual situation E70 electrodes are used for steels with F_y values from 36 to 60 ksi, while E80 is used when F_y is 65 ksi.

In addition to the nominal stresses given in Table 14-1 there are several other LRFD provisions applying to welding. Among the more important are the following:

1. The minimum length of a fillet weld may not be less than four times the nominal leg size of the weld. Should its length actually be less than this value, the weld size considered effective must be reduced to one-quarter of the weld length.

2. The maximum size of a fillet weld along edges of material less than $\frac{1}{4}$ in. thick equals the material thickness. For thicker material it may not be larger than the material thickness less $\frac{1}{16}$ in., unless the weld is specially built out to give a full throat thickness. For a plate with a thickness of $\frac{1}{4}$ in. or more it is desirable to keep the weld back at least $\frac{1}{16}$ in. from the edge so that the inspector can clearly see the edge of the plate and thus accurately determine the dimensions of the weld throat.

   As a general statement, the weldability of a material improves as the thickness to be welded decreases. The problem with thicker material is that thick plates take heat from welds more rapidly than thin plates even if the same weld sizes are used. (The problem can be somewhat alleviated by preheating the metal to be welded to a few hundred degrees Fahrenheit and holding it there during the welding operation.)

3. The minimum permissible size fillet welds of the LRFD Specification are given in Table 14-2 (Table J2.4 of the LRFD Specification). They vary from $\frac{1}{8}$ in. for $\frac{1}{4}$ in. or thinner material up to $\frac{5}{16}$ in. for material over $\frac{3}{4}$ in. in thickness. The smallest practical weld size is about $\frac{1}{8}$ in, and the most economical size is probably about $\frac{1}{4}$ or $\frac{5}{16}$ in. The $\frac{5}{16}$ in. weld is about the largest size that can be made in one pass with the shielded metal arc welded process (SMAW) and $\frac{1}{2}$ in. with the submerged arc process (SAW).

   These minimum sizes were not developed on the basis of strength considerations, but rather upon the fact that thick materials have a quenching or rapid cooling effect on small welds. If this happens the result is often a loss in weld ductility. In addition, the thicker material tends to restrain the weld material from shrinking as it cools, with the result that weld cracking can be a problem.

TABLE 14-2   MINIMUM SIZE OF FILLET WELDS

Material thickness of thicker part joined (in.)	Minimum size of fillet weld[a] (in.)
To $\frac{1}{4}$ inclusive	$\frac{1}{8}$
Over $\frac{1}{4}$ to $\frac{1}{2}$	$\frac{3}{16}$
Over $\frac{1}{2}$ to $\frac{3}{4}$	$\frac{1}{4}$
Over $\frac{3}{4}$	$\frac{5}{16}$

[a] Leg dimension of fillet welds. Single pass welds must be used.

Note that the minimum sizes given in Table 14-2 are dependent on the thicker of the two parts being joined. Regardless of the value given in the table, the minimum size does not have to exceed the thickness of the thinner part. It may be larger, however, if so required by the calculated strength.

4. When practical *end returns* (or *boxing*) should be used for fillet welds, as shown in Fig. 14-11. LRFD Specification J2.2b provides specific requirements. These are summarized in this paragraph. The length of a return should not be less than twice the nominal weld size. If end returns are used for connections that depend upon the flexibility of the outstanding legs (such as framing angles and simple end plate connections that are discussed in Chapter 15), their lengths may not exceed 4 times the nominal weld sizes. For connections such as beam seats, brackets, framing angles, and so on subjected to fatigue type loads that tend to initiate progressive failures of the welds, returns *must* be used around sides or ends for distances no less than 2 times nominal weld sizes.

5. The LRFD Specification J2.2b states that fillet welds shall be terminated at the extreme ends or sides of member parts. They must either be turned continuously around the corners for a distance of not less than two times the nominal weld sizes or they should terminate at a distance no less than the nominal weld size from the end. When returns are used they will strengthen welds at their most stressed points thus inhibiting cracks.

   When fillet welds are used on the opposite sides of a common plane they must be interrupted at corners which are common to the welds. Should the welder try to weld around such corners there will be a problem of melting the corners with consequent reduction in thickness. Figs. 8-40 and 8-41 in Part 8 of the Manual illustrate cases of this type.

6. When longitudinal fillet welds are used for the connection of plates or bars, their length may not be less than the perpendicular distance between them because of shear lag, discussed in Chapter 3. Furthermore, the distance between fillet welds may not be greater than 8 in. for end connections, unless end transverse welds or intermediate plug or slot welds are used (AWS Specification 8.8.1.)

7. For lap joints, the minimum amount of lap permitted is equal to 5 times the thickness of the thinner part joined, but may not be less than 1 in. The

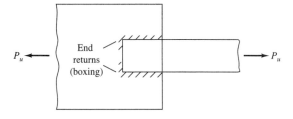

Figure 14-11 End returns.

purpose of this minimum lap is to keep the joint from rotating excessively [See Fig. 12.1(a)] when the connected parts are loaded. (AWS Specification 8.8.3).

14-13 DESIGN OF SIMPLE FILLET WELDS

Examples 14-1 through 14-3 illustrate the calculations necessary to determine the strength of various fillet welded connections, while Example 14-4 presents the design of such a connection. In these and other problems weld lengths are selected no closer than to the nearest $\frac{1}{4}$ in., because closer work cannot be expected in shop or field.

■ EXAMPLE 14-1

Determine the design strength of a 1-in. length of $\frac{5}{16}$-in. fillet weld using (a) the shielded metal arc process (SMAW) and (b) the submerged arc process (SAW). Use E70 electrodes with a minimum tensile strength of 70 ksi $= F_{EXX}$ and assume load parallel to weld.

Solution.

(a) Shielded metal arc process

Effective throat thickness of weld $= (0.707)(\frac{5}{16}) = 0.221$ in.

$$
\begin{aligned}
\text{Design strength} &= \phi F_w \\
&= (\phi)(\text{nominal strength of weld } 0.6 \, F_{EXX}) \\
&\quad \times (\text{throat } t)(\text{weld length}) \\
&= (0.75)(0.60 \times 70)(0.221)(1.0) = 6.96 \text{ k/in.}
\end{aligned}
$$

(b) Submerged arc process

From LRFD Section J2.2a the effective throat thickness of weld $= \frac{5}{16}$ in.

$$
\begin{aligned}
\text{Design strength} &= (0.75)(0.60 \times 70)(\tfrac{5}{16})(1.0) \\
&= 9.84 \text{ k/in.} \qquad\qquad ■
\end{aligned}
$$

Fillet welds may not be designed with a stress that is greater than the design stress on the adjacent members being connected. If the external force applied to the member (tensile or compressive) is parallel to the axis of the weld metal the design strength may not exceed the axial design strength of the member.

Examples 14-2 and 14-3 illustrate the calculations necessary to determine the design strength of plates connected with SMAW and SAW fillet welds. In each of these examples the shearing strength per inch of the welds controls and is multiplied by the total length of the welds to give the total capacity of the connections.

■ **EXAMPLE 14-2**

What is the design strength of the connection shown in Fig. 14-12 if $F_y = 50$ ksi and E70 electrodes are used? The $\frac{7}{16}$-in. fillet welds shown were made by the SMAW process.

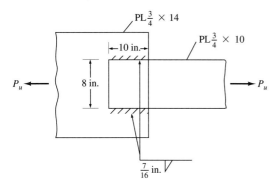

Figure 14-12

Solution.

$$\text{Effective throat thickness} = (0.707)(\tfrac{7}{16}) = 0.309 \text{ in.}$$
$$\text{Capacity of weld/in.} = \phi F_w = (0.75)(0.60 \times 70)(0.309)(1.0)$$
$$= 9.73 \text{ k/in.} \leftarrow$$
$$\text{Total capacity of weld} = (9.73)(20) = 194.65 \text{ k} \leftarrow$$
$$\text{Design strength of plate} = \phi F_y A_g = (0.90)(50)(\tfrac{3}{4} \times 10)$$
$$= 337.5 \text{ k}$$
$$\underline{\underline{\text{Design capacity} = 194.6 \text{ k}}}$$ ■

■ **EXAMPLE 14-3**

Repeat Example 14-2 if SAW welds are used.

Solution.

$$\text{Effective throat thickness} = (0.707)(\tfrac{7}{16}) + 0.11 = 0.419 \text{ in.}$$
$$\text{Capacity of weld/in.} = \phi F_w = (0.75)(0.60 \times 70)(0.419)(1.0)$$
$$= 13.20 \text{ k/in.}$$
$$\text{Total capacity of weld} = (13.20)(20) = 264 \text{ k} \leftarrow$$
$$\text{Design strength of plate} = \phi F_y A_g = (0.90)(50)(\tfrac{3}{4} \times 10)$$
$$= 337.5 \text{ k}$$
$$\underline{\underline{\text{Design capacity} = 264 \text{ k}}}$$ ■

■ EXAMPLE 14-4

Using A36 steel and E70 electrodes, design SMAW fillet welds to resist a full-capacity load on the $\frac{3}{8} \times 6$ in. member shown in Fig. 14-13.

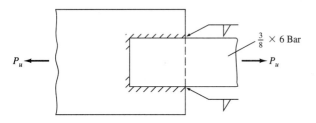

Figure 14-13

Solution.

$$P_u = \phi_t F_y A_g = (0.90)(36)(\tfrac{3}{8} \times 6) = 72.9 \text{ k}$$

Maximum weld size $= \frac{3}{8} - \frac{1}{16} = \frac{5}{16}$ in. (LRFD Section J2.2b)

Minimum weld size $= \frac{3}{16}$ in. (Table 14.2)

Use $\frac{5}{16}$ in. weld

Effective throat thickness of weld $= (0.707)(\frac{5}{16}) = 0.221$ in.

Capacity of weld per in. $= \phi F_W = (0.75)(0.60 \times 70)(0.221)(1.0)$

$$= 6.96 \text{ k/in.}$$

Length required $= \dfrac{72.9}{6.96} = 10.47$ in.

Use end returns not less than $2 \times \frac{5}{16} = \frac{10}{16}$ in. (Say 1 in.)

Leaving $10.47 - 2.0 = 8.47$ in. or $4\frac{1}{2}$ in. each side

However, must use $6 - 1 = 5$-in.-long welds each side as required by LRFD Specification J2.2b.   ■

Appendix J2.4 states that the strength of fillet welds loaded transversely in a plane through their centers of gravity may be determined with the following equation in which $\phi = 0.75$ and θ is the angle between the line of action of the load and the longitudinal axis of the weld.

$$\phi F_W = \phi(0.6 \, F_{EXX})(1.0 + 0.50 \sin^{1.5} \theta)(\text{throat } t)(\text{weld length})$$

As the angle θ is increased the strength of the weld increases. Should the load be perpendicular to the longitudinal axis of the weld, the result will be a 50 percent increase in the computed weld strength. Example 14-5 illustrates the application of this Appendix expression.

■ EXAMPLE 14-5

Determine the strength per in. of the $\frac{1}{4}$-in. SMAW fillets welds formed with E70 electrodes shown in part (a) of Fig. 14-14 if the load is applied parallel to the longitudinal axis of the welds. Then determine the strength per in. of the welds shown in part (b) of the figure where the load is applied at a 45° angle with the longitudinal axis of the welds.

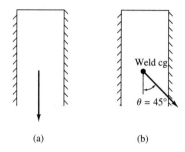

(a)   (b)

Figure 14-14

Solution.

(a) Effective throat thickness $= (0.707)(\frac{1}{4}) = 0.177$ in.
   $\phi F_w = (0.75)(0.6 \times 70)(0.177)(1.0) = 5.58$ k/in.

(b) $\phi F_w = (\phi)(0.6F_{EXX})(1.0 + 0.50 \sin^{1.5} \theta)$(effective throat t)(length)
   $= (0.75)(0.6)(70)(1.0 + 0.50 \sin^{1.5} 45°)(0.177)(1.0)$
   $= 7.23$ k/in., which is almost a 30 percent increase in strength over the longitudinally loaded case.   ■

On some occasions the lengths available for the usual longitudinal fillet welds are not sufficient for the load to be resisted. For the situation shown in Fig. 14-15 it may be possible to develop sufficient strength by welding along the back of the channel at the edge of the plate if sufficient space is available. The dashed lines in the figure show this weld.

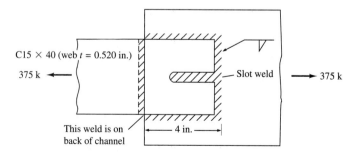

C15 × 40 (web $t = 0.520$ in.)

375 k

Slot weld

375 k

This weld is on back of channel

← 4 in. →

Figure 14-15

Another possibility is the use of slot welds as illustrated in Example 14-2. There are several LRFD requirements pertaining to slot welds that need to be mentioned here. This specification (J2.3) says that the width of a slot may not be less than the member thickness $+ \frac{5}{16}$ in. (rounded off to the next greater odd $\frac{1}{16}$ in. since structural punches are made in odd 16ths diameters), nor may it be greater than $2\frac{1}{4}$ times the weld thickness. For members up to $\frac{5}{8}$ in. thickness the weld thickness must equal the plate thickness, and for members greater than $\frac{5}{8}$ in. thickness the weld thickness may not be less than one-half the member thickness or $\frac{5}{8}$ in. The maximum length permitted for slot welds is 10 times the weld thickness. The limitations given in specifications for the maximum sizes of plug or slot welds are caused by the detrimental shrinkage that occurs around these types of welds when they exceed certain sizes. Should holes or slots larger than those specified be used, it is desirable to use fillet welds around the borders of the holes or slots rather than using a slot or plug weld. Slot and plug welds are normally used in conjunction with fillet welds in lap joints. Sometimes plug welds are used to fill in the holes temporarily used for erection bolts for beam and column connections. They may or may not be included in the calculated strength of these joints.

The strength of a plug or slot weld is equal to its design stress ϕF_W times its nominal area in the shearing plane. This area is equal to the area of contact at the base of the plug or slot. The length of a slot weld can then be determined from the following.

$$L = \frac{\text{load}}{(\text{width})(\text{design stress})}$$

Example 14-6 illustrates the design of the welds necessary to connect a channel to a plate. The calculations show that the ordinary side and end fillet welds do not provide sufficient strength in this case because of the limited space available. It is decided to use a slot weld to provide the remaining load resistance needed.

■ EXAMPLE 14-6

Design SMAW fillet welds to connect a C15 × 40 to the plate shown in Fig. 14-15. The load to be resisted is 375 k and E70 electrodes are to be used. As shown in the figure the channel may lap over the plate by only 4 in. due to space limitations. Space is not available for welding on back of channel. Notice the load is applied transversely to the end weld.

Solution.   Because of limited space use

$$\text{Max weld size} = \text{web } t - \tfrac{1}{16} = \tfrac{1}{2} - \tfrac{1}{16} = \tfrac{7}{16} \text{ in.}$$

Effective throat thickness $= (0.707)(\tfrac{7}{16}) = 0.309$ in.

Capacity of end and side welds

$$= (0.75)(0.60 \times 70)(0.309)(1 + 0.5 \sin^{1.5} 90°)(15)$$
$$+ (0.75)(0.60 \times 70)(0.309)(2 \times 4)$$
$$= 219 + 77.9 = 296.9 \text{ k} < 375 \text{ k}$$

Therefore try a slot weld

$$\text{Min width of slot} = 0.520 + \tfrac{5}{16} = \tfrac{13}{16} \text{ in.}$$
$$\text{Max width} = 2\tfrac{1}{4} \times \text{weld thickness}$$
$$= (2\tfrac{1}{4})(\text{web } t \text{ of channel}) = (2\tfrac{1}{4})(\tfrac{1}{2})$$
$$= 1\tfrac{1}{8} \text{ say } \tfrac{17}{16} \text{ (to odd } \tfrac{1}{16})$$

Use $\tfrac{15}{16}$ in.

$$\text{Capacity of } \tfrac{7}{16}\text{- in. fillet weld} = \left(\frac{15 - \tfrac{15}{16}}{15}\right)(219) + 77.9 = 283.2 \text{ k}$$

$$\text{Load to be resisted by slot weld} = 375 - 283.2 = 91.8 \text{ k}$$

$$\text{Length required for slot weld} = \frac{91.8}{(\tfrac{15}{16})(0.75)(0.60 \times 70)} = 3.11 \text{ in.}$$

Say $3\tfrac{1}{2}$ in.

$$\text{Max length permitted by LRFD} = (10)(\tfrac{1}{2})$$
$$= 5.00 \text{ in.} > 4\tfrac{1}{2} \text{ in.} \qquad \text{(OK)}$$
$$\text{Use } \tfrac{15}{16} \times 3\tfrac{1}{2} \text{ in. slot weld} \qquad ■$$

Alternate Solution Should space have been available on the back of the channel next to the plate, a $\tfrac{7}{16}$-in. fillet weld would carry 219 k > 91.8 k (OK)

14-14 DESIGN OF FILLET WELDS FOR TRUSS MEMBERS

Should the members of a welded truss consist of single angles, double angles, or similar shapes and be subjected to static axial loads only, the LRFD Specification (J1.8) permits the connections to be designed by the same procedures described in the preceding section. The designers can select the weld size, calculate the total length of the weld required, and place the welds around the member ends as they see fit. (It would not be logical, of course, to place the weld all on one side of a member, such as for the angle of Fig. 14-16, because of the rotation possibility.) Example 14-7 illustrates the simple calculations involved in designing the welds for the ends of a truss member.

■ **EXAMPLE 14-7**
Using A36 steel and E70 electrodes, design side and end fillet SMAW welds for the full capacity of a 6 × 4 × $\tfrac{1}{2}$-in. angle tension member with the long leg connected. Assume static load.

Figure 14-16

Solution.

Tensile capacity of angle $= \phi_t F_y A_g = (0.90)(36)(4.75) = 153.9$ k

Maximum weld size $= \frac{1}{2} - \frac{1}{16} - \frac{7}{16}$ in.

Minimum weld size $= \frac{3}{16}$ in. (from Table 14-2).

Use $\frac{5}{16}$-in. weld as maximum size which can be made in one pass

Effective throat thickness of weld $= (0.707)(\frac{5}{16}) = 0.221$ in.

Design strength of end weld $= (0.75)(0.60 \times 70)(0.221)$
$$\times (1 + 0.5 \sin^{1.5} 90°)(6)$$
$$= 62.7 \text{ k}$$

Load to be carried by side welds $= 153.9 - 62.7 = 91.2$ k

Design strength/in. of side welds $= (0.75)(0.60 \times 70)$
$$\times (0.221)(1) = 6.96 \text{ k/in.}$$

Length Reqd. $= \dfrac{91.2}{6.96} = 13.10$ in. say $6\frac{1}{2}$ in. each side

Place welds as shown in Fig. 14-16 ∎

The student should carefully note that the centroid of the welds and the centroid of the statically loaded angle do not coincide in the connection selected in Example 14-7 and shown in Fig. 14-16. Should a welded connection be subjected to varying stresses (such as those occurring in a bridge member), it is considered desirable to place the welds so that their centroid will coincide with the centroid of the member (or the resulting torsion must be accounted for in design). If the member being connected is symmetrical the welds will be placed

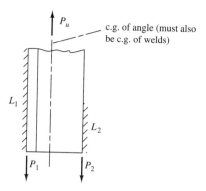

Figure 14-17

symmetrically, but if the member is not symmetrical the welds will not be symmetrical.

The force in an angle, such as the one shown in Fig. 14-17, is assumed to act along its center of gravity. If the center of gravity of weld resistance is to coincide with the angle force the weld must be asymmetrically placed, or in this figure L_1 must be longer than L_2. (When angles are connected by bolts there is usually an appreciable amount of eccentricity, but in a welded joint eccentricity can be fairly well eliminated.) The information necessary to handle this type of weld design can be easily expressed in equation form, but only the theory behind the equations is presented here.

For the angle shown in Fig. 14-17 the force acting along line L_2 (designated here as P_2) can be determined by taking moments about L_1. The member force and the weld resistance are to coincide and the moments of the two about any point must be zero. If moments are taken about L_1 the force P_1 (which acts along line L_1) will be eliminated from the equation and P_2 can be determined. In a similar manner P_1 can be determined by taking moments along L_2 or by $\Sigma V = 0$. Example 14-8 illustrates the design of fillet welds of this type. A similar problem is handled in Example 14-9 except an end fillet weld is included, thus permitting a shorter connection. The center of gravity and resistance of the end weld are known and can be easily included in the moment equations.

There are other possible solutions for the design of the welds for the angle considered in these two examples. Although the $\frac{7}{16}$-in. weld is the largest one permitted at the edges of the $\frac{1}{2}$-in. angle and at its end, a larger weld could be used on the other side next to the outstanding leg. From a practical point of view, however, the welds should be the same size because different size welds slow the welder down owing to the need to change electrodes to make different sizes.

It is also to be remembered from our fatigue discussion in Chapter 4 that should the estimated number of loading cycles during the structure's estimated life exceed 20,000, it will be necessary to study the stress range of the connection at service loads. This study may very well result in larger connections as required by Appendix K of the LRFD Specification.

■ EXAMPLE 14-8

Use A36 steel, E70 electrodes, and the SMAW process to design side fillet welds for the full capacity of the $5 \times 3 \times \frac{1}{2}$-in. angle tension member shown in Fig. 14-18. Assume the member is subjected to repeated stress variations, making any connection eccentricity undesirable. Check block shear strength of member.

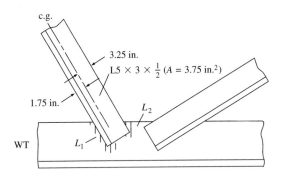

Figure 14-18

Solution.   Tensile capacity of angle

$$P_u = \phi_t F_y A_g = (0.9)(36)(3.75) = 121.5 \text{ k} \leftarrow$$

or

$$P_u = \phi_t F_u A_g \text{ assuming } U = 0.87$$
$$= (0.75)(58)(0.87)(3.75) = 141.9 \text{ k}$$

Maximum weld size $= \frac{1}{2} - \frac{1}{16} = \frac{7}{16}$ in.

Use $\frac{5}{16}$-in. weld

Effective throat thickness of weld $= (0.707)(\frac{5}{16}) = 0.221$ in.

Capacity of weld/in. $= \phi F_w = (0.75)(0.60 \times 70)(0.221)(1.0)$
$$= 6.96 \text{ k/in.}$$

Total weld length required $= \dfrac{121.5}{6.96} = 17.46$ in.

Taking moments about L_1 (see Fig. 4.18) to determine force P_2

$$(121.5)(1.75) - 5.00 \, P_2 = 0$$

$P_2 = 42.53$ k

$P_1 = P - P_2 = 121.5 - 42.53 = 78.97$ k

$L_1 = \dfrac{78.97}{6.96} = 11.35$ in.   say $11\frac{1}{2}$ in.

$L_2 = \dfrac{42.53}{6.96} = 6.11$ in.   say $6\frac{1}{2}$ in.   ■

Use end returns $2 \times \frac{5}{16} = \frac{10}{16}$ (say 1 in.). The end return lengths may be subtracted from the side weld lengths, making them $10\frac{1}{2}$ in. and $5\frac{1}{2}$ in., respectively.

Check block shearing strength assuming these dimensions

$$0.6F_u A_{nv} = (0.6)(58)(18 \times \tfrac{1}{2}) = 313.2 \text{ k}$$
$$> F_u A_{nt} = (58)(5 \times \tfrac{1}{2}) = 145 \text{ k}$$
$$\therefore \phi R_n = 0.75[(0.6)(58)(18 \times \tfrac{1}{2}) + (36)(3 \times \tfrac{1}{2})]$$
$$= 302.4 \text{ k} > 121.5 \text{ k} \qquad\qquad\qquad \text{(OK)}$$

■ **EXAMPLE 14-9**

Rework Example 14-8 using fillet welds along the sides and end of the angle.

Solution. Assuming $\frac{5}{16}$-in. welds (capacity of side welds = 6.96 k/in.)

Total strength of end weld
$$= (0.75)(0.60 \times 70)(0.707 \times \tfrac{5}{16})(1 + 0.5 \sin^{1.5} 90°)(5) = 52.2 \text{ k}$$
Taking moments about L_1 to determine force P_2,
$$(121.5)(1.75) - (2.50)(52.2) - 5.00 P_2 = 0$$
$$P_2 = 16.5 \text{ k}$$
$$P_1 = 121.5 - 52.2 - 25.12 = 52.8 \text{ k}$$
$$L_1 = \frac{52.8}{6.96} = 7.59 \text{ in.} \quad \underline{\text{say 8 in.}}$$
$$L_2 = \frac{16.5}{6.96} = 2.37 \text{ in.} \quad \underline{\text{say } 2\frac{1}{2} \text{ in.}} \qquad ■$$

It is rather convenient for design purposes to know the strength of a $\frac{1}{16}$-in. fillet weld 1 in. long. Though this size is below the minimum permissible size given in Table 14-2, the calculated strength of such a weld is useful for determining weld sizes for calculated forces. For a 1-in.-long SMAW weld with the load parallel to weld axis we have

$$\phi F_w = (0.75)(0.707 \times \tfrac{1}{16} \times 1.0)(0.60 F_{EXX}) = 0.02 F_{EXX}$$

For an E70 electrode ϕF_w is $(0.02)(70) = 1.4$ k/in. for a $\frac{1}{16}$-in. weld. If we are designing a fillet weld to carry a factored force of 6.5 k/in, the required weld size will be $6.5/1.4 = 4.64$ sixteenths of an inch, or, say, $\frac{5}{16}$ in.

14-15 SHEAR AND TORSION

Fillet welds are frequently loaded with eccentrically applied loads with the result that the welds are subjected to either shear and torsion or to shear and bending. Figure 14-19 is presented in an attempt to show the difference between the two situations. Shear and torsion, shown in part (a) of the figure, is the subject of this section, while shear and bending—shown in part (b) of the figure—are discussed in Sections 14-16 and 14-7.

As is the case for eccentrically loaded bolt groups (Section 13-1), the LRFD Specification provides the permissible design strength of welds, but does not specify a method of analysis for eccentrically loaded welds. The method to be used is left up to the designer.

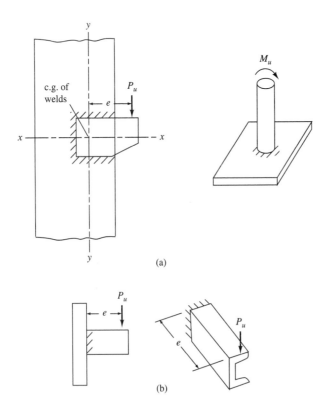

Figure 14-19 (a) Welds subjected to shear and torsion. (b) Welds subjected to shear and bending.

Elastic Method Initially the very conservative elastic method is presented. In this method friction or slip resistance between the connected parts is neglected and the connected parts are assumed to be perfectly rigid.

For this discussion the welded bracket of part (a) of Fig. 14-20 is considered. The pieces being connected are assumed to be completely rigid as they were in bolted connections. The effect of this assumption is that all deformation occurs in the weld. The weld is subjected to a combination of shear and torsion as was the eccentrically loaded bolt group considered in Section 13-1. The force caused by torsion can be computed from the following familiar expression.

$$f = \frac{Td}{J}$$

In this expression T is the torsion, d is the distance from the center of gravity of the weld to the point being considered, and J is the polar moment of inertia of the weld. It is usually more convenient to break the force down into its vertical and horizontal components. In the expressions to follow, h and v are the horizontal and vertical components of the distance d. (The formulas are almost identical to those used for determining stresses in bolt groups subject to torsion.)

$$f_h = \frac{Tv}{J} \qquad f_v = \frac{Th}{J}$$

These components are combined with the usual direct shearing stress which is assumed to equal the reaction divided by the total length of the welds. For design of a weld subject to shear and torsion it is convenient to assume a 1-in. weld and to compute the stresses on a weld of that size. Should the assumed weld be overstressed, a larger weld is required; if understressed, a smaller one is desirable.

Although the calculations will probably show the weld to be overstressed or understressed, the math does not have to be repeated because a ratio can be set up to give the weld size for which the load would produce a computed stress exactly equal to the design stress. The student should note that the use of a 1-in. weld simplifies the units because 1 in. of length of weld is 1 in.^{2} of weld, and the computed stresses can be said to be either kips per square inch or kips per inch of length. Should the calculations be based on some size other than a 1-in. weld, the student will have to be very careful in keeping the units straight, particularly in obtaining the final weld size. To further simplify the calculations the welds are assumed to be located at the edges along which the fillet welds are placed, rather than at the centers of their effective throats. As the throat dimensions are rather small, this assumption changes the results very little. Example 14-10 illustrates the calculations involved in determining the weld size required for a connection subjected to a combination of shear and torsion.

■ EXAMPLE 14-10

For the bracket shown in Fig. 14-20(a), determine the fillet weld size required if E70 electrodes, the LRFD Specification, and the SMAW process are used.

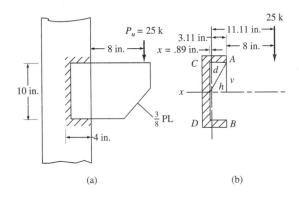

Solution.   Assuming a 1-in. weld as shown in part (b) of Fig. 14-20,

$$A = 18 \text{ in.}^2$$

$$\bar{x} = \frac{(4)(2)(2)}{18} = 0.89 \text{ in.}$$

$$I_x = \left(\frac{1}{12}\right)(1)(10)^3 + (2)(4)(5)^2 = 283.3 \text{ in.}^4$$

$$I_y = (2)\left(\frac{1}{3}\right)(0.89^3 + 3.11^3) + (10)(0.89)^2 = 28.5 \text{ in.}^4$$

$$J = 283.3 + 28.5 = 311.8 \text{ in.}^4$$

The 1-in. vertical weld has a design strength $= (0.75)(0.60 \times 70) \times (0.707 \times 1.0)(1.0) = 22.27$ k/in. while the horizontal weld has a design strength $= (0.75)(0.60 \times 70)(0.707 \times 1.0)(1 + 0.5 \ \sin^{1.5} \ 90°)(1.0) = 33.41$ k/in. Therefore, stress is computed at points A and B in the horizontal weld and at C and D in the vertical weld (see Fig. 14-20) and required weld size determined.

Horizontal weld	*Vertical weld*
$f_h = \dfrac{(25 \times 11.11)(5)}{311.8} = 4.45$ k/in.	$f_h = \dfrac{(25 \times 11.11)(5)}{311.8} = 4.45$ k/in.
$f_v = \dfrac{(25 \times 11.11)(3.11)}{311.8} = 2.77$ k/in.	$f_v = \dfrac{(25 \times 11.11)(0.89)}{311.8} = 0.793$ k/in.
$f_s = \dfrac{25}{18} = 1.39$ k/in.	$f_s = \dfrac{25}{18} = 1.39$ k/in.
$f_r = \sqrt{(2.77 + 1.39)^2 + (4.45)^2}$ $= 6.09$ k/in.	$f_r = \sqrt{(0.793 + 1.39)^2 + (4.45)^2}$ $= 4.96$ k/in.
Size $= \dfrac{6.09}{33.41} = 0.182$ in.	Size $= \dfrac{4.96}{22.27} = 0.233$ in. ←

<u>Use $\frac{1}{4}$ - in. fillet welds</u>

Ultimate Strength Method   An ultimate strength analysis of eccentrically loaded welded connections is much more realistic than is the very conservative elastic procedure just described. For the discussion to follow, the eccentrically loaded fillet weld of Fig. 14.21 is considered. As for eccentrically loaded bolted connections, the load tends to cause a relative rotation and translation between the parts connected by the weld.

Even if the eccentric load is of such a magnitude as to cause the most stressed part of the weld to yield, the entire connection will not yield. The load may be increased, the less stressed fibers will begin to resist more of the load, and failure will not occur until all the weld fibers yield. The weld will tend to rotate about its instantaneous center of rotation. The location of this point (which is indicated by the letter O in the figure) is dependent upon the location of the eccentric load, the geometry of the weld, and the deformations of the different elements of the weld.

If the eccentric load P_u is vertical and if the weld is symmetrical about a horizontal axis through its center of gravity, the instantaneous center will fall somewhere on the horizontal x axis. Each differential element of the weld will provide a resisting force R. As shown in Fig. 14.21, each of these resisting forces is assumed to act perpendicular to a ray drawn from the instantaneous center to the center of gravity of the weld element in question.

Studies have been made to determine the maximum shear forces that eccentrically loaded weld elements can withstand.[4,5] The results, which depend on the load deformation relationship of the weld elements, may be represented either with curves or in formula fashion. The ductility of the entire weld is governed

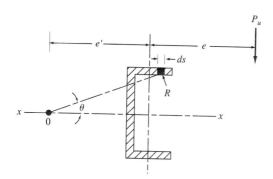

Figure 14-21

[4] L. J. Butler, S. Pal, and G. L. Kulak, "Eccentrically Loaded Weld Connections," *Journal of the Structural Division,* vol. 98, no. ST5, May 1972, pp. 989–1005.

[5] G. L. Kulak and P. A. Timler, "Tests on Eccentrically Loaded Fillet Welds," Dept of Civil Engineering, University of Alberta, Edmonton, December 1984.

by the maximum deformation of the weld element that first reaches its limit. (The element the greatest distance from the weld's instantaneous center probably reaches its limit first.)

Just as in eccentrically loaded bolted connections, the location of the instantaneous center of rotation is determined by trial and error. Unlike eccentrically loaded bolt groups, the strength and deformation of welds are dependent on the angle θ that the force in each element makes with the axis of that element. The deformation of each element is proportional to its distance from the instantaneous center. At maximum stress the weld deformation is Δ_m, which is determined with the following expression in which D is the leg size of the fillet weld.

$$\Delta_m = 0.209 \, (\theta + 2)^{-0.32} D$$

The deformation in a particular element is assumed to vary directly in proportion to its distance from the instantaneous center

$$\Delta_r = \frac{l_r}{l_m} \Delta_m$$

The nominal stress in a particular weld element is computed as

$$R = 0.60 F_{EXX}(1.0 + 0.50 \sin^{1.5} \theta) \, [p(1.9 - 0.9p)]^{0.3}$$

where θ = the angle of loading measured from the weld longitudinal axis and p = the ratio of the deformation of an element to its deformation at ultimate stress.

We can assume a location of the instantaneous center, determine R values for the different elements of the weld, and compute ΣR_x and ΣR_y. The three equations of equilibrium ($\Sigma M = 0$, $\Sigma R_x = 0$, and $\Sigma R_y = 0$) will be satisfied if we have the correct location of the instantaneous center. If they are not satisfied we will try another location and so on. Finally when the equations are satisfied the value of P_u can be computed as $\sqrt{(\Sigma R_x)^2 + (\Sigma R_y)^2}$.

Detailed information on the use of these expressions is given in Part 8 of the LRFD Manual. The LRFD Specification permits weld strengths in eccentrically loaded connections to exceed $0.6 \, F_{EXX}$. The solution of these types of problems is completely impractical without the use of computers or of computer generated tables such as those provided in Part 8 of the LRFD Manual. Example 14-11 illustrates the use of the LRFD tables to solve a problem of this type while Example 14-18 at the end of the chapter makes use of the computer program INSTEP for a solution.

The values given in the tables of Part 8 of the LRFD Manual were developed by this procedure. Using the tables, the ultimate strength P_u of a particular connection can be determined from the expression to follow in which C is a tabular coefficient that includes a ϕ of 0.75, C_1 is a coefficient depending on the

TABLE 14-3 ELECTRODE STRENGTH COEFFICIENT

Electrode	F_{EXX} (ksi)	C_1
E60	60	0.857
E70	70	1.00
E80	80	1.03
E90	90	1.16
E100	100	1.21
E110	110	1.34

Source: American Institute of Steel Construction, *Manual of Steel Construction Load & Resistance Factor Design,* 2nd ed. (Chicago: AISC, 1994), p. 8-158. Reprinted with the permission of the AISC.

electrode number and given in Table 14-3, D is the weld size in sixteenths of an inch, and l is the length of the vertical weld.

$$P_u = CC_1 Dl$$

The Manual includes tables for both vertical and inclined loads (at angles from the vertical of 0°, 45°, and 75°). The user is warned not to interpolate for angles in between these values because the results may be quite unconservative. Therefore, he or she is advised to use the value given for the next lower angle. Should the connection arrangement being considered not be covered by the tables, it is suggested that the conservative elastic procedure previously described be used. Examples 14-11 and 14-12 illustrate the use of these ultimate strength tables.

■ EXAMPLE 14-11

Repeat Example 14-10 using the LRFD tables that are based on an ultimate strength analysis. The connection is redrawn in Fig. 14-22.

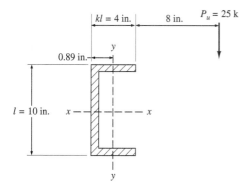

Figure 14-22

Solution.

$$e = 11.11 \text{ in.}$$

$$l = 10 \text{ in.}$$

$$a = \frac{11.11}{10} = 1.111$$

$$k = \frac{4}{10} = 0.4$$

$C = 0.986$ from Table 8-42, in the LRFD Manual

$C_1 = 1.0$ from same table

$$D = \text{weld size required} = \frac{P_u}{CC_1 l}$$

$$= \frac{25}{(0.986)(1.0)(10)} = 2.53 \text{ sixteenths}$$

$$= 0.158 \text{ in. (as compared with 0.233 in}$$

by the elastic method)

Use $\frac{3}{16}$ in.   ■

■ EXAMPLE 14-12

Determine the weld size required for the situation shown in Fig. 14-23 using the LRFD tables that are based on an ultimate strength analysis, (A36 steel, E70 electrodes).

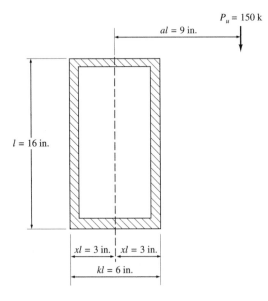

Figure 14-23

Solution.

$$k = \frac{6}{16} = 0.375$$

$$a = \frac{9}{16} = 0.5625$$

C from table 8-40 by interpolation $= 2.49$

$C_1 = 1.0$ for E70 electrodes (see Table 14-3)

Required weld size $= D = \dfrac{P_u}{CC_1 l} = \dfrac{150}{(2.49)(1.0)(16)}$

$$= 3.77 \text{ sixteenths}$$

Use $\frac{1}{4}$-in. weld   ■

14-16 SHEAR AND BENDING

The welds shown in Fig. 14-19(b) and in Fig. 14-24 are subjected to a combination of shear and bending.

For short welds of this type the usual practice is to consider a uniform variation of shearing stress. If, however, the bending stress is assumed to be given by the flexure formula, the shear does not vary uniformly for vertical welds but as a parabola with a maximum value $1\frac{1}{2}$ times the average value. These stress and shear variations are shown in Fig. 14-25.

The student should carefully note that the maximum shearing stresses and the maximum bending stresses occur at different locations. It is, therefore, probably not necessary to combine the two stresses at any one point. If the weld is capable of withstanding the worst shear and the worst moment individually it is probably satisfactory. In Example 14-13, however, a welded connection subjected to shear and bending is designed by the usual practice of assuming a uniform shear distribution in the weld and combining the value vectorially with the maximum bending stress.

Figure 14-24

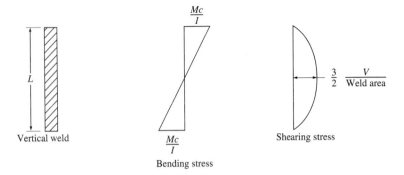

Figure 14-25

■ EXAMPLE 14-13

Using E70 electrodes, the SMAW process, and the LRFD Specification determine the weld size required for the connection of Fig. 14-24 if $P_u = 45$ k, $e = 2\frac{1}{2}$ in., and $L = 8$ in. Assume that the member thicknesses do not control weld size.

Solution.

$$f_s = \frac{45}{(2)(8)} = 2.81 \text{ k/in.}$$

$$f = \frac{(45 \times 2.5)(4)}{(\frac{1}{12})(1)(8)^3(2)} = 5.27 \text{ k/in.}$$

$$f_r = \sqrt{(2.81)^2 + (5.27)^2} = 5.97 \text{ k/in.}$$

$$\frac{\text{Weld size}}{\text{required}} = \frac{5.97}{(0.707)(1)(0.75)(0.60 \times 70)} = 0.268 \text{ in.} \quad (\text{say } \tfrac{5}{16} \text{ in.}) \quad ■$$

The subject of shear and bending is a very practical one, as it is the situation commonly faced in moment resisting connections. This topic is continued at length in Section 14.17.

14-17 DESIGN OF MOMENT-RESISTING CONNECTIONS

In this section a brief introduction to moment-resisting connections is presented. It is not the intention of the author to describe in detail all of the possible arrangements of bolted and welded moment-resisting connections at this time, nor to provide a complete design. Instead he attempts to provide the basic theory of transferring shear and moment from a beam to another member. One numerical example is included. This theory is very easy to understand and should

Possible column stiffener

Full-penetration groove welds

Fillet weld
for shear

Shear tab

Backing bar

Figure 14-26 A moment-resisting connection.

enable the reader to design other moment-resisting connections—whatever their configurations. This topic is continued in the next chapter.

One moment-resisting connection that is popular with many fabricators is shown in Fig. 14.26. There the flanges are groove-welded to the column, while the shear is carried separately by a single-plate or shear tab connection (to be described in detail in Section 15-7). The shear also could be carried by a seat angle or by web angles. Several other feasible moment-resisting connections are shown in Fig. 15-5.

The first step in the design is to compute the magnitude of the internal compression and tension forces T and C. These forces are assumed to be concentrated at the center of each flange as shown in Fig. 14-27

$$C = T = \frac{M}{d - t_f}$$

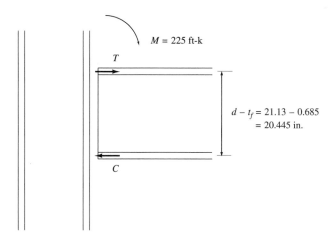

$M = 225$ ft-k

T

$d - t_f = 21.13 - 0.685$
$= 20.445$ in.

C

Figure 14-27

Next the areas of the full-penetration welds up against the column are determined. They equal the magnitude of C or T divided by the design stress of a full penetration groove weld as provided in Table 14-1 with $\phi = 0.9$.

$$\text{Area reqd.} = \frac{C \text{ or } T}{\phi F_y}$$

Using this procedure it is theoretically possible to have a weld area larger than the cross-sectional area of the flange. It would then be theoretically necessary to use an auxiliary plate on the flange to resist the extra force. (We may just transfer all of the forces with plates on the flanges. Sometimes the beam flanges are groove-welded flush with the column on one end and connected to the beam on the other end with the auxiliary plates just described. This could help us regarding our problems of fit. The forces are transferred from the beam to the plate with fillet welds and by groove welds from the plate to the column.)

Recent research at the University of California and Lehigh University has shown that the full plastic moment capacity of a beam can be developed with full-penetration welds made only to the flanges.

Example 14-14 illustrates the design of a moment-resisting connection with flange full-penetration groove welds. The reader should understand that this example is not quite complete. As described in the next chapter, it is necessary also to design the shear plate, seat angle, or whatever is used to transfer the shear, and also to check the column for the concentrated T or C force.

■ EXAMPLE 14-14

Design a moment-resisting connection for the W21 × 68 beam shown in Fig. 14-27 with the flanges groove-welded to a column. The beam, which consists of A36 steel, has a 60-k factored end reaction and an end moment of 225 ft-k. Use E70 electrodes.

Solution.

Using a W21 × 68 ($d = 21.13$ in., $b_f = 8.270$ in., $t_f = 0.685$ in.)

Design of moment weld

$$C = T = \frac{(12)(225)}{20.445} = 132.1 \text{ k}$$

$$\text{Area of groove weld reqd.} = \frac{132.1}{(0.9)(36)} = 4.08 \text{ in.}^2$$

$$\text{Width reqd.} = \frac{4.08}{t_f} = \frac{4.08}{0.685} = 5.96 \text{ in.}$$

Use 6-in. wide welds

Design of shear weld

Try $\frac{1}{4}$-in. fillet weld on shear tab (or on seat L or web L_s)

$$\text{Length reqd.} = \frac{60}{(0.707)(\frac{1}{4})(0.75)(0.60 \times 70)} = 10.78 \text{ in.}$$

Use 11 in. ∎

Note The design is incomplete, as it is necessary to design the shear tab, seat L, or whatever is used, and to check the column for the calculated shear force (C or T) from the beam. We will examine these tasks in the next chapter.

14-18 FULL-PENETRATION AND PARTIAL-PENETRATION GROOVE WELDS

Full-Penetration Groove Welds

When plates with different thicknesses are joined, the strength of a full-penetration groove weld is based on the strength of the thinner plate. Similarly, if plates of different strengths are joined, the strength of a full-penetration weld is based on the strength of the weaker plate. Notice that no allowances are made for the presence of reinforcement, or, that is, for any extra weld thickness.

Full-penetration groove welds are the best type of weld for resisting fatigue failures. In fact, in some specifications they are the only groove welds permitted if fatigue is possible. Furthermore, a study of Appendix K of the LRFD Specification will show that allowable stresses for fatigue situations are increased if the crowns or reinforcement of the groove welds have been ground flush.

Partial Penetration Groove Welds

When we have groove welds which do not extend completely through the full thickness of the parts being joined, they are referred to as *partial-penetration groove welds.* Such welds can be made from one or both sides with or without preparation of the edges (such as bevels). Partial-penetration welds are shown in Fig. 14-28.

Figure 14-28 Partial-penetration groove welds.

Partial-penetration groove welds often are economical where the welds are not required to develop large forces in the connected materials, such as for column splices and for the connecting together of the various parts of built-up members.

In Table 14-1 we can see that the design stresses are the same as for full-penetration welds when we have compression or tension parallel to the axis of the welds. When we have tension transverse to the weld axis there is a substantial strength reduction because of the possiblity of high stress concentrations.

Examples 14-15 and 14-16 illustrate the calculations needed to determine the strength of full-penetration and partial-penetration groove welds. The design strengths of both full-penetration and partial-penetration groove welds are given in table 14-1 of this chapter (LRFD Table J2.5).

In part (b) of Example 14-16 it is assumed that two W sections are spliced together with a partial-penetration groove weld and the design shear strength of the member is determined. To do this it is necessary to compute the following: (1) the shear rupture strength of the base material as per LRFD Section J4.1, (2) the shear yielding strength of the connecting elements as per LRFD Section J5.3, and (3) the shear strength of the weld as per LRFD Section J2.2 and LRFD Table J2.5 (Table 14-1 in this chapter). The design shear strength of the member is the least of these three values which are described as follows:

1. Shear fracture of base material $= \phi F_n A_{ns}$ with $\phi = 0.75$, $F_n = 0.6 F_u$, and $A_{ns} =$ net area subject to shear.
2. Shear yielding of connecting elements $= \phi R_n = \phi(0.60 A_{vg})F_y$ with $\phi = 0.90$ and $A_{vg} =$ gross area subjected to shear.
3. Shear yielding of the weld $= \phi F_w = \phi(0.60\ F_{EXX})A_w$ with $\phi = 0.75$ and $A_w = A_{eff} =$ area of weld.

■ EXAMPLE 14-15

(a) Determine the design tensile strength of a full-penetration groove weld for the plates shown in Fig. 14-29. Use A36 steel and E70 electrodes.

(b) Determine the design tensile strength of the plates shown in Fig. 14-29 if a $\frac{1}{4}$-in. partial-penetration groove weld is used on one side.

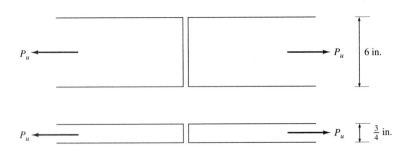

Figure 14-29

Solution.

(a) $P_u = \phi F_y A = (0.90)(36)(6 \times \frac{3}{4}) = \underline{\underline{145.8 \text{ k}}}$

(b) $P_u = \phi F_y A = (0.90)(36)(6 \times \frac{1}{4}) = \underline{\underline{48.6 \text{ k}}} \leftarrow$

   or

   $P_u = (\phi)(0.60 F_{EXX})(A) = (0.80)(0.60 \times 70)(6 \times \frac{1}{4}) = 50.4 \text{ k}$ ∎

■ EXAMPLE 14-16

(a) A full-penetration groove weld made with E70 electrodes is used to splice together the two halves of an A36 W21 × 166. Determine the shear design strength of the splice.

(b) Repeat part (a) if two vertical partial-penetration welds (A36 steel, E70 electrodes) with throat thicknesses of $\frac{1}{4}$ in. are used.

Solution.

Using a W21 × 166 ($d = 22.48$ in., $t_w = 0.750$ in.)

(a) $V_u = (0.90)(0.60F_y)A_w$

   $= (0.90)(0.60 \times 36)(22.48 \times 0.750) = \underline{\underline{327.8 \text{ k}}} \leftarrow$

   or

   $V_u = (0.80)(0.60 F_{EXX})(A_w)$

   $= (0.80)(0.60 \times 70)(22.48 \times 0.750) = 566.5 \text{ k}$

(b) shear fracture strength

   of base material $= \phi F_w A_{nv}$

   $= (0.75)(0.6 \times 58)(22.48 \times 0.750) = 440 \text{ k}$

   shear yielding of

   connecting elements $= \phi(0.60 A_g)F_y$

   $= (0.90)(0.60)(22.48 \times 0.750)(36)$

   $= 327.8 \text{ k} \leftarrow$

   shear yielding of weld $= \phi(0.60 F_{EXX}) A_w$

   $= (0.75)(0.60 \times 70)(22.48 \times 2 \times \frac{1}{4})$

   $= 354.1 \text{ k}$ ∎

14-19 COMPUTER EXAMPLES

Examples 14-17 and 14-18 illustrate the application of the computer program INSTEP for welded connections. In supplying the data requested you should notice that for the letter *B* (electrode number), the letter *C* (welding process as SAW or SMAW) and the letter *G* (weld style as load passing through c.g. of weld or eccentric shear) we don't actually type in the required information, we merely repeatedly press the key in question until the desired data appears. When the letter H is pressed a sketch showing possible weld configurations appears and it is necessary to carefully supply all the dimensions requested. After all the necessary information is supplied, the F1 key is pressed and the design strength of the weld will be furnished. *Should an eccentric load be located to the left of the weld c.g., it is necessary to use a minus sign on P_u in using INSTEP.*

■ EXAMPLE 14-17

Using the computer program INSTEP determine the design strength P_u of the $\frac{5}{16}$-in. E70 fillet welded connection shown in Fig. 14-30. The steel is A572, Grade 50 and the SMAW process was used.

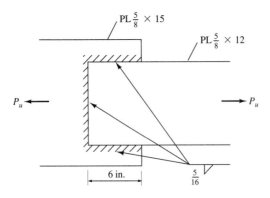

Figure 14-30

Solution. ■

```
                    INPUT
              A   Base Steel: A572 Grade 50
              B   Electrode: E70xx
              C   Process: Shielded Metal Arc

         CONNECTION DETAILS
              E   Weld Size: 0.3125 inches
              F   Weld Type: Fillet Weld
              G   Weld Style: Shear W/Wo Tension
              H   Specify Weld Geometry
```

JOINT FORCES ACTING

K $\quad$ P = 0 kips

L $\quad$ 0 = 0 degrees from horizontal

M $\quad$ e = 0 inches from CG

OUTPUT

EFFECTIVE WELD THROAT

t_e = (0.707).3125 =.220938 inches

COMPUTE WELD CAPACITY

.75 F_w=0.75(0.69)(70)(.220938)

.75 F_w=6.95953 kip/in

P_u = 208.786 kips

■ EXAMPLE 14-18

Using the computer program INSTEP determine the design strength P_u of the $\frac{3}{8}$-in. eccentrically loaded E70 fillet welded connection of Fig. 14-31. The steel is A572, Grade 50 and the SMAW process was used.

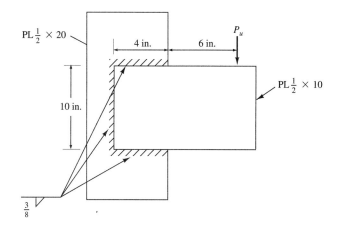

Figure 14-31

INPUT

A	Base Steel: A572 Grade 50
B	Electrode: E70xx
C	Process: Shielded Metal Arc

CONNECTION DETAILS

E	Weld Size: 0.3125 inches
F	Weld Type: Fillet Weld
G	Weld Style: Electric Shear only
H	Specify Weld Geometry

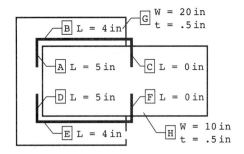

JOINT FORCES ACTING

K	P = 0 kips
L	0 = 0 degrees from horizontal
M	e = 9.11 inches from CG

OUTPUT

SUMMARY OF FINAL RESULTS

CG_x = .888889 inches
CG_y = 0 inches
P_n = 93.6681 kips
P_u = 0.75P_n = 70.2511 kips
e'= 1.60764 inches from CG
Interations: 10

PROBLEMS

14-1. A $\frac{1}{4}$-in. fillet weld SMAW process is used to connect the members shown in the accompanying illustration. Determine the design load that can be applied to this connection according to the LRFD Specification if the steel is A36 and E70 electrodes are used. (*Ans.* P_u = 97.2 k)

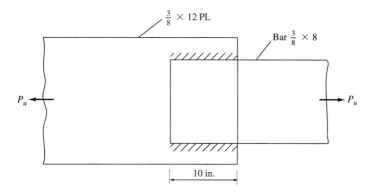

Problem 14-1

14-2. Rework Prob. 14-1 if the SAW process is used.

14-3. Rework Prob. 14-1 if A572 grade 65 steel and E80 electrodes are used. (*Ans. $P_u = 127.4$ k*)

14-4. Determine the strength of the $\frac{5}{16}$-in. fillet welds shown if E70 electrodes and A572 Grade 50 steel are used (SMAW Process).

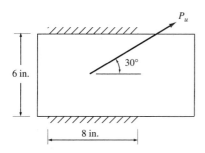

Problem 14-4

14-5. (a) Repeat Prob. 14-4 if $\frac{1}{4}$-in. welds are used and if $\theta = 20°$. (*Ans. $P_u = 98.1$ k*)
   (b) Repeat part (a) if $\theta = 10°$. (*Ans. $P_u = 92.4$ k*)

14-6. Design maximum size fillet welds if $P_u = 129.6$ k. Use E70 electrodes and the SMAW process.

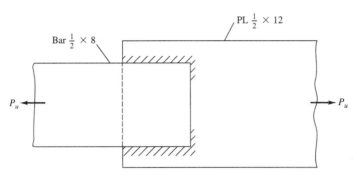

Problem 14-6

14-7. Rework Prob. 14-6 if the SAW process is used. (*Ans.* Use $\frac{7}{16}$-in. welds 7 in. long each side as required by LRFD Section J2.2b plus 1-in. end returns)

14-8. Rework Prob. 14-6 using side welds and a vertical end weld at the end of the $\frac{1}{2} \times 8$ bar. Also use A572 grade 65 steel and E80 electrodes.

14-9. Rework Prob. 14-6 using side welds and welds at the end of the $\frac{1}{2} \times$ 8 PL at the connection. (*Ans.* 8-in. end weld and 1-in. side welds)

14-10. The $\frac{5}{8} \times$ 8-in. PL shown in the accompanying illustration consists of A36 steel and is to be connected to a gusset plate with $\frac{5}{16}$-in. SMAW fillet welds. Determine the length L required to develop the full strength of the bar if E70 electrodes are used.

Problem 14-10

14-11. Design maximum size side SMAW fillet welds to develop a $P_u =$ 153.9 k for an L6 × 4 × $\frac{1}{2}$ using E70 electrodes and A36 steel. The member is connected on the sides of the 6-in. leg and is subject to alternating loads. (*Ans.* Use $5\frac{1}{2}$-in. and 11-in. side welds)

14-12. Rework Prob. 14-11 using side welds and a weld at the end of the angle.

14-13. Rework Prob. 14-12 using A588 Grade 50 steel and E80 electrodes. (*Ans.* 2-in. and $8\frac{1}{2}$-in. side welds and 6-in. end weld)

14-14. One leg of an $8 \times 8 \times \frac{3}{4}$ angle is to be connected with side welds and a weld at the end of the angle to a plate behind to develop a load $P_u = 370$ k (A36). Balance the SMAW fillet welds around the center of gravity of the angle. Use maximum weld size and assume E70 electrodes.

14-15. It is desired to design $\frac{5}{16}$-in. SMAW fillet welds necessary to connect a C10 $\times$ 30 made from A36 steel to a $\frac{3}{8}$-in. gusset plate. End, side, and slot welds may be used to develop a $P_u = 285.8$ k. No welding is permitted on the back of the channel. Use E70 electrodes. It is assumed that due to space limitations the channel can lap over the gusset plate by a maximum of 8 in. (*Ans.* $\frac{5}{16}$-in. fillet welds and one $1\frac{1}{16} \times 2\frac{1}{2}$ in. slot weld)

14-16. Rework Prob. 14-15 using A572 grade 60 steel and E80 electrodes and $\frac{3}{8}$-in. fillet welds.

14-17. Using the elastic method determine the maximum force per inch to be resisted by the fillet weld shown in the accompanying illustration. (*Ans.* 8.83 k/in.)

$P_u = 18$ k

8 in.

10 in.

Problem 14-17

14-18. Using the elastic method determine the maximum force to be resisted per inch by the fillet weld shown in the accompanying illustration.

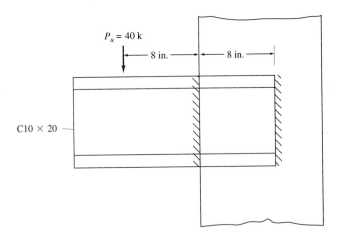

Problem 14-18

14-19. Using the elastic method rework Prob. 14-18 if welds are used on the top and bottom of the channel in addition to those shown in the figure. (*Ans.* 3.94 k/in.)

14-20. Using the elastic method determine the maximum force per inch to be resisted by the fillet welds shown in the accompanying illustration. Also determine the required weld thickness using A36 steel, E70 electrodes, and SMAW welds.

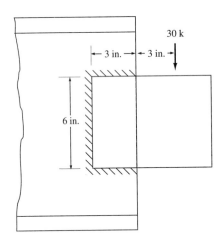

Problem 14-20

14-21. Determine the maximum eccentric load P_u that can be applied to the connection shown in the accompanying illustration if $\frac{1}{4}$-in. SMAW fillet welds are used. Assume plate thickness $= \frac{1}{2}$ in. E70. A36. (a) Use elastic method. (b) Use LRFD tables and ultimate strength method. [*Ans.* (a) $P_u = 25.3$ k (b) $P_u = 35.8$ k]

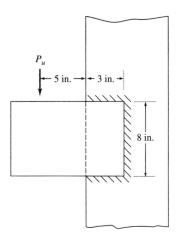

Problem 14-21

14-22. Rework Prob. 14-21 if $\frac{3}{8}$-in. fillet welds are used and the vertical weld is 10 in. high.

14-23. Determine the SMAW fillet weld size required for the connection of Prob. 14-15 if the load is increased to 30 k and the height of the weld is 12 in. A36. E70. (a) Use elastic method (b) use LRFD tables and ultimate strength method. (*Ans.* (a) 0.463 in., say $\frac{1}{2}$ in. (b) 0.227 in., say $\frac{1}{4}$ in.)

14-24. Using E70 electrodes, the SMAW process, and A36 steel determine the fillet weld size required for the bracket shown in the accompanying illustration. (a) Use elastic method. (b) Use LRFD Tables and ultimate strength method.

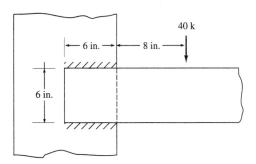

Problem 14-24

14-25. Rework Prob. 14-24 if the load is increased from 40 to 60 k and the horizontal weld lengths are increased from 6 to 8 in. [*Ans.* (a) 0.564 in., say $\frac{5}{8}$ in. (b) 0.489 in,. say $\frac{1}{2}$ in.]

14-26. Determine the SMAW fillet weld size required for the connection shown in the accompanying illustration. E70. A36. (a) Use elastic method. (b) Use LRFD tables and ultimate strength method.

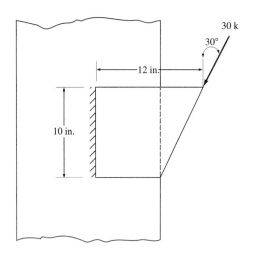

Problem 14-26

14-27. Determine the SMAW fillet weld size required for the connection shown in the accompanying illustration. E70. A36. What angle thickness should be used? (a) Use elastic method. (b) Use LRFD tables and ultimate strength method. [*Ans.* (a) 0.158 in., say $\frac{3}{16}$ in., $\frac{1}{4}$ in. angle *t*. (b) 0.118 in., say $\frac{3}{16}$ in., $\frac{1}{4}$ in. angle *t*.]

Problem 14-27

14-28. Assuming the SMAW process is to be used, determine the fillet weld size required for the connection shown in the accompanying illustration. Use A36 steel, E70 electrodes, and the elastic method.

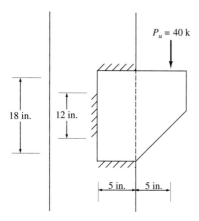

$P_u = 40$ k

18 in.   12 in.

5 in.   5 in.

Problem 14-28

14-29. Determine the fillet weld size required for the connection shown in the accompanying illustration. A36. E70. The SMAW process is to be used. (a) Use elastic method. (b) Use LRFD tables and ultimate strength method. [*Ans.* (a) 0.512 in., say $\frac{5}{8}$ in. (b) 0.381 in., say $\frac{7}{16}$ in.]

W14 × 82   Beam reaction = 45 k

6 in.

L6 × 4 × $\frac{3}{4}$

6 in.   8 in.

Problem 14-29

14-30. Determine the value of the load P_u that can be applied to the connection shown in the accompanying illustration if $\frac{3}{8}$-in. fillet welds are used. E70. A36. SAW. Use the elastic method.

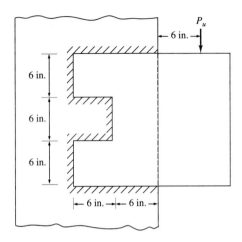

Problem 14-30

14-31. Determine the length of $\frac{1}{4}$-in. SMAW fillet welds 12 in. on center required to connect the cover plates for the section shown in the accompanying illustration at a point where the external shear V_u is 180 k. A36. E70. (*Ans.* 6.08 in., say 6-in. welds. 12 in. o.c.)

PL$\frac{3}{4}$ × 16

W21 × 62

PL$\frac{3}{4}$ × 16   **Problem 14-31**

14-32. The welded plate girder shown in the accompanying illustration has an external shear V_u of 720 k at a particular section. Determine the fillet weld size required to fasten the plates to the web if the SMAW process is used. E70. A36.

PL1 × 16

Web $\frac{1}{2}$ × 48

PL1 × 16

Problem 14-32

14-33. Design SMAW welded moment-resisting connections for the ends of a W24 × 76 to resist an R_u of 120 k and an M_u of 250 ft-k. Use A36 steel and E70 electrodes. Assume the column flange is 16 in. wide. The moment is to be resisted by full-penetration groove welds in the flanges and the shear is to be resisted by welded clip angles along the web. Assume the beam was selected for bending with 0.9 F_y. (*Ans.* $\frac{1}{4}$-in. fillet welds 11 in. high on web, 6 in. wide full-penetration welds in flanges)

14-34. The beam shown in the accompanying illustration is assumed to be attached at its ends with moment-resisting connections. Select the beam assuming full lateral support, A36 steel, and E70 SMAW electrodes. Use a connection of the type used in Prob. 14-33.

w_u = 3 k/ft (not including beam weight)

|← 30 ft →|

Problem 14-34

14-35 The two W16 × 40 beams shown in the accompanying illustration are to be made continuous across the supporting W24 × 84 girder. If the factored end reactions are each 80 k and their factored end moments are 70 ft-k, design the connections using E70 electrodes, SMAW welds, and A36 steel. (*Ans.* $\frac{5}{16}$-in. welds 6 in. long on web, $\frac{5}{16}$-in. welds 8 in. long each flange)

W16 × 40   W16 × 40

Seat L

W24 × 84

Problem 14-35

14-36. (a) Assuming the plates in Fig. 14-29 are 8 in. wide and $\frac{5}{8}$ in. thick, determine their design tensile strength if a full-penetration groove weld is used. Use A36 steel and E70 electrodes.

(b) Repeat part (a) if a $\frac{5}{16}$-in. partial-penetration groove weld is used on one side.

14-37. (a) If full-penetration groove welds formed with E70 electrodes are used to splice together the two halves of an A24 × 117, determine the shear design strength of the splice. (*Ans.* 259.4 k)

(b) Repeat part (a) it two vertical partial-penetration groove welds with $\frac{1}{4}$-in. throat thickness are used. (*Ans.* 259.4 k)

14-38. If the flanges of the spliced beam of Prob. 14-37 are joined with partial-penetration groove welds with $\frac{5}{16}$-in. throat thicknesses, determine the design shear strength at the splice if A572 Grade 50 steel is used.

14-39 to 14-41. *Solve the problems shown using the enclosed computer diskette and E70 electrodes.*

14-39. Determine the design strength of the welds shown. (*Ans.* 66.8 k)

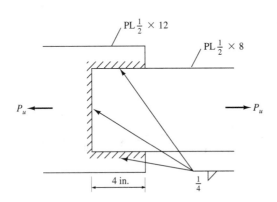

Problem 14-39

14-40. Calculate the fillet weld size required for the connection of Prob. 12-39 if the load P_u is equal to 125 k.

14-41. Determine the eccentric load P_u that the connection shown with its $\frac{3}{8}$-in. fillet weld can support. (*Ans.* 34.2 k)

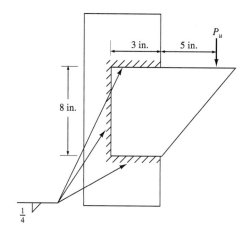

Problem 14-41

Chapter **15**

Building Connections

15-1 SELECTION OF TYPE OF FASTENER

This chapter is concerned with the actual beam-to-beam and beam-to-column connections commonly used in steel buildings. Under present-day steel specifications four types of fasteners are permitted for these connections. These are welds, unfinished bolts, high-strength bolts, and rivets.

Selection of the type of fastener or fasteners to be used for a particular structure usually involves many factors, including requirements of local building codes, relative economy, preference of designer, availability of good welders or riveters, loading conditions (as static or fatigue loadings), preference of fabricator, and equipment available. It is impossible to list a definite set of rules from which the best type of fastener can be selected for any given structure. One can only give a few general statements that may be helpful in making a decision. These are listed as follows.

1. Unfinished bolts are often economical for light structures subject to small static loads and for secondary members (such as purlins, girts, bracing, etc.) in larger structures.
2. Field bolting is very rapid and involves less skilled labor than welding. The purchase price of high-strength bolts, however, is rather high.
3. If a structure is later to be disassembled, welding probably is ruled out, leaving the job open to bolts.
4. For fatigue loadings, fully tensioned high-strength bolts and welds are very good.
5. Notice that special care has to be taken to properly install high-strength, slip-critical bolts.
6. Welding requires the smallest amounts of steel, probably provides the most attractive-looking joints, and also has the widest range of application to different types of connections.

7. When continuous and rigid fully moment-resisting joints are desired, welding probably will be selected.

8. Welding is almost universally accepted as being satisfactory for shop-work. For fieldwork it is very popular in some areas of the United States, while in a few others it is stymied by the fear that field inspection is rather questionable.

9. To use welds for very thick members requires a great deal of extra care, and bolted connections may very well be used instead. Furthermore, such bolted connections are far less susceptible to brittle fractures.

A most interesting article entitled "Choosing the Best Structural Fastener" by Henry J. Stetina appeared in the November 1963 issue of *Civil Engineering*. It would be well worth the student's time to read this article. In addition, a valuable reference on the relative economy of various types of beam-to-beam and beam-to-column connections is given in a handbook published by the United States Steel Corporation in October 1963 entitled *Building Design Data*.

15-2 TYPES OF BEAM CONNECTIONS

All connections have some restraint, that is, some resistance to changes of the original angles between intersecting members when loads are applied. Depending on the amount of restraint the LRFD Specification (A2.2) classifies connections as being fully restrained (Type FR) and partially restrained (Type PR). These two types of connections are described in more detail as follows:

1. Type FR connections are commonly referred to as rigid or continuous frame connections. They are assumed to be sufficiently rigid or restrained to keep the original angles between members virtually unchanged under load.

2. Type PR connections are those that have insufficient rigidity to keep the original angles virtually unchanged under load. Included in this classification are simple and semirigid connections as described in detail in this section.

A *simple connection* is a Type PR connection for which restraint is ignored. It is assumed to be completely flexible and free to rotate, thus having no moment resistance. A *semirigid connection* is a Type PR connection whose resistance to angle change falls somewhere between the simple and rigid types.

As there are no perfectly rigid connections nor completely flexible ones, all connections really are partly restrained or PR to one degree or another. The usual practice in the past was to classify connections based on a ratio of the moment developed for a particular connection to the moment that would be developed by a completely rigid connection. A rough rule was that simple connections had 0–20 percent rigidity, semirigid connections had 20–90

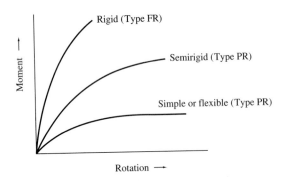

Figure 15-1 Typical moment-rotation curves for connections.

percent rigidity, and rigid connections had 90–100 percent rigidity. Fig. 15-1 shows a set of typical moment-rotation curves for these connections. Notice that the lines are curved because as the moments become larger, the rotations increase at a faster rate.

During the past few years quite a few investigators around the world have been trying to develop empirical formulas for describing the rotational characteristics of connections.[1,2,3,4] Though they have made some progress, the only accurate method for developing such information today involves the actual fabrication of connections followed by load testing. It is very difficult to include in a formula the effects of such things as poor fit, improper tightening of bolts, and so on.

Each of these three general types of connections is briefly discussed in this section, with little mention of the specific types of connectors used. The remainder of the chapter is concerned with detailed designs of these connections using specific types of fasteners. *In this discussion the author probably overemphasizes the semirigid and rigid type connections, because a very large percentage of the building designs with which the average designer works will be assumed to have simple connections.* A few descriptive comments are given in the following paragraphs concerning each of these three types of connections.

[1] R. M. Richard, "A Study of Systems Having Conservative Non-Linearity," Ph.D. Thesis, Purdue University, 1961.

[2] N. Kishi and W. F. Chen, "Data Base of Steel Beam-to-Column Connections," no. CE-STR-86-26, Vols. I and II (West Lafayette, IN: Purdue University, School of Engineering, July, 1986).

[3] L. F. Geschwindner, "A Simplified Look at Partially Restrained Beams," *Engineering Journal,* AISC, 28, no. 2 (2nd quarter, 1991), pp. 73–78.

[4] Wai-Fah Chen et al., "Semi-rigid Connections in Steel Frames," Council on Tall Buildings and Urban Habitat, Committee 43 (McGraw Hill, 1992).

A typical framing angle shop-welded to a beam. It will be field-bolted to another member. (Courtesy American Institute of Steel Construction, Inc.)

Simple connections (Type PR) are quite flexible and are assumed to allow the beam ends to be substantially free to rotate downward under load as true simple beams should. Although simple connections do have some moment resistance (or resistance to end rotation), it is assumed to be negligible and they are assumed to be able to resist shear only. Several types of simple connections are shown in Fig. 15-2. More detailed descriptions of each of these connections and their assumed behavior under load are given in later sections of this chapter. In this figure most of the connections are shown as being made entirely with the same type of fastener, while in actual practice two types of fasteners are often used for the same connection. For example, a very common practice is to shopweld the web angles to the beam web and field-bolt them to the column or girder.

Semirigid connections (Type PR) are those that have appreciable resistance to end rotation, thus developing appreciable end moments. In design practice it is quite common for the designer to assume all connections are either simple or rigid, with no consideration given to those situations in between, thereby simplifying the analysis. Should he or she make such an assumption for a true semirigid connection, he or she may miss an opportunity for appreciable moment reductions. To understand this possibility, the reader is referred to the moment diagrams shown in Fig. 15-3 for a group of uniformly loaded beams supported with connections having different percentages of rigidity. This figure shows that the maximum moments in a beam vary greatly with different types of end connections. For example, the maximum moment in the semirigid connection of part (d) of the figure is only 50 percent of the maximum moment in the simply supported beam of part (a) and only 75 percent of the maximum moment in the rigidly supported beam of part (b).

(a) Framed simple connection

(b) Seated simple connection

(c) Framed simple connection

(d) Single-plate or shear tab simple connection

Figure 15-2 Some simple connections. (a) Framed simple connection. (b) Seated simple connection. (c) Framed simple connection. (d) Single-plate or shear tab simple connection.

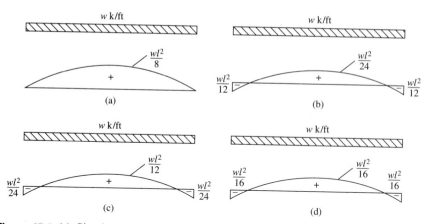

Figure 15-3 (a) Simple connections (0 percent). (b) Rigid connections (100 percent). (c) Semirigid connections (50 percent). (d) Semirigid connections (75 percent).

Actual semirigid connections are used fairly often, but usually no advantage is taken of their moment-reducing possibilities in the calculations. Perhaps one factor that keeps the design profession from taking advantage of them more often is the LRFD Specification (Section A2), which states that consideration of a connection as being semirigid is permitted only upon presentation of evidence that it is capable of providing a certain percentage of the end restraint furnished by a completely rigid connection. This evidence must consist of documentation in the technical literature, or must be established by analytical or empirical means.

Three practical semirigid or PR connections capable of providing considerable moment resistance are shown in Fig. 15-4. If the end-plate connection shown in part (a) of the figure is extended above the beam and more bolts are installed, the moment resistance of the connection can be appreciably increased. Part (c) of the figure shows a semirigid connection that is proving to be quite satisfactory for steel-concrete composite floors. Moment resistance in this connection is provided by reinforcing bars placed in the concrete slab above the beam and by the horizontal leg of the seat angle.[5] Another type of semirigid connection is illustrated in an accompanying photograph from the Lincoln Electric Company.

[5] D. J. Ammerman and R. T. Leon, "Unbraced Frames with Semirigid Composite Connections," *Engineering Journal*, AISC, 27, no. 1 (1st quarter 1990), pp. 12–21.

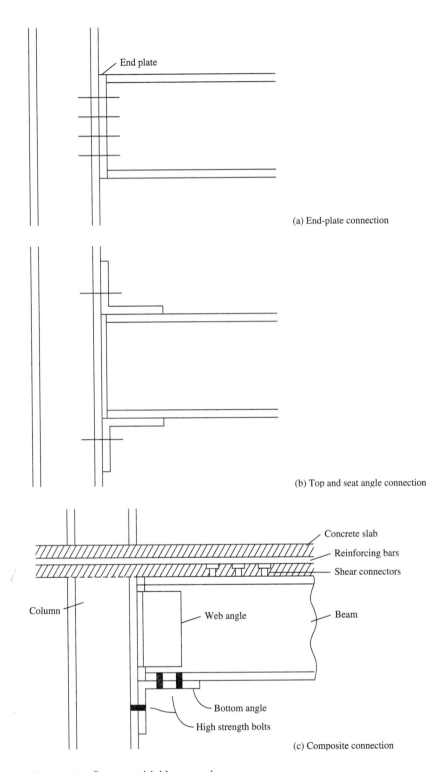

(a) End-plate connection

(b) Top and seat angle connection

Concrete slab

Reinforcing bars

Shear connectors

Column

Web angle

Beam

Bottom angle

High strength bolts

(c) Composite connection

Figure 15-4 Some semirigid connections.

A semirigid beam-to-column connection, Ainsley Building, Miami, Fla. (Courtesy of the Lincoln Electric Company.)

The use of partially restrained connections with roughly 60 to 75 percent rigidities is gradually increasing. When it becomes possible to accurately predict the percents of rigidity for various connections and when better design procedures are available, this type of design will probably become very common.

Rigid connections (Type FR) are those which theoretically allow no rotation at the beam ends and thus transfer close to 100 percent of the moment of a fixed end. Connections of this type may be used for tall buildings in which wind resistance is developed by providing continuity between the members of the building frame. Several Type FR connections that provide almost 100 percent restraint are shown in Fig. 15-5. It will be noticed in the figure that column web stiffeners may be required for some of these connections to provide sufficient resistance to rotation. The design of these stiffeners is discussed in Section 15-12.

The moment connection shown in part (d) is rather popular with steel fabricators, and the end-plate connection of part (e) also has been frequently used in recent years.[6]

---

[6] J. D. Griffiths, "End-Plate Moment Connections—Their Use and Misuse," *Engineering Journal,* AISC, vol. 21, no. 1 (first quarter, 1984) pp. 32–34.

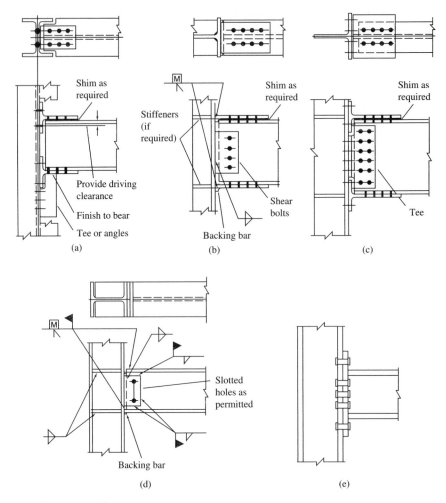

Figure 15-5 Moment-resisting connections.

You will note the use of *shims* in parts (a) to (c) of Fig. 15-5. Shims are thin strips of steel that are used to adjust the fit at connections. They can be one of two types: conventional or finger shims. *Conventional shims* are those that are installed with the bolts passing through them, while *finger shims* can be installed after the bolts are in place. You should realize that there is some variation in the depths of beams as they come from the steel mills (see page 1-188 in the LRFD Manual for permissible tolerances). To provide for such variations, it is common to make the distance between flange plates or angles larger than the nominal beam depths given in the manual.[7]

---

[7] W. T. Segui, *Fundamentals of Structural Steel Design,* 2nd. ed., (Boston: PWS-Kent, 1994), pp. 355–356.

15-3 STANDARD BOLTED BEAM CONNECTIONS

Several types of standard bolted connections are shown in Fig. 15-6. These connections are usually designed to resist shear only, as testing has proved this practice to be quite satisfactory. Part (a) of the figure shows a connection between beams with the so-called *framed connection.* This type of connection consists of a pair of flexible web angles probably shop-connected to the web of the supported beam and field-connected to the supporting beam or column. When two beams are being connected, usually it is necessary to keep their top

Figure 15-6 (a) Framed connection. (b) Framed connection. (c) Framed connection. (d) Seated connection. (e) Seated connection. (f) Seated connection with stiffener angles.

The beam on the right has a coped top flange, while the beam on the left has both flanges coped because its depth is very close to that of the supporting girder. (Courtesy American Institute of Steel Construction, Inc.)

flanges at the same elevation, with the result that the top flange of one will have to be cut back (called *coping*) as shown in part (b) of the figure. For such connections we must check block shear as discussed in Section 3-7 of this text.

Simple connections of beams to columns can be either framed or seated as shown in Fig. 15-6. In part (c) of the figure a framed connection is shown in which two web angles are connected to the beam web in the shop, after which bolts are placed through the angles and column in the field. It is often convenient to have an angle, called an *erection seat,* to support the beam during erection. Such an angle is shown in the figure.

The seated connection has an angle under the beam similar to the erection seat just mentioned, which is shop-connected to the column. In addition, there is another angle—probably on top of the beam—that is field-connected to the beam and column. A seated connection of this type is shown in part (d) of the figure. Should space limitation prove a problem above the beam, the top angle may be placed in the optional location shown in part (e) of the figure. The top angle, at either of the locations mentioned, is very helpful in keeping the top flange of the beam from being accidentally twisted out of place during construction.

The amount of load that can be supported by the types of connections shown in parts (c), (d), and (e) of Fig. 15-6 is severely limited by the flexibility or bending strength of the horizontal legs of the seat angles. For heavier loads it is necessary to use stiffened seats such as the one shown in part (f) of the figure.

The majority of these connections are selected by referring to standard tables. The LRFD Manual has excellent tables for selecting bolted or welded beam connections of the types shown in Fig. 15-6. After a rolled-beam section

(a)

(b)

Figure 15-7 (a) Bending of framed-beam connection. (b) Bending of seated-beam connection.

has been selected, it is quite convenient to refer to these tables and select one of the standard connections, which will be suitable for the vast majority of cases.

In order to make these standard connections have as little moment resistance as possible, the angles used in making up the connections are usually light and flexible. To qualify as simple end supports the ends of the beams should be as free as possible to rotate downward. Figure 15-7 shows the manner in which framed and seated end connections will theoretically deform as the ends of the beams rotate downward. The designer does not want to do anything that will hamper these deformations.

For the rotations shown in Fig. 15-7 to occur, there must be some deformation of the angles. As a matter of fact, if end slopes of the magnitudes that are computed for simple ends are to occur, the angles will actually bend enough to be stressed beyond their yield points. If this situation occurs, they will be permanently bent and the connections will quite closely approach true simple ends. The student should now see why it is desirable to use rather thin angles and large gages for the bolt spacing if flexible simple end connections are the goal of the designer.

These connections do have some resistance to moment. When the ends of the beam begin to rotate downward, the rotation is certainly resisted to some

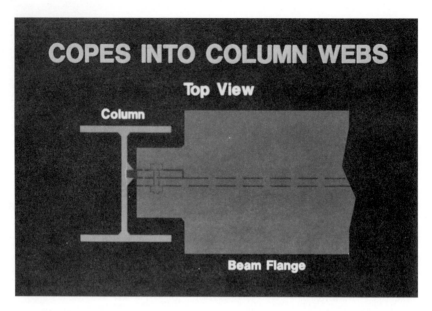

Top view showing a beam framing into a column web where the beam flange is wider than the opening between the column flanges. Thus, both top and bottom flanges are coped by flame-cutting.

extent by the tension in the top bolts, even if the angles are quite thin and flexible. Neglecting the moment resistance of these connections will cause conservative beam sizes. If moments of any significance are to be resisted, more rigid-type joints need to be provided than are available with the framed and seated connections.

15-4 LRFD MANUAL STANDARD CONNECTION TABLES

In Part 9 of the LRFD Manual a series of tables are presented that the designer may use to select several different types of standard connections. There are tables for bolted or welded two-angle framed connections, seated beam connections, stiffened seated beam connections, eccentrically loaded connections, single-angle framed connections, and others.

In the next few sections of this chapter (15-5 through 15-8) a few standard connections are selected using the tables in the Manual. The author hopes that these will be sufficient to introduce the reader to the LRFD tables and to enable him or her to make designs using the other tables with little difficulty.

Sections 15-9 to 15-11 present design information for some other types of connections.

15-5 DESIGNS OF STANDARD BOLTED FRAMED CONNECTIONS

For small and low-rise buildings (and that means most buildings) simple framed connections of the types previously shown in parts (a) and (b) of Fig. 15-6 are usually used to connect beams to girders or to columns. The angles used are rather thin ($\frac{1}{2}$ in. is the arbitrary maximum thickness used in the LRFD Manual) so they will have the necessary flexibility shown in Fig. 15-7. The angles will develop some small moments (supposedly not more than 20 percent of full fixed-end conditions), but they are neglected in design.

The framing angles extend out from the beam web by $\frac{1}{2}$ in. as shown in Fig. 15-8. This protrusion, which is often referred to as the *setback,* is quite useful in fitting members together during steel erection.

Several standard bolted framed connections for simple beams are designed in this section with the tables provided in Part 9 of the LRFD Manual. In these tables the following abbreviations for different bolt conditions are used:

1. A325-SC and A490-SC (slip-critical connections)
2. A325-N and A490-N (bearing-type connections with threads included in the shear planes)
3. A325-X and A490-X (bearing-type connections with threads excluded from shear planes)

It is thought that the minimum depth of framing angles should be at least equal to one-half of the distance between the web toes of the beam fillets (called the *T* distances and given in the properties tables of Part 1 of the Manual). This minimum depth is used so as to provide sufficient stability during steel erection.

Example 15-1 presents the design of standard framing angles for a simply supported beam using bearing-type bolts in standard-size holes. In this example the design strengths of the bolts and the angles are taken from the appropriate tables. The author shows how the table values can be checked with the procedures presented in Chapters 12 to 14.

Example 15-1 presents the design of a standard double-angle connection for a simply supported uncoped beam using bearing type bolts (threads included in shear plane) in standard size holes. This problem is handled with the tables in Part 9 of the Manual.

■ EXAMPLE 15-1

Select an all-bolted double-angle framed simple end connection for the W30 × 108 (t_w = 0.545 in.) shown in Fig. 15-8 if the factored end reaction R_u = 175 k and if the connection frames into the flange of a W14 × 61 column (t_f = 0.645 in.). Assume that F_y = 36 ksi and F_u = 58 ksi for the framing angles and 50 ksi and 65 ksi respectively for the beam and column. Use $\frac{3}{4}$-in. A325-N bolts (bearing type threads included in shear plane) in standard size holes.

Figure 15-8

Solution.   Using double-angle connection with F_y = 36 ksi and F_u = 58 ksi, we look in the tables for the least number of rows of bolts that can be used with a W30 section. It's 5 rows (page 9-36 Part 9 of the LRFD Manual), but the connection for $\frac{3}{4}$-in. A325-N bolts will not support a reaction that large. Thus we move to 6 rows on page 9-34 and try a connection with an angle $t = \frac{5}{16}$ in. and with a length $L = 15 + (2)(1\frac{1}{4}) = 17\frac{1}{2}$ in. (This length seems satisfactory when compared to the T of $26\frac{3}{4}$ in. for this shape.)

$$\phi R_n = 191 \text{ k} > 175 \text{ k} \tag{OK}$$

(Checking the preceding value: the 6 bolts in the beam web are in double shear and bearing on $\frac{1}{2}$ in. The shear controls and equals $(0.75)(2 \times 0.4418)$ $(48)(6) = 191$ k.)

Checking supported beam web (bottom of page 9-35 in Part 9 of the LRFD Manual)

With $L_{eh} = 1\frac{3}{4}$ in. and $L_{ev} = 1\frac{1}{4}$ in. (see Fig. 15-8)

ϕR_n for uncoped 0.545-in. web = $(0.545)(527) = 287.2$ k > 175 k (OK)

Checking supported column flange (bottom of page 9-35 in Part 9 of the LRFD Manual)

Noting there are two angle legs attached to flange

$$\phi R_n = (2)(527)(0.645) = 679.8 \text{ k} > 175 \text{ k} \tag{OK}$$

To select the lengths of the angle legs it is necessary to study the dimensions given in Fig. 15-9, which is a view along Section A-A in Fig. 15-8. On the lower right part of the figure are shown the minimum clearances needed for insertion and tightening of the bolts. These are the H_2 and C_1 distances shown and are obtained for $\frac{3}{4}$-in. bolts from Table 8-4 in Part 8 of the LRFD Manual.

Figure 15-9

For the legs bolted to the beam web a $2\frac{1}{2}$-in. gage is used. Using a minimum edge distance of 1 in. here we will make these angle legs $3\frac{1}{2}$ in. For the outstanding legs the minimum gage is $\frac{5}{16} + 1\frac{3}{8} + 1\frac{1}{4} = 2\frac{15}{16}$ in., say 3 in. We will make this a 4-in. angle leg.

Use 2Ls $4 \times 3\frac{1}{2} \times \frac{5}{16} \times 1$ ft $- 5\frac{1}{2}$ in.   ■

Using the tables of Part 9 of the LRFD Manual, framed bolted connections can be easily selected where one or both flanges of a beam are coped. Coping may substantially reduce the design strength of beams and require the use of larger members or the addition of web reinforcing. The design strength of coped members for flexural yielding, local buckling, and lateral torsional buckling is described in Part 9 of the LRFD Manual. Example 9-1 on pages 9-16 to 9-18 illustrates the procedure.

15-6 DESIGNS OF STANDARD WELDED FRAMED CONNECTIONS

Table 9-3 in Part 9 of the LRFD Manual includes the information necessary to use welds instead of bolts as used in Example 15-1. The values in the table are based on E70 electrodes. The table is normally used where the angles are attached to the beams in the shop and then field-bolted to the other member. Should the framing angles be welded to both members the weld values provided in Table 9-4 in Part 9 of the LRFD Manual are used.

A weld used to connect the angles to the beam web is called Weld A as shown in Fig. 15-10. If a weld is used to connect the beam to another member, that weld is called Weld B.

For the usual situations $4 \times 3\frac{1}{2}$ in. angles are used with the $3\frac{1}{2}$-in. legs connected to the beam webs. The 4-in. outstanding legs will usually accommodate the standard gages for the bolts going into the other members. The angle

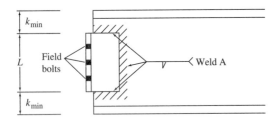

Figure 15-10

thickness selected equals the weld size plus $\frac{1}{16}$ in. or the minimum value given in Table II-A for the bolts. The angle lengths are the same as those used for the nonstaggered bolt cases (that is, $5\frac{1}{2}$ through $35\frac{1}{2}$ in.).

The design strengths of the welds to the beam webs (Weld A) given in Table 9-3 in Part 9 of the LRFD Manual were computed using the instantaneous center of rotation method which was briefly described in Chapter 14. To select a connection of this type the designer picks a weld size from Table 9-3 and then goes to Table 9-2 to determine the number of bolts required for connection to the other member. This procedure is illustrated in Example 15-2.

■ EXAMPLE 15-2

Design a framed beam connection to be welded (SMAW) to a W30 × 90 beam ($t_w = 0.470$ in. and $T = 26\frac{3}{4}$ in.) and then bolted to another member. The factored beam reaction R_u is 210 k, the steel is A36, the weld is E70, and the bolts are $\frac{3}{4}$-in. A325-N.

Solution.

 Table 9-3 Weld A Design Strength of Bolts.

 From this table one of several possibilities is a $\frac{3}{16}$-in. weld $20\frac{1}{2}$ in. long with a design strength of 228 k. The depth or length is compatible with the T value of $26\frac{3}{4}$ in. For a $\frac{3}{16}$-in. weld the minimum angle thickness is $\frac{1}{4}$ in. as per LRFD Specification J2.2b. The minimum web thickness for shear given in Table 9-3 is 0.43 in., but this is less than the 0.470 in. furnished.

 Table 9-2 Selection of Bolts.

 The angle length selected is $20\frac{1}{2}$ in. and this corresponds to a 7-row bolted connection in Table 9-2. From this latter table we find that 7 rows of $\frac{3}{4}$-in. A325-N bolts will support 188 k if the angle thickness is $\frac{1}{4}$ in., but R_u is 210 k. If we use a $\frac{5}{16}$-in. angle thickness the bolts will support 223 k, which is satisfactory.

 Use 2Ls 4 × $3\frac{1}{2}$ × $\frac{5}{16}$ × 1ft $8\frac{1}{2}$ in. ■

Table 9-4 in Part 9 of the LRFD Manual provides the information necessary for designing all-welded standard framed connections; that is, with Welds A and

Figure 15-11

B as shown in Fig. 15-11. The design strengths for Weld A were determined by the ultimate strength or instantaneous center or rotation method, while the values for Weld B were determined using the elastic method. A sample design is presented in Example 15-3.

■ EXAMPLE 15-3

Select an all-welded double-angle framed simple end connection for a W30 × 108 beam ($t_w = 0.545$ in.) to a W14 × 61 column ($t_f = 0.645$ in.) Assume $F_y = 50$ ksi and $F_u = 58$ ksi for the members and $R_u = 200$ k.

Solution.

Select Weld A for beam

From Table 9-4 one possibility is a $\frac{3}{16}$-in. weld 18 in. long. For a beam with $F_y = 50$ ksi the minimum web thickness is 0.31 in., which is $< t_w$ of 0.545 in. of the beam.

$$\phi R_n = 209 \text{ k} > 200 \text{ k} \qquad\qquad\qquad (\text{OK})$$

Select Weld B for column

The $\frac{1}{4}$- and $\frac{5}{16}$-in. welds will not provide sufficient capacity. ∴Use $\frac{3}{8}$ in. For a column with $F_y = 50$ ksi the minimum flange thickness is 0.31 in., which is $<t_f$ of 0.645 in. of the column.

Check minimum angle thickness

$$\min t = \tfrac{3}{8} + \tfrac{1}{16} = \tfrac{7}{16} \text{ in.}$$

$$\underline{\underline{\text{Use 2Ls } 4 \times 3 \times \tfrac{7}{16} \times 1 \text{ ft 6 in.}}} \qquad ■$$

15-7 SINGLE-PLATE OR SHEAR TAB FRAMING CONNECTIONS

A rather economical type of flexible connection for light loads that is being used more and more frequently is the single-plate framing connection that was illustrated in Fig. 15-2(d). The bolt holes are prepunched in the plate and the web of the beam. Then the plate is shop-welded to the supporting beam or column, and lastly the beam is bolted to the plate in the field. Steel erectors like

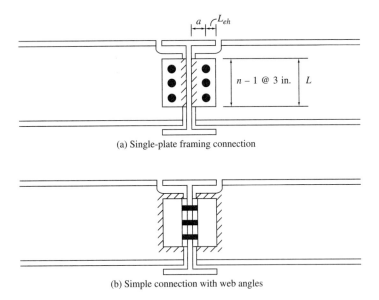

(a) Single-plate framing connection

(b) Simple connection with web angles

Figure 15-12 (a) Single-plate framing connection. (b) Simple connection with web angles.

this kind of connection because of its simplicity. They are particularly pleased with it when there is a beam connected to each side of a girder, as shown in Fig. 15-12(a). All they have to do is bolt the beam webs to the single plate on each side of the girder. Should web clip angles be used for such a connection, the bolts will have to pass through the angles on each side of the girder as well as the girder web, as shown in part (b) of the figure. This is a little more difficult field operation.

With the single-plate connection the reaction or shear load is assumed to be distributed equally among the bolts passing through the beam web. It is also assumed that relatively free rotation occurs between the end of the member and the supporting beam or column. Because of these assumptions this type of connection often is referred to as a "shear tab" connection. Various studies and tests have shown that these connections can develop some end moments depending on the number and size of the bolts and their arrangement, the thicknesses of the plate and beam web, the span-to-depth ratio of the beam, the type of loading, and the flexibility of the supporting element.

Tables for shear tab connections are presented in the LRFD Manual. An empirical design procedure using service loads is presented by Richard.[8]

---

[8] R. M. Richard et al. "The Analysis and Design of Single-Plate Framing Connections," *Engineering Journal,* AISC, vol. 17, no. 2 (second quarter, 1980), pp. 38–52.

A shear tab or single-plate framing connection. (Courtesy of American Institute of Steel Construction, Inc.)

■ EXAMPLE 15-4

Design a single-plate shear connection for the W16 × 50 beam to the W14 × 90 column shown in Fig. 15-13 if $R_u = 55$ k. Use $\frac{3}{4}$-in. A325-N high-strength bolts and E70 electrodes. The beam and column are to have $F_y = 50$ ksi and $F_u = 65$ ksi, while the plate is to have $F_y = 36$ ksi and $F_u = 58$ ksi.

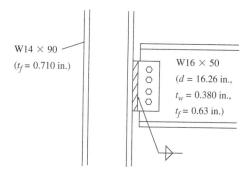

W14 × 90
($t_f = 0.710$ in.)

W16 × 50
($d = 16.26$ in.,
$t_w = 0.380$ in.,
$t_f = 0.63$ in.)

Figure 15-13

Solution.   Assuming the column provides rigid support: 4 rows of bolts, a $\frac{1}{4}$-in. plate and $\frac{3}{16}$-in. fillet welds are selected from Table 9-10 in Part 9 of the LRFD Manual.

$\phi R_n = 55.5 \text{ k} > 55 \text{ k}$ \hfill (OK)

Checking supported beam web

From Table 9-2 for 4 rows of bolts with $L_{ev} = 1\frac{1}{2}$ in. and $L_{eh} = 1\frac{1}{2}$ in. and an uncoped beam.

$\phi R_n = (0.380)(351) = 133 \text{ k} > 55 \text{ k}$ \hfill (OK)

$$\underline{\text{Use PL}\frac{1}{4} \times 4\frac{1}{2} \times 1 \text{ ft } 0 \text{ in. with 4 rows of } \frac{3}{4}\text{-in. bolts}}$$

$$\underline{\text{and } \frac{3}{16}\text{-in. fillet welds}} \quad ■$$

Should the beam be coped it will be necessary to also check flexural yielding and local web buckling at the cope. This process is illustrated in Example 9-12 in the LRFD Manual, pages 9-150 to 9-151.

15-8 END-PLATE SHEAR CONNECTIONS

Another type of connection is the end-plate connection. It consists of a plate shop-welded flush against the end of a beam and field-bolted to a column or another beam. To use this type of connection it is necessary to carefully control the length of the beam and the squaring of its ends so that the end plates are vertical. Camber must also be considered in its effect on the position of the end plate. After a little practice in erecting members with end-plate connections, fabricators seem to like to use them. Nonetheless, there still is trouble in getting the dimensions just right, and they are not as commonly used as the si le-plate connectors.

Part (a) of Fig. 15-14 shows an end-plate connection which is satisfactory for PR situations. End-plate connections are illustrated in Part 9 of the LRFD Manual. Should the end plate be extended above and below the beam as shown in part (b) of Fig. 15-14, appreciable moment resistance will be achieved.

(a) End-plate PR connection   (b) Extended end-plate FR connection

Figure 15-14

Part 10 of the LRFD Manual provides tables and a procedure for designing extended end-plate connections. These connections may be designed as being FR ones only for statically loaded structures and for buildings in seismic zone 1 areas and for unimportant buildings in seismic zone 2 areas.

15-9 DESIGNS OF WELDED SEATED BEAM CONNECTIONS

Another type of fairly flexible beam connection is obtained by using a beam seat such as the one shown in Fig. 15-15. Beam seats are obviously of advantage to the worker performing the steel erection. The connections for these angles may be bolts or welds, but only the welded type is considered here because of lack of space. For such a situation the seat angle would usually be shop-welded to the column and field-welded to the beam. When welds are used, the seat angles, which are also called *shelf angles,* may be punched for erection bolts as shown in the figure. These holes can be slotted if desired to permit easy alignment of the members.

A seated connection may be used only when a top angle is used, as shown in Fig. 15-15. This angle, which provides lateral support for the beam, may be placed on top of the beam, or at the optional location shown on the side of the beam in part (a) of the figure. As the top angle is not usually assumed to resist any of the load, its size probably is selected by judgment. Fairly flexible angles are used that will bend away from the column or girder to which it is connected when the beam tends to rotate downward under load. This desired situation was illustrated in Fig. 15-7(b). A common top angle size selected is $4 \times 4 \times \frac{1}{4}$.

As will be seen in the LRFD tables, unstiffened seated beam connections of practical sizes can support only fairly light factored loads up to as high as 128 k if 1-in.-thick seat angles of A36 steel are used. For loads of this size two vertical end welds on the seat are sufficient. The top angle is welded only on its toes so that, as the beam tends to rotate, this thin flexible angle will be free to pull away from the column.

The seat design strengths given in Tables 9-6 and 9-7 of the Manual were developed for seat angles with either $3\frac{1}{2}$-in. or 4-in. outstanding legs. The steel used for the angles has $F_y = 36$ ksi and $F_u = 58$ ksi, while values are given in the table for beams with either $F_y = 36$ ksi or $F_y = 50$ ksi.

The design strengths in the tables were obtained giving consideration to both shear and flexural yielding of the outstanding leg of the seat angle and also crippling of the beam web. The values were calculated on the basis of a $\frac{3}{4}$-in. setback rather than the nominal $\frac{1}{2}$ in. used for web framing angles. This larger value was used to provide for possible mill underrun in the beam lengths. The weld design strengths provided were calculated using the elastic method. Example 15-5 illustrates the use of the Manual tables for designing a welded unstiffened seated beam connection. Other tables are included in the Manual for bolted seated connections as well as bolted or welded stiffened seated connections.

End return

Optional location
of top L

Erection bolts

End return

Weld
length

Angle length

(a)

Top angle

$\frac{1}{2}$-in. setback

Erection bolt

Seat or shelf
angle

(b)

Figure 15-15

■ EXAMPLE 15-5

Design an unstiffened beam connection to support a factored reaction of 60 k from a W24 × 55 beam ($d = 23.57$ in., $t_w = 0.395$ in. or $\frac{3}{8}$ in. as a fraction, $t_f = 0.505$ in.). The connection is to be made to the flange of a W14 × 68 column ($t_f = 0.720$ in.). Use $\frac{3}{4}$-in. A325-N bolts in standard size holes to connect the beam to the seat and top angle and E70 electrodes to connect those angles to the column flange. The angles are A36 while the beam and column have an F_y of 50 ksi and an F_u of 65 ksi.

Solution.

Select angle size and check design strength of outstanding angle leg with reference made to Table 9-7 in the Manual.

Select an 8-in. angle leg from the table. In this table you can see that for a 4-in. outstanding leg and a beam web thickness of 0.395 in. or $\frac{3}{8}$ in. to nearest fraction ($F_y = 50$ ksi), a $\frac{3}{4}$-in. thick angle will support

$$\phi R_n = 68.2 \text{ k} > 60 \text{ k} \qquad \text{(OK)}$$

Select weld for seat angle

Using an 8×4 angle, a $\frac{5}{16}$-in. fillet weld will support

$$\phi R_n = 66.8 \text{ k} > 60 \text{ k} \qquad \text{(OK)}$$

Checking web crippling

$N/d = 4/23.57 = 0.170 < 0.2$ ∴ Making use of ϕR_3 and ϕR_4 values for a W24 $\times$ 55 ($F_y = 50$ ksi) in the Factored Uniform Load Tables of Part 4 of the Manual

$$\phi R_n = \phi R_3 + \phi R_4 = 63.6 + (4)(5.60) = 86 \text{ k} > 60 \text{ k} \qquad \text{(OK)}$$

Select bolts to connect beam to seat angle

Use two $\frac{3}{4}$-in. A325-N bolts (no calculated force)

Select top angle and bolts for connecting top angle to beam (no calculated forces)

Use top L4 $\times$ 4 $\times$ $\frac{1}{2}$ with two $\frac{3}{4}$-in. A325-N bolts between the beam and angle.

Use $\frac{3}{16}$-in. minimum fillet weld size along toe of the angle.   ∎

15-10 STIFFENED SEATED BEAM CONNECTIONS

When beams are supported by seated connections and when the factored reactions become fairly large (say above 128 k, depending on the steel grade), it is necessary to stiffen the seats. These larger reactions cause moments in the outstanding or horizontal legs of the seat angles that cannot be supported by standard thickness angles unless they are stiffened in some manner. Typical stiffened seated connections are shown in Fig. 15-6(f) and in Fig. 15-16.

Stiffened seats may be bolted or welded. Bolted seats may be stiffened with a pair of angles as shown in part (a) of Fig. 15-16. Structural tee stiffeners either bolted or welded may be used. A welded one is shown in part (b) of the same figure. Also commonly used are welded two-plate stiffeners such as the one shown in part (c).

Examples 9-9 and 9-10 in Part 9 of the LRFD Manual illustrate the design of stiffened seated beam connections. These designs make use of Tables 9-8 and 9-9 in the Manual.

(a) Bolted stiffened seat
with angles

(b) Welded stiffened seat
with structural tee

(c) Welded two-plate stiffened seat

Figure 15-16 (a) Bolted stiffened seat with angles. (b) Welded stiffened seat with structural tee. (c) Welded two-plate stiffened seat.

15-11 DESIGN OF MOMENT-RESISTING CONNECTIONS

In Chapter 14 a moment-resisting connection was designed (Example 14-14) where the flanges were welded to a column and a shear tab was used to transfer shear between the beam web and the column. Several other types of moment-resisting connections were shown in Fig. 15-5.

As the general procedure used in Example 14-15 is applicable to all moment-resisting connections, no additional numerical examples are presented in this text. The moment to be resisted is divided by the distance between the cgs. of the top and bottom parts of the couple (C and T), and then welds or bolts are selected that will provide the necessary design strengths so determined. Next, a shear tab, or a pair of framing angles, or a beam seat are selected to resist the shear force. Finally, it may be necessary, as described in the next section, to provide stiffeners for the column web, or to select a larger column section.

Another moment-resisting connection is presented in Fig. 15-17. In this particular connection the T and C values are transferred by fillet welds into the

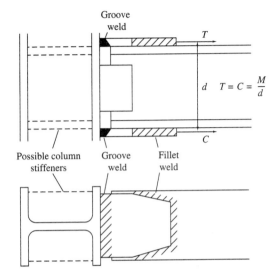

d $T = C = \dfrac{M}{d}$

Groove
weld

T

C

Possible column
stiffeners

Groove
weld

Fillet
weld

Figure 15-17

plates and by groove welds from the plates to the columns. For easier welding these plates may be tapered as shown in the bottom of the figure. The student may have noticed such tapered plates used for facilitating welding in other situations.

For a rigid or continuous connection of the type shown in Fig. 15-17 we must be careful to check the strength of the top and bottom plates. Should the plates be bolted, this involves the tensile strength of the top plate including the effect of the bolt holes and block shear. The design compressive strength of the other plate must also be checked. Examples 10-1 and 10-2 in the Manual illustrate such calculations.

15-12 COLUMN WEB STIFFENERS

If a column to which a beam is being connected bends appreciably at the connection, the moment resistance of the connection will be reduced no matter how good the connection may be. Furthermore, if the top connection plate in pulling away from the column tends to bend the column flange as shown in part (a) of Fig. 15-18, the middle part of the weld may be greatly overstressed (like the prying action for bolts we discussed in Chapter 13).

When there is a danger of the column flange bending as described here, we must make sure that the desired moment resistance of the connection is provided. This may be done either by using a heavier column with stiffer flanges or by introducing column web stiffener plates as shown in part (b) of Fig. 15-18. *It is almost always desirable to use a heavier column because column web stiffener plates are quite expensive and a nuisance to use.*

Column web stiffener plates are somewhat objectionable to architects as they find it convenient to run pipes and conduits inside their columns, but this objection can easily be overcome. First, if the connection is to only one column flange, the stiffener does not have to run for more than half the column depth as

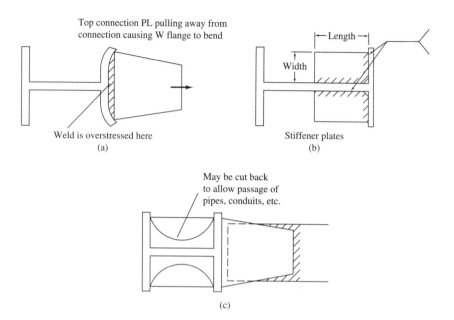

Top connection PL pulling away from
connection causing W flange to bend

Weld is overstressed here
(a)

Length

Width

Stiffener plates
(b)

May be cut back
to allow passage of
pipes, conduits, etc.

(c)

Figure 15-18

shown in part (b) of Fig. 15-18. If connections are made to both column flanges the column stiffener plates may be cut back to allow the passage of pipes, conduits, etc., as shown in part (c) of the figure.

For this discussion the factored force applied from the beam flange to the column is referred to as P_{bf}. Should its value be larger than any one of the following resisting forces, it will be necessary to provide column web stiffeners, says LRFD Specification K1. In the expressions to follow h is the depth of the column web between the web toes of the fillets and t_f is the thickness of the beam flange or connection plate from which the concentrated force is being applied.

$$R_n = P_{fb} = \phi = 0.9 \qquad \text{(LRFD Equation K1.1)}$$

$$R_n = 6.25t_f^2 F_{yf}$$

$$R_n = P_{wi}t_b + P_{wo} = \begin{array}{l}\text{local web yielding strength}\\ \text{considered in Chapter 10}\end{array}$$

$$\phi = 1.0$$

$$R_n = (5 k + N)F_{yw}t_w \qquad \text{(LRFD Equation K1.2)}$$

$$P_{wb} = \begin{array}{l}\text{compression buckling strength of an}\\ \text{unstiffened member web}\end{array}$$

$$\phi = 0.9$$

$$R_n = \frac{4100 \, t_w^3 \sqrt{F_{yw}}}{h} \qquad \text{(LRFD Equation K1.8)}$$

The values of P_{wb}, P_{fb}, and $P_{wi}t_b + P_{wo}$ for the W shapes normally used as columns have been computed and recorded in the column tables of Part 2 of the LRFD Manual for steels with $F_y = 36$ ksi and 50 ksi.

The LRFD Manual presents a set of suggested rules for the design of column web stiffeners. These are as follows:

1. The width of the stiffener plus one-half of the column web thickness should not be less than one-half the width of the beam flange or of the moment connection plate that applies the concentrated force.
2. The stiffener thickness should not be less than $t_b/2$.
3. If there is a moment connection applied to only one flange of the column, the length of the stiffener plate does not have to exceed one-half the column depth.
4. The stiffener plate should be welded to the column web with a sufficient strength to carry the force caused by the unbalanced moment on the opposite sides of the column.

For the column given in Example 15-7, which follows, it is necessary to use column web stiffeners or to select a larger column. These alternatives are considered in the solution.

■ EXAMPLE 15-7

It is assumed that a particular column is a W12 × 87 consisting of A36 steel and subjected to 171.9-k C and T forces transferred by an FR type connection through the flanges of a W18 × 46 beam. It will be found that this column is not satisfactory to resist these forces. (a) Select a larger W12 column section that will be satisfactory. (b) Using a W12 × 87 column, design column web stiffeners including the stiffener connections using E70 SMAW welds.

Solution.

Beam is a W18 × 46(b_f = 6.06 in., t_b = t_f = 0.605 in.)
Column is a W12 × 87(d = 12.53 in., t_w = 0.515 in., t_f = 0.810 in.)

Checking to see if forces transferred to column are too large. Using LRFD column tables for the W12 × 87,

$$P_{fb} = 133 \text{ k} < 171.9 \text{ k} \qquad \text{(NG)}$$
$$P_{wi}t_b + P_{wo} = (19)(0.605) + 139 = 150.5 \text{ k} < 171.9 \text{ k} \qquad \text{(NG)}$$
$$P_{wb} = 311 \text{ k} > 171.9 \text{ k} \qquad \text{(OK)}$$

∴ Either a larger column or column web stiffeners are required.
(a) Selecting a larger column
By examination a W12 × 96 will not do because P_{fb} in the table is < 171.9 k.
Try W12 × 106

$$P_{fb} = 198 \text{ k} > 171.9 \text{ k} \qquad \text{(OK)}$$
$$P_{wi}t_b + P_{wo} = (22)(0.605) + 185 = 198.3 \text{ k} > 171.9 \text{ k} \qquad \text{(OK)}$$
$$P_{wb} = 518 \text{ k} > 171.9 \text{ k} \qquad \text{(OK)}$$

Use W12 × 106

(b) Design of web stiffeners using a W12 × 87 column and the suggested LRFD rules presented before this example.

$$\text{Required stiffener area} = \frac{171.9 - 133}{36} = 1.08 \text{ in.}^2$$

$$\text{Min. width of stiffener} = \tfrac{1}{3}b_f - \frac{t_w}{2} = \frac{6.06}{3} - \frac{0.515}{2} = 1.76 \text{ in.}$$

$$\text{Min. } t \text{ of stiffeners} = \frac{t_b}{2} = \frac{0.605}{2} = 0.3025 \text{ in.}$$

$$\text{Required } t \text{ of stiffeners} = \frac{1.08}{1.76} = 0.614 \text{ in.} \quad \underline{\text{say } \tfrac{5}{8} \text{ in.}}$$

$$\text{Required width} = \frac{1.08}{0.625} = 1.728 \text{ in.} \quad \underline{\text{say 4 in. for practical}}$$
$$\underline{\text{conditions}}$$

$$\text{Min. length} = \frac{d}{2} - t_f = \frac{12.53}{2} - 0.810 = 5.46 \text{ in.} \quad \underline{\text{say 6 in.}}$$

$$\underline{\underline{\text{Use 2 PLs } \tfrac{5}{8} \times 4 \times 0 \text{ ft 6 in.}}}$$

Design of welds for stiffener plates

Min. weld size as required by LRFD Specification Table J2.4 $= \tfrac{1}{4}$ in.
based on the column web t_w of 0.515 in.

Required length of weld $=$

$$= \frac{171.9 - 133}{(0.75 \times 0.60 \times 70)(0.707)(\tfrac{1}{4})}$$

$$= 6.99 \text{ in.} \quad \underline{\text{say 7 in.}} \quad \blacksquare$$

15-13 CONNECTION DESIGN AIDS—HANDBOOKS AND COMPUTER PROGRAMS

Several computer programs are available for the design and checking of connections. They are economical to buy, save a great deal of time, and reduce the possibility of numerical mistakes. Anyone who has struggled through connection designs, particularly the selection of dimensions, will appreciate these programs.

One very satisfactory program is CONXPRT. It was developed by Structural Engineers Inc., of Radford, Virginia, and is marketed by the AISC. It is applicable to double-angle, shear end-plate, and single-plate simple framing connections. Not only is a program such as CONXPRT useful for developing designs, it also is very good for checking existing designs. Just imagine that you are working for a consulting engineer who receives the detailed drawings for a large steel building from the steel fabricator and asks you to check the adequacy of all the connections. If you check them with the handbook and a pocket calculator you are facing a monumental job—but with a computer program the work can be handled quickly and accurately.

PROBLEMS

15-1. Determine the maximum end reaction that can be transferred through the web angle connection shown in the accompanying illustration. The steel is A36 and the bolts are $\frac{3}{4}$-in. A325-N and are used in standard-size holes. The beam is connected to the web of a W30 × 90 girder. (*Ans.* $R_u = 122$ k)

W18 × 55

Angle $t = \frac{5}{16}$ in.

$l_v = 1\frac{1}{4}$ in.

3 @ 3 in.

$1\frac{1}{4}$ in.

$\frac{1}{2}$ in.

$l_{eh} = 1\frac{1}{2}$ in.

Problem 15-1

15-2. Repeat Prob. 15-1 if the bolts are $\frac{7}{8}$-in. A325-N.

15-3. Repeat Prob. 15-1 if the bolts are 1-in. A325-X. (*Ans.* $R_u = 114$ k)

15-4. Using the LRFD Manual select a pair of bolted standard web angles for a W33 × 141 connected to the flange of a W14 × 120 column with a dead load service reaction of 150 k and a live load service reaction of 100 k. The bolts are to be $\frac{7}{8}$-in. A325-N in standard-size holes and the steel is A36.

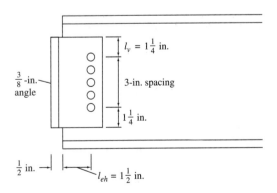

$l_v = 1\frac{1}{4}$ in.

$\frac{3}{8}$-in. angle

3-in. spacing

$1\frac{1}{4}$ in.

$\frac{1}{2}$ in.

$l_{eh} = 1\frac{1}{2}$ in.

Problem 15-4

15-5. Repeat Prob. 15-4 if 1-in. A325-N bolts are to be used. (*Ans.* 8-row connection 2Ls 5 × $3\frac{1}{2}$ × $\frac{1}{2}$ × 1 ft 11 in.)

15-6. Repeat Prob. 15-1 if $\frac{3}{4}$-in. A325 S.C. bolts are to be used.

15-7. Design a framed beam connection for a W27 × 84 connected to a W30 × 116 girder web to support a factored beam reaction of 250 k. The bolts are to be 1-in. A325 S.C. Class A in standard-size holes, and A36 steel is to be used. The edge distances and bolt spacings are the same as those shown in the sketch for Prob. 15-4. (*Ans.* 6-row connection 2Ls 5 × $3\frac{1}{2}$ × $\frac{1}{2}$ × 1 ft $5\frac{1}{2}$ in.)

15-8. Repeat Prob. 15-1 if the bolts are 1-in. A325-S.C. and are used in $1\frac{1}{16}$ × $1\frac{5}{16}$ in. short slots with long axes perpendicular to the transmitted force. Angle *t* is $\frac{1}{2}$ in.

15-9. Design a framed beam connection for a W18 × 50 to support a factored beam reaction of 65 k. The beam's top flange is to be coped for a 2-in. depth, and $\frac{7}{8}$-in. A325-X bolts in standard-size holes are to be used. A36 steel. The beam is connected to a W27 × 146 girder. (*Ans.* Use 4-row connection with 5 × $3\frac{1}{2}$ × 4 × 0 ft $11\frac{1}{2}$ in. angles)

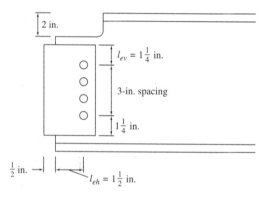

Problem 15-9

15-10. Repeat Prob. 15-7 if the factored beam reaction is to be 285 k and if 1-in. A325 N bolts and A572 steel (F_y = 50 ksi and F_u = 65 ksi) are to be used.

15-11. Select a framed beam connection for a W33 × 130 beam (A36 steel) with a factored beam reaction of 200 k. It is to be connected to the flange of a W36 × 50 column. The web angles are to be welded with E70 electrodes (weld A in the Manual) and are to be field-connected to the girder with $\frac{7}{8}$-in. A325-N bolts. (*One ans.* 2Ls 4 × $3\frac{1}{2}$ × $\frac{5}{16}$ × 1 ft $8\frac{1}{2}$ in., $\frac{1}{4}$-in. weld A and 7-row bolt connection to girder)

15-12. Repeat Prob. 15-11 using SMAW shop and field welds (welds A and B in the Manual).

15-13. Select a framed beam connection from the LRFD Manual for a W30 $\times$ 124 consisting of A36 steel ($F_y = 36$ ksi) using SMAW E70 shop and field welds. The factored beam reaction is 190 k. The beam is to be connected to the flange of a W14 $\times$ 145 column. (*Ans.* 2Ls $4 \times 3 \times \frac{3}{8} \times 1$ ft 6 in. with $\frac{3}{16}$-in. weld A and $\frac{5}{16}$-in. weld B)

15-14. Repeat Prob. 15-13 if $R_u = 290$ k.

15-15. Select an unstiffened seated beam connection bolted with $\frac{7}{8}$-in. A325-N bolts in standard-size holes for the following data: Beam is W16 $\times$ 67, column is W14 $\times$ 82, both consisting of A36 steel, R_u is 55 k, and the column gage is $5\frac{1}{2}$ in. (*Ans.* Use L7 $\times$ 4 $\times \frac{3}{4} \times 0$ ft 8 in.)

Chapter **16**

Composite Beams

16-1 COMPOSITE CONSTRUCTION

When a concrete slab is supported by steel beams and there is no provision for shear transfer between the two, the result is a noncomposite section. There is no doubt in noncomposite construction that the load causes the slab to deflect along with the beam, with the result that some of the load is carried by the slab. Unless there is a great deal of bond between the two (as would be the case where the steel beam is completely encased in the concrete, or where a system of mechanical shear connectors is provided), however, the load carried by the slab is small and may be neglected.

For many years steel beams and reinforced-concrete slabs were used together with no consideration being made for any composite effect. In recent decades, however, it has been shown that a great strengthening effect can be obtained by tying the two together to act as a unit in resisting loads. Steel beams and concrete slabs joined together compositely can often support a one-third, one-half, or even greater increase in load than could the steel beams alone in noncomposite action.

Composite construction for highway bridges was given the green light by the adoption of the 1944 AASHTO Specifications, which approved the method. Since about 1950 the use of composite bridge floors has rapidly increased until today they are commonplace all over the United States. In these bridges the longitudinal shears are transferred from the stringers to the reinforced-concrete slab or deck with shear connectors (to be described in Section 16.5), causing the

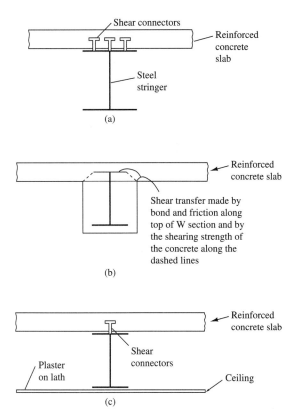

Figure 16-1 (a) Composite bridge floor with shear connectors. (b) Encased section for building floors. (c) Building floors with shear connectors.

slab or deck to assist in carrying the bending moments. This type of section is shown in part (a) of Fig. 16-1.

The first approval for composite building floors was given by the 1952 AISC Specification, and today they are very common. These floors may either be encased in concrete (very rare due to expense) as shown in part (b) of Fig. 16-1, or be nonencased with shear connectors as shown in part (c) of the figure. Almost all composite building floors being built today are of the nonencased type. If the steel sections are encased in concrete, the shear transfer is made by bond and friction between the beam and the concrete and by the shearing strength of the concrete along the dashed lines shown in part (b) of Fig. 16-1.

Today formed steel deck (illustrated in Fig. 16-2) is used for almost all composite building floors. The initial examples in this chapter, however, pertain to the calculations for composite sections where formed steel deck is not used. Sections that make use of formed steel deck are described later in the chapter.

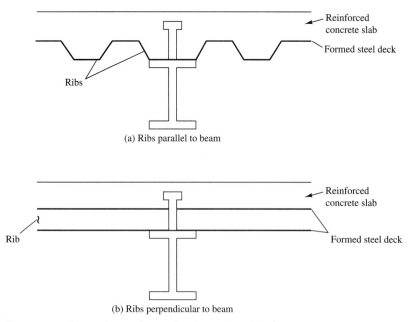

(a) Ribs parallel to beam

(b) Ribs perpendicular to beam

Figure 16-2 Composite sections using formed steel deck.

16-2 ADVANTAGES OF COMPOSITE CONSTRUCTION

The floor slab in composite construction acts not only as a slab for resisting the live loads but also as an integral part of the beam. It actually serves as a large cover plate for the upper flange of the steel beam, appreciably increasing the beam's strength.

A particular advantage of composite floors is that they make use of concrete's high compressive strength by putting a large part of the slab in compression. At the same time a larger percentage of the steel is kept in tension (also advantageous) than is normally the case in steel-frame structures. The result is less steel tonnage required for the same loads and spans (or longer spans for the same sections). Composite sections have greater stiffness than noncomposite sections and they have smaller deflections—perhaps only 20 to 30 percent as large. Furthermore, tests have shown that the ability of a composite structure to take overload is decidedly greater than for a noncomposite structure.

An additional advantage of composite construction is the possibility of having smaller overall floor depths—a fact of particular importance for tall buildings. Smaller floor depths permit reduced building heights with the consequent advantages of smaller costs for walls, plumbing, wiring, ducts, elevators, and foundations. Another important advantage available with reduced beam

Bridge on John F. Kennedy Expressway, Chicago, Ill. (Courtesy of Robert McCullough, Chicago, Ill.)

depths is a saving in fireproofing costs, because a coat of fireproofing material is provided on smaller and shallower steel shapes.

It frequently is necessary to increase the load-carrying capacity of an existing floor system. Often, this can be handled quite easily for composite floors merely by welding cover plates onto the bottom flanges of the beams.

A disadvantage for composite construction, however, is the cost of furnishing and installing the shear connectors. This extra cost usually will exceed the cost reductions mentioned when spans are short and lightly loaded.

16-3 DISCUSSION OF SHORING

After the steel beams are erected the concrete slab is placed on them. The formwork, wet concrete, and other construction loads must, therefore, be supported by the beams or by temporary shoring. Should no shoring be used, the steel beams must support all of these loads as well as their own weights. Most specifications say that after the concrete has gained 75 percent of its 28-day strength the section has become composite and all loads applied thereafter may be considered to be supported by the composite section. When shoring is used it supports the wet concrete and the other construction loads. It does not really support the weight of the steel beams unless they are given an initial upward deflection (probably impractical). When the shoring is removed (after the concrete gains at least 75 percent of its 28-day strength) the weight of the slab is transferred to the composite section, not just to the steel beams. The student can see that if shoring is used it will be possible to use lighter and thus cheaper steel beams. The question then arises, "Will the saving in steel cost be greater than the extra cost of shoring?" Probably the answer is "No." The usual decision is to use heavier steel beams and to do without shoring for several reasons. These include the following:

1. Apart from reasons of economy, the use of shoring is a tricky operation, particularly where settlement of the shoring is possible, as is often the case in bridge construction.

2. Both theory and load tests show that the ultimate strengths of composite sections of the same sizes are the same whether shoring is used or not. If lighter steel beams are selected for a particular span because shoring is used, the result is a smaller ultimate strength.

3. Another disadvantage of shoring is that after the concrete hardens and the shoring is removed, the slab will participate in composite action in supporting the dead loads. The slab will be placed in compression by these long-term loads and will have substantial creep and shrinkage parallel to the beams. The result will be a great decrease in the stress in the slab with a corresponding increase in the steel stresses. The probable consequence is that most of the dead load will be supported by the steel beams anyway and composite action will really apply only to the live loads as though shoring had not been used.

4. Also, in shored construction cracks occur over the steel girders, necessitating the use of reinforcing bars. In fact, we should use reinforcing over the girders in unshored construction, too. Although cracks will be smaller there, they are going to be present nonetheless, and we need to keep them as small as possible.

Despite these facts, shored construction does possess some advantages as compared to unshored construction. First, deflections are smaller because they are all based on the properties of the composite section. (In other words, the initial wet concrete loads are not applied to the steel beams alone, but rather to the whole composite section.) Second, it is not necessary to make a strength check for the steel beams for this wet load condition. This is sometimes quite important for situations in which we have low ratios of live to dead loads.

The deflections of unshored floors due to the wet concrete sometimes can be quite large. If the beams are not cambered, additional concrete (perhaps as much as 10 percent or more) will be used to even up the floors. If, on the other hand, too much camber is specified, we may end up with slabs that are too thin in those areas where wet concrete deflections aren't as large as the camber.

16-4 EFFECTIVE FLANGE WIDTHS

There is a problem involved in estimating how much of the slab acts as part of the beam. Should the beams be rather closely spaced, the bending stresses in the slab will be fairly uniformly distributed across the compression zone. If, however, the distances between beams are large, bending stresses will vary quite a bit nonlinearly across the flange. The further a particular part of the slab or flange is away from the steel beam, the smaller will be its bending stress. Specifications attempt to handle this problem by replacing the actual slab with a narrower or effective slab that has a constant stress. This equivalent slab is deemed to support the same total compression as is supported by the actual slab. The effective width of the slab b_e is shown in Fig. 16-3.

Figure 16-3

The portion of the slab or flange that can be considered to participate in the composite beam action is controlled by the specifications. LRFD Specification I3.1 states that the effective width of the concrete slab on each side of the beam center line shall not exceed the least of the values to follow. This same set of rules applies whether the slab exists on one or both sides of the beam.

1. One-eighth of the span of the beam measured center-to-center of supports for both simple and continuous spans.
2. One-half of the distance from the beam center line to the center line of the adjacent beam.
3. The distance from the beam center line to the edge of the slab.

The AASHTO requirements for determining effective flange widths are somewhat different. The maximum total flange width may not exceed one-fourth of the beam span, twelve times the least thickness of the slab, or the distance center-to-center of the beams. Should the slab exist on only one side of the beam, its effective width may not exceed one-twelfth of the beam span, six times the slab thickness, or one-half of the distance from the center line of the beam to the center line of the adjacent beam.

16-5 SHEAR TRANSFER

The concrete slabs may rest directly on top of the steel beams or the beams may be completely encased in concrete for fireproofing purposes. This latter case, however, is very expensive and is rarely used. The longitudinal shear can be transferred between the two by bond and shear and possibly some type of shear reinforcing if needed when the beams are encased. When not encased, mechanical connectors must transfer the load. Fireproofing is not necessary for bridges, and the slab is placed on top of the steel beams. Bridges are subject to heavy impactive loads and the bond between the beams and the deck, which is easily broken, is considered negligible. For this reason shear connectors are designed to resist all of the shear between bridge slabs and beams.

Various types of shear connectors have been tried including spiral bars, channels, zees, angles, and studs. Several of these types of connectors are shown in Fig. 16-4. Economic considerations have usually led to the use of round studs welded to the top flanges of the beams. These studs are available in diameters

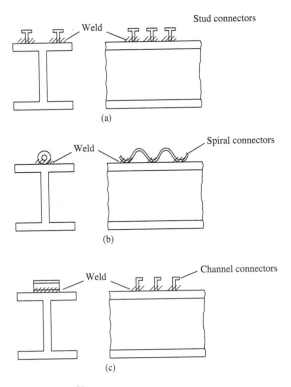

Figure 16-4 Shear connectors.

from $\frac{1}{2}$ to 1 in. and in lengths from 2 to 8 in., but the LRFD Specification (I5.1) states that their length may not be less than 4 stud diameters.

They actually consist of rounded steel bars welded on one end to the steel beams. The other end is upset or headed to prevent vertical separation of the slab from the beam. These studs can be quickly attached to the steel beams through the steel decks with stud-welding guns by semiskilled workers. The AISC Commentary (I3.5) describes special procedures needed for 16 gage and thicker decks and for decks with heavy galvanized coatings (>1.25 ounces per sq ft).

Shop installation of shear connectors is initially more economical, but there is a growing tendency to use field installation. There are two major reasons for this trend: the connectors may easily be damaged during transportation and setting of the beams, and they serve as a hindrance to the workers walking along the top flanges during the early phases of construction.

When a composite beam is being tested, failure will probably occur with a crushing of the concrete. At that time it seems reasonable to assume that the concrete and steel will both have reached a plastic condition.

For this discussion, reference is made to Fig. 16-5. Should the neutral axis fall in the slab, the maximum horizontal shear (or horizontal force on the plane between the concrete and the steel) is said to be $A_s F_y$; and if the neutral axis is in the steel section, the maximum horizontal shear is considered to equal

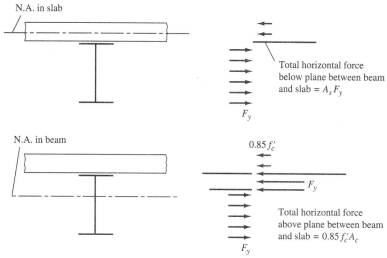

N.A. in slab

Total horizontal force
below plane between beam
and slab = $A_s F_y$

F_y

N.A. in beam

$0.85 f_c'$

F_y

Total horizontal force
above plane between beam
and slab = $0.85 f_c' A_c$

F_y

Figure 16-5

$0.85 f_c' A_c$ where A_c is the effective area of the concrete slab. (For the student unfamiliar with the strength design theory for reinforced concrete, the average stress at failure on the compression side of a reinforced concrete beam is usually assumed to be $0.85 f_c'$.)

Channel-section shear connectors, Grand Rapids, Mich. (Courtesy of the Lincoln Electric Company.)

From this information expressions for ΣQ_n (the shear to be taken by the connectors) can be determined. The LRFD (I5.2) says that for composite action the total horizontal shear between the points of maximum positive moment and zero moment is to be taken as the least of the following where ΣQ_n is the total nominal strength of the shear connectors provided as described in Section 16-7.

(a) $0.85 f'_c A_c$

(b) $A_s F_y$ (for hybrid beams this yield force must be calculated separately for each of the components of the cross section)

(c) ΣQ_n

16-6 PARTIALLY COMPOSITE BEAMS

For this discussion it is assumed that we need to select a steel section which when made composite with the concrete slab will have a design strength of 450 ft-k. It is further assumed that when we select a section from the Manual it has (when made composite with the slab) an M_u equal to 510 ft-k. If we now provide shear connectors for full composite action, the section will have a design strength of 510 ft-k. But we need only 450 ft-k.

It seems very logical to assume that we may decide to provide only a sufficient number of connectors to develop a design strength of 450 ft-k. In this way we can reduce the number of connectors and save some money (perhaps a good bit if we repeat this section a good many times in the structure). The resulting section is a *partially composite section,* one that does not have a sufficient number of connectors to develop the full flexural strength of the composite beam. We will encounter this situation in Examples 16-3 and 16-4.

It is usually felt that the total strength of the shear connectors used in a particular beam should not be less than 25 percent of the shearing strength required for full composite action ($A_s F_y$). Otherwise, our calculations may not accurately depict the stiffness and strength of a composite section.

16-7 STRENGTH OF SHEAR CONNECTORS

For composite sections it is permissible to use normal weight stone concrete (made with aggregates conforming to ASTM C33) or lightweight concrete weighing not less than 90 lb/ft^3 (made with rotary kiln-produced aggregates conforming to ASTM C330).

The LRFD Specification provides strength values for headed steel studs not less than 4 diameters in length after installation and for hot-rolled steel channels. *They do not, however, give resistance factors for the strength of shear connectors.* This is because they feel that the factor used for determining the flexural strength of the concrete is sufficient to account for variations in concrete strength, including those variations that are associated with shear connectors.

Stud Shear Connectors

The nominal shear strength in kips of one stud shear connector embedded in a solid concrete slab is to be determined with the following expression from LRFD Specification I5.3. In this expression A_{sc} is the cross-sectional area of the shank of the connector in square inches and f'_c is the specified compressive stress of the concrete in ksi. E_c is the modulus of elasticity of the concrete in ksi and equals $w^{1.5}\sqrt{f'_c}$ in which w is the unit weight of the concrete in lb/ft³. Finally, F_u is the specified minimum tensile strength of the steel stud in ksi.

$$Q_n = 0.5\,A_{sc}\sqrt{f'_c E_c} \leq A_{sc}F_u \qquad \text{(LRFD Equation I5-1)}$$

Table 16-1 (which is Table 5-1 from Part 5 of the LRFD Manual) shows a set of Q_n values calculated with the preceding equation for 3/4-in. headed studs made from A36 steel and embedded in concrete slabs with several different f'_c values and concrete weights of 115 and 145 lb/ft³.

TABLE 16-1	NOMINAL STUD SHEAR STRENGTH, Q_n (KIPS) FOR $\frac{3}{4}$-IN. HEADED STUDS	
f'_c ksi	w lb/ft³	Q_n kips
3.0	115	17.7
3.0	145	21.0
3.5	115	19.8
3.5	145	23.6
4.0	115	21.9
4.0	145	26.1

Channel Shear Connectors

The nominal shear strength in kips for one channel shear connector is to be determined from the following expression from LRFD Specification I5.4 in which t_f and t_w are respectively the flange and web thicknesses of the channel and L_c is its length. All of these values are to be used in inches.

$$Q_n = 0.3\,(t_f + 0.5t_w)\,L_c\,\sqrt{f'_c E_c} \qquad \text{(LRFD Equation I5-2)}$$

Other Connectors

Should other types of shear connectors be used, the LRFD Specification (I6) states that their nominal strengths are to be determined with suitable tests.

Stud Shear Connectors in the Ribs of Steel Decks

When shear connectors are placed in the ribs of steel decks, the LRFD Specification (13.5) states that their nominal capacity as determined by the appropriate Q_n expression may have to be reduced. This topic is addressed in Section 16-11.

16-8 NUMBER, SPACING, AND COVER REQUIREMENTS FOR SHEAR CONNECTORS

The number of shear connectors to be used between the point of maximum moment and each adjacent point of zero moment equals the horizontal force to be resisted divided by the nominal strength of one connector Q_n.

Spacing of Connectors

Tests of composite beams with shear connectors spaced uniformly, and of composite beams with the same number of connectors spaced in variation with the statical shear, show little difference as to ultimate strengths and as to deflections at working loads. These results are the case as long as the total number of connectors is sufficient to develop the shear on both sides of the point of maximum moment. As a result the LRFD Specification (I5.6) permits uniform spacings of connectors on each side of the maximum moment point, except that the number of connectors placed between a concentrated load and the nearest point of zero moment must be sufficient to develop the maximum moment at the concentrated load.

Maximum and Minimum Spacings

Except for formed steel decks, the minimum center-to-center spacing of shear connectors along the longitudinal axes of composite beams permitted by the LRFD Specification (I5.6) is 6 diameters, while the minimum value transverse to the longitudinal axis is 4 diameters. Within the ribs of formed steel decks, the minimum permissible spacing is 4 diameters in any direction.

When the flanges of steel beams are rather narrow, it may be difficult to achieve the minimum transverse spacings described here. For such situations the studs may be staggered. Figure 16-6 shows possible arrangements.

It the deck ribs are parallel to the axis of the steel beam and more connectors are required than can be placed within the rib, the LRFD Commentary (I5.6) permits the splitting of the deck so that adequate room is made available.

Shear connectors must be capable of resisting both horizontal and vertical movement because there is a tendency for the slab and beam to separate vertically as well as to slip horizontally. The upset heads of stud shear connectors help to prevent vertical separation. The LRFD Specification (I5.6) states that the maximum spacing of shear connectors may not exceed 8 times the total slab thickness.

Figure 16-6 Connector arrangements.

Cover Requirements

The LRFD Specification (I5.6) requires that there must be at least 1 in. of lateral concrete cover provided for shear connectors. This rule does not apply to connectors used within the ribs of formed steel decks because tests have shown that strengths are not reduced even when studs are placed as close as possible to the ribs.

When studs are not placed directly over beam webs, they have a tendency to tear out of the beam flanges before their full shear capacity is reached. To keep this situation from occurring, the LRFD Specification (I5.6) requires that the diameter of the studs may not be greater than 2.5 times the flange thickness of the beam to which they are welded unless they are located over the web.

When formed steel deck is used, the steel beam must be connected to the concrete slab with stud shear connectors with diameters not larger than $\frac{3}{4}$ in. These may be welded through the deck or directly to the steel beam. After their installation they must extend for at least $1\frac{1}{2}$ in. above the top of the steel deck and the concrete slab thickness above the steel deck may not be less than 2 in. (LRFD Specification I3.5a).

16-9 MOMENT CAPACITY OF COMPOSITE SECTIONS

The nominal flexural strength of a composite beam in the positive moment region may be controlled by the plastic strength of the section or by the strength of the concrete slab or by the strength of the shear connectors. Furthermore, if the web is very slender and if a large portion of the web is in compression, web buckling may possibly limit the nominal strength of the member.

Little research has been done on the subject of web buckling for composite sections, and for this reason the LRFD Specification (I3.2) has conservatively applied the same rules to composite section webs as to plain steel webs. The positive flexural strength ($\phi_b M_n$ with $\phi_b = 0.85$) of a composite section is to be determined assuming a plastic stress distribution if $h/t_w \leq 640/\sqrt{F_{yf}}$. In this expression h is the distance between the web toes of the fillet, that is, $d - 2k$, t_w is the web thickness, and F_{yf} is the yield stress of the beam flange. All of the rolled W, S, M, HP, and C shapes in the Manual meet this requirement for F_y values up to 65 ksi. (For built-up sections h is the distance between adjacent lines of fasteners or the clear distance between flanges when welds are used.)

Should h/t_w be greater than $640/\sqrt{F_{yf}}$, the value of $\phi_b M_n$ with $\phi_b = 0.90$ is to be determined by superimposing the elastic stresses. The effects of shoring must be considered for these calculations.

The nominal moment capacity of composite sections as determined by load tests can be estimated very accurately with the plastic theory. With this theory the steel section at failure is assumed to be fully yielded, and the part of the concrete slab which is on the compression side of the neutral axis assumed to be stressed to $0.85 f_c'$. If any part of the slab is on the tensile side of the neutral axis, it is assumed to be cracked and incapable of carrying stress.

The plastic neutral axis (PNA) may fall in the slab or in the flange of the steel section or in its web. Each of these cases is discussed in this section.

Neutral Axis in Concrete Slab

The concrete slab compression stresses vary somewhat from the PNA out to the top of the slab. For convenience in calculations, however, they are assumed to be uniform with a value of $0.85 f_c'$ over an area of depth a and width b_e determined as described in Section 16-4. (This distribution is selected to provide a stress block having the same total compression C and the same center of gravity for the total force as we have in the actual slab.)

The value of a can be determined from the following expression where the total tension in the steel section is set equal to the total compression in the slab.

$$A_s F_y = 0.85 f_c' a b_e$$

$$a = \frac{A_s F_y}{0.85 f_c' b_e}$$

If a is equal to or less than the slab thickness, the PNA will fall in the slab and the nominal or plastic moment capacity of the composite section may be written as the total tension T or the total compression times the distance between their centers of gravity. Reference is made here to Fig. 16-7.

Example 16-1, which follows, illustrates the calculation of $M_u = \phi_b M_p = \phi_b M_n$ for a composite section where the PNA falls within the slab.

Figure 16-7 Plastic neutral axis (PNA) in the slab.

■ **EXAMPLE 16-1**

Compute $M_u = \phi_b M_p = \phi_b M_n$ for the composite section shown in Fig. 16-8 if $f'_c = 4$ ksi and $F_y = 36$ ksi.

Figure 16-8

Solution.

$$a = \frac{A_s F_y}{0.85 f'_c b_e} = \frac{(31.7)(36)}{(0.85)(4)(84)} = 4.00 \text{ in.} < t$$

$$M_n = M_p = A_s F_y \left(\frac{d}{2} + t - \frac{a}{2} \right)$$

$$= (31.7)(36) \left(\frac{29.83}{2} + 5 - \frac{4.0}{2} \right)$$

$$= 20{,}445 \text{ in.-k} = 1704 \text{ ft-k}$$

$$M_u = \phi_b M_n = (0.85)(1704) = 1448.4 \text{ ft-k}$$

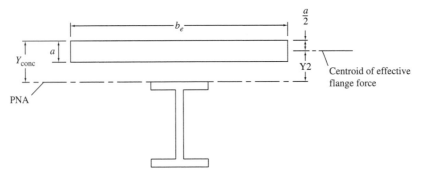

Figure 16-9

Note: If the reader refers to Part 5 of the LRFD Manual he or she can determine the M_u value for this composite beam with reference being made to Fig. 16-9. For this discussion it is assumed that the PNA is located at the top of the steel flange (TFL) or down in the steel shape. In the LRFD tables Y1 represents the distance from the PNA to the top of the beam flange while Y2 represents the distance from the centroid of the effective concrete flange force to the top flange of the beam $(t - a/2)$. ■

With the PNA for the preceding example being located at the top of the beam flange, from page 5-24 of the Manual with Y2 $= 5 - \frac{4}{2} = 3$ in. and Y1 $= 0$ and for a W30 $\times$ 108 the value of $M_u = \phi_b M_p = 1450$ ft-k.

Neutral Axis in Top Flange of Steel Beam

If a is calculated as previously described and is greater than the slab thickness t the PNA will fall down in the steel section. If this happens it will be necessary to find out whether the PNA is in the flange or below the flange. Suppose we assume that it's at the base of the flange. We can calculate the total compressive force C above $= 0.85 f'_c b_e t + A_f F_y$ where A_f is the area of the flange and the total tensile force below $T = F_y(A_s - A_f)$. If $C > T$ the PNA will be in the flange. If $C < T$ the PNA is below the flange.

Assuming we find the PNA is in the flange we can find its location letting $\bar{y}$ be the distance to the PNA measured from the top of the top flange by equating C and T as follows:

$$0.85 f'_c b_e t + F_y b_f \bar{y} = F_y A_s - F_y b_f \bar{y}$$

From which $\bar{y}$ is

$$\bar{y} = \frac{F_y A_s - 0.85 f'_c b_e t}{2 F_y b_f}$$

Figure 16-10

Then the nominal or plastic moment capacity of the section can be determined from the expression to follow with reference being made to Fig. 16-10. Taking moments about the PNA we get:

$$M_p = M_n = 0.85 f_c' b_e t \left(\frac{t}{2} + \bar{y}\right) + 2F_y b_f \bar{y}\left(\frac{\bar{y}}{2}\right) + F_y A_s\left(\frac{d}{2} - \bar{y}\right)$$

Example 16-2, which follows, illustrates the calculation of $M_u = \phi_b M_p = \phi_b M_n$ for a composite section in which the PNA falls in the flange.

■ **EXAMPLE 16-2**

Compute $M_u = \phi_b M_p = \phi_b M_n$ for the composite section shown in Fig. 16-11 if A36 steel is used and if f_c' is 4 ksi.

Figure 16-11

Solution.   Is PNA located at top of steel flange?

$$a = \frac{(36)(34.2)}{(0.85)(4)(80)} = 4.53 \text{ in.} > 4.00 \text{ in.}$$

∴ PNA is located down in steel section.

Is PNA in flange or in web? Here we assume it is at base of steel flange.

$C = (0.85)(4)(80)(4) + (36)(10.495)(0.85) = 1409.1 \text{ k}$

$T = (36)(34.2 - 10.495 \times 0.85) = 910 \text{ k}$

Since $C > T$, the PNA falls in the steel flange and can be located as follows:

$$\bar{y} = \frac{(36)(34.2) - (0.85)(4)(80)(4)}{(2)(36)(10.495)} = 0.190 \text{ in.}$$

Then

$$M_n = M_p = (0.85)(4)(80)(4)\left(\frac{4}{2} + 0.190\right)$$

$$+ (2)(36)(10.495)(0.190)\left(\frac{0.190}{2}\right)$$

$$+ (36)(34.2)\left(\frac{30.01}{2} - 0.190\right)$$

$$= 20{,}637 \text{ in.-k} = 1719.7 \text{ ft-k}$$

$$M_u = \phi_b M_p = \phi_b M_n = (0.85)(1719.7) = \underline{\underline{1461.8 \text{ ft-k}}}$$

By interpolation in the LRFD Composite Design tables with Y1 = 0.19 in. and Y2 = 2.0 in. we get $1480 - (0.19/0.21)(1480 - 1460) = 1461.9$ ft-k.

∎

If we have a partially composite section with ΣQ_n less than $A_s F_y$, the PNA will be down in the shape; if in the flange, the value of M_u can be determined with the equation used in Example 16-2. In the composite design tables presented in the Manual values of ΣQ_n and M_u are shown for seven different PNA locations—top of the flange, quarter points in the flange, bottom of the flange, and two points down in the web. Straight-line interpolation may be used for numbers in between the tabulated values.

Neutral Axis in Web of Steel Section

If for a particular composite section we find that a is larger than the slab thickness, and if we then assume the PNA is located at the bottom of the steel flange and we calculate C and T and find T is larger than C, the PNA will fall in the web. We can go through calculations similar to the ones we used for the case in which the PNA was located in the flange. Space is not taken to show such calculations because the Composite Design tables in Part 5 of the Manual cover most common cases.

16-10 DEFLECTIONS

Deflections for composite beams may be calculated by the same methods used for other types of beams. The student must be careful to compute deflections for the various types of loads separately. For example, there are dead loads applied to the steel section alone (if no shoring is used), dead loads applied to the composite section, and live loads applied to the composite section.

The long-term creep effect in the concrete in compression causes deflections to increase with time. These increases, however, are usually not considered significant for the average composite beam. If the designer feels that it is a matter of importance, he or she should compute long-term deflections using values of about $2n$ for the modular ratio when calculating the composite section properties to be used for deflections.

Should lightweight concrete be used, the actual modulus of elasticity of that concrete E_c (which may be rather small) should be used in calculating I_{tr} for deflection computations. For stress calculations we use E_c for normal-weight concrete.

Generally speaking, shear deflections are neglected, although on occasion they can be quite large.[1] The steel beams can be cambered for all or some portion of deflections. It may be feasible in some situations to make a floor slab a little thicker in the middle than on the edges to compensate for deflections.

The designer may want to control vibrations in composite floors subject to pedestrian traffic or other moving loads. This may be the case where we have large open floor areas with no damping furnished by partitions, as in shopping malls. For such cases dynamic analyses should be made.[2]

When the LRFD Specification is used to select steel beams for composite sections, the results often will be some rather small steel beams, and thus some quite shallow floors. Such floors when unshored frequently will have large deflections when the concrete is placed. As a result designers usually will require cambering of the beams. Other alternatives include the selection of larger beams or the use of shoring.[3,4]

The beams selected must, of course, have sufficient $\phi_b M_n$ values to support themselves and the wet concrete. Nevertheless, their sizes probably are dictated more by wet concrete deflections than by moment considerations.

An alternative solution for these problems involves the use of partly restrained or PR connections (discussed in Chapter 15). When these connections are used the midspan deflections and moments are appreciably reduced, enabling us to use smaller girders. Furthermore, there are reductions in the annoying vibrations that are a problem in shallow composite floors.

When PR connections are used, negative moments will develop at the supports. In Section 16-12 of this chapter it is shown that the LRFD

---

[1] L. S. Beedle et al., *Structural Steel Design* (New York: Ronald Press, 1964), p. 452.

[2] Thomas M. Murray, "Design to Prevent Floor Vibrations," *Engineering Journal,* AISC, 12, no. 3 (3d quarter, 1975), pp. 82–87.

[3] R. Leon, "Composite Semi-Rigid Connections," *Modern Steel Construction,* Volume 32, no. 9 (AISC, Chicago: October 1992), pp. 18–23.

[4] "Innovative Design Cuts Costs," *Modern Steel Construction,* Volume 33, no. 4 (AISC, Chicago: April 1993), pp. 18–21.

Specification permits the use of negative design moment strength for composite floors, provided certain requirements are met as to shear connectors and development of slab reinforcing in the negative moment region.

16-11 DESIGN OF COMPOSITE SECTIONS

Composite construction is of particular advantage economically when loads are heavy, spans are long, and beams are spaced at fairly large intervals. For steel building frames, composite construction is economical for spans varying roughly from 25 to 50 ft, with particular advantage in the longer spans. For bridges, simple spans have been economically constructed up to approximately 120 ft and continuous spans 50 or 60 ft longer. Composite bridges are generally economical for simple spans greater than about 40 ft and for continuous spans greater than about 60 ft.

Occasionally cover plates are welded to the bottom flanges of steel beams with improved economy. The student can see that with the slab acting as part of the beam there is a very large compressive area available, and by adding cover plates to the tensile flange a little better balance is obtained.

In tall buildings where headroom is a problem it is desirable to use the minimum overall floor thicknesses possible. For buildings, minimum depth-span ratios of approximately $\frac{1}{24}$ are recommended if the loads are fairly static and $\frac{1}{20}$ if the loads are of such a nature as to cause appreciable vibration. The thicknesses of the floor slabs are known (from the concrete design), and the depths of the steel beams can be fairly well estimated from these ratios.

Before we attempt some composite designs, several additional points relating to lateral bracing, shoring, estimated steel beam weights, and lower bound moments of inertia are discussed in the paragraphs to follow.

Lateral Bracing

After the concrete slab hardens it will provide sufficient lateral bracing for the compression flange of the steel beam. However, during the construction phase before the concrete hardens, lateral bracing may be insufficient and its design strength may have to be reduced depending on the estimated unbraced length. When steel-formed decking or concrete forms are attached to the beam's compression flange they usually will provide sufficient lateral bracing. The designer must be very careful in consideration of lateral bracing for fully encased beams.

Beams with Shoring

If beams are shored during construction we will assume that all loads are resisted by the composite section after the shoring is removed.

Beams without Shoring

If temporary shoring is not used during construction, the steel beam alone must be able to support all the loads before the concrete is sufficiently hardened to provide composite action.

Without shoring the wet concrete loads tend to cause large beam deflections which may lead us to build thicker slabs where the beam deflections are large. This situation can be counteracted by cambering the beams.

The LRFD Specification does not provide any extra margin against yield stresses occurring in beams during construction of unshored composite floors. Assuming satisfactory lateral bracing is provided, the Specification (F1.2) states that the maximum factored moment may not exceed $0.90F_yZ$. The 0.90 in effect limits the maximum factored moment to a value about equal to the yield moment F_yS.

To calculate the moment to be resisted during construction, it seems logical to count the wet concrete as a live load and to also include some extra live load as perhaps 20 psf to account for construction activities.

Estimated Steel Beam Weight

As illustrated in Example 16-3 it may be useful sometimes to make an estimate of the weight of the steel beam. The LRFD Manual provides the following empirical formula for this purpose in its Part 5.

$$\text{Estimated beam weight} = \left[\frac{12\,M_u}{(d/2 + Y_{\text{con}} - a/2)\phi F_y}\right]3.4$$

where M_u = required flexural strength of composite section
   d = nominal steel beam depth
   Y_{con} = distance from top of steel beam to top of concrete slab
   a = effective concrete slab thickness (which can be conservatively estimated as somewhere in the range of about 2 in.)
   ϕ = 0.85

Lower Bound Moment of Inertia

To calculate the service load deflections for composite sections, a table of lower bound moment of inertia values is presented in Part 5 of the Manual. These values are computed from the area of the steel beam and an equivalent concrete area of $\Sigma Q_n/F_y$. The remainder of the concrete flange is not used in these calculations. This means that if we have partially composite sections the value of the lower bound moment of inertia will reflect this situation because ΣQ_n will be smaller. The lower bound moment of inertia is computed with the following expression, reference being made to Fig. 16-12 where Y_{ENA} is the distance from the bottom of the beam to the elastic neutral axis (ENA).

$$I = I_x + A_s\left(Y_{ENA} - \frac{d}{2}\right)^2 + \left(\frac{\Sigma Q_n}{F_y}\right)(d + Y_2 - Y_{ENA})^2$$

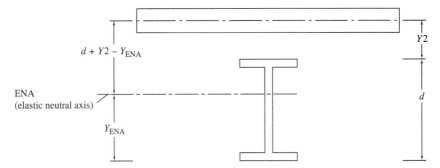

Figure 16-12

Extra Reinforcing

For building design calculations the spans are often considered to be simply supported, but the steel beams do not generally have perfectly simple ends. The result of this situation is that some negative moment may occur at the beam ends, with possible cracking of the slab above. To prevent or minimize cracking, some extra steel can be placed in the top of the slab extending 2 or 3 ft out into the slab. The amount of steel added is in addition to the temperature and shrinkage requirements specified by the American Concrete Institute.[5] See LRFD Commentary I3.2 for detailed comments on reinforcing.

Example Problems

Examples 16-3 and 16-4 illustrate the designs of two unshored composite sections.

■ EXAMPLE 16-3

Beams 10 ft on center with 36-ft simple spans are to be selected to support a 4-in.-deep lightweight concrete slab on a 3-in.-deep formed steel deck with no shoring. The ribs for the steel deck, which are perpendicular to the beam center lines, have average widths of 6 in. If the service dead load (including the beam weight) is to be 0.78 k/ft of length of the beams and the service live load is 1.2 k/ft, (a) select the beams, (b) determine the number of $\frac{3}{4}$-in.-diameter headed studs required, (c) compute the service live load deflection, and (d) check the beam shear. Other data: A36 steel, $f'_c = 3$ ksi, and concrete weight 115 lb/ft³.

---

[5] *Building Code Requirements for Reinforced Concrete* (ACI std. 318-89). Detroit: American Concrete Institute, 1989, Section 7.12.

Solution.

Loads and moments

$$w_u = (1.2)(0.78) + (1.6)(1.2) = 2.856 \text{ k/ft}$$

$$M_u = \frac{(2.856)(36)^2}{8} = 462.7 \text{ ft-k}$$

Effective flange width b_e

$$b_e = (2)(\tfrac{1}{8} \times 36 \times 12) = 108 \text{ in.} \leftarrow$$

$$b_e = (2)(5 \times 12) = 120 \text{ in.}$$

(a) Select W section

$Y_{con} = $ distance from top of slab to top of steel flange $= 4 + 3 = 7$ in.

Assume $a = 2$ in. (It's usually quite small, particularly for relatively light sections.)

Y1 is distance from PNA to top flange $= 0$ in.

Y2 is the distance from the center of gravity of the concrete flange force to the top flange of the beam $= 7 - a/2 = 7 - 2/2 = 6$ in.

Looking through the composite tables of the Manual with $M_u = 462.7$ ft-k, $Y1 = 0$, and $Y2 = 6$ in., we can see that several W18s (55 lb, 50 lb, and 46 lb) seem reasonable.

We can use LRFD estimated weight formula for an 18-in.-deep section.

$$wt = \left[\frac{(12)(462.7)}{(18/2 + 7 - 2/2)(0.85)(36)} \right] 3.4 = 41.12 \text{ lb}$$

Try W18 × 40 ($A = 11.8$ in.^{2}), though it seems a little small compared with estimated weight.

Assume $\Sigma Q_n = A_s F_y = (11.8)(36) = 424.8$ k

$$\text{Thus } a \text{ required} = \frac{\Sigma Q_n}{0.85 f'_c b_e} = \frac{424.8}{(0.85)(3)(108)} = 1.54 \text{ in.}$$

$$Y1 = 0$$

$$Y2 = 7.0 - \frac{1.54}{2} = 6.23 \text{ in.}$$

$$M_u \text{ from Manual by interpolation} = 450 + \left(\frac{0.23}{0.50}\right)(465 - 450)$$

$$= 456.9 \text{ ft-k} < 462.7 \text{ ft-k} \qquad \text{(NG)}$$

Try W18 × 46($A = 13.5$ in.^{2}, $d = 18.06$ in., $t_w = 0.360$ in.)

Now this time we will go to the case where Y1 is the largest possible to provide an M_u of about 462.7 ft-k with Y2 $= 6$ in., because ΣQ_n will be smaller and fewer shear connectors will be required. This will occur with Y1 $= 0.30$ in. where $M_u = 460$ ft-k if Y2 $= 6$ in.

$$\Sigma Q_n = 354 \text{ k}$$

$$a = \frac{354}{(0.85)(3)(108)} = 1.285 \text{ in.}$$

$$Y2 = 7 - \frac{1.285}{2} = 6.36 \text{ in.}$$

$$M_u = 460 + \left(\frac{0.36}{0.50}\right)(472 - 460)$$

$$= 468.6 \text{ ft-k} > 462.7 \text{ ft-k} \qquad \text{(OK)}$$

Use W18 × 46

(b) Design of studs

Since we have steel-formed deck we must compute a stud reduction factor (SRF) as required by Specification I3.5(b).

$h_r = 3$ in.

Assume $H_s = 5.0$ in. (can't be $> h_r + 3 = 6$ in.)
Assume 1 stud in a rib at beam intersection $= N_r$

$w_r = 6$ in.

$$\text{SRF} = \frac{0.85}{\sqrt{1}}\left(\frac{6}{3}\right)\left[\frac{5.0}{3} - 1.0\right] = 1.13 > 1.0 \qquad \text{(LRFD Equation I3-1)}$$

∴ No reduction is necessary.

Q_n from Table 16.1 for $\frac{3}{4}$-in. headed studs, $f'_c = 3$ ksi, and concrete weight of 115 lb/ft^3 = 17.7 k

ΣQ_n for W18 × 46 with Y1 of 0.30 in. = 354 k

$$\text{Total no. of connectors required} = \frac{2\Sigma Q_n}{Q_n} = \frac{(2)(354)}{17.7} = 40$$

Use 40 -$\frac{3}{4}$ in. studs (20 on each side of point of maximum moment which is ℄ here).

(c) Compute LL deflection

Assume maximum permissible LL deflection

$$\frac{1}{360}\text{span} = \left(\frac{1}{360}\right)(12 \times 36) = 1.2 \text{ in.}$$

$C_1 = 161$ from Fig. 10-8 of textbook (or Fig. 4-2 of Manual)

$$M_{LL} = \frac{(1.2)(36)^2}{8} = 194.4 \text{ ft-k}$$

I_x = lower bound moment of inertia from tables in Part 5 of Manual

$$= 2000 + \left(\frac{0.36}{0.50}\right)(2080 - 2000) = 2057.6 \text{ in.}^4$$

$$\Delta_{LL} = \frac{ML^2}{C_1 I_X} = \frac{(194.4)(36)^2}{(161)(2057.6)} = 0.760 \text{ in.} < 1.2 \text{ in.} \tag{OK}$$

(d) Check beam shear

$$V_u = \frac{(36)(2.856)}{2} = 51.4 \text{ k}$$

$$\phi V_n = \phi 0.6 F_y A_w \text{ for W18} \times 46$$
$$= (0.9)(0.6)(36)(18.06)(0.360)$$
$$= 126.4 \text{ k} > 51.4 \text{ k} \tag{OK}$$

Use W18 × 46 with forty $\frac{3}{4}$-in. headed studs

■ EXAMPLE 16-4

Using the same data as for Example 16-3 except that $F_y = 50$ ksi, concrete 145 lbs/ft^3 and unshored construction is to be used, perform the following tasks:

(a) Select steel beam.
(b) If the stud reduction factor for metal decks = 1.0, determine the number of $\frac{3}{4}$-in. headed studs required.
(c) Check the beam strength before the concrete hardens.
(d) Compute service load deflection before concrete hardens. Assume a construction live load of 20 psf.
(e) Determine the service live load deflection after composite action is available.
(f) Check shear.
(g) Select a steel section to carry all the loads if no shear connectors are used and compute its total service load deflection.

Solution. From Example 16-3

$$M_u = 462.7 \text{ ft-k}$$
$$b_e = 108 \text{ in.}$$

(a) Select W section

$$Y_{con} = 4 + 3 = 7 \text{ in.}$$

Assume $a = 2$ in.

$$Y1 = 0$$
$$Y2 = 7 - \tfrac{2}{2} = 6 \text{ in.}$$

Try W18 × 35 ($A = 10.3$ in.^{2}, $d = 17.70$ in., $t_w = 0.30$ in.)

Assume $\Sigma Q_n = (10.3)(50) = 515$ k

$$a = \frac{515}{(0.85)(3)(108)} = 1.87 \text{ in.}$$

$$Y2 = 7 - \frac{1.87}{2} = 6.065 \text{ in.}$$

M_u from Manual by interpolation

$$= 542 + \left(\frac{0.065}{0.50}\right)(560 - 542) = 544.3 \text{ ft-k} > 462.7 \text{ k} \qquad \text{(OK)}$$

(b) Design of studs

Q_n from stud table $= 17.7$ k

We can go down in W18 × 35 table to $Y1 = 0.21$ in. and still obtain the M_u required (it's between 487 and 500 ft-k)

$$\Sigma Q_n = 388 \text{ k}$$

$$\text{Number of connectors required} = \frac{2 \times 388}{17.7} = 43.8$$

Use 22 connectors each side of ℂ

(c) Check strength of W section before concrete hardens

Assume the wet concrete is an LL during construction and also add a 20 psf construction live load.

$$\text{Concrete wt} = (\tfrac{4}{12})(145)(10) = 483 \text{ lb/ft}$$

Other dead loads $= 0.780 - 0.483 = 0.297$ k/ft

$w_u = (1.2)(0.297) + (1.6)(0.020 \times 10 + 0.483) = 1.449$ k/ft

$$M_u = \frac{(1.449)(36)^2}{8} = 234.7 \text{ ft-k}$$

$\phi_b M_p$ from beam tables for W18 × 35 $= 249$ ft-k > 234.7 ft-k   (OK)

(d) Service load deflection before concrete hardens

$$C_1 = 161$$

I_x for W18 × 35 $= 510$ in.[4] (not lower bound I)

Use $w_D = 0.78 + (10)(0.02) = 0.98$ k/ft

$$M_{DL} = \frac{(0.98)(36)^2}{8} = 158.8 \text{ ft-k}$$

$$\Delta_{DL} = \frac{(158.8)(36)^2}{(161)(510)} = 2.51 \text{ in.}$$

(We might camber beam for this deflection and/or use PR connections.)

(e) Service LL deflection after composite action is available

$$M_L = \frac{(1.2)(36)^2}{8} = 194.4 \text{ ft-k}$$

Lower bound I with Y1 = 0.21 in. and Y2 = 6.065 in.

$$I = 1490 + \left(\frac{0.065}{0.50}\right)(1550 - 1490) = 1498 \text{ in.}^4$$

$$\Delta_{LL} = \frac{(194.4)(36)^2}{(161)(1498)} = 1.04 \text{ in.} < \frac{L}{360} = 1.2 \text{ in. (OK)}$$

(f) Check shear

$$V_u = \frac{(36)(2.856)}{2} = 51.4 \text{ k}$$

$$\phi V_n = (0.9)(0.6)(50)(17.70)(0.30)$$
$$= 143.4 \text{ k} > 51.4 \text{ k} \qquad\qquad\qquad \text{(OK)}$$

(g) Select section if we have no composite action

$$M_u = 462.7 \text{ ft-k} = \phi_b M_p$$

Requires a W24 × 55 (I_x = 1350 in.^{4}) from Load Factor Design Selection Table in Manual

$$\Delta = \frac{(462.7)(36)^2}{(161)(1350)} = 2.76 \text{ in.} > \frac{L}{360} = 1.2 \text{ in.}$$

Min. I to limit deflection to 1.2 in. $= \left(\frac{2.76}{1.2}\right)(1350) = 3105 \text{ in.}^4$

Use W30 × 90 (as compared with W18 × 35 if a composite section is used). ∎

16-12 CONTINUOUS COMPOSITE SECTIONS

The LRFD Specification (I3.2) permits the use of continuous composite sections. The flexural strength of a composite section in a negative moment region may be considered to equal $\phi_b M_n$ for the steel section alone, or it may be based upon the plastic strength of a composite section considered to be made up of the steel beam and the longitudinal reinforcement in the slab. For this latter method to be used the following conditions must be met:

1. The steel section must be compact and adequately braced.
2. The slab must be connected to the steel beams in the negative moment region with shear connectors.

Continuous-welded plate girders in Henry Jefferson County, Iowa. (Courtesy of the Lincoln Electric Company.)

3. The longitudinal reinforcing in the slab parallel to the steel beam and within the effective width of the slab must have adequate development lengths. (*Development length* is a term used in reinforced-concrete design and refers to the length that reinforcing bars have to be extended or embedded in the concrete to properly anchor them or develop their stresses by means of bond between the bars and the concrete.)

For a particular beam the total horizontal shear force between the point of zero moment and the point of maximum negative moments is to be taken as the smaller of $A_r F_{yr}$ and ΣQ_n where A_r is the cross-sectional area of the properly developed reinforcing and F_{yr} is the yield stress of the bars. The plastic stress distribution for negative moment in a composite section is illustrated in Fig. 16-13.

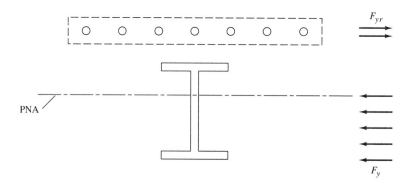

Figure 16-13 Stress distribution in negative moment range.

16-13 DESIGN OF CONCRETE-ENCASED SECTIONS

For fireproofing purposes, it is possible to completely encase in concrete the steel beams used for building floors. Such a practice, however, is definitely not economical because lightweight spray-on fire protection is so much cheaper. Furthermore, encased beams may increase the floor system dead load by as much as 15 percent.

For the rare situation in which encased beams are used, the horizontal shears between the slabs and beams can be considered to be transferred by natural bond and friction if certain conditions are met. The LRFD Specification (I1) states that for the transfer to be permissible the encasing concrete must be placed integrally with the slab concrete and must cover the steel by at least 2 in. on the sides and bottom (or soffit). It is further required that the top of the steel section be at least $1\frac{1}{2}$ in. below the top of the slab and 2 in. above the bottom of the slab. Finally, the encasing concrete must have adequate mesh or other reinforcing for its full depth and across the soffit of the beam to prevent spalling of the concrete. The exact amount, which is not specified by the LRFD, can be very nominal in size (from "chicken wire size" on up).

1. With one method, the design strength of the encased section may be based on the plastic moment capacity $\phi_b M_p$ of the steel section alone.
2. By another method, the design strength is based on the first yield of the tension flange, assuming composite action between the concrete which is in compression and the steel section.

If the second method is used and we have unshored construction, the stresses in the steel section caused by the wet concrete and other construction loads are calculated. Then the stresses in the composite section caused by loads applied after the concrete hardens are computed. These stresses are superimposed on the first set of stresses. The total stresses so computed may not exceed $\phi_b F_y$ with $\phi_b = 0.90$. If we have shored construction all of the loads may be assumed to be supported by the composite section and the stresses computed accordingly. For stress calculations, the properties of a composite section are computed with the transformed area method. In this method the cross-sectional area of one of the two materials is replaced or transformed into an equivalent area of the other. For composite design, it is customary to replace the concrete with an equivalent area of steel, whereas the reverse procedure is used in the working stress design method for reinforced-concrete design.

In the transformed area procedure the concrete and steel are assumed to be bonded tightly together so that their strains will be the same at equal distances from the neutral axis. The unit stress in either material can then be said to equal its strain times its modulus of elasticity (ϵE_c for the concrete or ϵE_s for the steel).

The unit stress in the steel is then $\epsilon E_s / \epsilon E_c = E_s / E_c$ times as great as the corresponding unit stress in the concrete. The E_s / E_c ratio is referred to as the modular ratio n; therefore, n in.^{2} of concrete are required to resist the same total stress as 1 in.^{2} of steel; and the cross-sectional area of the slab (A_c) is replaced with a transformed or equivalent area of steel equal to A_c / n.

The American Concrete Institute (ACI) Building Code suggests that the following expression may be used for calculating the modulus of elasticity of concrete weighing from 90 to 155 lb/ft.^{3}

$$E_c = w^{1.5} 33 \sqrt{f_c'}$$

In this expression w is the weight of the concrete in pounds per cubic foot and f_c' is the 28-day compressive strength in pounds per square inch.

There are no slenderness limitations required by the LRFD Specification for either of the two methods because the encasement is effective in preventing both local and lateral buckling.

In Example 16-5, which follows, stresses are computed with the elastic theory assuming composite action as described for the second method. Notice that the author has divided the effective width of the slab by n to transform the concrete slab into an equivalent area of steel.

■ EXAMPLE 16-5

Review the encased beam section shown in Fig. 16-14 if no shoring is used and the following data are assumed:

$$\text{Simple span} = 36 \text{ ft}$$

$$\text{Service dead load} = 0.50 \text{ k/ft before concrete hardens plus an}$$
$$\text{additional } 0.25 \text{ k/ft after concrete hardens}$$

$$\text{Construction live loads} = 0.2 \text{ k/ft}$$

$$\text{Service live load} = 1.0 \text{ k/ft after concrete hardens}$$

$$\text{Effective flange width } b_e = 60 \text{ in. and } n = 9$$

$$F_y = 50 \text{ ksi}$$

Figure 16-14

Solution.   Calculated Properties of Composite Section: Neglecting concrete area below flange,

$$A = 13.3 + \frac{(4)(60)}{9} = 39.96 \text{ in.}^2$$

$$y_b = \frac{(13.3)(10.44) + (26.66)(18)}{39.96} = 15.50 \text{ in.}$$

$$I = 586 + (13.3)(5.06)^2 + (\tfrac{1}{12})(\tfrac{60}{9})(4)^3 + (26.66)(2.5)^2$$
$$= 1129 \text{ in.}^4$$

Stresses before concrete hardens
  Assume wet concrete is a live load

$$w_u = (1.6)(0.5 + 0.2) = 1.12 \text{ k/ft}$$

$$M_u = \frac{(1.12)(36)^2}{8} = 181.4 \text{ ft-k}$$

$$f_t = \frac{(12)(181.4)(8.065)}{586} = 29.96 \text{ ksi}$$

$$< \phi_b F_y = (0.9)(50) = 45 \text{ ksi} \qquad\qquad\qquad \text{(OK)}$$

Stresses after concrete hardens

$$w_u = (1.2)(0.25) + (1.6)(1.0) = 1.9 \text{ k/ft}$$

$$M_u = \frac{(1.9)(36)^2}{8} = 307.8 \text{ ft-k}$$

$$f_t = \frac{(12)(307.8)(15.50 - 2.37)}{1129} = 42.96 \text{ ksi}$$

Total $f_t = 29.96 + 42.96 = 72.92 \text{ ksi} > 45 \text{ ksi} > 0.9F_y = 45 \text{ ksi}$ (NG)

∎

For buildings, continuous composite construction with encased sections is permissible. For continuous construction the positive moments are handled exactly as has been illustrated by the preceding examples. For negative moments, however, the transformed section is taken as shown in Fig. 16-15. The crosshatched area represents the concrete in compression, and all concrete on the tensile side of the neutral axis (that is, above the axis) is neglected.

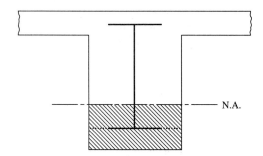

N.A.

Figure 16-15

PROBLEMS

16-1. Determine M_u for the section shown assuming sufficient shear connectors are provided to ensure full composite section. Solve using the procedure presented in Section 16-9 and check answer with tables in Manual. A36 steel. $f_c' = 3$ ksi. (*Ans.* 522.4 ft-k)

Problem 16-1

16-2. Repeat Prob. 16-1 if $F_y = 50$ ksi.

16-3. Repeat Prob. 16-1 using tables in Manual if it is considered to be partially composite and if ΣQ_n is 350 k. (*Ans.* 482.8 ft-k)

16-4. Determine M_u for the section shown if A36 steel and sufficient shear connectors are used to guarantee full composite action. Use formulas and check with Manual. $f_c' = 4$ ksi

Problem 16-4

16-5. Repeat Prob. 16-4 if $F_y = 50$ ksi (*Ans.* 678 ft-k)

16-6. Compute M_u for the composite section shown if A36 steel is used and sufficient shear connectors are used to provide full composite action. A

$3\frac{1}{4}$-in. concrete slab is supported by 3-in.-deep composite metal deck ribs perpendicular to beam. $f'_c = 4$ ksi. Check answer with Manual.

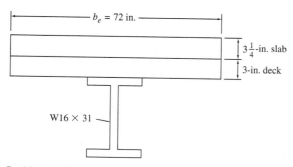

$b_e = 72$ in.

$3\frac{1}{4}$-in. slab

3-in. deck

W16 × 31

Problem 16-6

16-7. Repeat Prob. 16-6 using Manual tables if ΣQ_n of the connectors is 241 k. (*Ans.* 282.2 ft-k)

16-8. Using the Composite Design Tables of the LRFD Manual, A36 steel, a 145 lb/ft³ concrete slab with $f'_c = 3$ ksi and shored construction, select the steel section, design $\frac{3}{4}$-in. headed studs, calculate live load service deflection, and check the shear if the service live load is 100 psf.

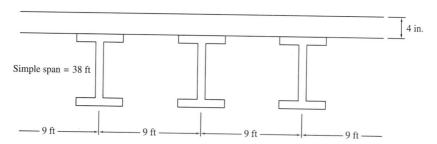

4 in.

Simple span = 38 ft

—— 9 ft ——   —— 9 ft ——   —— 9 ft ——   —— 9 ft ——

Problem 16-8

16-9. Repeat Prob. 16-8 if $F_y = 50$ ksi. (*Ans.* W16 × 36 with 30 studs)

16-10. For Prob. 16-9 calculate deflection during construction for wet concrete plus 20 psf live load for construction activities.

16-11. Select an A36 steel section to support a service dead load of 200 psf and a service live load of 100 psf. The beams are to have 36-ft simple spans and are to be spaced 8 ft 6 in. on center. Construction is shored, concrete weighs 115 lb/ft³, f'_c is 3.5 ksi, a metal deck with ribs perpendicular to the steel beams is used together with a 4-in. concrete slab. The ribs are 3 in. deep and have average widths of 6 in. Design $\frac{3}{4}$-in. headed studs and calculate live load deflection. (*Ans.* W21 × 57 with 46 studs)

16-12. Using the LRFD Manual and A36 steel, design a nonencased unshored composite section for the simple span beams shown in the accompanying figure if a 4-in. concrete slab (145 lb/ft³) with $f_c' = 3$ ksi is used. The total service dead load is 0.6 k/ft of length of the beam and the service live load is 1.25 k/ft.

(a) Select the beams.

(b) Determine the number of 3/4-in.-diameter headed studs required.

(c) Compute the service live load deflection.

(d) Check the beam shear.

Problem 16-12

16-13. Repeat Prob. 16-12 if $F_y = 50$ ksi and $f_c' = 4$ ksi. (*Ans.* W14 × 26 with 22 studs)

16-14. A36 beams 9 ft on center and spanning 40 ft are to be selected to support a 4-in.-deep lightweight concrete slab ($f_c' = 3$ ksi, weight 115 lb/ft³) on a 3-in.-deep formed steel deck with shoring. The ribs for the steel deck, which are perpendicular to the steel beams, have average widths of 6 in. If the total service dead load is to be 0.80 k/ft of length of the beams and the service live load is 1.25 k/ft, (a) select the beams, (b) determine the number of 3/4-in.-diameter headed studs required, (c) compute the service live load deflection, and (d) check the beam shear.

16-15. Repeat Prob. 16-14 if $F_y = 50$ ksi. (*Ans.* W18 × 40 with 58 studs)

16-16. Repeat Prob. 16-14 if spans are 34 ft.

16-17. Repeat Prob. 16-16 if $F_y = 50$ ksi. (*Ans.* W16 × 31 with 52 studs)

16-18. Using the same data as for Prob. 16-14 except that unshored construction is to be used, perform the following tasks:

(a) Select the steel beam.

(b) If the stud reduction factor for metal decks is 1.0, determine the number of $\frac{3}{4}$-in. headed studs required.

(c) Check the beam strength before the concrete hardens.

(d) Compute service load deflection before the concrete hardens assuming a construction live load of 25 psf.

(e) Determine the service load deflection after composite action is available.

(f) Check shear.

16-19. Repeat Prob. 16-18 if $F_y = 50$ ksi. (*Ans.* W18 × 40 with 58 studs. But deflection is high before concrete hardens.)

16-20. Using the transformed area method compute the stresses in the encased section shown in the accompanying illustration if no shoring is used. The section is assumed to be used for a simple span of 30 ft and to have a service dead uniform load of 30 psf applied after composite action develops and a service live uniform load of 120 psf. Assume $n = 9$, A36 steel, $f_c' = 3$ ksi, and concrete weighing 150 lb/ft³.

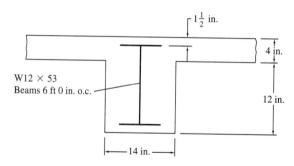

W12 × 53
Beams 6 ft 0 in. o.c.

Problem 16-20

Chapter 17

Composite Columns

17-1 INTRODUCTION

Composite columns are constructed with rolled or built-up steel shapes encased in concrete, or with concrete placed inside steel pipes or tubes. The resulting members are able to support significantly higher loads than can reinforced-concrete columns of the same sizes.

Several composite columns are shown in Fig. 17-1. In part (a) of the figure a W shape embedded in concrete is shown. The cross sections, which usually are square or rectangular, have one or more longitudinal bars placed in each corner. In addition, lateral ties are wrapped around the longitudinal bars at frequent vertical intervals. Ties are effective in increasing column strengths. They prevent the longitudinal bars from being displaced during construction and they resist the tendency of these same bars to buckle outward under load, which would cause breaking or spalling off of the outer concrete cover. It will be noted that these ties are all open and U-shaped. Otherwise they could not be installed, because the steel column shapes will have always been erected at an earlier time. In parts (b) and (c) of the figure, steel pipe and steel tubing sections filled with concrete are shown.

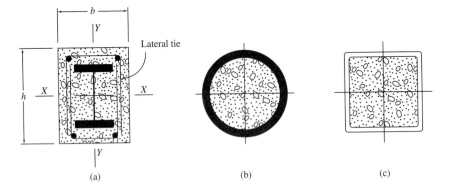

(a) (b) (c)

Figure 17-1 Composite columns.

17-2 **ADVANTAGES OF COMPOSITE COLUMNS**

For quite a few decades structural steel shapes have been used in combination with plain or reinforced concrete. Originally, the encasing concrete was used to provide only fire and corrosion protection for the steel, with no consideration given to its strengthening effects. During the last 20 to 30 years, however, the development and increasing popularity of composite frame construction has encouraged designers to include the strength of the concrete in their calculations.[1,2]

Composite columns may be practically used for low-rise and high-rise buildings. For the low-rise types such as warehouses, parking garages, and so on the steel columns are often encased in concrete for the sake of appearance or for protection of the steel from fire, corrosion, and from vehicles in garages. If, for structures like these, we are going to encase the steel in concrete anyway, we may as well take advantage of the concrete and use smaller steel shapes.

For high-rise buildings the sizes of composite columns are often considerably smaller than is required for reinforced-concrete columns to support the same loads. The results with composite designs are appreciable savings of valuable floor space. Closely spaced composite steel-concrete columns connected with spandrel beams may be used around the outsides of high-rise buildings to resist lateral loads by the tubular concept (to be described in Chapter 19). Very large composite columns are sometimes placed on the corners of high-rise buildings to increase lateral resisting moments. Also, steel sections embedded within reinforced-concrete shear walls may be used in the central core of high-rise buildings. This also insures a greater degree of precision in the construction of the core.

With composite construction the bare steel sections support the initial loads including the weights of the structure, the gravity and lateral loads occurring during construction, and the concrete later cast around the W shapes or inside the tube shapes. The concrete and steel are combined in such a fashion that the advantages of both materials are used in the composite sections. For instance, the reinforced concrete enables the building frame to more easily limit swaying or lateral deflections. At the same time the light weight and strength of the steel shapes permit the use of smaller and lighter foundations.[3]

---

[1] D. Belford, "Composite Steel Concrete Building Frame," *Civil Engineering* (New York: ASCE, July 1972), pp. 61–65.

[2] Fazlur R. Kahn, "Recent Structural Systems in Steel for High Rise Buildings," BCSA Conference on Steel in Architecture, November 24–26, 1969.

[3] L. G. Griffis, "Design of Encased W-shape Composite Columns," *Proceedings 1988 National Steel Construction Conference* (AISC, Chicago, June 8–11, 1988), pp. 20-1–20-28.

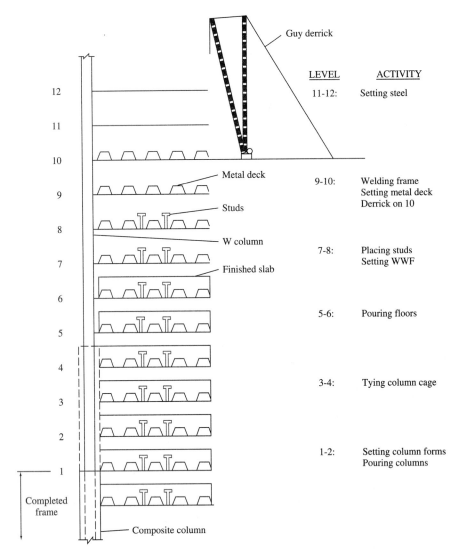

Figure 17-2 Sequence of construction operations for a composite building frame. (Courtesy of AISC.)

Composite high-rise structures are erected in a rather efficient manner. There is quite a vertical spread of construction going on at any one time with numerous trades working simultaneously. This situation, which is pictured in Fig. 17-2, is briefly described here.[4]

---

[4] Griffis, op. cit.

1. One group of workers may be erecting the steel beams and columns for one or two stories on top of the frame.
2. Two or three stories below another group will be setting the metal decking for the floors.
3. A few stories below another group may be placing the concrete for the floor slabs.
4. This continues as we go down the building with one group, tying the column reinforcing bars in cages, while below them others are placing the column forms, placing the column concrete, and so on.

17-3 DISADVANTAGES OF COMPOSITE COLUMNS

As described in the preceding section composite sections have several important advantages. They, however, also have a few disadvantages. One particular problem with their use in high-rise buildings is the difficulty of controlling their rates and amounts of shortening in relation to shear walls and perhaps adjacent plain steel columns. The accurate estimation of these items is made quite difficult by the different types and stages of construction activities going on simultaneously over a large number of building stories.

If composite columns are used around the outside of a high-rise building and plain steel sections are used in the building core (or if we have shear walls), the creep in the composite sections can be a problem. The results may be concrete floors that are not very level. Some erectors make very careful elevation measurements at column splices and then try to make appropriate adjustments with steel shims to try to even out the differences between measured elevations and computed elevations.

Another problem with composite columns is the lack of knowledge available concerning the mechanical bond between the concrete and the steel shapes. This is particularly important for the transfer of moments through beam-column joints. It is feared that if large cyclical strain reversals were to occur at such a joint (as in a seismic area) there could be a severe breakdown of the joint.[5]

17-4 LATERAL BRACING

Resistance to lateral loads for the usual structural steel or reinforced-concrete high-rise building is provided as the floors are being constructed. For instance, diagonal bracing or moment-resisting joints may be provided for each floor as a structural steel building frame is being constructed. In a similar manner the needed lateral strength of a reinforced-concrete frame may be provided by the

---

[5] Griffis, op. cit.

moment resistance obtained with monolithic construction of its members and/or by shear walls.

For composite construction the desired lateral strength of a building is not obtained until the concrete has been placed around or inside the erected steel members and has sufficiently hardened. This situation is probably being achieved 10 to 18 stories behind the steel erection (see Fig. 17-2).

As we have mentioned, the steel fabricator is used to erecting a steel frame and providing the necessary wind bracing as the floors are erected. The steel frames used for high-rise composite buildings, however, do not usually have such bracing and the frames will not have the desired lateral strength. This strength will be obtained only after the concrete is placed and cured for many building stories. Thus the engineer of record for a composite high-rise building must clearly state the lateral force conditions and what is to be done about them during erection.[6]

17-5 SPECIFICATIONS FOR COMPOSITE COLUMNS

Composite columns theoretically can be constructed with cross sections that are square, rectangular, round, triangular, or any other shape. Practically, however, they usually are square or rectangular, with one reinforcing bar in each column corner. This arrangement enables us to use reasonably simple connections from the exterior spandrel beams and floor beams to the steel shapes in the columns without unduly interfering with the vertical reinforcing.

The LRFD Specification does not provide detailed requirements for reinforcing bar spacings, splices, and so on. It therefore seems logical that the requirements in this regard of the ACI 318 Code[7] should be followed for situations not clearly covered by the LRFD Specification.

Section I2.1 of the LRFD Specification does provide detailed requirements pertaining to cross-sectional areas of steel shapes, concrete strengths, tie areas, and spacings for the vertical reinforcing bars, and so on. This information is listed and briefly discussed in the paragraphs to follow:

1. *The total cross-sectional area of the steel section or sections may not be less than 4 percent of the gross column area.* If the steel percentage is less than 4 percent, the member is classified as a reinforced-concrete column, and its design must be handled by the *Building Code Requirements for Reinforced Concrete* of the American Concrete Institute.

2. *When a steel core is encased in concrete the encasement must be reinforced with longitudinal load-carrying bars (which must be run continuously at framed levels) and with lateral ties spaced no farther apart than*

---

[6] Griffis, op. cit.

[7] American Concrete Institute, *Building Code Requirements for Reinforced Concrete,* ACI 318-89 (Detroit: 1989).

$\frac{2}{3}$ *times the least dimension of the composite member. The area of the ties must not be less than 0.007 sq. in. per inch of bar spacing. There must be at least $1\frac{1}{2}$ in. clear cover of concrete outside of any steel (ties or longitudinal bars).* The cover is needed for the protection of the steel from fire or corrosion. The amount of longitudinal and transverse reinforcing required in the encasement is thought to be sufficient to prevent severe spalling of the concrete surface from occurring during a fire.

3. *The specified compression strength of the concrete f'_c must be at least 3 ksi but not more than 8 ksi if normal weight concrete is used. For lightweight concrete it may not be less than 4 ksi or more than 8 ksi.* The upper limit of 8 ksi is provided because sufficient test data are not available for composite columns with higher-strength concretes at this time. The lower limit of f'_c was specified for the purpose of ensuring the use of good quality but readily available concrete and for the purpose of making sure that adequate quality control is used. This might not be the case if a lower grade of concrete were specified.

4. *The yield stresses of the steel sections and reinforcing bars used may not be greater than 55 ksi. If a steel with a yield stress greater than 55 ksi is actually used in a composite column, only 55 ksi may be used in the calculations.* One major objective in composite design is to prevent local buckling of the longitudinal reinforcing bars and the contained steel section. To achieve this objective the covering concrete must not be allowed to break or spall. It was assumed by the writers of the LRFD Specification that this concrete is in danger of breaking or spalling if its strain reaches 0.0018. If we take this strain and multiply it by E_s we get $(0.0018)(29,000) \approx 55$ ksi. Thus 55 ksi is a limit state for stress in the reinforcing.

5. *The minimum permissible wall thickness of steel tubing filled with concrete is equal to $b\sqrt{F_y/3E}$ for each face of width b of rectangular sections. The minimum thickness for circular sections of outside diameter D is $D\sqrt{F_y/8E}$.* These values are the same as those given in the 1989 ACI Code. It is desired that steel pipe or tubing of sufficient thickness be used so that they will not buckle before yielding.

6. *When composite columns contain more than one steel shape, those shapes have to be connected with lacing, tie plates, or batten plates so that local buckling of the individual shapes is not possible before the concrete hardens.* After the concrete hardens, it is assumed that all the parts of the column act together as a unit in resisting load.

7. It is necessary to avoid the overstressing of either the concrete or the structural steel at connections. As a consequence, LRFD Specification I2.4 requires that *the part of the design strength of axially loaded composite columns resisted by concrete must be developed by direct bearing at connections. Should the supporting concrete be wider on one or more sides than the loaded area and be otherwise restrained against lateral*

expansion on the remaining side or sides, the design compressive strength of the composite column resisted by the concrete $\phi_c P_{nc}$ is to be computed as being equal to $1.7\phi_c f_c' A_B$ with $\phi_c = 0.60$ for bearing on concrete where A_B is the loaded area.

17-6 AXIAL DESIGN STRENGTHS OF COMPOSITE COLUMNS

The contribution of each component of a composite column to its overall strength is difficult if not impossible to determine. The amount of flexural concrete cracking varies throughout the height of the column. The concrete is not nearly as homogeneous as is the steel, and, furthermore, the modulus of elasticity of the concrete varies with time and under the action of long-term or sustained loads. The effective lengths of composite columns in the rigid mono-lithic structures in which they are frequently used cannot be determined very well. The contribution of the concrete to the total stiffness of a composite column varies depending on whether it is placed inside a tube or whether it is on the outside of a W section where its stiffness contribution is less.[8]

The preceding paragraph presented some of the reasons it is difficult to develop a useful theoretical formula for the design of composite columns. As a result, a set of empirical formulas is presented in the LRFD Specification for the design of composite columns.

The design strengths of composite columns ($\phi_c P_n$ with $\phi_c = 0.85$ and $P_n = A_g F_{cr}$) are determined much as are the design strengths of plain steel columns. *The formulas to be used for composite columns for F_{cr}, the critical stress, are the same, except that the areas, radii of gyration, yield stresses, and moduli of elasticity are modified in an attempt to account for composite behavior.* The column expressions given in Section E2 of the Specification and previously described in Chapter 5 of this text are listed as follows:
If $\lambda_c \leq 1.5$

$$F_{cr} = (0.658^{\lambda_c^2})F_y \qquad \text{(LRFD Equation E2-2)}$$

If $\lambda_c > 1.5$

$$F_{cr} = \left(\frac{0.877}{\lambda_c^2}\right)F_y \qquad \text{(LRFD Equation E2-3)}$$

where

$$\lambda_c = \frac{KL}{r\pi}\sqrt{\frac{F_y}{E}} \qquad \text{(LRFD Equation E2-4)}$$

---

[8] Task Group 20, Structural Stability Council, "A Specification for the Design of Steel-Concrete Composite Columns," *Engineering Journal,* AISC 16, no. 4 (fourth quarter, 1979), pp. 101–115.

The modifications made in these formulas are as follows:

1. *Replace A_g with A_s where A_s is the area of the steel shape, tube, or pipe,* but not including any reinforcing bars.
2. *Replace r with r_m where r_m is the radius of gyration of the steel shapes,* pipes, or tubes. For steel shapes encased in concrete it may not be less than 0.3 times the overall thickness of the composite member in the plane of buckling.
3. *Replace F_y with the modified yield stress F_{my} and E with the modified modulus of elasticity E_m.* These values follow:

$$F_{my} = F_y + c_1 F_{yr}\left(\frac{A_r}{A_s}\right) + c_2 f_c'\left(\frac{A_c}{A_s}\right) \qquad \text{(LRFD Equation I2-1)}$$

$$E_m = E + c_3 E_c\left(\frac{A_c}{A_s}\right) \qquad \text{(LRFD Equation I2-2)}$$

In these expressions for F_{my} and E_m the following abbreviations are used:

1. A_c, A_s, and A_r are, respectively, the areas of the concrete, steel section, and reinforcing bars.
2. E and E_c are, respectively, the moduli of elasticity of the steel and the concrete. LRFD Specification I2.2 states that E_c, the modulus of elasticity of the concrete in ksi, may be computed from $w_c^{1.5}\sqrt{f_c'}$ where w_c is the unit weight of the concrete in lbs/ft and f_c' is the specified compression strength of the concrete in ksi.
3. F_y and F_{yr} are the specified minimum yield stresses of the steel section and the reinforcing bars.
4. c_1, c_2, and c_3 are numerical coefficients. For concrete-filled pipes and tubing $c_1 = 1.0$, $c_2 = 0.85$, and $c_3 = 0.4$. For concrete-encased shapes $c_1 = 0.7$, $c_2 = 0.6$, and $c_3 = 0.2$.

The value of $\phi_c P_n$ for a composite column with a W section embedded in concrete is calculated in Example 17-1 in accordance with these LRFD requirements.

■ EXAMPLE 17-1

Compute the value of $\phi_c P_n$ for the composite column shown in Fig. 17-3 if A36 steel, 3.5 ksi concrete with a weight of 145 lb/ft³, and KL of 12 ft are used.

No. 9 bars Grade 60 (1 sq. in. each)

W12 × 72 (A = 21.1 in.²)

Ties No. 3 bars spaced 13 in. o.c.

20 in.

20 in.

Figure 17-3

Solution.   Note: As required by LRFD Specification (I2.1) F_{yr} is used as 55 ksi and not 60 ksi.

$$F_{my} = F_y + c_1 F_{yr}\left(\frac{A_r}{A_s}\right) + c_2 f_c'\left(\frac{A_c}{A_s}\right)$$

$$= 36 + (0.7)(55)\left(\frac{4}{21.1}\right) + (0.6)(3.5)\left(\frac{400 - 25.1}{21.1}\right)$$

$$= 80.61 \text{ ksi}$$

$$E_c = w^{1.5}\sqrt{f_c'} = 145^{1.5}\sqrt{3.5} = 3266.5 \text{ ksi}$$

$$E_m = E + c_3 E_c\left(\frac{A_c}{A_s}\right)$$

$$= 29,000 + (0.2)(3266.5)\left(\frac{400 - 25.1}{21.1}\right)$$

$$= 29,000 + 11,608 = 40,608 \text{ ksi}$$

Replace F_y with 80.61 ksi, E with 40,608 ksi, and A_g with $A_s = 21.1$ in.^{2}

r_y of W12 × 72 = 3.04 in., may not be less than $(0.3)(20) = 6.0$ in.

$$\lambda_c = \frac{KL}{r\pi}\sqrt{\frac{F_y}{E}}$$

$$= \frac{(12)(12)}{6.0\pi}\sqrt{\frac{80.61}{40,608}} = 0.340$$

$$F_{cr} = (0.658^{\lambda_c^2})F_y = (0.658^{0.340^2})80.61 = 76.80 \text{ ksi}$$

$$\phi_c P_n = \phi_c F_{cr} A_s = (0.85)(76.80)(21.1) = 1377.4$$

(LRFD table described in Section 17-7 gives 1380 k)   ■

17-7 LRFD TABLES

The LRFD Manual in Part 5 presents a series of tables for determining the design axial strengths of various square and rectangular encased W shapes, as well as for numerous pipes and structural tubes filled with concrete. The tables are set up in exactly the same fashion as they were for axially loaded plain steel columns in Part 3 of the Manual. The strengths are given with respect to the minor axes for a range of $K_y L_y$ values.

The tables were prepared for normal weight concretes and include encased W shapes with F_y values of 36 and 50 ksi and f_c' values of 3.5, 5, and 8 ksi; 36-ksi pipes filled with 3.5- and 5-ksi concrete; and 46-ksi tubing filled with 3.5 and 5-ksi concrete. The reinforcing bars used for the encased sections are all grade 60 (or 60 ksi).

Examples 17-2 and 17-3, which follow, illustrate the applications of these tables.

■ **EXAMPLE 17-2**

(a) Using A36 steel and 3.5-ksi concrete select a square composite column with a W shape encased in concrete to support a factored load $P_u = 940$ k if $KL = 14$ ft.

(b) Repeat part (a) if a plain W shape is used instead of a composite column.

Solution.

(a) Possible selections from Part 5 of the Manual include:

   18 in. × 18 in. column with W10 × 45 (971 k)

   16 in. × 16 in. column with W8 × 67 (1000 k)

   Use 18 in. × 18 in. column with W10 × 45

(b) Possible selections from Part 3 of Manual include:

   W14 × 120

   W12 × 136

   Use W14 × 120   ■

■ **EXAMPLE 17-3**

An 18-in. × 22-in. composite column with an encased W14 × 61 ($F_y = 36$ ksi, $f'_c = 5$ ksi) has a $K_x L_x = 26$ ft and a $K_y L_y = 16$ ft. Using the composite column tables determine $P_u = \phi_c P_n$.

Solution.   Entering the tables with $K_y L_y = 16$ ft, $P_u = \phi_c P_n$ is found to equal 1440 k. However, $K_x L_x = 26$ ft and $r_x/r_y = 1.22$.

$$\frac{K_x L_x}{r_x/r_y} = \frac{26}{1.22} = 21.31 \text{ ft} > K_y L_y$$

$\therefore K_x L_x$ controls

Reentering tables with $K_y L_y = 21.31$ ft we find P_u is found by interpolation to equal

$$1340 - \left(\frac{1.31}{2}\right)(1340 - 1280) = 1301 \text{ k} \qquad ■$$

It has been shown that the axial load strength of composite columns using W sections greatly exceeds the axial load strengths of plain W sections. The longer the columns become, the greater is the ratio of the strength of composite columns to that of noncomposite ones.

TABLE 17-1   AXIAL DESIGN STRENGTHS

[An A36 W14 × 90 column compared with strengths of that same section composite with a 22 in. × 22 in. reinforced concrete section (f'_c = 3.5 ksi, Grade 60 reinforcing bars)]

Effective Length KL (ft)	Axial design strength of composite section (kips)	Axial design strength of W14 × 90 (kips)	Ratio of strength of composite Section to that of W14 × 90
0	1780	811	2.19
10	1740	767	2.27
20	1600	650	2.46
30	1390	493	2.82
40	1150	334	3.44

The increasing strength advantage of longer composite columns over that of plain steel columns is clearly shown in Table 17-1. In this table the axial design strength of a 22 in. × 22 in. column (with f'_c = 3.5 ksi and Grade 60 reinforcing bars) composite with an A36 W14 × 90 is compared with the axial design strength of a plain A36 W14 × 90 column. The ratio of strength of the composite section to that of the plain section goes from 2.27 for an effective length of 10 ft to 3.44 for an effective length of 40 ft. *Thus we can see that the strength of a composite column falls off at an appreciably lower rate than does that of a plain W column as the effective length increases.*

Example 17-4 presents the design of a pipe column filled with concrete. After a little preliminary scratchwork, the author chooses a trial section and computes its axial load strength using the necessary LRFD expressions. The results are checked with the Part 5 LRFD Manual tables for composite columns.

■ **EXAMPLE 17-4**

Select an A36 pipe column filled with 3.5 ksi concrete (145 lbs/ft³) to support a load P_u = 320 k if KL = 12 ft.

Solution. Assume an 8-in. standard steel pipe. Its properties are shown in Fig. 17-4.

Minimum wall t as required by LRFD I2.1 $= D\sqrt{\dfrac{F_y}{8E}}$

$$= 8.625\sqrt{\dfrac{36}{(8)(29 \times 10^3)}} = 0.107 \text{ in.} < 0.322 \text{ in.} \qquad \text{(OK)}$$

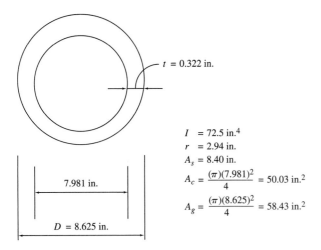

$I = 72.5$ in.^{4}
$r = 2.94$ in.
$A_s = 8.40$ in.
$A_c = \dfrac{(\pi)(7.981)^2}{4} = 50.03$ in.^{2}
$A_g = \dfrac{(\pi)(8.625)^2}{4} = 58.43$ in.^{2}

$t = 0.322$ in.

7.981 in.

$D = 8.625$ in.

Figure 17-4 Properties of 8-in. standard steel pipe.

Checking cross-sectional area of pipe column as a percentage of total composite section area

$$= \frac{8.40}{58.43} = 0.144 > 0.04 \text{ reqd. by LRFD I2.1} \qquad \text{(OK)}$$

Computing modified values of yield stress (F_{my}), modulus of elasticity (F_m), and radius of gyration (r_m). (Note that the second part of LRFD Equation I2-1, which follows, equals zero since there are no longitudinal reinforcing bars and thus $A_r = 0$.)

$$F_{my} = F_y + c_1 F_{yr}\left(\frac{A_r}{A_s}\right) + c_2 f'_c\left(\frac{A_c}{A_s}\right)$$

$$= 36 + 0 + (0.85)(3.5)\left(\frac{50.03}{8.40}\right) = 53.72 \text{ ksi}$$

$$E_c = w_c^{1.5}\sqrt{f'_c} = 145^{1.5}\sqrt{3.5} = 3267 \text{ ksi}$$

$$E_m = E + c_3 E_c\left(\frac{A_c}{A_s}\right) = 29{,}000 + (0.4)(3267)\left(\frac{50.03}{8.40}\right) = 36{,}783 \text{ ksi}$$

$r_m = r$ of pipe $= 2.94$ in., but not less than
$(0.3D) = (0.3)(8.625) = 2.59$ in. per LRFD I2.2.

Slenderness parameter

$$\lambda_c = \frac{KL}{r_m \pi}\sqrt{\frac{F_{my}}{E_m}} = \frac{(12)(12)}{2.94\pi}\sqrt{\frac{53.72}{36{,}783}} = 0.596 < 1.5$$

$$\therefore F_{cr} = (0.658^{\lambda_c^2})F_{my} = 0.658^{(0.596)^2}(53.72) = 46.30 \text{ ksi}$$

Computing $\phi_c P_n$

$$\phi_c P_n = \phi_c A_s F_{cr} = (0.85)(8.40)(46.30) = \underline{330.6 \text{ k}} > 320 \text{ k} \qquad \text{(OK)}$$

Note: LRFD Manual in Part 5 gives 330 k   ■

Section I2.4 of the LRFD Specification requires that the part of the design strength of a composite column resisted by the concrete ($\phi_c P_n - \phi_c P_{ns}$) must be resisted by direct bearing at connections. This supporting strength is equal to $0.85 \, \phi_c f'_c A_B$ if the supporting concrete area is the same size as the column. In this expression A_B is the loaded area and $\phi_c = 0.60$. If the supporting concrete is wider than the loaded area on one or more sides and restrained against lateral expansion on the other sides, the supporting strength is equal to $1.7 \, \phi_c f'_c A_B$.

In Example 17-5 the author computes the bearing area A_B required for load transfer for the composite column of Example 17-1.

■ EXAMPLE 17-5

It is assumed that all of the load for the composite column of Example 17-1 is applied to the column at one level. Determine the bearing area A_B of the concrete as required by Section I2.4 of the LRFD Specification. The supporting concrete is wider than the loaded area on all sides.

Solution.   From the Prob. 17-1 solution $\phi_c P_n = 1377.4$ k

From the LRFD column tables a W12 × 72 (A36 steel) with $KL = 12$ ft can alone support a value $\phi_c P_{ns} = 574$ k.

The design compressive load resisted by the concrete $\phi_c P_{nc}$ must be developed by direct bearing at the connection

$$\phi_c P_{nc} = \phi_c P_n - \phi_c P_{ns} = 1377.4 - 574 = 803.4 \text{ k}$$

Section I2.4 of the LRFD Specification requires that

$$\phi_c P_{nc} \le 1.7 \phi_c f'_c A_B$$

$$\therefore A_B \ge \frac{\phi_c P_{nc}}{1.7 \phi_c f'_c} = \frac{803.4}{(1.7)(0.6)(3.5)} = 225.04 \text{ in.}^2$$

Our composite column has an area of 20 × 20 = 400 in.^{2} > 225.04 in.^{2}
                                                                        (OK)   ■

17-8 FLEXURAL DESIGN STRENGTHS OF COMPOSITE COLUMNS

The nominal flexural strength of composite columns is computed assuming a plastic distribution of stresses. We can locate the plastic neutral axis by equating the tensile forces on one side of the member to the compression force on the other side. On the tensile side there will be reinforcing bars and part of the embedded steel section stressed to their yield stresses. On the compression side there will be a compressive force equal to $0.85 f_c'$ times the area of an equivalent stress block. The equivalent stress block will have a width equal to the column width and a depth equal to β_1 times the distance to the PNA. (The value of β_1 is provided by the ACI Code.) The nominal flexural strength M_n' then equals the sum of the moments of the axial forces about the PNA.

Values of $\phi_b M_{nx}$ and $\phi_b M_{ny}$ are shown for each of the composite columns in Part 5 of the Manual. These values will be needed for analyzing beam-columns as described in the next section.

17-9 AXIAL LOAD AND BENDING EQUATION

The following interaction formulas are used to check plain steel members subject to axial load and bending.
If $P_u/\phi P_n \geq 0.2$

$$\frac{P_u}{\phi P_n} + \frac{8}{9}\left(\frac{M_{ux}}{\phi_b M_{nx}} + \frac{M_{uy}}{\phi_b M_{ny}}\right) \leq 1.0 \qquad \text{(LRFD Equation H1-1a)}$$

If $P_u/\phi P_n < 0.2$

$$\frac{P_u}{2\phi P_n} + \left(\frac{M_{ux}}{\phi_b M_{nx}} + \frac{M_{uy}}{\phi_b M_{ny}}\right) \leq 1.0 \qquad \text{(LRFD Equation H1-1b)}$$

These equations and their application were discussed in Chapter 11. That presentation included definitions of the various values needed to calculate M_{ux} and $M_{uy}(B_1, M_{nt}, B_2,$ and $M_{1t})$.

These same interaction equations are used to determine the adequacy of composite beam-columns except that some of the terms are modified. These modifications follow:

1. The Euler elastic buckling loads P_{ex} and P_{ey} that are used in the calculations of the bending factors B_1 and B_2 are to be determined with the following expression in which F_{my} is the modified yield stress that was defined in Section 17-6. The values of P_{ex} and P_{ey} times the square of the appropriate effective length in feet divided by 10^4 are given in the tables for each of the composite columns.

$$P_e = \frac{A_s F_{my}}{\lambda_c^2}$$

2. The resistance factor ϕ_b is to be used as it is in composite beams where it equals 0.85 if $h/t_w \leq 640/\sqrt{F_{yf}}$ and a plastic stress distribution is used to compute M_n; or it is taken as 0.9 if $h/t_w > 640/\sqrt{F_{yf}}$ and M_n determined by superimposing the elastic stresses.

3. The column slenderness parameter λ_c is to be modified as it was for determining the design strengths of axially loaded composite columns in Section 17-6.

17-10 DESIGN OF COMPOSITE COLUMNS SUBJECT TO AXIAL LOAD AND BENDING

This section is devoted to the design of composite columns to resist axial loads and moments. The procedure is a trial-and-error one involving the selection of a trial section, the application of the appropriate interaction formula, probably the selection of another trial section, the application of the formula, and so on until a satisfactory column is obtained.

A perfectly satisfactory design can be made as described above, but the process can involve quite a few trials if the first estimate is not too good. For this reason a rough method of estimating sizes is presented in the next few paragraphs. This method usually will enable the designer to make a fairly good first size estimate and thus reduce the number of trial designs that have to be made.

For this discussion it is assumed that a composite column is to be designed to support a certain P_u and a certain M_{ux} with M_{uy} equal to zero. It is further assumed that LRFD Equation H1-1a applies to the member. If M_{uy} is zero the formula becomes

$$\frac{P_u}{\phi P_n} + \frac{8}{9} \frac{M_{ux}}{\phi_b M_{nx}} \leq 1.0$$

The designer may estimate the final values of the two parts of this equation. He or she may very well assume the two parts are equal.

$$\frac{P_u}{\phi P_n} = 0.5 \quad \text{and} \quad \frac{8}{9} \frac{M_{ux}}{\phi_b M_{nx}} = 0.5$$

For this discussion it is assumed that a composite column with an encased section with $KL = 12$ ft consisting of A36 steel and 3.5-ksi concrete is to be selected to resist $P_u = 500$ k and $M_{ux} = 100$ ft-k. Estimated values of ϕP_n and $\phi_b M_{nx}$ can be determined as follows:

$$\frac{P_u}{\phi P_n} = 0.5 \qquad \frac{8}{9} \frac{M_{ux}}{\phi_b M_{nx}} = 0.5$$

$$\frac{500}{\phi P_n} = 0.5 \qquad \frac{8}{9} \frac{100}{\phi_b M_{nx}} = 0.5$$

$$\phi P_n = 1000 \text{ k} \qquad \phi_b M_{nx} = 177.8 \text{ ft-k}$$

The designer may then go to the composite tables and try a section that has ϕP_n and $\phi_b M_{nx}$ values somewhere in the range of these values. For instance, an 18 in. × 18 in. encased section with a W10 × 49 has a ϕP_n of 1020 k while a 16 in. × 16 in. encased section with a W8 × 35 has a $\phi_b M_{nx}$ equal to 174 ft-k. As a first trial the designer may like to try a section somewhere in between these two in the tables, as perhaps the 16 in. × 16 in. with a W8 × 48 or W8 × 58. Example 17-6 presents the complete design of this particular member.

■ **EXAMPLE 17-6**

Select a composite column with an encased W section to resist $P_u = 500$ k and $M_{ux} = 100$ ft-k. The column which is braced against sidesway or lateral translation at its ends is to consist of A36 steel and 3.5-ksi concrete. $KL = 12$ ft and $C_m = 0.85$.

Solution.   As described in the paragraphs before this example, assume rough values of $\phi P_n = 1000$ k and $\phi_b M_{nx} = 177.8$ ft-k and try an 16 in. × 16 in. composite column with an encased W8 × 48 (for which $\phi P_n = 870$ k, $\phi_b M_{nx} = 223$ ft-k, and $P_e(K_x L_x)^2/10^4 = 89.3$ ft²-k).

$$P_e = \frac{(89.3)(10)^4}{(12)^2} = 6201.4 \text{ k}$$

with LRFD Equation C1-2

$$B_1 = \frac{C_m}{1 - P_u/P_e} = \frac{0.85}{1 - 1000/6201.4} = 1.013$$

and

$$M_{ux} = (1.013)(100) = 101.3 \text{ ft-k}$$

$$\frac{P_u}{\phi P_n} = \frac{500}{870} > 0.2$$

∴ Must use LRFD Equation H1-1a

$$\frac{P_u}{\phi P_n} + \frac{8}{9}\frac{M_{ux}}{\phi_b M_{nx}} \leq 1.0$$

$$\frac{500}{870} + \frac{8}{9}\frac{100}{223} = 0.575 + 0.399 = 0.974 < 1.0 \qquad \text{(OK)}$$

Use 16 in. × 16 in. column with encased W8 × 48, 4 No. 7 longitudinal bars and No. 3 ties spaced 10 in. o.c.   ■

For Example 17-6 the proportions between the axial load P_u and the bending moment M_{ux} seemed fairly normal. Should the ratio between the two be somewhat different, as where we have a rather large moment in comparison with the axial load, we might change the estimated values in the interaction equation and have a little better luck in the first trials. For instance, if the moment M_{ux} is rather large in comparison with P_u we might assume the following approximate values for the interaction formula and backfigure the values of ϕP_n and $\phi_b M_{nx}$ to look up in the tables.

$$\frac{P_u}{\phi P_n} = 0.3 \qquad \frac{8}{9}\frac{M_{ux}}{\phi_b M_{nx}} = 0.7$$

The rough procedure used here to estimate beam-column sizes doesn't always work as well as it does for the author's two enclosed examples (17-6 and 17-7). Frequently, several trials will still have to be made.

■ **EXAMPLE 17-7**

It is desired to design an encased composite column with A36 steel and 5-ksi concrete to resist a P_u of 600 k and an M_{ux} of 500 ft-k. Assume $KL = 14$ ft and $C_m = 0.85$.

Solution. Assuming that Equation H1-1a will apply. As the moment is quite large in comparison with the axial load, the following values are assumed:

$$\frac{P_u}{\phi P_n} = 0.3 \qquad \frac{8}{9}\frac{M_{ux}}{\phi_b M_{nx}} = 0.7$$

$$\frac{600}{\phi P_n} = 0.3 \qquad \frac{8}{9}\frac{500}{\phi_b M_{nx}} = 0.7$$

$$\phi P_n = 2000 \text{ k} \qquad \phi_b M_{nx} = 634.9 \text{ ft-k}$$

Try a 22 in. × 22 in. section with an encased W14 × 90 ($\phi P_n = 2000$ k, $\phi_b M_{nx} = 691$ ft-k, and $[P_{ex}(K_x L_x)^2]/10^4 = 335$ ft²-k)

$$P_e = \frac{(335)(10)^4}{(14)^2} = 17{,}092 \text{ k}$$

$$B_1 = \frac{0.85}{1 - 600/17{,}092} = 0.881 \quad \underline{\text{Use } 1.0}$$

$$M_{ux} = (1.0)(500) = 500 \text{ ft-k}$$

$$\frac{P_u}{\phi P_n} = \frac{600}{2000} > 0.2$$

∴ Must use LRFD Equation H1-1a

$$\frac{600}{2000} + \frac{8}{9}\frac{500}{691} = 0.3 + 0.643 = 0.943 \tag{OK}$$

Use 22 in. × 22 in. section with an encased W14 × 90, 4 No. 10 longitudinal bars and No. 3 ties at 14 in. o.c. ■

17-11 LOAD TRANSFER AT FOOTINGS AND OTHER CONNECTIONS

A small steel base plate usually is provided at the base of a composite column. Its purpose is to accommodate the anchor bolts needed to anchor the embedded steel shape to the footing for the loads occurring during the erection of the structure before the encasing concrete hardens and composite action is developed. This plate desirably will be sufficiently small so as to be out of the way of the dowels needed for the reinforced-concrete part of the column.[9]

The LRFD Specification does not provide details for the design of these dowels, but a procedure similar to the one provided by the ACI 318-89 Code seems to be a good one. If the column P_u exceeds $1.7 \, \phi_c f_c' A_B$ the excess load should be resisted by dowels. If P_u does not exceed $1.7 \, \phi_c f_c' A_B$ it would appear that no dowels are needed. For such a situation the ACI Code (Sections 15.8.2.1 and 15.8.2.3) states that a minimum area of dowels equal to 0.005 times the cross-sectional area of the column must be used and those dowels may not be larger than No. 11 bars. This diameter requirement ensures sufficient tying together of the column and footing over the whole contact area. The use of a very few large dowels spaced far apart might not do this very well.

PROBLEMS

For all problems grade 60 reinforcing bars and 145 lb/ft³ concrete are used.

17-1 to 17-3. *Determine $\phi_c P_n$ for the composite columns shown using the appropriate LRFD formulas. The steel shapes are A36 and f_c' is 3.5 ksi for the concrete. Column effective lengths, W section sizes, and reinforcing bar numbers are shown with the figures.*

No. 10 bar (1.27 in.²)

W14 × 90
KL = 12 ft

No. 3 bars 14 in. c. to c.

22 in.

22 in.

Problem 17-1 (*Ans.* 1715 k)

---

[9] Griffis, op. cit.

No. 10 bar (1.27 in.²)

W12 × 230
KL = 14 ft

No. 3 bars 14 in. c. to c.

24 in.

22 in.

Problem 17-2

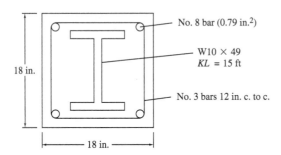

No. 8 bar (0.79 in.²)

W10 × 49
KL = 15 ft

No. 3 bars 12 in. c. to c.

18 in.

18 in.

Problem 17-3 (*Ans.* 989 k)

17-4. Repeat Prob. 17-2 using F_y = 50 ksi.

17-5. Repeat Prob. 17-3 using F_y = 50 ksi and f'_c = 5 ksi. (*Ans.* 1319.5 k)

17-6. Repeat Probs. 17-1, 17-2, and 17-3 using the tables in the LRFD Manual.

17-7 and 17-8. *Compute* $\phi_c P_n$ *for the concrete-filled steel pipe composite columns shown using the appropriate LRFD formulas. The pipes are made from A36 steel and f'_c is 3.5 ksi for the concrete. The section sizes and effective lengths are shown with the figures.*

12 in. pipe (65.42 lb/ft, t_{well} = 0.500 in.)
KL = 16 ft

Problem 17-7 (*Ans.* 763 k)

6 in. pipe (53.16 lb/ft, t_{well} = 0.864 in.)
KL = 15 ft

Problem 17-8

17-9. Repeat Probs. 17-7 and 17-8 using the LRFD tables. (*Ans.* 764 k and 345 k, respectively)

17-10 and 17-11. *Determine* $\phi_c P_n$ *for the concrete-filled square structural tubes shown using the appropriate LRFD formulas. The tubes are made from steel with* F_y = 46 *ksi and concrete with* f'_c = 3.5 *ksi.*

16 in.

16 in.

16 × 16 tube (103.30 lb/ft,
$t_{wall} = \frac{1}{2}$ in.)
KL = 32 ft

Problem 17-10

5 in.

5 in.

5 × 5 tube (22.37 lb/ft,
$t_{wall} = \frac{3}{8}$ in.)
KL = 18 ft

Problem 17-11 (*Ans.* 117 k)

17-12. Repeat Probs. 17-10 and 17-11 using the tables in the LRFD Manual.

17-13. A 20 in. × 20 in. composite column (f'_c = 3.5 ksi) with an encased W12 × 58 (A36) and with one No. 9 bar (1.00 in.^{2}) in each corner is subjected to P_u = 1200 k and M_{ux} = 125 ft-k. If the member has a KL = 14 ft and is braced against sidesway or lateral translation at its ends and if C_m = 0.85, is it satisfactory? (*Ans.* 1.24 > 1.00 NG)

17-14. A 16 in. × 16 in. tube (103.30 lb/ft, $t = \frac{1}{2}$ in., and F_y = 46 ksi) is encased with 3.5-ksi concrete. If the member has a KL of 16 ft and is braced against sidesway with C_m = 0.85 and if P_u is 800 k and M_{ux} is 100 ft-k, is it satisfactory?

17-15. Select a composite column (f'_c = 3.5 ksi) with an encased W section (A36 steel) to resist P_u = 1500 k and M_{ux} = 150 ft-k. The column which is to have a KL = 15 ft is to be braced against lateral translation at its ends and is to have C_m = 0.85. (*One ans:* 22 in. × 22 in. column with a W14 × 109, 4 No. 10 longitudinal bars and No. 3 ties, 14 in. o.c.)

17-16. Select a composite column (f'_c = 5 ksi) with an encased W14 section (A36 steel) to resist P_u = 1800 k and M_{ux} = 200 ft-k. Assume KL = 12 ft and C_m = 0.85. The column is braced against sidesway at its ends.

Chapter **18**

Built-up Beams, Built-up Wide-Flange Sections, and Plate Girders

18-1 COVER-PLATED BEAMS

Should the largest available W section be insufficient to support the loads anticipated for a certain span, several possible alternatives may be taken. Perhaps the most economical solution involves the use of a higher-strength steel W section. If this is not feasible, we may make use of one of the following: (1) two or more regular W sections side-by-side (an expensive solution), (2) a cover-plated beam, (3) a built-up wide-flange section, (4) a plate girder, or (5) a steel truss. This section is concerned with the cover-plated beam alternative. The built-up wide flange is presented in Section 18-2, and the remainder of the chapter is devoted to plate girders.

In addition to being practical for cases where the moments to be resisted are slightly in excess of those that can be supported by the deepest W sections, there are other useful applications for cover-plated beams. On some occasions the total depth may be so limited that the resisting moments of W sections of the specified depth are too small. For instance, the architect may show a certain maximum depth for beams in his or her drawings for a building. In a bridge, beam depths may be limited by clearance requirements. Cover-plated beams will frequently prove to be a very satisfactory solution for situations like these. Furthermore, there may be economical uses for cover-plated beams where the depth is not limited and where there are standard W sections available to support the loads. A smaller W section than required by the maximum moment can be selected and have cover plates attached to its flanges. These plates can be cut off

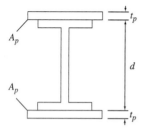

Figure 18-1

where the moments are smaller, with resulting saving of steel. Applications of this type are quite common for continuous beams.

Should the depth be fixed and a cover-plated beam seem to be a feasible solution, the usual procedure will be to select a standard section with a depth that leaves room for top and bottom cover plates. Then the cover plate sizes can be selected.

For this discussion reference is made to Fig. 18-1. For the derivation to follow, Z is the plastic modulus for the entire built-up section, Z_W is the plastic modulus for the W section, d is the depth of the W section, t_p is the thickness of one cover plate, and A_p is the area of one cover plate. An expression for the required area of one plate can be developed as follows:

$$Z \text{ required} = \frac{M_u}{\phi_b F_y}$$

The total Z of the built-up section must at least equal the Z required. It will be furnished by the W shape and the cover plates as follows:

$$Z \text{ required} = Z_W + Z_{\text{plates}}$$
$$= Z_W + 2 A_p\left(\frac{d}{2} + \frac{t_p}{2}\right)$$
$$A_p = \frac{Z \text{ required} - Z_W}{d + t_p}$$

Example 18-1 illustrates the design of a cover-plated beam. Quite a few satisfactory solutions involving different W sections and varying size cover plates are available other than the one made in this problem.

■ **EXAMPLE 18-1**

Select a beam limited to a maximum depth of 29.50 in. for the loads and span of Fig. 18-2. A 50 ksi steel is used and the beam is assumed to have full lateral bracing for its compression flange.

Figure 18-2

Solution.

$$\text{Assume beam wt} = 350 \text{ lb/ft}$$

$$w_u = (1.2)(6.35) + (1.6)(9.5) = 22.82 \text{ k/ft}$$

$$M_u = \frac{(22.82)(40)^2}{8} = 4564 \text{ ft-k}$$

$$Z \text{ required} = \frac{M_u}{\phi_b F_y} = \frac{(12)(4564)}{(0.9)(50)} = 1217.1 \text{ in.}^3$$

The W24 × 408 is the only W section with a depth of 29.50 in. or less that will provide the needed Z. As it is a very heavy section, the author decided to use a cover-plated beam.

Try a W27 × 146 (d = 27.38 in., Z_x = 461 in.^{3}, b_f = 13.965 in.)

Assume 1-in.-thick plates

$$A_p = \frac{Z \text{ required} - Z_W}{d + t_p} = \frac{1217.1 - 461}{27.38 + 1.0} = 26.64 \text{ in.}^2$$

Use W27 × 146 with one PL1 × 28 each flange   ■

18-2 BUILT-UP WIDE-FLANGE SECTIONS

In this section the author initially defines the terms *built-up wide-flange sections* and *plate girders* because the reader may quite understandably find it difficult to distinguish between the two.

In Chapter G of the LRFD Specification a clear distinction is made between beams (whether rolled shapes or built-up wide-flange sections) and plate girders. It is stated that plate girders are to be distinguished from beams on the basis of web slenderness. This slenderness is measured by the ratio h/t_w. The letter h represents (1) the clear distance between flanges less two times the fillet or corner radius for rolled shapes, (2) the distance between adjacent lines of fasteners for built-up sections, or (3) the clear distance between flanges for built-up sections when welds are used. These values are illustrated in Fig. 18-3. Should h/t_w be greater than λ_r (that is, the width-thickness ratio above which a web is classified as being non-compact in LRFD Section B5), the member is classified as a plate girder and the provisions of Appendix G of the LRFD Specification apply. Reference is made here to Fig. 18-4.

A section is classified as a plate girder if h/t_w for its web is $> 970/\sqrt{F_{yf}}$. (Other values apply for webs that are in combined flexural and axial compres-

(a) Rolled section   (b) Bolted built-up section   (c) Welded built-up section
(seldom used today)

Figure 18-3 Values of h.

sion or for members that have different flange sizes. For such situations it is necessary to see LRFD Section B5 and Appendix B5.1.)

You will note in applying the $970/\sqrt{F_{yf}}$ ratio that we use the yield stress of the flange F_{yf} instead of the yield stress of the web F_{yw} because for hybrid girders inelastic buckling of the web due to bending is dependent upon the strain in the flange.

With the LRFD Specification overall economy often can be obtained when built-up wide-flange sections are used instead of plate girders. With these sections the webs selected are sufficiently thick to carry shear without buckling.

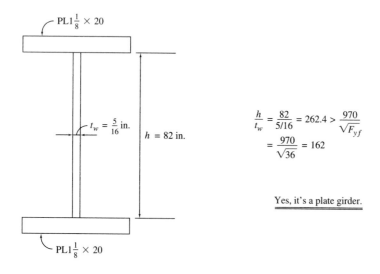

$$\frac{h}{t_w} = \frac{82}{5/16} = 262.4 > \frac{970}{\sqrt{F_{yf}}}$$

$$= \frac{970}{\sqrt{36}} = 162$$

Yes, it's a plate girder.

Figure 18-4 Is it a plate girder if $F_{yf} = 36$ ksi?

Even though these sections with their unstiffened webs will be heavier than plate girders for the same spans and loads, their overall costs will often be less because of their smaller fabrication costs. In addition, design calculations will be appreciably reduced.

According to LRFD Specification F1.2d plastic analysis is permitted for compact beams and girders if they meet certain conditions. They must be singly or doubly symmetric and loaded in the plane of symmetry. When bent about their major axes the laterally unbraced lengths of their compression flanges at plastic hinge locations associated with failure mechanisms must not exceed certain values as given in Chapter F of the Specification.

According to Table B5-1 of the LRFD Specification, a section is compact if

$$\frac{b_f}{2t_f} \le \frac{65}{\sqrt{F_y}} \quad \text{and} \quad \frac{h}{t_w} \le \frac{640}{\sqrt{F_{yf}}}$$

After a compact web size is selected for a built-up wide-flange member, the next step is to select the flange size. For this discussion reference is made to Fig. 18-5. The total design bending strength of the girder shown equals the bending strength of its web plus the bending strength of its flanges.

The plastic modulus of the entire girder equals the statical moment of the compression and tension areas of the web about the neutral axis plus the statical moment of the areas of both flanges about the neutral axis. After such an expression is written it can be solved for the required area of one of the flanges.

$$\text{Total } Z \text{ required for girder} = \frac{M_u}{\phi_b F_y}$$

$$Z \text{ furnished} = (t_w)\left(\frac{h}{2}\right)\left(\frac{h}{4}\right)(2) + (A_p)\left(\frac{h}{2} + \frac{t_f}{2}\right)(2)$$

$$\frac{M_u}{\phi_b F_y} = \frac{t_w h^2}{4} + A_p(h + t_f)$$

$$A_p \text{ required} = \frac{M_u}{\phi_b F_y(h + t_f)} - \frac{t_w h^2}{4(h + t_f)}$$

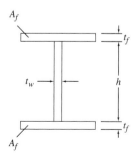

Figure 18-5

■ **EXAMPLE 18-2**

Design a 60-in.-deep welded built-up wide-flange section with no intermediate stiffeners for a 70-ft simple span to support a service dead load of 1 k/ft and a service live load of 2 k/ft. The section is to be framed between columns (thus no end-bearing stiffeners are required as described in Section 18-8) and is to have full lateral bracing for its compression flange. The design is to be made with A36 steel, SMAW fillet welds, and E70 electrodes.

Solution.   Maximum shear and moment

$$\text{Assume beam wt} = 200 \text{ lb/ft}$$
$$w_u = (1.2)(1.2) + (1.6)(2) = 4.64 \text{ k/ft}$$
$$M_u = \frac{(4.64)(70)^2}{8} = 2842 \text{ ft-k}$$
$$V_u = (35)(4.64) = 162.4 \text{ k}$$

Design for a compact web and flange

$$Z \text{ required} = \frac{M_u}{\phi F_y} = \frac{(12)(2842)}{(0.9)(36)} = 1052.6 \text{ in.}^3$$

Web size required

For web to be compact h/t_w must be

$$\leq \frac{640}{\sqrt{F_{yf}}} = \frac{640}{\sqrt{36}} = 107$$

Assume $h = 60 - 2 = 58$ in.

$$\text{Min. } t_w = \frac{58}{107} = 0.542 \text{ in.} \quad \underline{\text{say } \tfrac{9}{16} \text{ in.}}$$

Try $\tfrac{9}{16} \times 58$ web ($A_w = 32.62$ in.²)

$$\frac{h}{t_w} = \frac{58}{9/16} = 103.1$$

Since this value is $>418/\sqrt{F_{yf}} = 70$ (see LRFD Appendix G-4), transverse stiffeners may be needed. However, if $V_u \leq \phi V_n$ as determined from LRFD Appendix Equation A-F2-3 and A-G3-3, stiffeners will not be necessary. (This topic is discussed in Section 18-5.)

Instead of substituting into the above-mentioned formulas it is possible to determine the value of $\phi_v V_n/A_w$ from Table 10-36 in Part 6 of the Manual with $a/h > 3.0$ and $h/t_w = 103.1$ (a is the clear distance between transverse stiffeners).

$$\frac{\phi_v V_n}{A_w} = 11.2 \text{ ksi}$$

$$\frac{V_u}{A_w} = \frac{162.4}{32.62} = 4.98 \text{ ksi} < 11.2 \text{ ksi}$$

$\therefore$ Intermediate stiffeners are not required

Flange area

$$A_f = \frac{M_u}{\phi_b F_y (h + t_p)} - \frac{t_w h^2}{4(h + t_p)}$$

Assume $\frac{3}{4}$-in. plates

$$A_f = \frac{(12)(2842)}{(0.9)(36)(58 + 0.75)} - \frac{(9/16)(58)^2}{(4)(58 + 0.75)} = 9.87 \text{ in.}^2$$

Try $\frac{3}{4} \times 14$ plate each flange (10.5 in.^{2}). Is flange compact?

$$\frac{b_f}{2 t_f} = \frac{14}{(2)(0.75)} = 9.33 < \frac{65}{\sqrt{36}} = 10.8 \qquad\qquad \text{(YES)}$$

Check section

$$Z = (29)(\tfrac{9}{16})(14.5)(2) + (14)(0.75)(29.375)(2)$$
$$= 1090 \text{ in.}^3 > 1052.6 \text{ in.}^3 \qquad\qquad \text{(OK)}$$

Use $\frac{9}{16} \times 58$ web and one $\frac{3}{4} \times 14$ plate each flange   ∎

Comments on Solution

1. On some occasions the tables entitled "Built-Up Wide-Flange Sections" in Part 4 of the Manual may be very useful in designing these types of sections.
2. Should the section selected be noncompact, the beam will be designed as per the requirements of Appendix F1 of the LRFD Specification.

18-3 INTRODUCTION TO PLATE GIRDERS

Plate girders are large I-shaped sections built up from plates and perhaps rolled sections. (Also see Fig. 18-4 of this chapter.) They usually have design strengths somewhere between those of rolled beams and steel trusses. Several possible arrangements are shown in Fig. 18-6. Rather obsolete bolted girders are shown in parts (a) and (b) of the figure, while several welded types are shown in parts (c) through (f). *Since nearly all plate girders constructed today are welded (although they may make use of bolted field splices) this chapter is devoted almost exclusively to welded girders.*

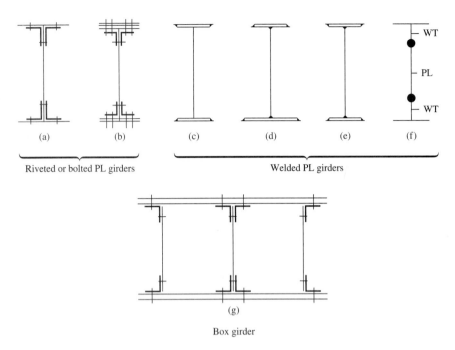

Figure 18-6

The welded girder of part (d) of Fig. 18-6 is arranged to reduce overhead welding as compared with the girder of part (c), but in so doing may be creating a little worse corrosion situation if the girder is exposed to the weather. A box girder, illustrated in part (g), occasionally is used where moments are large and depths are quite limited. Box girders also have great resistance to torsion and lateral buckling. In addition, they make very efficient curved members because of their high torsional strengths.

Plates and shapes can be arranged to form plate girders of almost any reasonable proportions. This fact may seem to give them a great advantage for all situations, but for the smaller sizes the advantage is usually canceled by the higher fabrication costs. For example, it is possible to replace a W36 with a plate girder roughly twice as deep that will require considerably less steel and will have much smaller deflections; however, the higher fabrication costs will almost always rule out such a possibility.

Most steel highway bridges built today for spans less than about 80 ft are steel-beam bridges. For longer spans the plate girder begins to compete very well economically. Where loads are extremely large, as for railroad bridges, plate girders are competitive for spans as low as 45 or 50 ft.

The upper economical limits of plate girder spans depend on several factors, including whether the bridge is simple or continuous, whether a highway or railroad bridge is involved, the largest section that can be shipped in one piece,

Buffalo Bayou Bridge, Houston Tex.—a 270-ft span. (Courtesy of the Lincoln Electric Company.)

etc. Generally speaking, plate girders are very economical for railroad bridges for spans from 50 to 130 ft (15 to 40 m) and for highway bridges for spans from 80 to 150 ft (24 to 46 m). However, they are often very competitive for much longer spans, particularly when continuous. In fact, they are actually common for 200-ft (61-m) spans and have been used for many spans in excess of 400 ft (122 m). The main span of the continuous Bonn-Beuel plate girder bridge over the Rhine River in Germany is 643 ft.

Plate girders are not used only for bridges. They also are fairly common in various types of buildings where they are called upon to support heavy concentrated loads. Frequently, a large ballroom or dining room with no interfering columns is desired on a lower floor of a multistory building. Such a situation is shown in Fig. 18-7. The plate girder shown must support some tremendous column loads for many stories above. The usual building plate girder of this type is simple to analyze because it probably does not have moving loads, although some building girders may be called upon to support traveling cranes.

The usual practical alternative to plate girders in the spans for which they are economical is the truss. In general, plate girders have the following advantages with particular comparison made to trusses.

1. The pound price for fabrication is lower than for trusses but higher than for rolled beam sections.
2. Erection is cheaper and faster than for trusses.
3. Due to their compactness, vibration and impact are not serious problems.

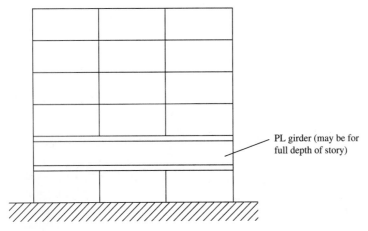

PL girder (may be for full depth of story)

Figure 18-7

4. Plate girders require smaller vertical clearances than trusses.
5. The plate girder has fewer critical points for stresses than do trusses.
6. A bad connection here or there is not as serious as in a truss where such a situation could spell disaster.
7. There is less danger of injury to plate girders in an accident as compared with trusses. Should a truck run into a bridge plate girder it would probably just bend it a little, but a similar accident with a bridge truss member could cause a broken member and perhaps failure.
8. A plate girder is more easily painted than a truss.

On the other hand, plate girders usually are heavier than trusses for the same spans and loads, and they have a further disadvantage in the large number of connections required between webs and flanges.

18-4 PLATE GIRDER PROPORTIONS

Depth

The depths of plate girders vary from about $\frac{1}{6}$ to $\frac{1}{15}$ of their spans, with average values of $\frac{1}{10}$ to $\frac{1}{12}$, depending on the particular conditions of the job. One condition that may limit the proportions of the girder is the largest size that can be fabricated in the shop and shipped to the job. There may be a transportation problem such as clearance requirements that limit maximum depths to 10 or 12 ft along the shipping route.

Web Size

After the total girder depth is estimated, the general proportions of the girder can be established from the maximum shear and the maximum moment. As previously described for I-shaped sections in Section 10-2, the web of a beam carries nearly all of the shearing stress; this shearing stress is assumed by the LRFD Specification to be uniformly distributed throughout the web. The web depth can be closely estimated by taking the total girder depth and subtracting a reasonable value for the depths of the flanges (roughly 1 to 2 in. each). The web depths usually are selected to the nearest even inch, because these plates are not stocked in fractional dimensions.

As a plate girder bends, its curvature creates vertical compression in the web, as illustrated in Fig. 18-8. This is due to the downward vertical component of the compression flange bending stress and the upward vertical component of the tension flange bending stress.

The web must have sufficient vertical buckling strength to withstand the squeezing effect shown in Fig. 18-8. This problem is handled in Appendix G1 of the LRFD Specification by providing maximum permissible web slenderness h/t_w values. The values are dependent upon stiffener spacings as represented by a/h ratios, where a is the clear distance between transverse stiffeners. The purpose of these limiting values is to prevent the vertical buckling of a girder flange into the web before the yield stress is reached in the flange.

If transverse stiffeners are used and spaced no more than $1\frac{1}{2}$ times h, the maximum ratio is

For $a/h \le 1.5$

$$\frac{h}{t_w} \le \frac{2000}{\sqrt{F_{yf}}}$$
(LRFD Equation A-G1-1)

Figure 18-8 Squeezing of plate girder web.

If no transverse stiffeners are used, or if they are used but are spaced at greater distances on center than $1\frac{1}{2}$ times h, the maximum ratio is

For $a/h > 1.5$

$$\frac{h}{t_w} \leq \frac{14{,}000}{\sqrt{F_{yf}(F_{yf} + 16.5)}} \qquad \text{(LRFD Equation A-G1-2)}$$

For this discussion, it is assumed that stiffener spacing is greater than 1.5 times the distance between the flanges. If we use A36 steel our web may not have a depth-thickness or h/t_w ratio less than $970/\sqrt{F_{yf}} = 970/\sqrt{36} = 162$ and still be classified as a plate girder, nor may it be larger than $14{,}000/\sqrt{36(36 + 16.5)} = 322$. Thus, for a trial girder section using A36 steel, a depth-thickness ratio for the web of somewhere between 162 and 322 is tried. From a corrosion standpoint the usual practice is to use some absolute minimum thickness. For bridge girders, $\frac{3}{8}$ in. is a common minimum, while $\frac{1}{4}$ or $\frac{5}{16}$ in. probably are the minimum values for the more sheltered building girders.

The shallower girders probably will be used when the loads are light, and the deeper ones when very large concentrated loads need to be supported, as from the columns in a tall building. If there are no depth restrictions for a

Indiana Harbor Works, East Chicago, Ind. (Courtesy of IR Construction Products Co.)

particular girder it will probably pay the designer to make rough designs and corresponding cost estimates to arrive at a depth decision. (Computer solutions will be very helpful in preparing these alternate designs.)

Flange Size

After the web dimensions are selected, the next step is to select an area of the flange so that it will not be overloaded in bending. The total bending strength of a plate girder equals the bending strength of the flange plus the bending strength of the web. As almost all of the bending strength is provided by the flange, an approximate expression can be developed to estimate the flange area as follows:

$$A_f F_y h \approx M_u$$

From which

$$A_f \approx \frac{M_u}{F_y h}$$

After A_f is computed and trial dimensions are selected, the overall bending strength of the girder is checked as will be described in Sections 18-6 and 18-7.

18-5 DETAILED PROPORTIONS OF WEBS

Buckling Considerations

The depths of plate girders have been previously discussed in Section 18-4 and were said to have average values varying from $\frac{1}{10}$ to $\frac{1}{12}$ of spans. After the total girder depth is assumed, the web depth can be estimated to be from 2 to 4 in. less than the total depth and selected to the nearest inch. A plate-girder web must have sufficient thickness to prevent *vertical buckling of the compression flange*. As the compression flange of a simply supported beam deflects or curves downward, it pushes against the web subjecting it to vertical compression. The amount of this downward force can be estimated to equal the total bending stress in the compression flange times the sine of the angle made by the curved flange with the horizontal.

The total load that can be applied to the web in this manner before the web buckles can be estimated by the Euler formula. Expressing this another way, the Euler formula can be used to determine a limiting height-to-thickness ratio to prevent buckling. Based on such a derivation (including an assumption for residual stress variation) the maximum ratios for webs given in the last section of this chapter were developed.

Web Shear

The LRFD Specification for plate girders permits their design on the basis of postbuckling strength. Designs on this basis provide a more realistic idea of the actual strength of a girder. (Such designs, however, do not necessarily result in better economy because more expensive stiffeners are required.) Should a girder be loaded until initial buckling occurs, it will not collapse because of a phenomenon known as *tension field action.*

After initial buckling a plate girder acts much like a truss. The web acts as a truss with tension diagonals and is able to resist additional shear. A diagonal strip of the web acts similarly to the diagonal of a parallel chorded truss (see Fig. 18-9). The stiffeners keep the flanges from coming together and the flanges keep the stiffeners from coming together. The intermediate stiffeners, which before initial buckling were assumed to resist no load, will after buckling resist compressive loads (or will serve as the compression verticals of a truss) due to diagonal tension. The result is that a plate-girder web probably can resist loads equal to two or three times those present at initial buckling before complete collapse will occur.

Until the web buckles initially, deflections are relatively small. However, after initial buckling the girder's stiffness decreases considerably, and deflections may increase to several times the values estimated by the usual deflection theory.

The estimated total or ultimate shear that a panel (a part of the girder between a pair of stiffeners) can withstand equals the shear initially causing web buckling plus the shear that can be resisted by tension field action. The amount of tension field action is dependent on the proportion of the panels.

The design shear strength of a plate girder is equal to $\phi_v V_n$ with $\phi_v = 0.90$ and V_n equal to the appropriate value determined from the various formulas given in LRFD Appendix G3 as follows:

If $h/t_w \leq 187 \sqrt{k_v/F_{yw}}$

$$V_n = 0.6 A_w F_{yw} \qquad\qquad\text{(LRFD Equation A-G3-1)}$$

If $h/t_w > 187 \sqrt{k_v/F_{yw}}$

$$V_n = 0.6 A_w F_{yw}\left[C_v + \frac{1 - C_v}{1.15\sqrt{1 + (a/h)^2}}\right] \qquad\text{(LRFD Equation A-G3-2)}$$

Figure 18-9 Tension field action in plate-girder web.

For the end panels of nonhybrid girders, for all the panels of hybrid girders, for web-tapered girders, and when $a/h > 3.0$ or $> [260(h/t_w)]^2$, tension field action is not permitted and

$$V_n = 0.6 A_w F_{yw} C_v \qquad \text{(LRFD Equation A-G3-3)}$$

The terms k_v and C_v used in the preceding formulas are defined as follows:

1. k_v is a web plate-buckling coefficient

$$= 5 + \frac{5}{(a/h)^2} \qquad \text{(LRFD Equation A-G3-4)}$$

except that it is to be used $= 5.0$ if $a/h > 3.0$ or $> [260/(h/t_w)]^2$

2. C_v is the ratio of the "critical" web stress (according to the linear buckling theory) to the shear yield stress of the web material as follows:
If $187 \sqrt{k_v/F_{yw}} \le h/t_w \le 234 \sqrt{k_v/F_{yw}}$

$$C_v = \frac{187\sqrt{k_v/F_{yw}}}{h/t_w} \qquad \text{(LRFD Equation A-G3-5)}$$

If $h/t_w > 234 \sqrt{k_v/F_{yw}}$

$$C_v = \frac{44,000 \, k_v}{(h/t_w)^2 \, F_{yw}} \qquad \text{(LRFD Equation A-G3-6)}$$

18-6 DESIGN OF PLATE GIRDERS WITH SLENDER WEBS, BUT WITH FULL LATERAL BRACING FOR THEIR COMPACT COMPRESSION FLANGES

In Example 18-3 a plate girder with full lateral bracing for its compression flange is selected. As the girder proportions are not compact, a plastic analysis is not permissible. Transverse stiffeners are considered in detail later in this chapter, and thus their design is not included in this example or the next.

If we have a plate girder with a slender web (that is, with $h/t_w > 970/\sqrt{F_{yf}}$), its flexural strength is the lesser value computed from the limit states of tension-flange yielding and buckling, as determined by the formulas to follow:

$$M_u = \phi_b M_n \quad \text{with} \quad \phi_b = 0.9$$

For tension-flange yielding

$$M_n = S_{xt} R_e F_{yt} \qquad \text{(LRFD Equation A-G2-1)}$$

For compression flange buckling

$$M_n = S_{xc} R_{PG} R_e F_{cr} \qquad \text{(LRFD Equation A-G2-2)}$$

Swinging a plate girder into place on Goat Island Bridge across the Niagara River, Niagara Falls, N.Y. (Courtesy of Bethlehem Steel Corporation.)

in which

$$R_{PG} = 1 - \frac{a_r}{1,200 + 300a_r}\left(\frac{h_c}{t_w} - \frac{970}{\sqrt{F_{cr}}}\right) \leq 1.0$$

(LRFD Equation A-G2.3)

R_e is a hybrid girder factor given in Appendix G2 of the Manual. It is to be taken as 1.0 for nonhybrid girders.

a_r is the ratio of the web area to the compression flange area ≤ 10.

F_{cr} is the critical compression flange stress as determined in Appendix G-2 of the Specification. It equals F_{yf} if the compression flange is fully braced laterally, that is, if $\lambda \leq \lambda_p$.

S_{xc} and S_{xt} are the section moduli values referred to the compression and tension flanges respectively.

h_c = two times the distance from the centroid of the girder to the nearest line of fasteners at the compression flange, or the inside face of the compression flange if welds are used.

■ EXAMPLE 18-3

Design a welded plate girder of A36 steel for a 70-ft span to support a uniform service dead load of 2 k/ft and a uniform service live load of 3.5 k/ft. The compression flange will be laterally braced for its entire length. The girder is to be simply supported at its ends and is not framed into columns. Do not consider stiffener design.

Solution.   Maximum shear and moment

Assume plate girder weight = 250 lb/ft

$w_u = (1.2)(2.25) + (1.6)(3.5) = 8.3 \text{ k/ft}$

$M_u = \dfrac{(8.3)(70)^2}{8} = 5084 \text{ ft-k}$

$V_u = (35)(8.3) = 290.5 \text{ k}$

Preliminary web design

Assume girder depth = $(\frac{1}{10})(12 \times 70) = 84$ in.

Assume web depth = $84 - 2 = 82$ in.

Minimum web t_w if we assume $a/h > 1.5$

$$= \frac{82}{322} = 0.255 \text{ in.}$$

Maximum t_w for noncompact web (to be classified as plate girder) = $82/162 = 0.506$ in. or 0.50 in.

Select minimum $t_w = \frac{5}{16}$ in. (rather than 0.25 in., because of corrosion minimum)

$$\frac{h}{t_w} = \frac{82}{5/16} = 262.4$$

Preliminary flange design

Use the following formula to approximate flange area

$A_f = \dfrac{M_u}{F_y h} \approx \dfrac{(12)(5084)}{(36)(82)} = 20.67 \text{ in.}^2$

Try $1\frac{1}{8} \times 20$ plate ($A_f = 22.5$ in.^{2}). Is it compact?

$$\frac{b_f}{2t_f} = \frac{20}{(2)(1\frac{1}{8})} = 8.89 < \frac{65}{\sqrt{36}} = 10.8 \qquad \text{(YES)}$$

Check trial girder section

$I = (\frac{1}{12})(\frac{5}{16})(82)^3 + (2)(1\frac{1}{8} \times 20)(41.56)^2 = 92{,}084 \text{ in.}^4$

$S_{xt} = \dfrac{92{,}084}{42.125} = 2186 \text{ in.}^3$

As $\dfrac{h}{t_w} = 262.4 > \dfrac{970}{\sqrt{36}} = 162$ Appendix G2 applies

$a_r = \dfrac{(\frac{5}{16})(82)}{(1\frac{1}{8})(20)} = 1.139$

$R_{PG} = 1 - \dfrac{1.139}{1{,}200 + (300)(1.139)}\left(262.4 - \dfrac{970}{\sqrt{36}}\right) = 0.926$

Design flexural strength for compression-flange buckling

$$M_n = S_{xt} R_{PG} R_e F_{yt}$$
$$= (2186)(0.926)(1.0)(36) = 72,872 \text{ in.-k} = 6073 \text{ ft-k}$$
$$M_u = \phi_b M_n = (0.9)(6073) = 5465 \text{ ft-k} > 5084 \text{ ft-k} \qquad \text{(OK)}$$

Use $\frac{5}{16} \times 82$ web with one $1\frac{1}{8} \times 20$ plate for each flange

Note: A check of shear values for this girder will show that transverse stiffeners are required. ∎

18-7 DESIGN OF PLATE GIRDERS WITH NONCOMPACT FLANGES AND WITHOUT FULL LATERAL BRACING FOR COMPRESSION FLANGES

As indicated in Section 18-6 the critical flange stress F_{cr} for plate girders with slender webs ($h/t_w > 970/\sqrt{F_{yf}}$), compact flanges, and with full lateral bracing supplied for the compression flange equals F_{yf}. Should these latter two conditions not be provided, it is necessary to consider the limit states of lateral-torsional buckling and local flange buckling as described below and to use *the lower value obtained.*

If $\lambda \le \lambda_p$

$$F_{cr} = F_{yf} \qquad \text{(LRFD Equation A-G2-4)}$$

If $\lambda_p < \lambda \le \lambda_r$

$$F_{cr} = C_b F_{yf}\left[1 - \frac{1}{2}\left(\frac{\lambda - \lambda_p}{\lambda_r - \lambda_p}\right)\right] \le F_{yf} \qquad \text{(LRFD Equation A-G2-5)}$$

If $\lambda > \lambda_r$

$$F_{cr} = \frac{C_{PG}}{\lambda^2} \qquad \text{(LRFD Equation A-G2-6)}$$

For the preceding equations it is necessary to determine the slenderness parameter for the limit state of lateral-torsional buckling as well as the limit state of local buckling. The slenderness parameter that results in the lowest value of F_{cr} governs.

For the limit state of lateral torsional buckling the following values are to be used in Formulas A-G2-5 and A-G2-6:

$$\lambda = \frac{L_b}{r_T} \qquad \text{(LRFD Equation A-G2-7)}$$

$$\lambda_p = \frac{300}{\sqrt{F_{yf}}} \qquad\qquad \text{(LRFD Equation A-G2-8)}$$

$$\lambda_r = \frac{756}{\sqrt{F_{yf}}} \qquad\qquad \text{(LRFD Equation A-G2-9)}$$

C_{PG} = plate girder coefficient

$\qquad = 286{,}000C_b \qquad\qquad$ (LRFD Equation A-G2-10)

$$C_b = \frac{12.5M_{max}}{2.5M_{max} + 3M_A + 4M_B \; 3M_C} \qquad \text{(LRFD Equation F1-3)}$$

r_T = radius of gyration of compression flange plus one-third of the compression part of the web.

For the limit state of flange local buckling the following values are to be used:

$$\lambda = \frac{b_f}{2t_f} \qquad\qquad \text{(LRFD Equation A-G2-11)}$$

$$\lambda_p = \frac{65}{\sqrt{F_{yf}}} \qquad\qquad \text{(LRFD Equation A-G2-12)}$$

$$\lambda_r = \frac{230}{\sqrt{F_{yf}/k_c}} \qquad\qquad \text{(LRFD Equation A-G2-13)}$$

$C_{PG} = 26{,}200 \; k_c \qquad\qquad$ (LRFD Equation A-G2-14)

$C_b = 1.0$

In these expressions $k_c = 4/\sqrt{h/t_w}$ and $0.35 \leq k_c \leq 0.763$.

Example 18-4 illustrates the selection of the cross section for a plate girder for which full lateral bracing is not provided for the compression flange.

■ EXAMPLE 18-4

The cross section for a welded plate girder is to be selected for a 54-ft span with simple end supports to support a uniform service dead load of 2 k/ft and concentrated service live loads of 150 k each at its one-third points. The design is to be made with A36 steel assuming that lateral bracing is provided for the compression flange only at the ends and one-third points. Do not design stiffeners.

Solution. Maximum shear and moment

Assume beam weight = 400 lb/ft

$\qquad W_u = (1.2)(2.4) = 2.88$ k/ft

$\qquad P_u = (1.6)(150) = 240$ k

Drawing shear and moment diagrams, (Fig 18-10)

Figure 18-10

Preliminary web design

Assume girder depth $= (\frac{1}{10})(12 \times 54) = 64.8$ in.   Say 64 in.

Assume web depth $= 64 - 2 = 62$ in.

Minimum web t_w if we assume $\dfrac{a}{h} > 1.5 = \dfrac{62}{322} = 0.193$ in.

Maximum t_w for noncompact web (to be classed as plate girder) $= 62/162 = 0.383$ in.

Select $t_w = \frac{3}{8}$ in. noting a corrosion minimum of about $\frac{5}{16}$ in.

$$\frac{h}{t_w} = \frac{62}{3/8} = 165.3$$

Preliminary flange design

Use the following formula to approximate flange area

$$A_f = \frac{M_u}{F_{yh}} = \frac{(12)(5369)}{(36)(62)} = 28.87 \text{ in.}^2$$

Try $1\frac{1}{4} \times 24$ plate ($A_f = 30$ in.^{2}). Is it compact?

$$\frac{b_f}{2t_f} = \frac{24}{(2)(1\frac{1}{4})} = 9.6 < \frac{65}{\sqrt{36}} = 10.8 \qquad\qquad \text{(YES)}$$

Check trial girder section

Computing section modulus

$$I = (\tfrac{1}{12})(3/8)(62)^3 + (2)(1\tfrac{1}{4})(24)(31.625)^2 = 67,456 \text{ in.}^4$$

Arc-welded plate girders, Louisiana State Fair Grounds, Shreveport, La. (Courtesy of the Lincoln Electric Company.)

$$S = \frac{67,456}{32.25} = 2091.7 \text{ in.}^3$$

$$\frac{h}{t_w} = \frac{62}{3/8} = 165.3 > \frac{970}{\sqrt{36}} = 162$$

$\therefore$ Appendix G applies

Properties of compression flange plus one-third of compression part of web

$$A_f + \tfrac{1}{6} A_w = (1.25)(24) + (\tfrac{1}{6})(\tfrac{3}{8} \times \tfrac{62}{2}) = 31.94 \text{ in.}^2$$

$$I_y = (\tfrac{1}{12})(1.25)(24)^3 = 1440 \text{ in.}^4$$

$$r_T = \sqrt{\frac{1440}{31.94}} = 6.71 \text{ in.}$$

Checking limitations of Appendix G. If $a/h \le 1.5$,

$$\frac{h}{t_w} \text{ maximum} = \frac{2000}{\sqrt{36}} = 333 > 165.3 \quad (\text{OK}) \quad (\text{LRFD Equation A-G1-1})$$

Checking bending design strength of center 18-ft section.

For the limit state of lateral-torsional buckling

$$\lambda = \frac{L_b}{r_T} = \frac{(12)(18)}{6.71} = 32.19 \qquad \text{(LRFD Equation A-G2-7)}$$

λ_p = limiting slenderness parameter for compact elements

$$= \frac{300}{\sqrt{F_{yf}}} = \frac{300}{\sqrt{36}} = 50 > 32.19 \qquad \text{(LRFD Equation A-G2-8)}$$

$$\therefore F_{cr} = F_{yf} = 36.0 \text{ ksi}$$

For the limit state of flange local buckling

$$\lambda = \frac{b_f}{2t_f} = \frac{24}{(2)(1.25)} = 9.6 \qquad \text{(LRFD Equation A-G2-11)}$$

$$\lambda_p = \frac{65}{\sqrt{F_{yf}}} = \frac{65}{\sqrt{36}} = 10.8 > 9.6 \qquad \text{(LRFD Equation A-G2-12)}$$

$$\therefore F_{cr} = F_{yf} = 36.0 \text{ ksi}$$

Design flexural strength

$$a_r = \frac{\text{web area}}{\text{compression flange area}} = \frac{(3/8)(62)}{(1.25)(24)} = 0.775$$

F_{cr} = 36 as per LRFD Equation A-G2-4

$$R_{PG} = 1 - \frac{0.775}{1200 + (300)(0.775)} \left(\frac{2 \times 31}{0.375} - \frac{970}{\sqrt{36}} \right) = 0.998$$

$$\text{(LRFD Equation A-G2-3)}$$

$$M_n = (2091.7)(0.998)(1.0)(36) = 75{,}151 \text{ in.-k} = 6263 \text{ ft-k}$$

$$\text{(LRFD Equation A-G2-2)}$$

$$M_u = \phi_b M_n = (0.9)(6263) = 5637 \text{ ft-k} > 5369 \text{ ft-k} \qquad \text{(OK)}$$

Check bending strength of end 18-ft section

$$M_{\max} = M @ 18 \text{ ft} = 5253 \text{ ft-k}$$
$$M_A = M @ 4.5 \text{ ft} = (317.76)(4.5) - (4.5)(2.88)(2.25) = 1401 \text{ ft-k}$$
$$M_B = M @ 9.0 \text{ ft} = (317.76)(9.0) - (9.0)(2.88)(4.5) = 2743 \text{ ft-k}$$
$$M_C = M @ 13.5 \text{ ft} = (317.76)(13.5) - (13.5)(2.88)(6.75) = 4027 \text{ ft-k}$$
$$C_b = \frac{(12.5)(5253)}{(2.5)(5253) + (3)(1401) + (4)(2743) + (3)(4027)}$$
$$= 1.62 \qquad \text{(LRFD Equation F1-3)}$$

Obviously, bending strength in this segment is satisfactory

For the limit state of lateral-torsional buckling

$$\lambda = \frac{L_b}{r_T} = \frac{(12)(18)}{6.71} = 33.19 \qquad \text{(LRFD Equation A-G2-7)}$$

$$\lambda_p = \frac{300}{\sqrt{F_{yf}}} = \frac{300}{\sqrt{36}} = 50 > 33.19 \qquad \text{(LRFD Equation A-G2-8)}$$

$\therefore F_{cr} = 36.0$ ksi

For the limit state of flange local buckling. As it was for the center 18-ft section $F_{cr} = F_{yf} = 36.0$ ksi and $R_{PG} = 0.998$.

$$\phi_b M_n = 5637 \text{ ft-k} > 5369 \text{ ft-k} \qquad \text{(OK)}$$

Use $\frac{3}{8} \times 62$ web and one $1\frac{1}{4} \times 24$ plate each flange ∎

18-8 DESIGN OF STIFFENERS

As previously indicated, it usually is necessary to stiffen the high thin webs of plate girders to keep them from buckling. If the girders are bolted the stiffeners probably will consist of a pair of angles bolted to the girder webs. If the girders are welded (the normal situation) the stiffeners probably will consist of a pair of plates welded to the girder webs. Figure 18-11 illustrates these types of stiffeners.

Stiffeners are divided into two groups: *bearing stiffeners,* which transfer heavy reactions or concentrated loads to the full depth of the web, and *intermediate or nonbearing stiffeners,* which are placed at various intervals along the

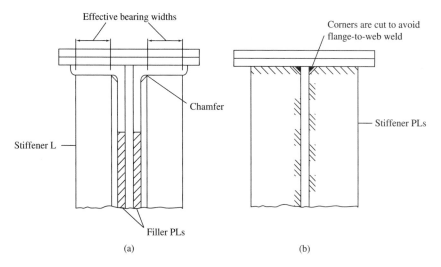

Figure 18-11 (a) Angle bearing stiffeners. (b) Plate-bearing stiffeners.

web to prevent buckling due to diagonal compression. An additional purpose of bearing stiffeners is the transfer of heavy loads to plate girder webs without putting all the loads on the flange connections.

As described in Appendix F2.3 of the LRFD Specification, transverse stiffeners may be either single or double. They do not have to be connected to the flanges except in the following situations:

1. Where bearing strength is needed to transmit concentrated loads or reactions.
2. Where single stiffeners are used and the flange of the girder consists of a rectangular plate. For such a situation the stiffener must be attached to the flange to resist any possible uplift tendency that may be caused by torsion in the flange.
3. Where lateral bracing is attached to a stiffener or stiffeners. Such a stiffener must be connected to the compression flange with a strength sufficient to transmit no less than 1 percent of the total flange stress unless the flange is composed only of angles.

Welds used to attach stiffeners to girder webs must be terminated at distances not less than 4 times the web thickness nor more than 6 times the web thickness from the near toe of the flange to the web weld.

If bolts are used to connect stiffeners to the web, they may not be spaced further apart than 12 inches. If intermittent fillet welds are used, the clear distance between them may not exceed the lesser of 16 times the web thickness or 10 inches.

Stiffeners that are located at concentrated loads or reactions have some special requirements because of the possibility of web crippling or compression buckling of the web. In these situations the stiffeners need to be designed as columns. Should the load or reaction be tensile, it will be necessary to weld the stiffeners to the loaded flange. If the force is compressive the stiffener may either bear against the loaded flange or be welded to it.

Bearing Stiffeners

Bearing stiffeners are placed in pairs on the webs of plate girders at unframed girder ends and where required for concentrated loads. They should fit tightly against the flanges being loaded and should extend out toward the edges of the flange plates or angles as far as possible. If the load normal to the girder flange is tensile, the stiffeners must be welded to the loaded flange. If the load is compressive it is necessary to obtain a snug fit, that is, a good bearing between the flange and the stiffeners. To accomplish this goal the stiffeners may be welded to the flange or the outstanding legs of the stiffeners may be milled.

A bearing stiffener is a special type of column that is difficult to accurately analyze because it must support the load in conjunction with the web. The amount of support provided by these two elements is difficult to estimate. The LRFD Specification (K1.9) states that the factored load or reaction may not exceed the design strength of a column consisting of the stiffener effective area plus a portion of the web equal to 12 t_w at girder ends and 25 t_w at interior concentrated loads. Only the part of the stiffeners outside of the fillets of the flange angles or the parts outside of the flange to web welds (see Fig. 18-11) are to be considered effective to support the bearing loads. The effective length of these "bearing stiffener columns" is assumed by the LRFD Specification (K1.9) as being equal to 0.75 h.

At an unframed girder end an end-bearing stiffener is required if the factored reaction R_u is larger than ϕR_n. If an interior load or reaction is larger than the same ϕR_n an interior bearing stiffener is needed.

If the concentrated force is applied at a distance from the member end greater than the member depth d

$$R_n = (5k + N)F_{yw}t_w \qquad \text{(LRFD Equation K1-2)}$$

If the concentrated force is applied at a distance from the member end less than or equal to the member depth d

$$R_n = (2.5k + N)F_{yw}t_w \qquad \text{(LRFD Equation K1-3)}$$

In these expressions

k = distance from outer face of flange to web toe of fillet, in.

N = length of bearing (not less than k for end reaction), in.

ϕ = 1.0

Intermediate Stiffeners

Intermediate or nonbearing stiffeners that also are called stability or transverse intermediate stiffeners are not required by LRFD Specification Appendix G4 if $h/t_w \leq 418/\sqrt{F_{yw}}$ or if the factored shear V_u is $\leq 0.6\phi_v A_w F_{yw} C_v$ where C_v is computed with the appropriate expression to follow, using a value of $k_v = 5.0$ and $\phi_v = 0.90$.

If $187\sqrt{\dfrac{k_v}{F_{yw}}} \leq \dfrac{h}{t_w} \leq 234\sqrt{\dfrac{k_v}{F_{yw}}}$

$$C_v = \frac{187\sqrt{k_v/F_{yw}}}{h/t_w} \qquad \text{(LRFD Equation A-G3-5)}$$

If $\dfrac{h}{t_w} > 234\sqrt{\dfrac{k_v}{F_{yw}}}$

$$C_v = \frac{44{,}000\,k_v}{(h/t_w)^2\,F_{yw}} \qquad \text{(LRFD Equation A-G3-6)}$$

The LRFD Specification also imposes some additional arbitrary limits on panel aspect ratios (a/h) for plate girders, even where shear stresses are small. The purpose of these limitations is to facilitate the handling of girders during fabrication and erection. The Specification (Appendix G3) states that tension field action is not permitted for end panels in nonhybrid plate girders, for all panels in hybrid and web tapered plate girders, and when $a/h > 3.0$ or $>[260/(h/t_w)]^2$ and that

$$V_n = 0.6 A_w F_{yw} C_v \qquad \text{(LRFD Equation A-G3-3)}$$

You should note here that as the intermediate stiffener spacing is decreased, C_v will become larger, as will the shear capacity of the girder.

The LRFD Specification (Appendix F2.3) states that the moment of inertia of a transverse intermediate stiffener about an axis at the center of the girder web if a pair of stiffeners is used, or about the face in contact with the web when single stiffeners are used, may not be less than

$$I_{st} \text{ minimum } = a t_w^3 j$$

where $j = \dfrac{2.5}{(a/h)^2} - 2 \geq 0.5$ (LRFD Equation A-F2-4)

The Appendix to the LRFD Specification (G4) states that, when designing for tension field action, the stiffener area A_{st} may not be less than the value given by the following formula:

$$A_{st} \text{ minimum } = \frac{F_{yw}}{F_{yst}}\left[0.15 \, Dht_w(1 - C_v)\frac{V_u}{\phi_v V_n} - 18 \, t_w^2 \right] \geq 0$$

$$\text{(LRFD Equation A-G4-1)}$$

In the preceding formula D is to be taken equal to 1.0 for stiffeners in pairs, 1.8 for single-angle stiffeners, and 2.4 for single-plate stiffeners. The terms C_v and V_n are defined in Appendix G3, while V_u is the factored shear at the section in question.

The values given for I_{st} minimum, j, and A_{st} minimum are the same as those required by the AASHTO in its specification for load factor design for steel bridges.[1]

Longitudinal Stiffeners

Longitudinal stiffeners, though not as effective as transverse ones, frequently are used for bridge plate girders because many designers feel they are more attractive. As such stiffeners are seldom used for plate girders in buildings, they are not covered in this textbook.

---

[1] *Standard Specifications for Highway Bridges,* 14th ed., 1989 (Washington, D. C.: American Association of State Highway and Transportation Officials), p. 145.

Highway bypass bridge, Stroudsburg, Pa. (Courtesy of Bethlehem Steel Corporation.)

18-9 FLEXURE-SHEAR INTERACTION

When a plate girder whose web depends on tension field action is subjected to relatively large shear and bending moment at the same location, the girder cannot provide its full capacity either in shear or in moment. As a result, an empirical interaction equation is used to check the adequacy of the girder. This equation, which follows, agrees quite well with test results.

If $0.6\ \phi V_n \le V_u \le \phi V_n$, and if $0.75\phi M_n \le M_u \le \phi M_n$ with $\phi = 0.90$, plate girders designed for tension field action also must satisfy the following interaction equation.

$$\frac{M_u}{\phi M_n} + 0.625\ \frac{V_u}{\phi V_n} \le 1.375 \qquad \text{(LRFD Equation A-G5-1)}$$

In the preceding equation M_n is the nominal flexural strength of the girder as determined by Appendix G2, ϕ is 0.90, and V_n is the nominal shear strength of the girder as determined by Appendix G3. The value of M_u may not exceed ϕM_n, nor may V_u exceed ϕV_n, with $\phi = 0.90$ in both cases.

Bearing and intermediate stiffeners are designed in Example 18-5 for the plate girder of Example 18-4. In addition, the girder is checked for flexure-shear interaction at several locations.

■ **EXAMPLE 18-5**

Design bearing and intermediate stiffeners for the plate girder of Example 18-4. Also check flexure-shear interaction at points 6, 12, and 18 ft from the girder ends.

Solution.

End Bearing Stiffeners

(a) Are end bearing stiffeners required?

Assume point bearing of the end reaction (that is, $N = 0$) and assume a $\frac{5}{16}$-in. web-to-flange fillet weld.

Check local web yielding

$$k = \text{distance from outer edge of flange to web end of fillet weld}$$
$$= 1\tfrac{1}{4} + \tfrac{5}{16} = 1.56 \text{ in.}$$
$$R_n = (5k + N)F_{yw}t_w \qquad\qquad \text{(LRFD Equation K1-2)}$$
$$\phi R_n = (1.0)(5 \times 1.56 + 0)(36)(\tfrac{3}{8}) = 105.3 \text{ k}$$
$$< 317.76 \text{ k} \therefore \text{ End bearing stiffeners are required}$$

If local web yielding check had been satisfactory (that is, $\phi R_n \geq 317.76$ k), it would have then been necessary also to check the criteria set forth in Sections K1.4 and K1.5 of the LRFD Specification before we could definitely say that end bearing stiffeners were not required.

(b) Design of end bearing stiffeners

Try two plate stiffeners $\frac{3}{4} \times 11$ as shown in Fig. 18-12.

Check width-thickness ratio (Table B5.1 Specification)

$$\frac{11}{3/4} = 14.67 < \frac{95}{\sqrt{36}} = 15.83 \qquad\qquad \text{(OK)}$$

Check column strength of stiffener (area is shown crosshatched in following sketch)

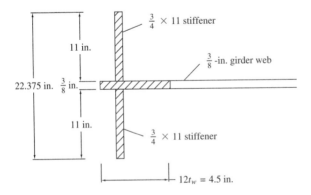

Figure 18-12 End bearing stiffeners.

$$I = (\tfrac{1}{12})(\tfrac{3}{4})(22.375)^3 = 700 \text{ in.}^4$$

(neglecting the almost infinitesimal contribution of the web)

$$A \text{ of column} = (2)(\tfrac{3}{4})(11) + (4.5)(\tfrac{3}{8}) = 18.19 \text{ in.}^2$$

$$r = \sqrt{\frac{700}{18.19}} = 6.20 \text{ in.}$$

$$KL = (0.75)(62) = 46.5 \text{ in.}$$

$$\frac{KL}{r} = \frac{46.5}{6.20} = 7.50$$

$\phi F_{cr} = 30.51$ ksi from Table 3-36 of Part 6 of Manual or can be determined using LRFD Formulas E2-2 or E2-3 as applicable.

$$\phi P_n = (30.51)(18.19) = 555.0 \text{ k} > 317.76 \text{ k} \tag{OK}$$

Check bearing criterion

$$\phi R_n = (0.75)(2.0 \, F_y A_{pd})$$

A_{pd} = projected bearing area of stiffeners subtracting $\tfrac{1}{2}$ in. from the 11-in. width of each stiffener for the cutouts for the web-to-flange welds as shown in Fig. 18-11 (b).

$$A_{pd} = (2)(11 - 0.5)(\tfrac{3}{4}) = 15.75 \text{ in.}$$

$$\phi R_n = (0.75)(2.0)(36)(15.75) = 850 \text{ k} > 317.76 \text{ k} \tag{OK}$$

Use two plates $\tfrac{3}{4} \times 11 \times 5$ ft $1\tfrac{3}{4}$ in. for bearing stiffeners (a depth of $61\tfrac{3}{4}$ in. is used instead of 62 in. for fitting purposes)

(c) Design of interior bearing stiffeners

The design procedure for interior bearing stiffeners is the same as for end bearing stiffeners, except a 25 t_w length of the web may be counted as part of the "column," as compared with the 12 t_w value used for the end stiffeners. Here the same size stiffeners are used at the concentrated loads as at the end reactions.

Intermediate Stiffeners

(a) Are intermediate stiffeners required?

$$\frac{h}{t_w} = \frac{62}{3/8} = 165.3 > \frac{418}{\sqrt{36}} = 70$$

$\therefore$ Intermediate stiffeners are needed.

(b) Is tension field action permitted?

Tension field action is not permitted in the end panels of nonhybrid girders according to LRFD Appendix G3.

(c) Spacing of intermediate stiffeners by trial and error

The author assumes a certain distance from the end of the girder to the first stiffener, computes ϕV_n to see if it's sufficiently large to resist the computed shear V_u, tries another spacing, and so on until $\phi V_n \geq V_u$.

After trials of 72 in., 54 in., 48 in., and 45 in. (all being unsatisfactory), a 42-in. distance is tried below.

Try 42 in. to first stiffener

$$\frac{a}{n} = \frac{42}{62} = 0.677$$

$$j = \frac{2.5}{(0.677)^2} - 2 = 3.45 > 0.5 \qquad \text{(OK)} \quad \text{(LRFD Equation A-F2-4)}$$

$$k_v = 5 + \frac{5}{(0.677)^2} = 15.91 \qquad \text{(LRFD Equation A-G3-4)}$$

$$C_v = \frac{(44,000)(15.91)}{(165.3)^2(36)} = 0.712 \qquad \text{(LRFD Equation A-G3-6)}$$

$$V_n = (0.6)(\tfrac{3}{8} \times 62)(36)(0.712) = 357.6 \text{ k} \quad \text{(LRFD Equation A-G3-3)}$$

$$\phi V_n = (0.90)(357.6) = 321.8 \text{ k} > 317.76 \text{ k} \qquad\qquad \text{(OK)}$$

Place first stiffener 42 in. from girder end

The shear V_u has fallen off slightly in this 42 in. from the girder end, as can be seen in Fig. 18-10. [It's $317.76 - (\frac{42}{12})(2.88) = 307.68$ k.] Thus, we can try a slightly larger spacing and go through these steps again to determine ϕV_n. It is much simpler, however, to use Table 9-50 in Part 6 of the LRFD Manual to determine stiffener spacings, and this is done below.

(d) Spacings of intermediate stiffeners using LRFD Manual

$$\frac{\phi_v V_n}{A_w} = \frac{V_u}{A_w} = \frac{317.76}{(3/8)(62)} = 13.67 \text{ ksi}$$

$$\frac{h}{t_w} = \frac{62}{3/8} = 165.3$$

From Table 9-50 in Manual, $\frac{a}{h}$ can be determined to equal 0.684 by double interpolation.

$$a_1 = (0.684)(62) = 42.41 \text{ in.}$$

Use 42 in. to first stiffener

Are additional stiffeners required?

V_u at first interior stiffener $= 317.76 - (\frac{42}{12})(2.88) = 307.68$ k

$$\frac{V_u}{A_w} = \frac{307.68}{3/8 \times 62} = 13.23 \text{ ksi} > 4.35 \text{ ksi}$$

$\frac{a_2}{h}$ by double interpolation $= 0.702$

$$a_2 = (0.702)(62) = 43.52 \text{ in.}$$

$\therefore$ Spacing continues to increase

Figure 18-13 Stiffener spacing diagram

Use 1 stiffener at 42 in. and remainder at $\dfrac{(12)(18) - 42}{4} = 43.5$ in. o.c. See Fig. 18-13.

(e) Check center 18-ft panel

$$\frac{h}{t_w} = 165.3$$

$$\frac{a}{h} = \frac{(12)(18)}{62} = 3.48 > \left[\frac{260}{h/t_w}\right]^2 = 2.47$$

$$k = 5.0; \; C_v = \frac{187\sqrt{5.0/36}}{165.3} = 0.422$$

$$V_n = (0.6)(\tfrac{3}{8} \times 62)(36)(0.422) = 211.9 \text{ k}$$

$$\frac{\phi V_n}{A_w} = \frac{(0.9)(211.9)}{3/8 \times 62} = 8.20 \text{ ksi}$$

$$\frac{V_u}{A_w} = \frac{25.92}{(3/8)(62)} = 1.11 \text{ ksi} < 8.20 \text{ ksi}$$

∴ Intermediate stiffeners not required

(f) Design of intermediate stiffeners

Stiffener design with no tension field action means that we have to comply with the minimum moment of inertia expression, $at_w^3 j$ given in Appendix F2.3.

$j = 3.45$ for 42 in. spacing

Min $I_{st} = at_w^3 j = (42)(\tfrac{3}{8})^3(3.45) = 7.64$ in.^{4}

There are many possible satisfactory stiffener sizes. We will try two of them by trial and error.

Try a $\frac{1}{2} \times 6$ single plate stiffener

$I_{st} = (\frac{1}{3})(\frac{1}{2})(6)^3 = 36$ in.$^4 > 7.64$ in.^{4} (OK)

Try a pair of $\frac{1}{2} \times 4\frac{1}{2}$ plate stiffeners

$I_{st} = (\frac{1}{12})(\frac{1}{2})(9.375)^3 = 34.33$ in.$^4 > 7.64$ in^4.   (OK)

Use $\frac{1}{2} \times 6$ single plate stiffeners

(g) Check flexure-shear interaction

Computing the values of V_u, M_u, and other values needed to substitute into interaction formula A-G5-1 at points 6, 12, and 18 ft from the end of the span.

Distance from end support (ft)	V_u (kips)	M_u (ft-k)	V_n (kips)	M_n (ft-k)	$\dfrac{0.6\ V_n}{M_n}$	$\dfrac{V_u}{M_u}$	$\dfrac{V_n}{0.75\ M_n}$
6	300.5	1854.7	353	6263	0.0338	0.162	0.0780
12	283.2	3605.8	353	6263	0.0338	0.079	0.0780
18	265.9	5253.1	353	6263	0.0338	0.051	0.0780

Since $0.6\ V_n/M_n < V_u/M_u > V_n/0.75\ M_n$, it is not necessary to check flexure-shear interaction.   ■

PROBLEMS

18-1. Select a cover-plated W section limited to a maximum depth of 18.00 in. to support the service loads shown in the accompanying figure. Use A36 steel and assume the beam has full lateral bracing for its compression flange. (*Ans.* One solution W16 $\times$ 36 with 1 PL $1\frac{1}{2} \times 14$ each flange)

D = 3 k/ft (not including beam weight)
L = 4 k/ft

30 ft

Problem 18-1

18-2. The architects specify that a beam no greater than 16.50 in. in depth be designed to support a dead service uniform load of 4 klf, not including the beam weight, and a live service uniform load of 6 klf for a 24-ft simple span. A572 steel with $F_y = 50$ ksi is to be used, and full lateral bracing for the compression flange is to be provided.

18-3. Design a 48-in.-deep welded built-up wide-flange section with no intermediate stiffeners for a 50-ft simple span to support a service dead load of 1 klf (not including the beam weight) and a service live load of 1.8 klf. The section is to be framed between columns and is to have full lateral bracing for its compression flange. The design is to be made with A36 steel. (*Ans.* One solution $\frac{7}{16} \times 46$ in. web with one $\frac{5}{8} \times 10$ in. PL each flange)

18-4. Repeat Prob. 18-3 if the span is to be 60 ft and if A572 steel ($F_y = 50$ ksi) is to be used.

18-5. Design a welded plate girder with A36 steel for a 60-ft span to support a uniform service dead load of 2.0 klf (not including the girder weight) and a service live load of 4.5 klf. The compression flange will be laterally braced for its entire length. The girder is to be simply supported at its ends and is not framed into the columns. Do not design stiffeners. (*Ans.* One solution $\frac{5}{16} \times 70$ web with one PL1 $\times$ 22 each flange)

18-6. Repeat Prob. 18-5 if A572 steel ($F_y = 50$ ksi) is used.

18-7. The cross section for a welded plate girder is to be selected for the span and service loads shown in the accompanying illustration. The design is to be made with A36 steel assuming that lateral bracing is provided for the compression flange at the ends and concentrated load points only. Do not design stiffeners. (*Ans.* One solution $\frac{5}{16} \times 58$ web and one 2 $\times$ 26 cover plate each flange)

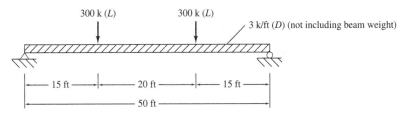

Problem 18-7

18-8. Repeat Prob. 18-7 using A572 steel ($F_y = 50$ ksi).

Chapter **19**

Design of Steel Buildings

19-1 INTRODUCTION TO LOW-RISE BUILDINGS

The material in this section and the next pertains to the design of low-rise steel buildings up to several stories in height, while Sections 19-3 to 19-10 provide information concerning common types of building floors and Section 19-11 presents common types of roof construction. Sections 19-12 and 19-13 are concerned with walls, partitions, and fireproofing. The sections thereafter present general information relating to high-rise or multistory buildings.

The low-rise buildings considered include apartment houses, office buildings, warehouses, schools, and institutional buildings that are not very tall with respect to their least lateral dimensions. The separating factor between the buildings discussed here and those mentioned in the multistory sections is the matter of wind forces and not the actual height of the building in question. The usual rule of thumb is that if the height of the building is not greater than twice its least lateral dimension, provision for wind forces is unnecessary. For buildings of these dimensions, the walls and partitions probably provide sufficient resistance to wind forces, except in unusual circumstances. Following such a rule of thumb, however, is a dangerous engineering practice. Each structure should be considered on the basis of its dimensions, location, surrounding structures, and so on. Then a decision can be made as to whether a definite system of lateral bracing needs to be used.

19-2 TYPES OF STEEL FRAMES USED FOR BUILDINGS

Steel buildings usually are classified as being in one of four groups according to their type of construction. These types are bearing-wall construction, skeleton construction, long-span construction, and combination steel and concrete framing. More than one of these construction types can be used in the same building. Each of these types is briefly discussed in the following paragraphs.

Coliseum in Spokane, Wash. (Courtesy of Bethlehem Steel Corporation.)

Bearing-Wall Construction

Bearing-wall construction is the most common type of single-story light commercial construction. The ends of beams or joists or light trusses are supported by the walls that transfer the loads to the foundation. The old practice was to rapidly thicken load-bearing walls as buildings became taller. For instance, the wall on the top floor of a building might be one or two bricks thick while the lower walls might be increased by one brick thickness for each story as we come down the building. As a result, this type of construction was usually thought to have an upper economical limit of about two or three stories. A great deal of research has been conducted for load-bearing construction in recent decades, and it has been discovered that thin load-bearing walls may be quite economical for many buildings up to 10 or 20 or even more stories.

The average engineer is not very well versed in the subject of bearing-wall construction, with the result that he or she may often specify complete steel or reinforced-concrete frames where bearing-wall construction might have been just as satisfactory and more economical. Bearing-wall construction is not very resistant to seismic loadings and does have an erection disadvantage for buildings of more than one story, however. For such cases, it is necessary to place the steel floor beams and trusses floor by floor as the masons complete their work below, thus requiring alternation of the masons and ironworkers.

Bearing plates usually are necessary under the ends of the beams or light trusses that are supported by the masonry walls, because of the relatively low bearing strength of the masonry. Although theoretically the beam flanges may

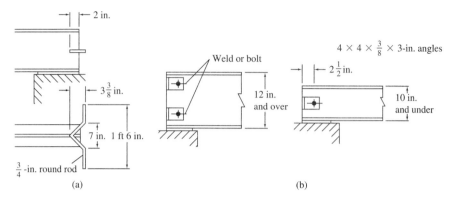

Figure 19-1 (a) Government anchor. (b) Angle wall anchors. From American Institute of Steel Construction, *Manual of Steel Construction Load & Resistance Factor Design,* 2 ed., (Chicago: AISC, 1994), p. 12-24. Reprinted with the permission of AISC.

on many occasions provide sufficient bearing without bearing plates, the plates almost always are used—particularly where the members are so large and heavy that they must be set by a steel erector. The plates usually are shipped loose and set in the walls by the masons. Setting them in their correct positions and at the correct elevation is a critical part of the construction. Should they not be properly set, there will be some delay in correcting their positions. If a steel erector is used, he or she probably will have to make an extra trip to the job.

When the ends of a beam are enclosed in a masonry wall, some type of wall anchor is desirable to prevent the beam from moving longitudinally with respect to the wall. The usual anchors consist of bent steel bars passing through beam webs. These so-called *government anchors* are shown in Fig. 19-1(a). Occasionally, clip angles attached to the web are used instead of government anchors. These are shown in Fig. 19-1(b). Should longitudinal loads of considerable magnitude be anticipated, regular vertical anchor bolts may be used at the beam ends.

For small commercial and industrial buildings bearing-wall construction is quite economical when the clear spans are not greater than roughly 35 or 40 ft. If the clear spans are much greater, it becomes necessary to thicken the walls and use pilasters to ensure stability. For these cases it may often be more economical to use intermediate columns if permissible.

Skeleton Construction

In skeleton construction the loads are transmitted to the foundations by a framework of steel beams and columns. The floor slabs, partitions, exterior walls, and so on, all are supported by the frame. This type of framing, which can be erected to tremendous heights, often is referred to as beam-and-column construction.

In beam-and-column construction the frame usually consists of columns spaced 20, 25, or 30 ft apart, with beams and girders framed into them from

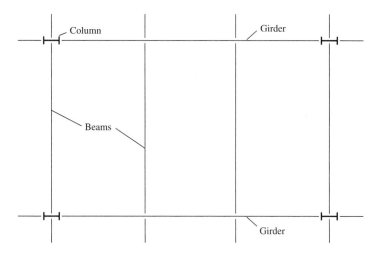

Figure 19-2 Beam-and-column construction.

both directions at each floor level. One very common method of arranging the members is shown in Fig. 19-2. The girders are placed in the longer direction between the columns, while the beams are framed between the girders in the short direction. With various types of floor construction, other arrangements of beams and girders may be used.

For skeleton framing the walls are supported by the steel frame and generally are referred to as *nonbearing* or *curtain walls.* The beams supporting the exterior walls are called *spandrel beams.* These beams, illustrated in Fig. 19-3, usually can be placed so that they will serve as the lintels for the windows.

Figure 19-3 Spandrel beams.

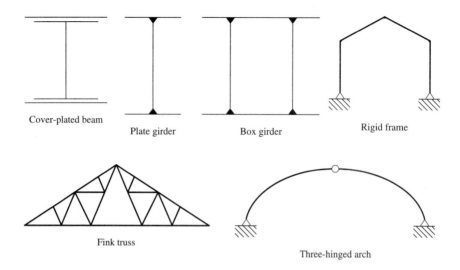

Cover-plated beam

Plate girder   Box girder   Rigid frame

Fink truss

Three-hinged arch

Figure 19-4 Long-span structures.

Long-Span Steel Structures

When it becomes necessary to use very large spans between columns, as for field houses, auditoriums, theaters, hangars, or hotel ballrooms, the usual skeleton construction may not be sufficient. Should the ordinary rolled W sections be insufficient, it may be necessary to use cover-plated beams, plate girders, box girders, large trusses, arches, rigid frames, and the like. When depth is limited, cover-plated beams, plate girders, or box girders may be called upon to do the job. Should depth not be so critical, trusses may be satisfactory. For very large spans, arches and rigid frames often are used. These various types of structures are referred to as *long-span structures*. Fig. 19-4 shows a few of these types of structures.

Combination Steel and Concrete Framing

A tremendous percentage of the buildings erected today make use of a combination of reinforced concrete and structural steel. If reinforced-concrete columns were used in very tall buildings, they would be rather large on the lower floors and take up considerable space. Steel column shapes surrounded by and bonded to reinforced concrete may be used and are referred to as combination or composite columns. These were discussed in Chapter 17.

19-3 COMMON TYPES OF FLOOR CONSTRUCTION

Concrete floor slabs of one type or another are used almost universally for steel-frame buildings. They are strong and have excellent fire ratings and good acoustic ratings. On the other hand, appreciable time and expense are required to provide the formwork necessary for most slabs. Concrete floors are heavy, they must include some type of reinforcing bars or mesh, and there may be a problem involved in making them watertight. Among the many types of concrete floors used today for steel-frame buildings are the following:

1. Concrete slabs supported with open-web steel joists (Section 19-4)
2. One-way and two-way reinforced-concrete slabs supported on steel beams (Section 19-5)
3. Concrete slab and steel beam composite floors (Section 19-6)
4. Concrete-pan floors (Section 19-7)
5. Steel-decking floors (Section 19-8)
6. Flat slab floors (Section 19-9)
7. Precast concrete slab floors (Section 19-10)

Among the several factors to be considered in selecting the type of floor system to be used for a particular building are loads to be supported; fire rating desired; sound and heat transmission; dead weight of floor; ceiling situation below (to be flat or have beams exposed); facility of floor for locating conduits, pipes, wiring, and so on; appearance; maintenance required; time required to construct; and depth available for floor.

The student can obtain a great deal of information about these and other construction practices by referring to various engineering magazines and catalogs, particularly *Sweet's Catalog File* published by McGraw-Hill Information Systems Company of New York. The author cannot make too strong a recommendation for the student to examine these books to see the tremendous amount of data available there that may be of use in his or her engineering practice. The sections to follow present brief descriptions of the floor systems mentioned in this section, along with some discussions of their advantages and chief uses.

19-4 CONCRETE SLABS ON OPEN-WEB STEEL JOISTS

Perhaps the most common type of floor slab in use for small steel-frame buildings is the slab supported by open-web steel joists. The joists are small parallel chord trusses whose members often are made from bars (hence the common name *bar joist*) or small angles or other rolled shapes. Steel forms or decks are

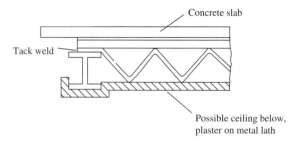

Figure 19-5 Open-web joists.

usually attached to the joists by welding or self-drilling or self-tapping screws, then concrete slabs are poured on top. This is one of the lightest types of concrete floors and also one of the most economical. A sketch of an open-web joist floor is shown in Fig. 19-5.

Open-web joists are particularly well suited to building floors with relatively light loads and for structures where there is not too much vibration. They have been used a good deal for fairly tall buildings, but generally speaking they are better suited for shorter buildings. They are very satisfactory for supporting floor and roof slabs for schools, apartment houses, hotels, office buildings, restaurant buildings, and similar low-level buildings. These joists are generally not suitable for supporting concentrated loads, however, unless they are especially detailed to carry those loads.

Open-web joists must be braced laterally to keep them from twisting or buckling and to keep the floors from being too springy. Lateral support is provided by *bridging,* which consists of continuous horizontal rods fastened to the top and bottom chords of the joists or of diagonal cross bracing. Bridging is desirably used at spaces not exceeding 7 ft on centers.

Open-web joists are easy to handle and are quickly erected. If desired, a ceiling can be attached to the bottom of the joists or suspended therefrom. The open spaces in the webs are admirably suited for placing conduits, ducts, wiring, piping, and so on. The joists should be either welded to supporting steel beams or well anchored in the masonry walls. When concrete slabs are placed on top of the joists, they are usually from 2 to $2\frac{1}{2}$ in. thick. Nearly all of the many concrete slabs cast in place or precast on the market today can be used successfully on top of open-web joists.

These joists are selected from manufacturers' catalogs. The student can learn a great deal about open-web joists by studying various manufacturers' literature. An exceptionally useful reference of this type is the previously mentioned *Sweet's Catalog File.*

The joists are designated as J-Series when the material used has a minimum yield point of 36 ksi (248 MPa), and as H-Series when the chord material consists of steel with a minimum yield point of 50 ksi (345 MPa). There are also long-span joists L-J Series and the deep long-span joists (DLJ Series with $F_y = 36$ ksi) intended for roof construction. The DLH Series are deep long-span joists with minimum yield points of 50 ksi. In the tables the number preceding the joist number is the nominal depth of the joist in inches, the letter or letters indicate the joist series, and the number to the right designates the chord section.

The tables provided in the manufacturers' catalogs give the safe uniform loads in pounds per foot that the joists can support. Also provided are the maximum end reaction in pounds that a joist can support from the standpoint of shear and the maximum uniform live load in pounds per foot that it can support without having a deflection larger than $\frac{1}{360}$ of the span.

19-5 ONE-WAY AND TWO-WAY REINFORCED-CONCRETE SLABS

One-Way Slabs

A very large number of concrete floor slabs in old office and industrial buildings consisted of one-way slabs about 4 in. thick supported by steel beams 6 to 8 ft on centers. These floors were often referred to as concrete arch floors because at one time brick or tile floors were constructed in approximately the same shape, that is, in the shape of arches with flat tops.

A one-way slab is shown in Fig. 19-6. The slab spans in the short direction shown by the arrows in the figure. One-way slabs usually are used when the long direction is two or more times the short direction. In such cases the short span is so much stiffer than the long span that almost all of the load is carried by the short span. The short direction is the main direction of bending and will be the direction of the main reinforcing bars in the concrete, but temperature and shrinkage steel is needed in the other direction.

Figure 19-6 One-way slab.

Welded wire mesh for bonding
to the fireproofing concrete.
Rarely are steel beams encased
in concrete as shown here due
to the great expense.

Figure 19-7 Cross section of one-way slab floor.

A typical cross section of a one-way slab floor with supporting steel beams
is shown in Fig. 19-7. When steel beams or joists are used to support reinforced-
concrete floors, it may be necessary to encase them in concrete or other materials to provide the required fire rating. Such a situation is shown in the figure.

It may be necessary to leave steel lath protruding from the bottom flanges
or soffits of the beam for the purposes of attaching plastered ceilings. Should

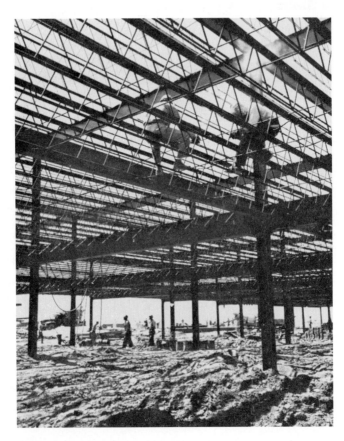

Short-span joists in the Univac Center of Sperry-Rand Corporation, Blue Bell, Pa. (Courtesy of Bethlehem Steel Corporation.)

such ceilings be required to cover the beam stems, this floor system will lose a great deal of its economy.

One-way slabs have an advantage when it comes to formwork in that the forms can be supported entirely by the steel beams with no vertical shoring needed. They have a disadvantage in that they are much heavier than most of the newer lightweight floor systems. The result is that they are not used as often as formerly for lightly loaded floors; but when a floor to support heavy loads, a rigid floor, or a very durable floor are desired, the one-way slab may be a very appropriate selection.

Two-Way Concrete Slabs

The two-way concrete slab is used when the slabs are square or nearly so, with supporting beams under all four edges. The main reinforcing runs in both directions. Other characteristics are similar to those of the one-way slab.

19-6 COMPOSITE FLOORS

Composite floors, previously discussed in Chapter 16, have steel beams (rolled sections, cover-plated beams, or built-up members) bonded together with concrete slabs in such a manner that the two act as a unit in resisting the total loads that the beam sections would otherwise have to resist alone. The steel beams can be smaller when composite floors are used, because the slabs act as part of the beams.

A particular advantage of composite floors is that they utilize concrete's high compressive strength by keeping all or nearly all of the concrete in compression, and at the same time stress a larger percentage of the steel in tension than is normally the case with steel-frame structures. The result is less steel tonnage in the structure. A further advantage of composite floors is an appreciable reduction in total floor thickness, which is particularly important in taller buildings.

Two types of composite floor systems are shown in Fig. 19-8. The steel beam can be completely encased in the concrete and the horizontal shear trans-

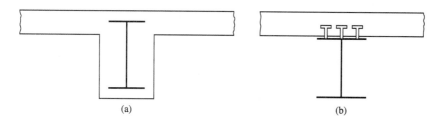

(a)   (b)

Figure 19-8 Composite floors. (a) Steel beam encased in concrete (very expensive), (b) Steel beam bonded to concrete slab with shear connectors.

The 104-ft joists for the Bethlehem Catholic High School, Bethlehem, Pa. (Courtesy of Bethlehem Steel Corporation.)

ferred by friction and bond (plus some shear reinforcement if necessary). *This type of composite floor is usually quite uneconomical.* The usual type of composite floor is shown in part (b), where the steel beam is bonded to the concrete slab with some type of shear connectors. Various types of shear connectors have been used during the past few decades, including spiral bars, channels, angles, studs, and so on, but economic considerations usually lead to the use of round studs welded to the top flanges of the beams in place of the other types mentioned. Typical studs are $\frac{1}{2}$ to $\frac{3}{4}$ in. in diameter and 2 to 4 in. in length.

Cover plates may occasionally be welded to the bottom flanges of rolled steel sections. The student can see that with the slab acting as a part of the beam there is quite a large area available on the compressive side of the beam. By adding plates to the tensile flange a little better balance is obtained.

19-7 CONCRETE-PAN FLOORS

There are several types of pan floors that are constructed by placing concrete in removable pan molds. (Some special light corrugated pans also are available that can be left in place.) Rows of the pans are arranged on wooden floor forms, and the concrete is placed over the top of them, producing a floor cross section

One-piece metal dome pans. (Courtesy of Gateway Erectors, Inc.)

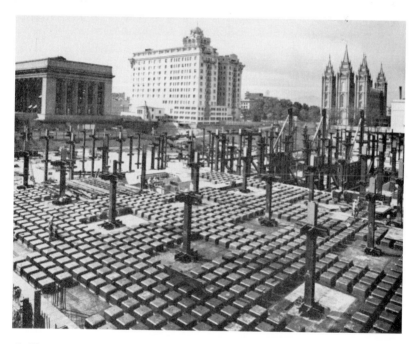

Temple Plaza parking facility, Salt Lake City, Utah. (Courtesy of Ceco Steel Products Corporation.)

Figure 19-9 Concrete-pan floor.

such as the one shown in Fig. 19-9. Joists are formed between the pans, giving a tee-beam-type floor.

These floors, which are suitable for fairly heavy loads, are appreciably lighter than the one-way and two-way concrete slab floors. They require a good deal of formwork, including appreciable shoring underneath the stems. Labor is thus higher than for many floors, but saving due to weight reduction and reuse of standard-size pans may make them competitive. If suspended ceilings are required, pan floors will have a decided economic disadvantage.

Two-way construction is available—that is, with ribs or stems running in both directions. Pans with closed ends are used, and the result is a waffle-type floor. This type of floor usually is used when the floor panels are square or nearly so. Two-way construction can be obtained for reasonably economical prices and makes a very attractive ceiling below with fairly good acoustical properties.

19-8 STEEL-DECKING FLOORS

Typical cross sections of steel-decking floors are shown in Fig. 19-10; several other variations are available. *Today, formed steel decking with a concrete topping is by far the most common type of floor system used for office and apartment buildings.* It also is popular for hotels and other buildings where the loads are not very large.

Figure 19-10 Steel-decking floors.

A particular advantage of steel-decking floors is that the decking immediately forms a working platform. The light steel sheets are quite strong and can span up to 20 ft or more. Due to the considerable strength of the decking the concrete does not have to be particularly strong. This permits the use of lightweight concrete as thin as 2 or $2\frac{1}{2}$ in.

The cells in the decking are convenient for conduits, pipes, and wiring. The steel usually is galvanized, and if exposed underneath can be left as it comes from the manufacturer or painted as desired. Should fire resistance be necessary, a suspended ceiling with metal lath and plaster may be used. The same is necessary if a flat ceiling is required below for the types of decking shown in Fig. 19-10, but decking is available with a flat soffit.

19-9 FLAT SLABS

Formerly, flat slab floors were limited to reinforced-concrete buildings, but today it is possible to use them in steel-frame buildings. A flat slab is reinforced in two or more directions and transfers its loads to the supporting columns without the use of beams and girders protruding below. The supporting concrete beams and girders are made so wide that they are the same depth as the slab.

Flat slabs are of great value when the panels are approximately square, when more headroom is desired than is provided with the normal beam and girder floors, when heavy loads are anticipated, and when it is desired to place the windows as near to the tops of the walls as possible. Another advantage is the flat ceiling produced for the floor below. Although the large amounts of reinforcing steel required cause additional costs, the simple formwork cuts expenses decidedly. The significance of simple formwork will be understood when it is realized that over one-half of the cost of the average poured reinforced-concrete floor slab is in the formwork.

For some reinforced-concrete frame buildings with flat slab floors, it is necessary to flare out the tops of the columns, forming column capitals, and perhaps thicken the slab around the column with the so-called drop panels. These items, which are shown in Fig. 19-11, may be necessary to prevent shear or punching failures in the slab around the column.

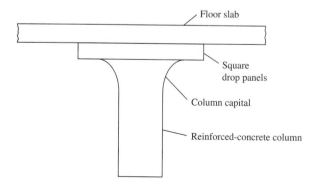

Figure 19-11 A flat slab floor for a reinforced-concrete building.

It is possible today in steel-frame buildings to use short steel cantilever beams connected to the steel columns and embedded in the slabs. These beams serve the purposes of the flared columns and drop panels in ordinary flat slab construction. This arrangement often is called a *steel grillage* or *column head.*

The flat slab is not a very satisfactory type of floor system for the usual tall building where lateral forces (wind or earthquake) are appreciable, because protruding beams and girders are desirable to serve as part of the lateral bracing system.

19-10 PRECAST CONCRETE FLOORS

Precast concrete sections are more commonly associated with roofs than they are with floor slabs, but their use for floors is increasing. They are quickly erected and reduce the need for formwork. Lightweight aggregates are often used in the concrete, making the sections light and easy to handle. Some of the aggregates used make the slabs nailable and easily cut and fitted on the job. For floor slabs with their fairly heavy loads, the aggregates should be of a quality that will not greatly reduce the strength of the resulting concrete.

101 Hudson St., Jersey City, N.J. (Courtesy of Owen Steel Company, Inc.)

Concrete plank   Hollow-cored slab   Prefabricated concrete block system   Channel slab

18–24 in.   5–9 in. $1\frac{1}{2}$ in.

Figure 19-12 Precast concrete roof and floor slabs.

The reader again is referred to *Sweet's Catalog File* at this point. In these catalogs a great amount of information is available on the various types of precast floor slabs on the market today. A few of the common types available are listed below, and a cross section of each type mentioned is presented in Fig. 19-12. Due to slight variations in the upper surfaces of precast sections, it usually is necessary to use a mortar topping of 1 to 2 in. before asphalt tile or other floor coverings can be installed.

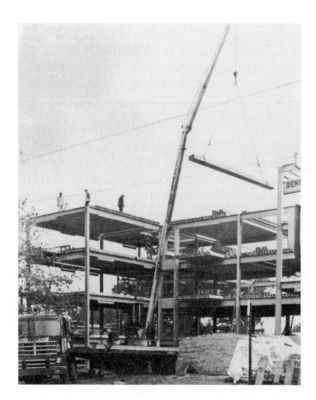

Interns' living quarters, Good Samaritan Hospital, Dayton, Ohio. (Courtesy of Flexicore Company, Inc.)

1. *Precast concrete planks* are roughly 2 to 3 in. thick and 12 to 24 in. wide and are placed on joists up to 5 or 6 ft on center. They usually have tongue-and-groove edges.
2. *Hollow-cored slabs* are sections roughly 6 to 8 in. deep and 12 to 18 in. wide and have hollow, perhaps circular, cores formed in a longitudinal direction, thus reducing their weights by approximately 50 percent (as compared with solid slabs of the same dimensions). These sections, which can be used for spans of from roughly 10 to 25 ft, may be prestressed.
3. *Prefabricated concrete block systems* are slabs made by tying together precast concrete blocks with steel rods (may also be prestressed). These seldom-used floors vary from 4 to 8 in. deep and can be used for spans of roughly 8 to 32 ft.
4. *Channel slabs* are used for spans of approximately 8 to 24 ft. Rough dimensions of these slabs are given in Fig. 19-12.

19-11 TYPES OF ROOF CONSTRUCTION

The types of roof construction commonly used for steel-frame buildings include concrete slabs on open-web joists, steel-decking roofs, and various types of precast concrete slabs. For industrial buildings the roof system in predominant use today is one that involves the use of a cold-rolled steel deck. In addition, the other types mentioned in the preceding sections on floor types are occasionally used, but they usually are unable to compete economically. Among the factors to be considered in selecting the specific types of roof construction are strength, weight, span, insulation, acoustics, appearance below, and type of roof covering to be used.

The major differences between floor-slab selection and roof-slab selection probably occur in the considerations of strength and insulation. The loads applied to roofs generally are much smaller than those applied to floor slabs, thus permitting the use of many types of lightweight aggregate concretes that may be appreciably weaker. Roof slabs should have good insulation properties, or they will have to have insulation materials placed on them and covered by the roofing. Among the many types of lightweight aggregates used are wood fibers, zonolite, foams, sawdust, gypsum, and expanded shale. Although some of these materials decidedly reduce concrete strengths, they provide very light roof decks with excellent insulating properties.

Precast slabs made with these aggregates are light, are quickly erected, have good insulating properties, and usually can be sawed and nailed. For poured concrete slabs on open-web joists, several lightweight aggregates work very well (zonolite, foams, gypsum, etc.), and the resulting concrete easily can be pumped up to the roofs, thereby facilitating construction. By replacing the aggregates with certain foams, concrete can be made so light it will float in water (for a while). Needless to say, the strength of the resulting concrete is quite low.

Steel decking with similar thin slabs of lightweight and insulating concrete placed on top make very good economical roof decks. A competitive variation consists of steel decking with rigid insulation board placed on top, followed by the regular roofing material. The other concrete-slab types are hard pressed to compete economically with these types for lightly loaded roofs. Other types of poured concrete decks require much more labor.

19-12 EXTERIOR WALLS AND INTERIOR PARTITIONS

Exterior Walls

The purposes of exterior walls are to provide resistance to atmospheric conditions, including insulation against heat and cold, satisfactory sound-absorption and light-refraction characteristics, sufficient strength, and fire ratings. They should look good and be reasonably economical.

For many years exterior walls were constructed of some type of masonry, glass, or corrugated sheeting. In recent decades, however, the number of satisfactory materials for exterior walls has increased tremendously. Available and commonly used today are precast concrete panels, insulated metal sheeting, and many other prefabricated units. Of increasing popularity are the light prefabricated sandwich panels which consist of three layers. The exterior surface is made from aluminum, stainless steel, ceramics, plastics, and others. The center of the panel consists of some type of insulating material such as fiberglass or fiberboard, while the interior surface is made from metal, plaster, masonry, or some other attractive material.

Interior Partitions

The main purpose of interior partitions is to divide the inside space of a building into rooms. Partitions are selected for appearance, fire rating, weight, or acoustical properties. Partitions may be bearing or nonbearing.

Bearing partitions support gravity loads in addition to their own weights and are thus permanently fixed in position. They can be constructed from wood or steel studs or from masonry units and faced with plywood, plaster, wallboard, or other material.

Nonbearing partitions do not support any loads in addition to their own weights and thus may be fixed or movable. Selection of the type of material to be used for a partition is based on the answers to the following questions: Is the partition to be fixed or movable? Is it to be transparent or opaque? Is it to extend all the way to the ceiling? Is it to be used to conceal piping and electrical conduits? Are there fire-rating and acoustical requirements? The more common types are made of metal, masonry, or concrete. For design of the floor slabs in a building that has movable partitions, some allowance should be given to the fact that the partitions may be moved. Probably the most common practice is to increase the floor-design live load by 15 or 20 psf over the entire floor area.

19-13 FIREPROOFING OF STRUCTURAL STEEL

Although structural steel members are incombustible, their strength is tremendously reduced at temperatures normally reached in fires when the other materials of a building burn. Many disastrous fires have occurred in empty buildings where the only fuel was the buildings themselves. Steel is an excellent heat conductor, and nonfireproofed steel members may transmit enough heat from one burning compartment of a building to ignite materials with which they are in contact in adjoining sections of the building.

The fire resistance of structural steel members can be greatly increased by applying to them fire-protective covers such as concrete, gypsum, mineral fiber sprays, special paints, and others. The thickness and kind of fireproofing used depends on the type of structure, the degree of fire hazard, and economics.

In the past, concrete was commonly used for fire protection. Although concrete is not a particularly good insulation material, it is very satisfactory when applied in thicknesses of $1\frac{1}{2}$ to 2 in. or more, due to its mass. Furthermore, the water in concrete (16 to 20 percent when fully hydrated) improves its fireproofing qualities appreciably. The boiling off of the water from the concrete requires a great deal of heat. As the water or steam escapes, it greatly reduces the concrete temperature. This is identical to the behavior of steel steam boilers, where the steel temperature is held to maximum values of only a few hundred degrees Fahrenheit due to the escape of the steam. It is true, however, that in very intense fires the boiling off of the water may cause severe cracking and spalling of the concrete.

Though concrete is an everyday construction material, and though in mass it is quite a satisfactory fireproofing material, its installation cost is extremely high and its weight is large. As a result, for most steel construction, spray-on fireproofing materials have almost completely replaced concrete.

The spray-on materials usually consist of either mineral fibers or cementious fireproofing materials. The mineral fibers formerly used were asbestos. Due to the health hazards associated with this material, its use has been discontinued, and other fibers now are used by manufacturers. The cementious fireproofing materials are composed of gypsum, perlite, vermiculite, and others. Sometimes, when plastered ceilings are required in a building, it is possible to hang the ceilings and light-gage furring channels by wires from the floor systems above and to use the plaster as the fireproofing.

The cost of fireproofing structural steel buildings is high and hurts steel in its economic competition with other materials. As a result, a great deal of research is being conducted by the steel industry on imaginative new fireproofing methods. Among these ideas is the coating of steel members with expansive and insulative paints. When heated to certain temperatures, these paints will char, foam, and expand, forming an insulative shield around the members. These swelling or extumescent paints are very expensive.

Other techniques involve the isolation of some steel members outside of the building, where they will not be subject to damaging fire exposure, and the circulation of liquid coolants inside box- or tube-shaped members in the building. It is probable that major advances will be made with these and other fireproofing ideas in the near future.[1]

19-14 INTRODUCTION TO HIGH-RISE BUILDINGS

Space is not taken in this introductory text to discuss the design of multistory buildings in detail. The material of the next few sections gives the student a general idea of the problems involved in the design of such buildings, but does not present elaborate design examples.

Office buildings, hotels, apartment houses, and other buildings of many stories are quite common in the United States, and the trend is toward an even larger number of tall buildings in the future. Available land for building in our heavily populated cities is becoming scarcer and scarcer, and costs are becoming higher and higher. Tall buildings require a smaller amount of this expensive land to provide required floor space. Other factors contributing to the increased number of multistory buildings are new and better materials and construction techniques.

There are, on the other hand, several factors that may limit the heights to which buildings will be erected.

1. Certain city building codes prescribe maximum heights for buildings.
2. Foundation conditions may not be satisfactory for supporting buildings of many stories.
3. Floor space may not be rentable above a certain height. Someone will always be available to rent the top floor or two of a 250-story building, but floor numbers 100 through 248 may not be so easy to rent.
4. There are several cost factors that tend to increase with taller buildings. Among these factors are elevators, plumbing, heating and air-conditioning, glazing, exterior walls, and wiring.

Whether a multistory building is used for an office building, a hospital, a school, an apartment house, or whatever, the problems of design generally are the same. The construction usually is of the skeleton type in which the loads are

---

[1] W. A. Rains, "A New Era in Fire Protective Coatings for Steel," *Civil Engineering* (New York: ASCE, September 1976), pp. 80–83.

transmitted to the foundation by a framework of steel beams and columns. The floor slabs, partitions, and exterior walls all are supported by the frame. This type of framing, which can be erected to tremendous heights, may also be referred to as beam-and-column construction. Bearing-wall construction is not often used for buildings of more than a few stories, although it has on occasion been used for buildings up to 20 or 25 stories. The columns of a skeleton frame usually are spaced 20, 25, or 30 ft on centers, with beams and girders framing into them from both directions (see Fig. 19-2). On some of the floors, however, it may be necessary to have much larger open areas between columns for dining rooms, ballrooms, and so on. For such cases very large beams (perhaps plate girders) may be needed to support column loads for many floors above.

It usually is necessary in multistory buildings to fireproof the members of the frame with concrete, gypsum, or some other material. The exterior walls often are constructed with concrete or masonry units, although an increasing number of modern buildings are being erected with large areas of glass in the exterior walls.

For these tall and heavy buildings the usual spread footings may not be sufficient to support the loads. If the bearing strength of the soil is high, steel grillage footings may be sufficient, but for poor soil conditions pile or pier foundations may be necessary.

For multistory buildings the beam-to-column systems are said to be superimposed on top of each other story by story or tier by tier. The column sections can be fabricated for one, two, or more floors, with the two-story lengths probably being the most common. Theoretically, column sizes can be changed

Broadview Apartments, Baltimore, Md. (Courtesy of Lincoln Electric Company.)

at each floor level, but the costs of the splices involved usually more than cancel any savings in column weights. Columns of three or more stories in height are difficult to erect. The two-story heights work out very well most of the time.

19-15 DISCUSSION OF LATERAL FORCES

For tall buildings, lateral forces must be considered as well as gravity forces. High wind pressures on the sides of tall buildings produce overturning moments. These moments probably are resisted without difficulty by the axial strengths of the columns, but the horizontal shears produced on each level may be of such magnitude as to require the use of special bracing or moment-resisting connections.

Unless they are fractured, the floors and walls provide a large part of the lateral stiffness of tall buildings. Although the amount of such resistance may be several times that provided by the lateral bracing, it is difficult to estimate and may not be reliable. Today so many modern buildings have light movable interior partitions, glass exterior walls, and lightweight floors that only the steel frame should be assumed to provide the required lateral stiffness.

A building must be not only sufficiently braced laterally to prevent failure but also prevented from deflecting so much as to injure its various parts. Another item of importance is the provision of sufficient bracing to give the occupants a feeling of safety. They might not have this feeling in tall buildings that have a great deal of lateral movement in times of high winds. There have actually been tales of occupants of the upper floors of tall buildings complaining of seasickness on very windy days.

The horizontal deflection of a multistory building due to wind or seismic loading is called *drift*. It is represented by Δ in Fig. 19-13. Drift is measured by the *drift index*, Δ/h, where h is the height or distance to the ground.

The usual practice in the design of multistory steel buildings is to provide a structure with sufficient lateral stiffness to keep the drift index between approximately 0.0015 and 0.0030 radians for the worst storms that occur in a period of approximately 10 years. These so-called 10-year storms might have wind in the range of about 90 mph, depending on location and weather records. It is felt

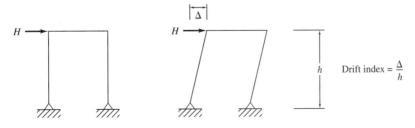

Figure 19-13 Lateral drift.

that when the index does not exceed these values, the users of the building will be reasonably comfortable.

In addition, multistory buildings should be designed to withstand safely 50-year storms. For such cases the drift index will be larger than the range mentioned, with the result that the occupants will suffer some discomfort. The 1450-ft World Trade Center in New York City deflects or sways up to about 3 ft in 10-year storms (drift index = 0.0021), while in hurricane winds it will sway up to about 7 ft (drift index = 0.0048).

Most buildings can be designed with little extra expense to withstand the forces caused during an earthquake of fairly severe intensity. On the other hand, earthquakes during recent years have clearly shown that the average building that is not designed for earthquake forces can be destroyed by earthquakes that are not particularly severe. The usual practice is to design buildings for additional lateral loads (representing the estimate of the earthquake forces) that are equal to some percentage (5 to 10) of the weight of the building and its contents.

The lateral forces require the use of more steel, even though load factors are reduced for wind and earthquake forces. Additional steel will be used for bracing or moment-resisting connections, as described in the next section.

19-16 TYPES OF LATERAL BRACING

A steel building frame with no lateral bracing is shown in Fig. 19-14(a). Should the beams and columns shown be connected with the standard ("simple beam") connections, the frame would have little resistance to the lateral forces shown. Assuming the joints to act as frictionless pins, the frame would be laterally deflected as shown in Fig. 19-14(b).

To resist these lateral deflections, the simplest method from a theoretical standpoint is the insertion of full diagonal bracing as shown in Fig. 19-14(c). From a practical standpoint, however, the student can easily see that in the average building full diagonal bracing would often be in the way of doors, windows, and other wall openings. Furthermore, many buildings have movable

Figure 19-14

Chase Manhattan Bank Building, New York City. (Courtesy of Bethlehem Steel Corporation.)

interior partitions, and the presence of interior cross bracing would greatly reduce this flexibility. Usually diagonal bracing is convenient in solid walls in and around elevator shafts, stairwells, and other walls in which few or no openings are planned.

Another method of providing resistance to lateral forces involves the use of moment-resisting connections as illustrated in Fig. 19-15. This *bracket-type*

Figure 19-15 Bracket-type bracing.

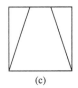

(a)   (b)   (c)

Figure 19-16 (a) X brace. (b) K brace. (c) Full-story knee brace.

bracing may be used economically to provide lateral bracing for lower-rise buildings. As buildings become taller, however, bracket-type bracing is not very economical, nor is it very satisfactory in limiting lateral deflections.

Some ways in which we can transmit lateral forces to the ground for buildings in the 20- to 60-story range are shown in Fig. 19-16. The X bracing system of part (a) works very well, except that it might get in the way. Furthermore, as compared with the systems shown in Fig. 19-16(b) and (c), the floor beams have longer spans and will have to be larger.

The K bracing system (b) provides more freedom for the placing of openings than does the full X bracing. K braces are supported at midspan, with the result that the floor beams will have smaller moments. The K bracing system (like the knee bracing system) uses less material than does X bracing. If we need more room than is available with the K bracing system, we can go to the full-story knee bracing system shown in Fig. 19-16(c).[2]

In the usual building the floor system (beams and slabs) is assumed to be rigid in the horizontal plane, and the lateral loads are assumed to be concentrated at the floor levels. Floor slabs and girders acting together provide considerable resistance to lateral forces. Investigation of steel buildings that have withstood high wind forces have shown that the floor slabs distribute the lateral forces so that all of the columns on a particular floor have equal deflections. When rigid floors are present, they spread the lateral shears to the columns or walls in the building. When lateral forces are particularly large, as in very tall buildings or where seismic forces are being considered, certain specially designed walls may be used to resist large parts of the lateral forces. These walls are called *shear walls.*

It is not necessary to brace every panel in a building. Usually the bracing can be placed in the outside walls with less interference than in the inside walls, where movable partitions may be desired. Probably the bracing of the outside panels alone is not enough, and some interior panels may need to be braced. It is assumed that the floors and beams are sufficiently rigid to transfer the lateral forces to the braced panels. Three possible arrangements of braced panels are

[2] E. H. Gaylord, Jr., and C. N. Gaylord, *Structural Engineering Handbook,* 2d ed. (New York: McGraw-Hill, 1979), pp. 19–77 to 19–111.

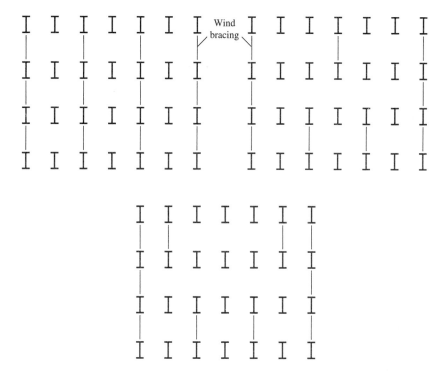

Figure 19-17

shown in Fig. 19-17. A symmetrical arrangement probably is desirable to prevent uneven lateral deflection in the building and thus torsion.

Bracing around elevator shafts usually is permissible, while for other locations it will interfere often with windows, doors, movable partitions, glass exterior walls, open spaces, and so on. Should bracing be used around an elevator shaft as shown in Fig. 19-18(a), and should calculations show that the drift index is too large, it may be possible to use a *hat truss* on the top floor, as shown in part (b). Such a truss will substantially reduce drift. If a hat truss is not feasible because of interference with other items, one or more *belt trusses* may be possible, as shown in Fig. 19-18(c). A belt truss will appreciably reduce drift, although not as much as a hat truss.

The bracing systems described so far are not efficient for buildings above approximately 60 stories. These taller buildings have very large lateral wind and perhaps seismic forces applied hundreds of feet above the ground. The designer needs to develop a system that will resist these loads without failure and in such a manner that lateral deflections are not so great as to frighten the occupants. The bracing methods used for these buildings usually are based on a tubular frame concept.

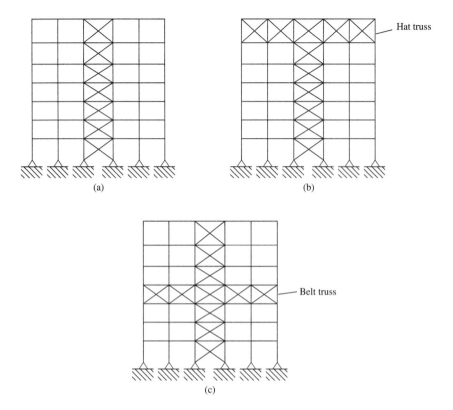

Figure 19-18

With the tubular system a vertical tubular cantilever frame is created much like the tube of Fig. 19-19. The tube consists of the building columns and girders in both the longitudinal and transverse directions of the building. The idea is to create a tube that will act like a continuous chimney or stack.

To build the tube, the exterior columns are spaced close together—from 3 or 4 ft up to 10 or 12 ft on center. They are connected with spandrel beams at the floor levels, as shown in Fig. 19-20(a).

Figure 19-19 Solid-wall tube.

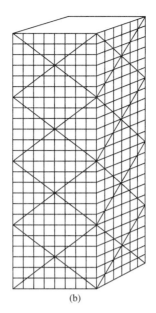

(a)   (b)

Figure 19-20 (a) Tubular frame. (b) Cross-braced tubular frame.

Further improvements of the tubular system can be made by cross bracing the frame with X bracing over many stories, as illustrated in Fig. 19-20(b). This latter system is very stiff and efficient, and is very helpful in distributing gravity loads rather uniformly over the exterior columns.

Another variation on this system is the tube-within-a-tube system. Interior columns and girders are used to create additional tubes. In tall buildings it is common to group elevator shafts, utility shafts, and stairs, and these systems can include shear walls and braced bents.[3]

19-17 ANALYSIS OF BUILDINGS WITH DIAGONAL WIND BRACING FOR LATERAL FORCES

Full diagonal cross bracing has been described as being an economical type of wind bracing for tall buildings. Where this type of bracing cannot practically be used due to interference with windows, docrs, or movable partitions, moment-resisting brackets often are used. However, should very tall narrow buildings (with height–least-width ratios of 5.0 or greater) be constructed, lateral deflections may become a problem with moment-resisting joints. The joints can satisfactorily resist the moments, but the deflections may be excessive.

[3] R. N. White, P. Gergely, and R. G. Sexsmith, *Structural Engineering,* vol. 3 (New York: Wiley 1974), pp. 537–546.

Transfer truss, 150 Federal Street, Boston, Mass. (Courtesy of Owen Steel Company, Inc.)

Should maximum wind deflections be kept under 0.002 times the building height, there is little chance of injury to the building says ASCE Subcommittee 31.[4] The deflection is to be computed neglecting any resistance supplied by floors and walls. To keep lateral deflections within this range, it is necessary to use deep knee bracing, K bracing, or full diagonal cross bracing when the height–least-width ratio is about 5.0; and for greater values of the ratio, full diagonal cross bracing or some other method such as the tubular frame is needed.[5]

The student often may see bracing used in buildings for which he or she would think wind stresses were negligible. Such bracing stiffens up a building appreciably and serves the useful purpose of plumbing the steel frame during erection. Before bracing is installed, the members of a steel building frame may be twisted in all sorts of directions. Connecting the diagonal bracing should pull members into their proper positions.

When diagonal cross bracing is used, it is desirable to introduce initial tension into the diagonals. This prestressing will make the building frame tight and reduce its lateral deflection. Furthermore, these light diagonal members can support compressive stress due to their pretensioning. Since the members can resist compression, the horizontal shear to be resisted will be assumed to split equally between the two diagonals. For buildings with several bay widths, equal shear distribution usually is assumed for each bay.

Should the diagonals not be initially tightened, rather stiff sections should be used so they will be able to resist appreciable compressive forces. The direct axial forces in the girders and columns can be found for each joint from the shear forces assumed in the diagonals. Usually, the girder axial forces so computed are too small to consider, but the values for columns can be quite important. The taller the building becomes, the more critical are the column axial forces caused by lateral forces.

For types of bracing other than cross bracing, similar assumptions can be made for analysis. The student is referred to pages 333–339 of *Theory of Modern Steel Structures* by L. E. Grinter (New York: Macmillan, 1962) for discussion of this subject.

19-18 MOMENT-RESISTING JOINTS

For a large percentage of buildings under eight to ten stories, the beams and girders are connected to each other and to the columns with simple end-framed connections of the types described in Chapter 15. As buildings become taller, it is absolutely necessary to use a definite wind-bracing system or moment-resisting joints. Moment-resisting joints also may be used in lower buildings where it is desired to take advantage of continuity and the consequent smaller

[4] "Wind Bracing in Steel Buildings," *Transactions ASCE* 105 (1940), pp. 1713–1739.

[5] L. E. Grinter, *Theory of Modern Steel Structures* (New York: Macmillan, 1962), p. 326.

(a) Shop-welded, field-bolted
moment connection

(b) Field-welded, field-bolted
moment connection

Figure 19-21 Popular moment-resisting connections.

beam sizes and depths and shallower floor construction. Moment-resisting brackets also may be necessary in some locations for loads that are applied eccentrically to columns.

Two types of moment-resisting connections that may be used as wind bracing are shown in Fig. 19-21. The design of connections of these types was also presented in Chapter 14. The average design company through the years probably will develop a file of moment-resisting connections from their previous designs. When they have a wind moment of such and such a value, they merely refer to their file and select one of their former designs that will provide the required moment resistance.

These are the two most commonly used moment-resisting connections today. Most fabricators select the type shown in 19-21(a) as being the most economical. Single-plate (or shear tab) connectors are shown, but framing angles may be used instead. Column stiffeners (shown as dashed lines in the figures) sometimes are required by LRFD Specification K1 for local flange bending, local web yielding, sidesway buckling of the web, and so on. Stiffeners are a nuisance and cost an appreciable amount of money. As a result, we will, when the specification shows they are needed, try to avoid them by increasing column sizes.

A very satisfactory variation of these last two connections involves an end plate as shown on pages 10-10 and 10-20 in Part 10 of the LRFD Manual. This type of connection may, however, be used only for static load situations.

19-19 ANALYSIS OF BUILDINGS WITH MOMENT-RESISTING JOINTS FOR LATERAL LOADS

Approximate methods have been very popular for many years for analyzing multistory buildings for lateral forces. Among the several reasons for this popularity have been the following:

The 20-story United Founders Life Tower in Oklahoma City, Okla. (Courtesy of Lincoln Electric Company.)

1. Large building frames are statically indeterminate to a very high degree, and their analysis by an "exact" method using a pocket calculator is a lengthy and difficult problem.
2. The resistance to lateral forces supplied by the walls and floors in a tall building is difficult to estimate accurately, and the results of analysis by an "exact" method therefore are not very precise.

3. Before the members of a statically indeterminate structure can be designed, their forces have to be determined, but these forces are dependent upon their sizes. Application of an approximate analysis should yield forces from which very good estimates of member sizes can be made for preliminary design.

The analysis of multistory buildings is a very difficult problem, and past practice has been to use one of the several approximate methods of analysis available. Today, however, with the availability of electronic computers it is feasible to make "exact" analyses in appreciably less time than required to make the approximate analyses (without the use of computers). The more accurate values obtained permit the use of smaller members. Computer usage saves money in analysis time and in the use of smaller members.

In the pages that follow, a brief review of the portal and cantilever approximate methods is presented. No consideration is given in either of these methods to the elastic properties of the members. These omissions can be very serious in unsymmetrical frames and in very tall buildings. For example, consider the changes in member sizes in a very tall building. In such a building there probably will not be a great deal of change in beam sizes from the top floor to the bottom floor. For the same loadings and spans, the changed sizes would be due to the large wind moments in the lower floors. The change, however, in column sizes from top to bottom would be tremendous. The result is that the relative sizes of columns and beams on the top floors are entirely different from their relative sizes on the lower floors. When this fact is not considered, it causes large errors in the analysis.

If the height of a building is roughly five or more times its least lateral dimension, generally it is felt that a more precise method of analysis should be used than the methods described here. There are several excellent approximate methods that make use of the elastic properties and that give values closely approaching the results of the "exact" methods. These include the Factor method, the Witmer method of K percentages, and the Spurr method. Should an exact method be desired, the slope-deflection and moment-distribution methods are available. If the slope-deflection procedure is used, the designer will have the problem of solving a large number of simultaneous equations. The problem is not so serious, however, if a digital computer is used.

The Portal Method

The most common approximate method of analyzing tall building frames for lateral loads is the portal method. The chief advantage of this method, which is said to be satisfactory for most buildings up to 25 stories,[6] is its simplicity. A detailed description of this method, including the basis for the assumptions

---

[6] "Wind Bracing in Steel Buildings," *Transactions ASCE* 105 (1940), p. 1723.

made, can be found in most structural analysis textbooks. Only a brief listing of the steps involved in its application is given here.

1. The horizontal shears on each level are arbitrarily distributed between the columns. One commonly used procedure is to assume the shear divides between the columns in the ratio of one part to exterior columns and two parts to interior columns. Another common distribution (and the one used in the illustrative problem to follow) is to assume the shear V taken by each column is in proportion to the floor area it supports.
2. The moment M in each column is assumed to equal the column shear times half the column height (thus assuming a point of contraflexure at middepth).
3. The girder moments M are determined by joints, by noting that the sum of the girder moments at any joint equals the sum of the column moments at that joint. These calculations are easily made by starting at the upper left joint and working joint by joint across to the right; after which the next level is handled left to right, and so on.
4. The shear V in each girder is assumed to equal its moment divided by half the girder length.
5. Finally, the column axial forces S are determined by summing up the beam shears and other column axial forces at each joint. These calculations again are handled conveniently by working from left to right and from the top floor down.

Fig. 19-22 shows a building frame that is to be analyzed by the portal method. The frames are assumed to be placed 20 ft 0 in. on centers and the exterior walls are assumed to be subjected to a wind pressure of 20 lb per vertical square foot. From these data the horizontal loads shown at each floor level are calculated.

Figure 19-22

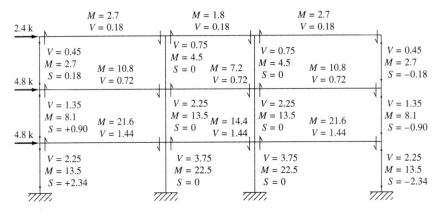

Figure 19-23 Frame analysis by the portal method.

This frame is analyzed in Fig. 19-23 by the portal method. The arrows shown on the figure give the direction of the girder shears and the column axial forces. The student can visualize the stress condition of the frame if he or she assumes the frame is tending to be pushed over from left to right by the wind, stretching the left exterior columns and compressing the right exterior columns.

The Cantilever Method

Another simple method of analyzing building frames for lateral forces is the cantilever method. This method is said to be a little more desirable for high narrow buildings than the portal method and may be used satisfactorily for buildings with heights not in excess of 25 to 35 stories.[7] It is not as popular as the portal method.

Instead of initially assuming the horizontal shears to be divided between the columns on each level in some proportion, the first assumption in the cantilever method pertains to the column axial forces. The axial force in a particular column is assumed to vary in direct proportion to its distance from the center of gravity of the group of columns on that level. With the wind blowing from left to right, the columns to the left of the center of gravity will be in tension, while those to the right will be in compression. The steps involved in analyzing a building frame for lateral forces by the cantilever method are as follows:

1. To obtain the column axial forces, moments are taken above an assumed plane of contraflexure through the middepth of the columns on each level. As an illustration of this procedure, moments are taken here to determine the axial forces in the columns on the top level of the building

---

[7] Ibid.

Figure 19-24

frame of Fig. 19-22. A free body is drawn in Fig. 19-24 of the part of the building above the assumed plane of contraflexure on the top level. The value of S is determined:

$$\Sigma M_x = 0$$
$$(2.4)(6) + (30)(S) - (50)(S) - (80)(4S) = 0$$
$$S = 0.0424 \text{ k}$$

2. The girder shears are determined joint by joint from the column axial forces.
3. The girder moments are determined by multiplying the girder shears by the half-girder lengths.
4. The column moments are found joint by joint from the girder moments.
5. The column shears are obtained by dividing the column moments by the half-column heights.

The analysis of the frame of Fig. 19-22 by the cantilever method is illustrated in Fig. 19-25. Arrows representing the directions of the column axial forces and the girder shears are shown again.

Figure 19-25 Frame analysis by the cantilever method.

19-20 ANALYSIS OF BUILDINGS FOR GRAVITY LOADS

Simple Framing

If simple framing is used, the design of the girders is quite simple because the shears and moments in each girder can be determined by statics. The gravity loads applied to the columns are relatively easy to estimate, but the column moments may be a little more difficult. If the girder reactions on each side of the interior column of Fig. 19-26 are equal, theoretically no moment will be produced in the column at that level. This situation probably is not realistic, however, because it is highly possible for the live load to be applied on one side of the column and not on the other (or at least be unequal in magnitude). The results will be column moments. If the reactions are unequal, the moment produced in the column will equal the difference between the reactions times the distances to the center of gravity of the column.

Exterior columns often may have moments due to spandrel beams opposing the moments caused by the floor loads on the inside of the column. Nevertheless, gravity loads generally will cause the exterior columns to have larger moments than the interior columns.

"Topping out" of Blue Cross–Blue Shield Building in Jacksonville, Fla. (Courtesy of Owen Steel Company, Inc.) The Christmas tree is an old north European custom used to ward off evil spirits. It also is used today to show that the steel frame was erected with no lost-time accidents to personnel.

Figure 19-26 Simple framing of interior column.

To estimate the moment applied to a column above a certain floor, it is probably reasonable to assume that the unbalanced moment at that level splits evenly between the column above and below. In fact, such an assumption is often on the conservative side, as the column below may be larger than the one above.

Rigid Framing

For buildings with moment-resisting joints it is a little more difficult to estimate the girder moments and make preliminary designs. If the ends of each girder are assumed to be completely fixed, the moments for uniform loads are as shown in Fig. 19-27(a). Conditions of complete fixity probably are not realized, with the result that the end moments are smaller than shown. There is a corresponding increase in the positive centerline moments approaching the simple moment ($wL^2/8$) shown in Fig. 19-27(b). Probably a moment diagram somewhere in between the two extremes is more realistic. Such a moment diagram is represented by the dotted line of Fig. 19-27(a). A reasonable procedure is to assume a moment in the range of $wL^2/10$, where L is the clear span.

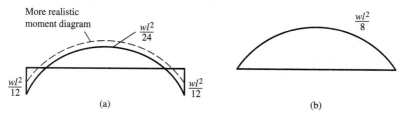

Figure 19-27 (a) Fixed-end beam. (b) Simple beam.

When the beams and girders of a building frame are rigidly connected to each other, a continuous frame is the result. From a theoretical standpoint, an accurate analysis of such a structure cannot be made unless the entire frame is handled as a unit.

Before this subject is pursued further, it should be realized that in the upper floors of tall buildings the wind moments will be small, and the framed and seated connections described in Chapter 15 will provide sufficient moment resistance. For this reason the beams and girders of the upper floors may very well be designed on the basis of simple beam moments, while those of the lower floors may be designed as continuous members with moment-resistant connections because of the larger wind moments.

From a strictly theoretical viewpoint there are several live-load conditions that need to be considered to obtain maximum shears and moments at various points in a continuous structure. For the building frame shown in Fig. 19-28 it is desired to place live loads to cause maximum positive moment in span *AB*. A qualitative influence line for positive moment at the centerline of this span is shown in part (a). This influence line shows that, to obtain maximum positive moment at the centerline of span *AB*, the live loads should be placed as shown in Fig. 19-28(b).

To obtain maximum negative moment at point *B* or maximum positive moment in span *BC*, other loading situations need to be considered. With the availability of computers, more of this detailed analysis is being done every day. The student can see, however, that unless the designer uses a computer, he or she probably will not have sufficient time to go through all of these theoretical situations. Furthermore, it is doubtful if the accuracy of our analysis methods

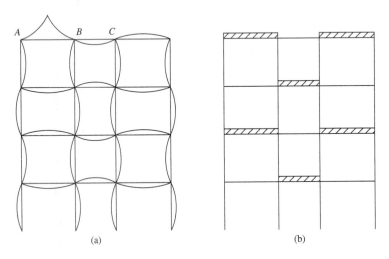

(a)   (b)

Figure 19-28

would justify all of the work anyway. Nonetheless, it often is feasible to take out two stories of the building at a time as a free body and analyze that part by one of the "exact" methods such as the successive correction method of moment distribution.

Before such an "exact" analysis can be made, it is necessary to make an estimate of the member sizes, probably based on the results of analysis by one of the approximate methods.

19-21 DESIGN OF MEMBERS

The girders can be designed for the load combinations mentioned in Section 19-20. For the top several floors the second condition will not control design, but farther down in the building it will control the sizes for the reasons described in the next paragraph.

Each of the members of a building frame must be designed for all of the dead loads that it supports, but it may be possible to design some members for lesser loads than their full theoretical live-load values. For example, it seems unlikely in a building frame of several stories that the absolutely maximum-design live load will occur on every floor at the same time. The lower columns in a building frame are designed for all of the dead loads above, but probably for a percentage of live loads appreciably less than 100 percent. Some specifications require that the beams supporting floor slabs be designed for full dead and live loads, but permit the main girders to be designed under certain conditions for reduced live loads. (This reduction is based on the improbability that a very large area of a floor would be loaded to its full live-load value at any one time.) The ASCE Code provides some very commonly used reduction expressions.[8]

Should simple framing be used, the girders will be proportioned for simple beam moments plus the moments caused by the lateral loads. For continuous framing the girders will be proportioned for $wL^2/10$ (for uniform loads) plus the moments caused by the lateral loads. An interesting comparison of designs by the two methods is shown in pages 717–719 of Beedle et al., *Structural Steel Design* (New York: Ronald, 1964). It may be necessary to draw the moment diagram for the two cases (gravity loads and lateral loads) and add them together to obtain the maximum positive moment out in the span. The value so computed may very well control the girder size in some of the members.

A major part of the design of a multistory building is involved in setting up a column schedule that shows the loads to be supported by the various columns story by story. The gravity forces can be estimated very well, while the shears,

---

[8] *American Society of Civil Engineers Minimum Design Loads for Buildings and Other Structures,* ASCE 7–93 (New York: ASCE), p. 6.

axial forces, and moments caused by lateral forces can be roughly approximated for the preliminary design from some approximate method (portal, cantilever, factor, or other).

If both axes of the columns are free to sway, the column effective lengths theoretically must be calculated for both axes as described in Chapter 5. If diagonal bracing is used in one direction, thus preventing sidesway, the K value will be less than 1.0 in that direction. When frames are braced in the narrow direction, it is reasonable to use $K = 1.0$ for both axes.

After the sizes of the girders and columns are tentatively selected for a two-story height, an "exact" analysis can be performed and the members re-designed. The two stories taken out for analysis often are referred to as a *tier*. This process can be continued tier by tier down through the building. Many tall buildings have been designed and are performing satisfactorily in which this last step (the two-story "exact" analysis) was omitted.

Appendix **A**

Allowable Stress Design

A-1 INTRODUCTION

In the pages to follow a very brief introduction to the allowable stress design (ASD) method is presented as it applies to tension members, columns, beams, and connections. It is assumed that the reader has previously studied these topics using the LRFD procedure. Therefore, the author does not repeat items that were involved in that prior study, such as net areas, effective net areas, effective column lengths, and so on.

With ASD the working or service loads are combined for the various members of a particular structure to obtain the maximum loads. Then members are selected such that their computed elastic stresses (when the maximum loads are applied) do not exceed certain allowable values. In structural steel design these stresses usually are given as being equal to some percentage of the yield stress, as, say, $0.60F_y$, $0.66F_y$, etc.

The ASD design procedure described in these pages is based upon the "Specification for Structural Steel Buildings—Allowable Stress Design and Plastic Design" and the *Manual of Steel Construction Allowable Stress Design* 9th ed., both published in 1989 by the AISC. The Specification is included in the Manual.

A-2 TENSION MEMBERS

The ASD Specification (D-1) provides an allowable tensile stress of $0.60F_y$ on the gross cross-sectional areas of members at sections where there are no holes. For sections where there are holes for bolts, the allowable tensile *stress* is $0.50F_u$, applicable to the effective net areas. Thus, a safety factor of 1.67 is provided against yielding of the gross section ($F_y/0.60F_y = 1.67$), and a value of 2.0 is provided against the fracture of the member through its smallest effective net area ($F_u/0.50F_u = 2.0$). Thus the allowable capacity of a tension member with bolt holes permitted by the ASD Specification is equal to the smaller of the following two values:

$$T = 0.60F_y A_g \quad \text{or} \quad T = 0.50F_u A_e$$

The above value of A_e is the effective net area of a steel section and equals UA_n. To compute the net area A_n of a bolted member, the diameter of holes are assumed to equal the bolt diameters plus 1/8 in.

The preceding values are not applicable to threaded steel rods or to members with pin holes (as in eyebars). For members with pin holes, an allowable tensile stress of $0.60F_y$ is permitted on the gross area of the member, while a value of $0.45F_y$ is permitted on the net area through the pin hole.

Example A-1 illustrates the calculations necessary for determining the effective net area of a bolted W section connected only to its flanges. In addition, the allowable tensile strength of the member is computed.

■ EXAMPLE A-1

Determine the allowable tensile load for a W10 × 45 with two lines of $\frac{3}{4}$-in.-diameter bolts in each flange, using A36 steel and the ASD Specification. There are at least three bolts in each line, and the bolts are not staggered with respect to each other. Select U from Table 3-2 on page 80.

Solution.   Using a W10 × 45 ($A_g = 13.3$ in.^{2}, $d = 10.10$ in., $b_f = 8.020$ in.)

(a) $T = 0.60F_y A_g = (0.60)(36)(13.3) = 287.3 \text{ k} \leftarrow$

(b) $A_n = 13.3 - (4)(\frac{7}{8})(0.620) = 11.13$ in.^{2}

$U = 0.90$ since $b_f > \frac{2}{3}d$

$A_e = UA_n = (0.90)(11.13) = 10.02$ in.^{2}

$T = 0.50F_u A_e = (0.50)(58)(10.02) = 290.6 \text{ k}$

Allowable $T = 287.3 \text{ k}$ ■

Example A-2 illustrates the design of a bolted tension W section. As previously stated, the tensile design strength T of a member is the least of (a) $0.60F_y A_g$ or (b) $0.50F_u A_e$.

To satisfy the first of these expressions, the minimum gross area must be at least

$$\min A_g = \frac{T}{0.60F_y} \tag{1}$$

To satisfy the second expression, the minimum value of A_e must be at least

$$\min A_e = \frac{T}{0.50F_u}$$

And since $A_e = UA_n$, the minimum value of A_n is

$$\min A_n = \frac{\min A_e}{U} = \frac{T}{0.50F_u U}$$

Then the minimum A_g for the second expression must at least equal the minimum value of A_n plus the estimated hole areas.

$$\min A_g = \frac{T}{0.50F_u U} + \text{estimated hole areas} \tag{2}$$

The designer can substitute into Equations (1) and (2), taking the larger value of A_g so obtained for an initial size estimate. However, notice that the maximum preferable slenderness ratio L/r is 300. From this value it is easy to compute the preferable minimum value of r for a particular design, that is, the value of r for which the slenderness ratio will be exactly 300. It is undesirable to consider a section whose least r is less than this value, because its slenderness ratio would exceed the preferable maximum value of 300.

Preferable min $r = \dfrac{L}{300}$

In Example A-2 a W section is selected for a given tensile load.

■ EXAMPLE A-2

Select a 30-ft-long W12 section of A36 steel to support a total tensile load of 220 k. As shown in Fig. A-1, the member is to have two lines of bolts in each flange for $\frac{7}{8}$-in. bolts (at least three in a line). Select U from Table 3-2 on page 80.

Figure A-1

Solution.   Computing the minimum A_g required:

$$\text{Min } A_g = \frac{T}{0.60F_y} = \frac{220}{(0.60)(36)} = 10.19 \text{ in.}^2$$

$$\text{Min } A_g = \frac{T}{0.50F_u U} + \text{estimated hole areas}$$

Assume $U = 0.90$ and assume flange thickness is about 0.515 in. after looking at W12 sections in the ASD Manual, which have areas of about 10.19 in.^{2}

$$\text{Min } A_g = \frac{220}{(0.50)(58)(0.90)} + (4)(\tfrac{7}{8} + \tfrac{1}{8})(0.515) = 10.49 \text{ in.}^2$$

$$\text{Preferable min } r = \frac{L}{300} = \frac{(12)(30)}{300} = 1.2 \text{ in.}$$

Try W12 × 40 ($A_g = 11.8$ in.^{2}, $d = 11.94$ in.,
$b_f = 8.005$ in., $t_f = 0.515$ in., $r_y = 1.93$ in.)

Checking:

$$T = 0.60F_y A_g = (0.60)(36)(11.8) = 254.9 \text{ k} > 220 \text{ k} \qquad \text{(OK)}$$

$$T = 0.50F_u UA_n, \text{ with } U = 0.90 \text{ since } \frac{b_f}{d} > \frac{2}{3},$$

$$\text{and } A_n = 11.8 - (4)(1.00)(0.515) = 9.74 \text{ in.}^2$$

$$T = (0.50)(58)(0.90)(9.74) = 254.2 \text{ k} > 220 \text{ k} \qquad \text{(OK)}$$

$$\frac{L}{r} = \frac{(12)(30)}{1.93} = 187 < 300 \qquad \text{(OK)}$$

■

A-3 COLUMNS

In the ASD Specification for columns it is assumed that because of residual stresses the upper limit of elastic buckling is defined by an average stress equal to one-half of the yield point ($\tfrac{1}{2}F_y$). If this stress is equated to the Euler expression, the value of the slenderness ratio at this upper limit can be determined for a particular steel. This value is referred to as C_c, the slenderness ratio dividing elastic from inelastic buckling, and is determined as follows:

$$\frac{1}{2}F_y = \frac{\pi^2 E}{(L/r)^2} = \frac{\pi^2 E}{C_c^2}$$

$$C_c = \sqrt{\frac{2\pi^2 E}{F_y}}$$

The values of C_c can be computed with little difficulty, but the ASD Manual gives its values for each steel (126.1 for A36, 116.7 for the 42,000 psi yield point steels, etc.). For columns with slenderness ratios less than C_c a parabolic formula, ASD Equation E2-1, is used.

$$F_a = \frac{\left[1 - \dfrac{(KL/r)^2}{2C_c^2}\right]F_y}{\dfrac{5}{3} + \dfrac{3(KL/r)}{8C_c} - \dfrac{(KL/r)^3}{8C_c^3}} \qquad \text{(ASD Equation E2-1)}$$

For values of KL/r greater than C_c the Euler formula is used. With a factor of safety of 1.92 (or 23/12), the expression becomes

$$F_a = \frac{12\pi^2 E}{23(KL/r)^2} \qquad \text{(ASD Equation E2-2)}$$

Figure A-2 shows the ranges in which the two ASD column expressions are used.

The denominator of Equation E2-1 is the factor of safety to be used and usually gives a value not much greater than the one used for axially loaded tension members.

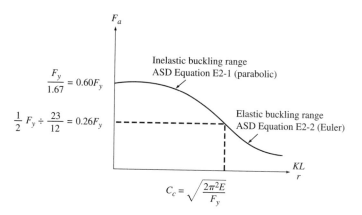

Figure A-2 ASD column allowable-stress curve.

In Example A-3 the allowable load that a particular column can support is calculated. The value of K, the effective length factor for the column, is determined in the usual manner. (Here, Table C-C2.1 in Part 5 of the ASD Manual is used.) Then the slenderness ratio is computed and the allowable design stress F_a is determined from the appropriate column formula as printed in Table 3-36 of Part 6 of the Manual. Finally, this value is multiplied by the cross-sectional area of the W section to determine the allowable column load.

■ **EXAMPLE A-3**

Using the column design stress values shown in Table 3-36 in the ASD Manual for A-36 steel, determine the allowable compression load P that the axially loaded column shown in Fig. A-3 can support.

Figure A-3

Solution. Using a W12 × 72 (A = 21.1 in.^{2}, r_y = 3.04 in.)

$K = 0.80$ from Table 5-1 in this text (page 141)

$$\frac{KL}{r} = \frac{(0.80)(12 \times 15)}{3.04} = 47.37$$

$F_a = 18.58$ ksi from Table 3-36 of ASD Manual

$P = F_a A_g = (18.58)(21.1) = \underline{\underline{392 \text{ k}}}$ ■

In Example A-4 a column with $KL = 14$ ft is selected using the ASD equations. An effective slenderness ratio of 50 is assumed, the allowable stress for that value is determined from Table C-50 of the ASD Manual, and the resulting stress is divided into the column load to obtain an estimated column area. After a trial section is selected with approximately that area, its actual slenderness ratio and allowable strength are computed. The first estimated size is too small, and the next larger section in that series of shapes is tried and found to be satisfactory.

■ EXAMPLE A-4

Select a W14 section for the column and load shown in Fig. A-4 using a steel with $F_y = 50$ ksi.

Figure A-4

Solution.

Assume $\dfrac{KL}{r} = 50$

$F_a = 24.35$ ksi from ASD Table C-50

A reqd. $= \dfrac{375}{24.35} = 15.40$ in.^{2}

Try W14 × 53 ($A = 15.6$ in.^{2}, $r_y = 1.92$ in.)

$\dfrac{KL}{r} = \dfrac{(1.0)(12 \times 14)}{1.92} = 87.5$

$F_a = 17.47$ ksi

Allow $P = (17.47)(15.6) = 272.5$ k $<$ 375 k   (NG)

Try W14 × 61 ($A = 17.9$ in.^{2}, $r_y = 2.45$ in.)

$\dfrac{KL}{r} = \dfrac{(1.0)(12 \times 14)}{2.45} = 68.57$

$F_a = 21.20$ ksi

Allow $P = (21.20)(17.9) = 379.5$ k $>$ 375 k   (OK)

Use W14 × 61

Note: The same answer can be obtained (and much more quickly) by using the ASD Column Tables in Part 3 of the Manual.   ■

A-4 BEAMS

Included in the items that need to be considered in beam design are moments, shears, crippling, buckling, lateral support, deflection, and perhaps fatigue. Beams probably will be selected that satisfactorily resist the bending moments and then checked to see if any of the other items are critical. To select a beam for a given situation the maximum moment is calculated for the assumed loading and a section having that much resisting moment is selected from the ASD Manual. It is usually more convenient, however, to work with the section modulus as described herein.

In the flexure formula f_b is the fiber stress in the outermost fiber at a distance c from the neutral axis, and I is the moment of inertia of the cross section. It should be remembered that this formula is limited to stress situations below the elastic limit because it is based on the usual elastic assumptions: a plane section before bending remains a plane section after bending; stress is proportional to strain; and so on.

If a beam is to be designed for a particular bending moment M and for a certain allowable stress F_b, the section modulus required to provide a beam of sufficient bending strength can be obtained from the flexure formula.

$$\frac{M}{F_b} = \frac{I}{c} = S$$

From the table in Part 2 of the ASD Manual entitled "Allowable Stress Design Selection Table," steel shapes having sufficient section moduli can be quickly selected.

The limiting moment permitted by the allowable-stress method is the moment at which the stress in the outermost fibers first reaches the yield point. The true bending strength of a beam, however, is larger than this commonly used value because the beam will not fail at this condition. The outermost fibers will yield, and the stress in the inner fibers will increase until they reach the yield stress, and so on, until the whole section is plastified.

The plastification process just mentioned is correct only if the beam remains stable in other ways; that is, it must have sufficient lateral support to prevent lateral buckling of the compression flange and it must have a sufficiently stocky profile to prevent local buckling.

The ASD Specification (F1) gives different allowable bending stresses for different conditions. For most cases the allowable bending stress is

$$F_b = 0.66F_y \qquad \text{(ASD Equation F1-1)}$$

This expression can be used to determine the allowable bending stress in the extreme fibers of compact hot-rolled shapes and built-up members (not including members consisting of steels with $F_y > 65$ ksi or hybrid girders built up with different yield stress steels) that are symmetrical about and loaded in the plane of their minor axis and that meet the other requirements of Section B5.1 of the ASD Specification for compact sections. One of these requirements is that the flange must be continuously attached to the web. A built-up section with its flanges intermittently welded to the web does not meet this requirement.

Example A-5, which follows, illustrates the design of a steel beam whose compression flange has full lateral support, thus permitting the use of the same allowable stress in both its tension and compression flanges.

■ EXAMPLE A-5

A 5-in. reinforced-concrete slab is to be supported with steel beams 8 ft 0 in. on centers. The beams, which will span 20 ft, are assumed to be simply supported. If the concrete slab is designed to support a live load of 100 psf, determine the lightest steel section required to support the slab. The compression flange of the beam will be incorporated in the concrete slab and is thus laterally supported. The concrete weighs 150 lb/ft^3, and the steel has an $F_y = 36$ ksi. Thus, $F_b = (0.66)(36) = 24$ ksi.

Solution.

$$\text{Dead load: slab} = (8)\left(\frac{5}{12}\right)(150) = 500 \text{ lb/ft}$$

$$\text{Estimated beam weight} = 26$$
$$\text{Live load: } 8 \times 100 = \underline{800}$$
$$\text{Total uniform load} = 1326 \text{ lb/ft}$$

$$M = \frac{(1.326)(20)^2}{8} = 66.3 \text{ ft-k}$$

$$S_{\text{reqd.}} = \frac{(12)(66.3)}{24} = 33.2 \text{ in.}^3$$

Use W12 × 26 $S_x = 33.4$ in.^{3} ■

The ASD Specification does not require a reduction in the allowable strength of a beam with bolts in either flange, as long as the tensile fracture strength of the net flange area divided by a factor of safety of 2.0 is at least as large as the tensile yield strength of the gross flange area divided by a factor of safety of 1.67. Expressed as an equation, no deduction for bolt or rivet holes in either flange is necessary if

$$0.5F_u A_{fn} \geq 0.6F_y A_{fg} \qquad \text{(ASD Equation B10-1)}$$

In this equation A_{fn} is the net flange area and A_{fg} is the gross flange area. Substituting into this expression we find that no deduction is necessary if the net flange area is equal to or greater than 75 percent of the gross flange area for A36 steel or 92 percent for A572 Grade 50 steel. These values are shown in Table A-1.

TABLE A-1 WHEN TO IGNORE BOLT
AND RIVET HOLES IN BEAM AND
GIRDER FLANGES

Steel	No reduction if $A_{fn}/A_{fg} \geq$
A36	0.75
A572 Grade 50	0.92
A588	0.86

Should $0.5F_u A_{fn}$ be less than $0.6F_y A_{fg}$, the ASD Specification requires that the flexural properties of the section be based on an effective tension flange area A_{fe}, determined as follows:

$$A_{fe} = \frac{5}{6}\frac{F_u}{F_y}A_{fn} \qquad\qquad \text{(ASD Equation B10-3)}$$

In other words, to calculate the moment of inertia for the flanges, the transfer expression Ad^2 becomes $A_{fe}d^2$. If the bolt or rivet holes are in the tension flange, we assume they are in the compression flange as well. If they are only in the compression flange, they are ignored because it is felt that the fasteners can adequately transmit compression through the holes.

A beam with holes in its tension flange is examined in Example A-6 in accordance with the provisions of ASD Specification B10.

■ EXAMPLE A-6

A W16 × 31 beam section (d = 15.88 in., b_f = 5.525 in., t_f = 0.440 in.) consisting of A572 Grade 50 steel has two holes for $\frac{7}{8}$-in. bolts in its tension flange. Does the flange area need to be reduced for calculating the beam properties? If so, how much should I_x be reduced?

Solution.

$A_{fg} = (5.525)(0.440) = 2.431$ in.^{2}

$A_{fn} = 2.431 - (2)(1.0)(0.440) = 1.551$ in.^{2}

$\dfrac{A_{fn}}{A_{fg}} = \dfrac{1.551}{2.431} = 0.638 < 0.92$ for A572 steel

∴ A reduction is necessary

$A_{fe} = \dfrac{5}{6}\dfrac{F_u}{F_y}A_{fn} = \left(\dfrac{5}{6}\right)\left(\dfrac{65}{50}\right)(1.551) = 1.680$ in.^{2}

Reduction in flange area = $2.431 - 1.680 = 0.751$ in.^{2}

Reduction in $I_x = 2Ad^2$

$$= (2)(0.751)\left(\frac{15.88}{2} - \frac{0.440}{2}\right)^2 = \underline{\underline{89.5 \text{ in.}^4}} \qquad ■$$

The ASD Specification (F4) states that the allowable shear stress F_v is $0.40F_y$ if $h/t_w \leq 380/\sqrt{F_y}$. In this expression h is the clear distance between the beam or girder flanges. Almost all rolled sections fall into this class, and the cross-sectional area effective in resisting shear is considered to be the overall depth of the member times its web thickness. For such cases the shear stress is calculated as follows:

$$f_v = \frac{V}{dt_w} \leq 0.40F_y$$

Should h/t_w be greater than $380/\sqrt{F_y}$, the allowable shear stress F_v becomes $F_y/2.89(C_v) \leq 0.40F_y$, where C_v is given in Section F4 of the Specification and is the ratio of the "critical" web stress to the shear yield stress of the web. This allowable shear stress is applicable to the clear distance between the flanges times the web thickness.

$$f_v = \frac{V}{ht_w}$$

The maximum permissible external shear that a particular beam can resist can be calculated by substituting the allowable shear stress F_v into the preceding expression and solving for V.

$$V = F_v dt_w \quad \text{or} \quad V = F_v ht_w$$

In the beam tables of the ASD Manual this value V is tabulated for each of the sections normally used as beams with yield stresses of 36 and 50 ksi. With a glance at these tables the designer can check the shear in a particular beam. If it is too high he or she can quickly select another section that has a satisfactory allowable V.

■ EXAMPLE A-7

Select a W section for the load and span shown in Fig. A-5. Use A36 steel and assume full lateral bracing for the compression flange. Also check shear.

Figure A-5

Solution.

Assume beam weight = 84 lb/ft

$$M = \frac{(2.084)(24)^2}{8} + (30)(8) = 390.0 \text{ ft-k}$$

$$S_{reqd.} = \frac{(12)(390.0)}{24} = 195.0 \text{ in.}^3$$

W24 × 84 (d = 24.10 in., t_w = 0.470 in.)

Check shear:

$$\frac{d}{t_w} = \frac{24.10}{0.470} = 51.28 < \frac{380}{\sqrt{36}} = 63.33$$

$$V = (12)(2.084) + 30 = 55.0 \text{ k}$$

$$f_v = \frac{55.0}{(24.10)(0.470)} = 4.86 \text{ ksi}$$

$$< 0.4 f_y = 14.4 \text{ ksi} \qquad \text{(OK)} \quad \blacksquare$$

A-5 BOLTED CONNECTIONS

Bearing-Type Bolts

The allowable shearing stresses for high-strength bolts are presented in Table J3.2 of the ASD Specification. On observing these values you will note that the allowable shearing stress is reduced when the bolt threads are included in the shear planes because of the reduced cross-sectional area through the threads. For example, the allowable shear stress for A325 bolts for a bearing-type connection and standard-size holes is 21 ksi if threads are not excluded from the shear plane and 30 ksi if threads are excluded. Should a bolt be in double shear, its shearing strength is considered to be twice its single shear value.

The allowable design strength of a bolt in bearing equals the allowable bearing stress of the connected part in kips per square inch times the diameter of the bolt times the thickness of the member that bears against the bolt. (For countersunk bolts and rivets, one-half of the depth of the countersink should be deducted, says ASD Specification J3.3.) When the distance L_e in the direction of the force from the center of a regular or oversized hole (or from the center of the end of a slotted hole) to the edge of a connected part is not less than $1\frac{1}{2}$ times the bolt diameter d, and the distance s center-to-center of the holes is not less than $3d$, and two or more bolts are used in the direction of the line of force, the bearing strength is

$F_p = 1.2F_u$ for standard or short-slotted holes   (ASD Equation J3-1)

$F_p = 1.0F_u$ for long-slotted holes perpendicular to the load

(ASD Equation J3-2)

Should deformations around the hole not be a design consideration, the two preceding expressions may be replaced with

$$F_p = 1.5F_u \qquad \text{(ASD Equation J3-4)}$$

For the examples and home problems in this appendix, F_p is assumed to equal $1.2F_u$. This also is the value used for the examples and tables of Part 4 of the ASD Manual.

Though the allowable bearing stress of a bolt with a small end distance is reduced, *the allowable values of the other bolts in the connection are not reduced.* The value of F_p for a single bolt or for two or more bolts in the line *each with an end distance less than* $1\frac{1}{2}d$ is to be determined with the following expression:

$$F_p = \frac{L_e F_u}{2d} \leq 1.2F_u \qquad \text{(ASD Equation J3-3)}$$

The allowable load on the edge bolt is the lesser of the allowable shear strength of the bolt on its cross-sectional area and the allowable bearing strength against the side of the hole. Should the allowable bearing strength control, we can increase the allowable strength of the connection by increasing the thickness of the parts being connected and by increasing the edge distance (if edge distance causes a reduction in F_p).

Example A-8 illustrates the calculations involved in determining the strength of a bearing-type connection. Using a similar procedure, the number of bolts required for a certain loading condition is calculated in Example A-9.

The values given for the strength of bolts in this chapter, whether bearing-type or slip-critical, can be obtained from the tables entitled "Bolts, Threaded Parts, and Rivets Shear Allowable Load in Kips" of Part 4 of the ASD Manual.

■ EXAMPLE A-8

Determine the allowable design strength P of the bearing-type connection shown in Fig. A-6. The steel is A36, the bolts are $\frac{7}{8}$-in. A325, the holes are standard sizes, the threads are excluded from the shear plane, edge distances are $>1\frac{1}{2}d$, and the distance center-to-center of the holes is $>3d$. $F_p = 1.2F_u$.

Figure A-6

Solution. Allowable tensile force:

$$A_g = (\tfrac{1}{2})(12) = 6.0 \text{ in.}^2$$
$$A_n = 6.00 - (2)(1.0)(\tfrac{1}{2}) = 5.0 \text{ in.}^2 = A_e$$
$$P = (0.60)(36)(6.0) = 129.6 \text{ k}$$
$$P = (0.50)(58)(5.0) = 145 \text{ k}$$

Bolts in single shear and bearing on $\tfrac{1}{2}$ in.:

$$P \text{ for single shear} = (0.6)(30)(4) = 72 \text{ k} \leftarrow$$
$$P \text{ for bearing} = (0.5)(\tfrac{7}{8})(1.2 \times 58)(4) = 121.8 \text{ k}$$
$$\underline{\underline{\text{Design } P = 72 \text{ k}}} \qquad \blacksquare$$

■ EXAMPLE A-9

How many $\tfrac{3}{4}$-in. A325 bolts in standard-size holes with threads excluded from the shear plane are required for the bearing-type connection shown in Fig. A-7? Use A36 steel and assume edge distances and spacing requirements are met, including those for $F_p = 1.2F_u$.

Figure A-7

Solution. Bolts in double shear and bearing on $\tfrac{3}{4}$ in.:

$$\text{Design shear strength per bolt} = (2)(0.44)(30) = 26.4 \text{ k} \leftarrow$$
$$\text{Design bearing strength per bolt} = (\tfrac{3}{4})(\tfrac{3}{4})(1.2 \times 58) = 39.1 \text{ k}$$
$$\text{Number of bolts required} = \frac{190}{26.4} = 7^+$$
$$\underline{\underline{\text{Use 8 or 9 bolts (depending on arrangement)}}} \qquad \blacksquare$$

Slip-Critical Connections

If bolts are tightened to their required tensions for slip-critical connections there is very little chance of their bearing against the plates which they are connecting. In fact, tests show that there is very little chance of slip occurring unless there is a calculated shear of at least 50 percent of the total bolt tension. This means that slip-critical bolts are not stressed in shear; however, the ASD Specification provides allowable shear strengths (they really are permissible friction values on the faying surfaces) so that the designer can handle the connections in just about the same manner he or she uses for bearing-type connections. The ASD

Specification assumes the bolts are in "shear" and no bearing, and the allowable shear stresses for high-strength bolts are given in their Table J3.2.

Example A-10 presents the design of a slip-critical connection for a lap joint.

■ EXAMPLE A-10

It is desired to design a slip-critical connection for the plates shown in Fig. A-8 to resist an 80-k load using 1-in. A325 high-strength bolts with threads excluded from the shear plane and meeting all edge and center-to-center distance requirements.

Figure A-8

Solution.   Slip-critical design, bolts in "single shear," and no bearing:

$$\text{"Single shear"} = (A_{\text{bolt}})(17.0) = (0.785)(17.0) = 13.35 \text{ k}$$

$$\text{Number of bolts required} = \frac{80}{13.35} = 5.99$$

Use 6 bolts   ■

A-6 WELDED CONNECTIONS

Table J2.5 of the ASD Specification provides allowable stresses for fillet welds, complete-penetration groove welds, and partial-penetration groove welds. Example A-11 illustrates the calculations necessary to determine the allowable strength of a fillet-welded connection. You will notice that the allowable stress in the weld equals 0.30 times the nominal tensile strength of the weld.

■ EXAMPLE A-11

Determine the allowable shear force that may be applied to a 1-in. length of a $\frac{5}{16}$-in. fillet weld using (a) the shielded metal arc process (SMAW) and (b) the submerged arc process (SAW). Use the ASD Specification and E70 electrodes with a minimum tensile strength of 70 ksi

Solution.

(a) SMAW process:

$$\text{Effective throat thickness of weld} = (0.707)(\tfrac{5}{16}) = 0.221 \text{ in.}$$

$$\text{Allowable capacity} = (0.221)(0.30 \times 70)$$

$$= 4.64 \text{ k/in.}$$

(b) SAW process:

From ASD (J2.2a) the effective throat thickness of weld $= \frac{5}{16}$ in.

Allowable capacity $= (\frac{5}{16})(0.30 \times 70) = \underline{\underline{6.56 \text{ k/in.}}}$ ■

Fillet welds may not be designed using an allowable stress that is greater than permitted on the adjacent members being connected. For instance, the allowable shear stress on the effective area of fillet welds is 0.30 times the tensile strength of the electrode, but may not exceed the allowable stress in the base material ($0.60F_y$ in tension). In the ASD Specification fillet welds are assumed to transmit loads by shear on the effective area regardless of the direction of the loads or of the position of the welds at the connections.

Examples A-11 and A-12 illustrate the calculations necessary to determine the allowable loads that can be applied to plates connected with SMAW and SAW fillet welds. In each of these examples the allowable strength per inch of the welds controls and is multiplied by the total length of the welds to give the total allowable capacity of the connections.

■ **EXAMPLE A-12**
What is the allowable capacity of the connection shown in Fig. A-9 if A36 steel, the ASD Specification, and E70 electrodes are used? The $\frac{7}{16}$-in. fillet welds shown were made by the SMAW process.

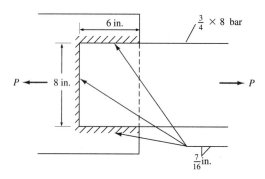

Figure A-9

Solution.

Effective throat thickness $= t_e = (0.707)(\frac{7}{16}) = 0.309$ in.

Allowable capacity of weld/in. $= (0.309)(0.30 \times 70) = 6.489$ k/in. ←

Allowable tensile capacity of weld $= (20)(6.489) = 129.8$ k

Allowable tensile capacity of PL $= (\frac{3}{4} \times 8)(0.60 \times 36) = 129.6$ k $\leftarrow$

$P = 129.6$ k   ■

■ **EXAMPLE A-13**

Repeat Example A-11 using SAW welds.

Solution.

Effective throat thickness $= t_e = (0.707)(\frac{7}{16}) + 0.11 = 0.419$ in.

Allowable shear capacity of weld/in. $= (0.419)(0.30 \times 70)$

$$= 8.799 \text{ k/in.} \leftarrow$$

Allowable shear capacity of weld $= (20)(8.799) = 176$ k

Allowable tensile capacity of PL $= (\frac{3}{4} \times 8)(0.60 \times 36) = 129.6$ k $\leftarrow$

$P = 129.6$ k   ■

Example A-14 illustrates the design of fillet welds for two plates.

■ **EXAMPLE A-14**

Using A36 steel, the ASD Specification, and E70 electrodes, design SMAW fillet welds to resist a full capacity load on the $\frac{3}{8} \times 6$-in. member shown in Fig. A-10.

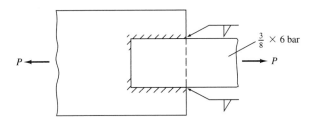

Figure A-10

Solution.

$P = (\frac{3}{8})(6)(22) = 49.5$ k

Max weld size $= \frac{3}{8} - \frac{1}{16} = \frac{5}{16}$ in.   (ASD Section J2-2b)

Min weld size $= \frac{3}{16}$ in. (Table 14-2 in this text)   (ASD Table J2.3)

Use $\frac{5}{16}$-in. weld

Effective thickness $= t_e = (0.707)(\tfrac{5}{16}) = 0.221$ in.

Allowable capacity of weld/in. $= (0.221)(0.30 \times 70) = 4.64$ k/in. $\leftarrow$

Length required $= \dfrac{49.5}{4.64} = 10.67$ in.

Use end return not less than $2 \times \tfrac{5}{16} = \tfrac{10}{6}$ (say 1 in.)

$10.67 - 2 = 8.67$ in. or $4\tfrac{1}{2}$ in. each side

However, use $6 - 1 = 5$-in. welds each side as required by ASD Section J2.2b plus end returns.   ∎

Appendix **B**

Derivation of the Euler Formula

The Euler formula is derived in this section for a straight, concentrically loaded, homogeneous, long, slender, elastic, and weightless column with rounded ends. It is assumed that this perfect column has been laterally deflected by some means as shown in Fig. B-1 and that, if the concentric load P was removed, the column would straighten out completely.

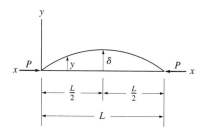

Figure B-1

The x and y axes are located as shown in the figure. As the bending moment at any point in the column is $-Py$, the equation of the elastic curve can be written as follows.

$$EI\frac{d^2y}{dx^2} = -Py$$

For convenience in integration both sides of the equation are multiplied by $2dy$ and the integration is performed.

$$EI\, 2\frac{dy}{dx}d\frac{dy}{dx} = -2Py\,dy$$

$$EI\left(\frac{dy}{dx}\right)^2 = -Py^2 + C_1$$

When $y = \delta$, $dy/dx = 0$, and the value of C_1 will equal $P\delta^2$ and

$$EI\left(\frac{dy}{dx}\right)^2 = -Py^2 + P\delta^2$$

The preceding expression is arranged more conveniently as follows.

$$\left(\frac{dy}{dx}\right)^2 = \frac{P}{EI}(\delta^2 - y^2)$$

$$\frac{dy}{dx} = \sqrt{\frac{P}{EI}}\sqrt{\delta^2 - y^2}$$

$$\frac{dy}{\sqrt{\delta^2 - y^2}} = \sqrt{\frac{P}{EI}}dx$$

Integrating this expression, the result is

$$\text{arc sin}\frac{y}{\delta} = \sqrt{\frac{P}{EI}}x + C_2$$

When $x = 0$ and $y = 0$, $C_2 = 0$. The column is bent into the shape of a sine curve expressed by the equation

$$\text{arc sin}\frac{y}{\delta} = \sqrt{\frac{P}{EI}}x$$

When $x = L/2$, $y = \delta$, resulting in

$$\frac{\pi}{2} = \frac{L}{2}\sqrt{\frac{P}{EI}}$$

In this expression P is the *critical buckling load* or the maximum load that the column can support before it becomes unstable. Solving for P

$$P = \frac{\pi^2 EI}{L^2}$$

This expression is the Euler formula, but usually it is written in a little different form involving the slenderness ratio. Since $r = \sqrt{I/A}$ and $r^2 = I/A$ and $I = r^2 A$, the Euler formula may be written as

$$\frac{P}{A} = \frac{\pi^2 E}{(L/r)^2} = F_e$$

Appendix **C**

Slender Compression Elements

Section B5 of the LRFD Specification is concerned with the local buckling of compression elements. In that section elements are classified as being compact, noncompact, or slender. As compact and noncompact elements have been discussed previously, this section is concerned only with the slender element. A brief summary of the LRFD method for determining design stresses for such members is presented in the next few paragraphs.

Should b/t ratios exceed the values given in LRFD Table B5.1 for noncompact elements, those elements will be classified as being slender, and their critical or F_{cr} stresses will have to be reduced. The design strength of an axially loaded compression member with slender elements will be reduced by multiplying it by a reduction factor Q.

The value of Q is equal to the product of two reduction factors Q_s and Q_a. Their values are dependent on whether the member consists of stiffened or unstiffened elements. Q_a is a reduction factor for slender stiffened compression elements, while Q_s is a reduction factor for slender unstiffened compression elements. Three cases are considered in Appendix B of the LRFD Specification.

1. For members consisting of unstiffened elements only, Q_s is to be determined with the appropriate formulas presented in this appendix, and $Q_a = 1.0$.
2. For members consisting of stiffened elements only, $Q_s = 1.0$, and Q_a is to be determined with the formulas to be presented in this appendix.
3. For members consisting of some stiffened elements and some unstiffened ones, Q_s and Q_a are determined with the appropriate formulas to be presented in this appendix.

Expressions for computing Q are given in LRFD Appendix B5.3a for the following types of members: (a) single angles; (b) flanges, angles, and plates projecting from rolled beams or columns or other compression members; (c) flanges, angles, and plates projecting from built-up columns or other compression members; and (d) stems of tees. In these equations the following terms are used:

b = width of unstiffened compression element, in.

t = thickness of unstiffened compression element, in.

F_y = specified minimum yield stress, ksi

Only the Q_s expressions for case (a) are presented here.
When $76.0/\sqrt{F_y} < b/t < 155/\sqrt{F_y}$

$$Q_s = 1.340 - 0.00447(b/t)\sqrt{F_y} \qquad \text{(LRFD Equation A-B5-3)}$$

When $b/t > 155/\sqrt{F_y}$

$$Q_s = 15{,}500/[F_y(b/t)^2] \qquad \text{(LRFD Equation A-B5-4)}$$

In Example C-1, the value of Q_s is computed for a pair of angles that are used as a compression member. As the member consists of stiffened elements, $Q_a = 1.0$. The computed value for Q_s can be checked in the double-angle tables of Part 1 of the Manual. This same double-angle member is further considered in Appendix D as to lateral-torsional buckling and the Q_s determined here is used there.

■ EXAMPLE C-1

The pair of L8 × 6 × $\frac{1}{2}$ s, long legs back-to-back separated by $\frac{3}{8}$ in., shown in Fig. C-1, is used as a compression member. Compute Q_s for this member, which is assumed to have an $F_y = 50$ ksi.

Figure C-1

Solution.   As shown in LRFD Appendix B5.2a, if

$$\frac{76}{\sqrt{F_y}} < \frac{b}{t} < \frac{155}{\sqrt{F_y}}$$

then LRFD Equation A-B5-3 is used to compute Q_s for angles projecting from a column.

$$\frac{76}{\sqrt{50}} = 10.75 < \frac{8}{\frac{1}{2}} = 16 < \frac{155}{\sqrt{50}} = 21.92$$

$$Q_s = 1.340 - 0.00447\left(\frac{b}{t}\right)\sqrt{F_y} \qquad \text{(LRFD Equation A-B5-3)}$$

$$= 1.340 - (0.00447)(16)(\sqrt{50})$$

$$= \underline{\underline{0.834}}$$

Checks with value in double angle tables.

If we have slender elements in a compression member, its design compression strength is to be computed as follows

For $\lambda_c\sqrt{Q} < 1.5$

$$F_{cr} = Q(0.658^{Q\lambda_c^2})F_y \qquad \text{(LRFD Equation A-B5-15)}$$

For $\lambda_c\sqrt{Q} > 1.5$

$$F_{cr} = \left[\frac{0.877}{\lambda_c^2}\right]F_y \qquad \text{(LRFD Equation A-B5-16)}$$

Appendix **D**

Flexural-Torsional Buckling of Compression Members

Usually symmetrical members such as W sections are used as columns. Torsion will not occur in such sections if the lines of action of the lateral loads pass through their shear centers. The *shear center* is that point in the cross section of a member through which the resultant of the transverse loads must pass so that no torsion will occur. The calculations necessary to locate shear centers are presented in Chapter 10. The shear centers of the commonly used doubly symmetrical sections occur at their centroids. This is not necessarily the case for other sections such as channels and angles. Shear center locations for several types of sections are shown in Fig D-1. Also shown in the figure are the

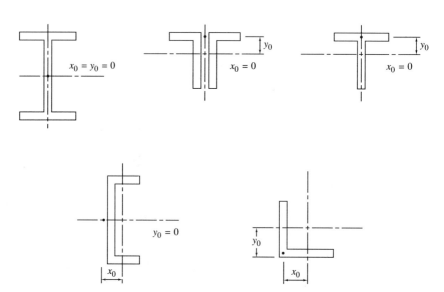

Figure D-1 Shear center locations for some common column sections.

coordinates x_0 and y_0 for the shear center of each section with respect to its centroid. These values are needed to solve the flexural-torsional formulas, presented later in this section.

Even though loads pass through shear centers, torsional buckling still may occur. If you load any section through its shear center no torsion will occur—but one still computes torsional buckling strength for these members, i.e., buckling load does not depend on the nature of the axial or transverse loading, but, rather, on the cross-section properties, column length, and support conditions.

The average designer does not consider the torsional buckling of symmetrical shapes or the flexural-torsional buckling of unsymmetrical shapes. Their usual feeling is that these conditions don't control the critical column loads, or at least don't affect them very much. Should we, however, have unsymmetrical columns or even symmetrical columns made up of thin plates, we will find that torsional buckling or flexural-torsional buckling may significantly reduce column capacities.

In Appendix E of the LRFD Specification a long list of formulas is presented for computing the flexural-torsional strength of column sections. The values given for column design strengths ($\phi_c P_n$ values) for double angles, single angles, and tees in Part 3 of the LRFD Manual make use of these formulas.

For flexural-torsion $P_u = \phi_c P_n = \phi_c A_g F_{cr}$ with $\phi_c = 0.85$ and F_{cr} to be determined from the formulas to follow from the specification. A list of definitions that are needed for using these formulas also is provided.

If $\lambda_e \sqrt{Q} \leq 1.5$

$$F_{cr} = Q(0.658^{Q\lambda_e^2})F_y \qquad \text{(LRFD Equation A-E3-2)}$$

If $\lambda_e \sqrt{Q} > 1.5$

$$F_{cr} = \left(\frac{0.877}{\lambda_e^2}\right)F_y \qquad \text{(LRFD Equation A-E3-3)}$$

in which $Q = 1.0$ for elements meeting the width-thickness ratios λ_r of LRFD section B5.1 and if not calculated as described in LRFD Appendixes E3 and B5.3.

$$\lambda_e = \sqrt{\frac{F_y}{F_e}} \qquad \text{(LRFD Equation A-E3-4)}$$

F_e = critical flexural-torsional elastic buckling stress

For doubly symmetric shapes

$$F_e = \left[\frac{\pi^2 E C_w}{(K_z L)^2} + GJ\right]\frac{1}{I_x + I_y} \qquad \text{(LRFD Equation A-E3-5)}$$

For singly symmetric shapes where y is the axis of symmetry

$$F_e = \frac{F_{ey} + F_{ez}}{2H}\left[1 - \sqrt{1 - \frac{4F_{ey}F_{ez}H}{(F_{ey} + F_{ez})^2}}\right] \qquad \text{(LRFD Equation A-E3-6)}$$

For unsymmetrical section F_e is the lowest root of the following cubic equation

$$(F_e - F_{ex})(F_e - F_{ey})(F_e - F_{ez}) - F_e^2(F_e - F_{ey})\left(\frac{x_0}{\bar{r}_0}\right)^2$$

$$- F_e^2(F_e - F_{ex})\left(\frac{y_0}{\bar{r}_0}\right)^2 = 0 \qquad \text{(LRFD Equation A-3E-7)}$$

A perhaps more convenient form of LRFD Equation A-E3-7 follows

$$HF_e^3 + \left[\frac{1}{\bar{r}_0^2}(y_0^2 F_{ex} + x_0^2 F_{ey}) - (F_{ex} + F_{ey} + F_{ez})\right]F_e^2$$

$$+ (F_{ex}F_{ey} + F_{ex}F_{ez} + F_{ey}F_{ez})F_e - F_{ex}F_{ey}F_{ez} = 0$$

K_z = effective length factor for torsional buckling

G = shear modulus (ksi)

C_w = warping constant (in.^{6})

J = torsional constant (in.^{4})

$$\bar{r}_0^2 = x_0^2 + y_0^2 + \frac{I_x + I_y}{A} \qquad \text{(LRFD Equation A-E3-8)}$$

$$H = 1 - \left(\frac{x_0^2 + y_0^2}{\bar{r}_0^2}\right) \qquad \text{(LRFD Equation A-E3-9)}$$

$$F_{ex} = \frac{\pi^2 E}{(KL/r)_x^2} \qquad \text{(LRFD Equation A-E3-10)}$$

$$F_{ey} = \frac{\pi^2 E}{(KL/r)_y^2} \qquad \text{(LRFD Equation A-E3-11)}$$

$$F_{ez} = \left[\frac{\pi^2 EC_w}{(K_z L)^2} + GJ\right]\frac{1}{A\bar{r}_0^2} \qquad \text{(LRFD Equation A-E3-12)}$$

The values of C_w, J, $\bar{r}_0$, and H are provided for many sections in the "Flexural-Torsional Properties" tables of Part 1 of the Manual.

In Example D-1 which follows, the author has gone through all of these formulas for a double-angle column. The resulting value for $\phi_c P_n$ is shown to coincide with the value given in the Manual.

■ EXAMPLE D-1

Determine (a) the flexural buckling strength and (b) the flexural-torsional buckling strength of an 18-ft pinned-end column consisting of two Ls 8 × 6 × $\frac{1}{2}$ long legs back to back with a $\frac{3}{8}$-in. gusset plate between them. The angles are made from A36 steel and are connected to each other at 6-ft intervals with two fully tensioned high-strength bolts as shown in Fig. D-2. The cross section of the member is shown in Fig. D-3. $G = 11,200$ ksi, and $K = 1.0$.

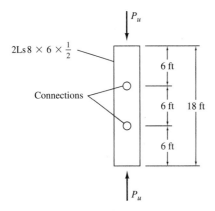

Figure D-2

Solution. Using two Ls 8 × 6 × $\frac{1}{2}$ long legs back to back, $\frac{3}{8}$-in. gusset plate in between ($A = 13.5$ in.^{2}, $r_x = 2.56$ in., r_z for one $L = 1.30$ in., and $r_y = 2.44$ in.)

(a) Flexural buckling

$$\left(\frac{KL}{r}\right)_y = \frac{(1.0)(12 \times 18)}{2.44} = 88.52$$

$$\phi_c F_{cr} = 20.26 \text{ ksi}$$

$$\phi_c P_n = (20.26)(13.5) = \underline{\underline{273.5 \text{ k}}}$$

(b) Flexural-torsional buckling

Cross section of member shown in Fig. D-2 together with some properties needed in the calculations.

Figure D-3

From flexural-torsional properties tables: $J = 0.584$ in.^{4} $\left.\right\}$ for 1 angle
$C_w = 2.280$ in.^{6}

$r_0 = 4.18$ in. $\left.\right\}$ for 2 angles
$H = 0.718$

Checking table values of r_0 and H

$$\bar{r}_0^2 = x_0^2 + y_0^2 + r_x^2 + r_y^2 = 0^2 + 2.22^2 + 2.56^2 + 2.44^2$$

(LRFD Equation A-E3-8)

$$\bar{r}_0 = 4.18 \text{ in.}$$

$$H = 1 - \frac{0^2 + 2.22^2}{(4.18)^2} = 0.718$$

(LRFD Equation A-E3-9)

Computing values of F_{ey}, F_{ez}, and F_e

$$\frac{a}{r_z} = \frac{72}{1.30} = 55.38 > 50 \ \therefore \text{ Use } \left(\frac{KL}{r}\right)_m \text{ rather than } \frac{K_y L_y}{r_y}$$

$$\left(\frac{KL}{r}\right)_y = \frac{(1.0)(12 \times 18)}{2.44} = 88.52$$

$$\left(\frac{KL}{r}\right)_m = \sqrt{(88.52)^2 + (55.38 - 50)^2} = 88.68$$

$$F_{ey} = \frac{(\pi^2)(29{,}000)}{(88.68)^2} = 36.40 \text{ ksi} \qquad \text{(LRFD Equation A-E3-11)}$$

$$F_{ez} = \left[\frac{(\pi)^2(29{,}000)(2 \times 2.28)}{(1.0 \times 12 \times 18)^2} + 11{,}200 \times 2 \times 0.584\right]$$

$$\times \frac{1}{(13.5)(4.18)^2}$$

$$= 55.58 \text{ ksi} \qquad \text{(LRFD Equation A-E3-12)}$$

$$F_e = \frac{36.40 + 55.58}{(2)(0.718)}\left[1 - \sqrt{1 - \frac{(4)(36.40)(55.58)(0.718)}{(36.40 + 55.58)^2}}\right]$$

$$= 28.20 \text{ ksi} \qquad \text{(LRFD Equation A-E3-6)}$$

$$\lambda_e = \sqrt{\frac{F_y}{F_e}} = \sqrt{\frac{36}{28.20}} = 1.130 < 1.5$$

$$\text{(LRFD Equation A-E3-4)}$$

$$\phi_c F_{cr} = (0.85)(0.911 \times 36)(0.658)^{1.130^2} = 16.34 \text{ ksi}$$

where F_y is multiplied by $Q_s = 0.911$ from Manual page 1-97 as a slender unstiffened element

$$\phi_c P_n = (16.34)(13.5) = \underline{220.6 \text{ k}}$$

which checks with LRFD Manual table value page 3-65. This is an 11.5 percent reduction in load from the flexural buckling value. ∎

Appendix E

Moment-Resisting Column Base Plates

Frequently, column bases are designed to resist bending moments as well as axial loads. An axial load causes compression between a base plate and the supporting footing, while a moment increases the compression on one side and decreases it on the other side. For average moments the forces may be transferred to the footing through flexure of the base plate, but when they are very large, stiffened or booted connections may be used. For a small moment the entire contact area between the plate and the supporting footing will remain in compression. This will be the case if the resultant load falls within the middle third of the plate length in the direction of bending.

Fig. E-1(a) and (b) shows base plates suitable for resisting relatively small moments. For these cases the moments are sufficiently small to permit their transfer to the footings by bending of the base plates. The anchor bolts may or may not have calculable stresses, but they are nevertheless considered necessary for good construction practice. They definitely are needed to hold the columns firmly in place and upright during the initial steel erection process. Temporary guy cables are also necessary during erection. The anchor bolts should be substantial and capable of resisting unforeseen erection forces. Sometimes these small plates are attached to the columns in the shop and sometimes they are shipped loose to the job and carefully set to the correct elevations in the field.

Should the eccentricity $(e = M/P)$ be sufficiently large so that the resultant falls outside the middle third of the plate, there will be an uplift on the other side of the column, putting the anchor bolts on that side in tension.

The moment will be transferred from the column into the footing by means of the anchor bolts, embedded a sufficient distance into the footing to develop the forces. The embedment should be calculated as required by reinforced-concrete design methods.[1] The booted connection shown in Fig. E-1(c) is assumed to be

---

[1] *Building Code Requirements for Reinforced Concrete (ACI 318-89) and Commentary (ACI 318R-89)* (Detroit: American Concrete Institute, 1989), pp. 188–189, 250–251.

(a)

(b)

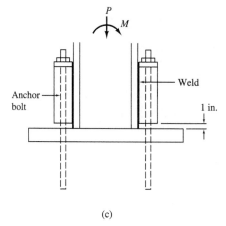

(c)

Figure E-1 Moment-resisting column bases.

welded to the column. The boots generally are made of angles or channels and are not, as a rule, connected directly to the base plate. Rather, the moment is transmitted from the column to the foundation by means of the anchor bolts. When booted connections are used, the base plates are normally shipped loose to the job and carefully set to the correct elevations in the field.

The capacity of these connections to resist rotation is dependent on the lengths of the anchor bolts, which are available to deform elastically. This capacity can be increased somewhat by pretensioning the anchor bolts. (This is similar to our prestress discussion for high-strength bolts in Section 13-3 in this text.) Actually, prestressing is not very dependable and usually is not done because of the long-term creep in the concrete.

When a moment-resisting or rigid connection between a column and its footing is used, it is absolutely necessary for the supporting soil or rock beneath the footing to be appreciably noncompressible, or the column base will rotate as shown in Fig. E-2. If this happens, the rigid connection between the column and the footing is useless. For the purpose of this section the subsoil is assumed to be capable of resisting the moment applied to it without appreciable rotation.

Figure E-2

Quite a few methods have been developed through the years for designing moment-resisting base plates. One rather simple procedure used by many designers is presented here by the author. As a first numerical example, a column base plate is designed for an axial load and a relatively small bending moment such that the resultant load falls between the column flanges. Assumptions are made for the width and length of the plate, after which the pressures underneath the plate are calculated and compared to the permissible value. If the pressures are unsatisfactory, the dimensions are changed and the pressures recalculated, until the values are satisfactory. The moment in the plate is calculated and the plate thickness determined. The critical section for bending is assumed to be at the center of the flange on the side where the compression is highest. Various designers will assume the point of maximum moment is located at some other point, as at the face of the flange or the center of the anchor bolt.

The moment is calculated for a 1-in.-wide strip of the plate and is equated to its resisting moment. The resulting expression is solved for the required thickness of the plate as follows:

$$M_u = \frac{\phi_b F_y I}{c} = \frac{\phi_b F_y (\frac{1}{12})(1)(t)^3}{t/2}$$

$$t = \sqrt{\frac{6M_u}{\phi_b F_y}} \quad \text{with} \quad \phi_b = 0.9$$

From Section J9 of the LRFD Specification

$$P_p = 0.85 f_c' A_1 \sqrt{\frac{A_2}{A_1}}$$

If we assume $\sqrt{\dfrac{A_2}{A_1}} \geq 2$

$$P_p = 1.7 f_c' A_1$$
$$\phi_c P_p = \phi_c\, 1.7 f_c' A_1 \quad \text{with} \quad \phi_c = 0.60$$

■ EXAMPLE E-1

Design a moment-resisting base plate to support a W14 × 120 column with an axial load of 620 k and a bending moment of 225 ft-k. Use A36 steel with $F_y = 36$ ksi and a concrete footing with $f_c' = 3.0$ ksi. $\phi_c F_p = (0.6)(1.7)(3.0) = 3.06$ ksi

Solution.   Using a W14 × 120 ($d = 14.48$ in., $t_w = 0.590$ in., $b_f = 14.670$ in., $t_f = 0.940$ in.)

$$e = \frac{(12)(225)}{620} = 4.35 \text{ in.}$$

∴ The resultant falls between the column flanges and within the middle third of the plate.

7.23 in. ⟶|⟵ 13.54 in. ⟶|⟵ 7.23 in.

⟵ 28 in. ⟶

0.075

(5.8)

(7.73)

1.606   2.139   **Figure E-3**

Try a 20 × 28 in. plate (after a few trials)

$$f = -\frac{P}{A} \pm \frac{P_u ec}{I} = -\frac{620}{(20)(28)} \pm \frac{(620)(4.35)(14)}{(\frac{1}{12})(20)(28)^3}$$

$$= -1.107 \pm 1.032 \begin{cases} -2.139 < 3.06 \text{ ksi} \\ -0.075 \text{ ksi} \end{cases}$$ (OK)

Taking moments to right at center of right flange (see Fig. E-3):

$$M_u = (5.81)\left(\frac{7.23}{3}\right) + (7.73)\left(\frac{2}{3} \times 7.23\right) = 51.26 \text{ in.-k}$$

$$t = \sqrt{\frac{6M_u}{\phi_b F_y}} = \sqrt{\frac{(6)(51.26)}{(0.9)(36)}} = 3.08 \text{ in.}$$

Use PL$3\frac{1}{4}$ × 20 × 2 ft 4 in.   ■

The moment considered in Example E-2 is of such a magnitude that the resultant load falls outside the column flange. As a result, there will be uplift on one side, and the anchor bolt will have to furnish the needed tensile force to provide equilibrium.

In this design the anchor bolts are assumed to have no significant tension due to tightening. As a result, they are assumed not to affect the force system. As the moment is applied to the column, the pressure shifts toward the flange on the compression side. It is assumed that the resultant of this compression is located at the center of the flange.

■ EXAMPLE E-2

Repeat Example E-1 with the same column and design stresses, but with the moment increased from 225 ft-k to 460 ft-k. Refer to Fig. E-4.

Figure E-4

Solution.   Using a W14 × 120 (d = 14.48 in., t_w = 0.590 in., b_f = 14.670 in., t_f = 0.940 in.)

$$e = \frac{(12)(460)}{620} = 8.90 \text{ in.}$$

∴ The resultant falls outside the column flange.
Taking moments about the center of the right flange:

$$(620)(8.90 - 6.77) - 15.26 \, T = 0$$

$$T = 86.54 \text{ k}$$

$$\text{Anchor bolt } A_{\text{reqd.}} = \frac{86.54}{\phi 0.75 F_u} = \frac{86.54}{(0.75)(0.75)(58)}$$

$$= 2.65 \text{ in.}^2$$

Use two $1\frac{3}{8}$-in. diameter bolts each side

Approximate plate size, assuming a triangular pressure distribution

$$R_u = 620 + 86.54 = 706.54 \text{ k}$$

A of plate reqd. $= \dfrac{R_u}{\text{Avg } \phi_c F_p} = \dfrac{R_u}{\phi_c F_p/2}$ with $\phi_c F_p$ given $= 3.06$ ksi in
statement of Example E-1.

$$A \text{ reqd.} = \frac{706.54}{3.06/2} = 461.79 \text{ in.}^2$$

Try a 24-in.-long plate
    The load is located 24 in. minus the distance from the column c.g. to the column flange c.g. $= \frac{24}{2} - 6.77 = 5.23$ in. from edge of plate. Thus the pressure triangle will be $3 \times 5.23 = 15.69$ in. long and the required plate length will equal

$$\frac{706.54}{\frac{1}{2} \times 3.06 \times 15.69} = 29.43 \text{ in.} \quad \text{(Say 30 in. long)}$$

If the plate is made 20 in. wide, the pressure zone will have an area $= 20 \times 29.43 = 588.6$ in.^{2} and the maximum pressure will be twice the average pressure, or

$$\frac{706.54}{588.6} \times 2 = 2.40 \text{ ksi} < 3.06 \text{ ksi} \qquad \text{(OK)}$$

Figure E-5

Figure E-6 Fillet welds from column to base plate.

Taking moments to right at center of right column flange (see Fig. E-5)

$$M = (5.16)\left(\frac{5.23}{3}\right) + (6.28)(\tfrac{2}{3} \times 5.23) = 30.89 \text{ in.-k}$$

$$t = \sqrt{\frac{(6)(30.89)}{(0.9)(36)}} = 2.39 \text{ in.}$$

Use PL$2\frac{1}{2}$ × 20 × 2 ft 6 in.

Design of weld from column to base PL (see Fig. E-6)

Total length of fillet weld each flange

$$= (2)(14.48) - 0.590 = 28.37 \text{ in.}$$

$$C = T = \frac{(12)(460)}{13.54} = 407.68 \text{ k}$$

Strength of 1-in. long 1-in. fillet weld using E70 electrodes

$$= (0.75)(0.6 \times 70)(1.0) = 31.5 \text{ k/in.}$$

$$\text{Weld size required} = \frac{407.68}{(28.37)(31.5)} = 0.456 \text{ in.}$$

Use $\frac{1}{2}$-in. weld   ■

In these examples the variation of stress in the concrete supporting the columns has been assumed to vary in a triangular or straight-line fashion. It is possible to work with an assumed ultimate concrete theory where the concrete in compression under the plate is assumed to fail at a stress of $0.85 f'_c$. Examples of such designs are available in several texts.[2] More detailed information concerning moment-resisting bases is available in several places.[3,4]

---

[2] W. McGuire, *Steel Structures* (Englewood Cliffs, N.J.: Prentice Hall, 1968), pp. 987–1004.

[3] C. G. Salmon, L. Schenker, and B. G. Johnston, "Moment-Rotation Characteristics of Column Anchorages," *Transactions ASCE,* 122 (1957), pp. 132–154.

[4] J. T. DeWolf and E. F. Sarisley, "Column Base Plates with Axial Loads and Moments," *Journal of Structural Division,* ASCE, 106, ST11 (November 1980), pp. 2167–2184.

Appendix F

Ponding

It has been claimed that almost 50 percent of the lawsuits faced by building designers are concerned with roofing systems.[1] Ponding, a problem with many flat roofs, is one of the common subjects of such litigation. If water accumulates more rapidly on a roof than it runs off, ponding results because the increased load causes the roof to deflect into a dish shape that can hold more water, which causes greater deflections, and so on. This process continues until equilibrium is reached, or until collapse occurs. Ponding can be caused by increasing deflections, clogged roof drains, settlement of footings, warped roof slabs, and so on.

The best way to prevent ponding is to use appreciable roof slopes ($\frac{1}{4}$ in. per ft or more) together with good drainage facilities. Supposedly more than two thirds of the flat roofs in the United States have slopes less than $\frac{1}{4}$ in. per ft. The construction of roofs with slopes this large will increase building costs by only a few percent as compared to perfectly flat roofs. The supporting girders for flat roofs with long spans should definitely be cambered to reduce the possibility of ponding (as well as the sagging that is so disturbing to the people occupying a building).

Section K2 of the LRFD Specification states that, unless roof surfaces have sufficient slopes to areas of free drainage or sufficient individual drains to prevent water accumulation, the strength and stability of the roof systems during

---

[1] Gary Van Ryzin, "Roof Design: Avoid Ponding by Sloping to Drain," *Civil Engineering* (New York: ASCE, January 1980), pp. 77–81.

ponding conditions must be investigated. The very detailed work of Marino[2] forms the basis of the ponding provisions of the LRFD Specification. Many other useful references are available.[3,4,5]

The amount of water that can be retained on a roof depends on the flexibility of the framing. The specifications state that a roof system can be considered stable and not requiring further investigation if the following expressions are satisfied:

$$C_p + 0.9C_s \leq 0.25 \qquad \text{(LRFD Equation K2-1)}$$

$$I_d \geq 25(S^4)10^{-6} \qquad \text{(LRFD Equation K2-2)}$$

where

$C_p = 32\, L_s L_p^4/10^7 I_p$

$C_s = 32\, SL_s^4/10^7 I_s$

L_p = length of primary members, ft

L_s = length of secondary members, ft

S = spacing of secondary members, ft

I_p = moment of inertia of primary members, in.^{4}

I_s = moment of inertia of secondary members, in.^{4}

I_d = moment of inertia of the steel deck (if one is used) supported on the secondary members, in.^{4} per ft

Should steel roof decks be used, their I_d must at least equal the value given by Equation K2-2. If the roof decking is the secondary system (i.e., no secondary beams, joists, etc.), it should be handled with Equation K2-1.

Some other LRFD requirements in applying these expressions follow:

1. The moment of inertia I_s must be decreased by 15 percent for trusses and steel joists.
2. Steel decking is considered to be a secondary member supported directly by the primary members.
3. Stresses caused by wind or seismic forces do not have to be considered in the ponding calculations.

Should moments of inertia be needed for open-web joists, they can be computed from the member cross sections, or perhaps more easily backfigured

[2] F. J. Marino, "Ponding of Two-Way Roof System," *Engineering Journal*, AISC, 3rd quarter, no. 3 (1966), pp. 93–100.

[3] L. B. Burgett, "Fast Check for Ponding," *Engineering Journal*, AISC, 10, no. 1 (1st quarter, 1973), pp. 26–28.

[4] J. Chinn, "Failure of Simply-Supported Flat Roofs by Ponding of Rain," *Engineering Journal*, AISC, no. 2, 2nd quarter (1965), pp. 38–41.

[5] J. L. Ruddy, "Ponding of Concrete Deck Floors," *Engineering Journal*, AISC, 23, no. 2 (3rd quarter, 1986), pp. 107–115.

from the resisting moments and allowable stresses given in the joist tables. (Since $M_R = FI/c$, we can compute $I = M_R c/F$.)

In effect these equations reflect *stress indexes* or percentage stress increases. For instance, here we are considering the percentage increase in stress in the steel members caused by ponding. If the stress in a member increases from $0.60F_y$ to $0.80F_y$, we say that the stress index U is

$$U = \frac{0.80F_y - 0.60F_y}{0.60F_y} = 0.33$$

The terms C_p and C_s are, respectively, the approximate stiffnesses of the primary and secondary support systems. LRFD Equation K2-1 ($C_p + 0.9C_s \leq 0.25$), which gives us an approximate stress index during ponding, is limited to a maximum value of 0.25. Should we substitute into this equation and obtain a value no greater than 0.25, ponding supposedly will not be a problem. Should the index be larger than 0.25, however, it will be necessary to conduct a further investigation. One method of doing this is presented in the LRFD Commentary and will be described later in this section.

Example F-1 presents the application of LRFD Equation K2-1 to a roofing system.

■ EXAMPLE F-1
Check the roof system shown in Fig. F-1 for ponding, using the LRFD Specification.

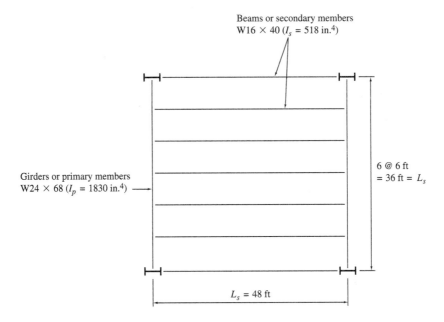

Beams or secondary members
W16 × 40 (I_s = 518 in.[4])

Girders or primary members
W24 × 68 (I_p = 1830 in.[4])

6 @ 6 ft
= 36 ft = L_s

L_s = 48 ft

Figure F-1

Solution.

$$C_p = \frac{32\ L_s L_p^4}{10^7 I_p} = \frac{(32)(48)(36)^4}{(10^7)(1830)} = 0.141$$

$$C_s = \frac{32\ S L_s^4}{10^7 I_s} = \frac{(32)(6)(48)^4}{(10^7)(518)} = 0.197$$

$$C_p + 0.9C_s = 0.141 + (0.9)(0.197) = 0.318 > 0.25$$

Indicates insufficient strength and stability, and thus a more precise method of checking should be used. ∎

The curves in the LRFD Commentary (K2) provide a design aid for use when we need to compute a more accurate flat-roof framing stiffness than is given by the specification provision that $C_p + 0.9C_s \leq 0.25$.

For this application the term f_o represents the computed bending stress due to the supported loads not including the ponding effect. The following stress indexes are computed for the primary and secondary members:

$$U_p = \left(\frac{F_y - f_o}{f_o}\right)_p$$

$$U_s = \left(\frac{F_y - f_o}{f_o}\right)_p$$

In these expressions f_o represents the stress due to $1.2D + 1.2R$ (where D = nominal dead load and R = nominal load due to rainwater or ice exclusive of ponding contribution). These loads should include any snow that is present, although most ponding failures have occurred during torrential summer rains.

We enter Fig. A-K2.1 in Section K of the LRFD Appendix with our computed U_p and move horizontally to the calculated C_s value of the secondary members. Then we go vertically to the abscissa scale and read the flexibility constant. If this value is more than our calculated C_p value computed for the primary members, the stiffness is sufficient. The same process is followed by Fig. A-K2.2, where we enter with our computed U_s and C_p values and pick from the abscissa the flexibility constant, which should be no less than our C_s value. This procedure is illustrated in Example F-2.

■ **EXAMPLE F-2**
Recheck the roof system considered in Example F-1 using the LRFD curves. Assume that as ponding begins, f_o is 20 ksi. in both girders and beams.

Solution.   Checking girders:

$$U_p = \frac{F_y - f_o}{f_o} = \frac{36 - 20}{20} = 0.80$$

For $U_p = 0.80$ and $C_s = 0.197$, we read in Fig. A-K2.1 $C_p = 0.28 >$ our calculated C_p of 0.141.

∴ Girders   OK

Checking beams:

$$U_s = \frac{F_y - f_o}{f_o} = \frac{36 - 20}{20} = 0.80$$

For $U_s = 0.80$ and $C_p = 0.141$, we read in Fig. A-K2.2 $C_s = 0.29 >$ our calculated C_s of 0.197.

∴ Beams   OK   ■

Glossary

Amplification Factor   A multiplier used to increase the computed moment or deflection in a member to account for the eccentricity of the load.

Aspect Ratio   The ratio of the lengths of the sides of a rectangular panel to each other.

Bar Joist   See *open web-joist*.

Bays   The areas between columns in a building.

Beam   A member that supports loads transverse to its longitudinal axis.

Beam-Column   A column that is subjected to axial compression loads as well as bending moments.

Bearing Wall Construction   Building construction where all the loads are transferred to the walls and thence down to the foundations.

Block Shear   A fracture type shear where fracture may occur on either the tension plane or the shear plane followed by yielding on the other plane (see Fig. 3-16).

Braced Frame   A frame that has resistance to lateral loads supplied by some type of auxiliary bracing.

Brittle Fracture   Abrupt fracture with little or no prior ductile deformation.

Buckling Load   The load at which a straight compression member assumes a deflected position.

Built-Up Member   A member made up of two or more steel elements bolted or welded together to form a single member.

Camber   The construction of a member bent or arched in one direction so that it won't look so bad when the service loads bend it in the opposite direction.

Cast Iron   An iron with a very low carbon content.

Charpy V-Notch Test   A test used for measuring the fracture toughness of steel by fracturing it with a pendulum swung from a certain height.

Cladding   The exterior covering of the structural parts of a building.

Cold-Formed Light-Gage Steel Shapes   Shapes made by cold bending thin sheets of carbon or low-alloy steels into desired cross sections.

Column   A structural member whose primary function is to support compressive loads.

Compact Section   A section that has a sufficiently stocky profile so that it is capable of developing a fully plastic stress distribution before buckling.

Composite Beam   A steel beam made composite with a concrete slab by providing shear transfer between the two (see Fig. 16-1).

694

Composite Column A column constructed with rolled or built-up steel shapes, encased in concrete or with concrete placed inside steel pipes or tubes (see Fig. 17-1).

Coping The cutting back of the flanges of a beam to facilitate its connection to another beam (see Fig. 10-7).

Dead loads Loads of constant magnitude that remain in one position. Examples: weights of walls, floors, roofs, fixtures, structural frames, and so on.

Drift Lateral deflection of a building.

Drift Index The ratio of lateral deflection of a building to its height.

Ductility The property of a material by which it can withstand extensive deformation without failure under high tensile stress.

Effective Length The distance between points of zero moment in a column; that is, the distance between its inflection points.

Effective Length Factor *K* A factor that when multiplied by the length of a column will provide its effective length.

Elastic Design A method of design that is based on certain allowable stresses.

Elastic Limit The largest stress that a material can withstand without being permanently deformed.

Elasticity The ability of a material to return to its original shape after it has been loaded and then unloaded.

Elastic Strain Strain that occurs in a member under load before its yield stress is reached.

Endurance Limit The maximum fatigue-type stress in a material for which the material seems to have an infinite life.

Euler Load The compression load at which a long and slender member will buckle elastically.

Eyebar A pin-connected tension member whose ends are enlarged with respect to the rest of the member so as to make the strength of the ends approximately equal to the strength of the rest of the member.

Factored Load A nominal load multiplied by a load factor.

Fasteners A term representing bolts, welds, rivets, or other connecting devices.

Fatigue A fracture situation caused by changing stresses.

Faying Surface The contact or shear area between members being connected.

Fillet Weld A weld placed in the corner formed by two overlapping parts in contact with each other (see Fig. 14-2).

First-Order Analysis Analysis of a structure in which equilibrium equations are written based on an assumed nondeformed structure.

Floor Beams The larger beams in many bridge floors that are perpendicular to the roadway of the bridge and that are used to transfer the floor loads from the stringers to the supporting girders or trusses.

Fracture Toughness The ability of a material to absorb energy in large amounts. For instance, steel members can be subjected to large deformations during erection and fabrication without failure, thus allowing them to be bent, hammered, sheared, and have holes punched in them.

Gage Transverse spacing of bolts measured perpendicular to the long direction of the member (see Fig. 12-4).

Girder A rather loosely used term usually indicating a large beam and perhaps one into which smaller beams are framed.

Girts Horizontal members running along the sides of industrial buildings used primarily to resist bending due to wind. They often are used to support corrugated siding.

Government Anchors Bent steel bars used when ends of steel beam are enclosed by concrete or masonry walls. The bars pass through the beam webs parallel to the walls and are enclosed in the walls. They keep beams from moving longitudinally with respect to the walls.

Groove Welds Welds made in grooves between members that are being joined. They may extend for the full thickness of the parts (complete-penetration groove welds) or they may extend for only a part of the member thickness (partial-penetration groove welds) (see Fig. 14-2).

Hybrid Member A structural steel member made from parts that have different yield stresses.

Impact Loads The difference between the magnitude of live loads actually caused and the magnitude of those loads had they been applied as dead loads.

Inelastic Action The deformation of a member that does not disappear when the loads are removed.

Influence Line A diagram whose ordinates show the magnitude and character of some function of a structure (shear, moment, etc.) as a unit load moves across the structure.

Instability A situation occurring in a member where increased deformation of that member causes a reduction in its load-carrying ability.

Intermediate Columns Columns that fail both by yielding and buckling. Their behavior is said to be inelastic. Most columns fall in this range, where some of the fibers reach the yield stress and some do not.

Ironworker A person performing steel erection (it's a name carried over from the days when iron structural members were used).

Joists The closely spaced beams supporting the floors and roofs of buildings.

Jumbo Sections Very heavy steel W sections (and structural tees cut from those sections) in Groups 4 and 5 of the LRFD Manual. Serious cracking problems sometimes occur in these sections when welding or thermal cutting is involved.

Killed Steel Steel that has been deoxidized to prevent gas bubbles and to reduce its nitrogen content.

Lamellar Tearing A separation in the layers of a highly restrained welded joint caused by "through the thickness" strains produced by shrinking of the weld metal.

Limit State A condition at which a structure or some point of the structure ceases to perform its intended function either as to strength or as to serviceability.

Lintels Beams over openings in masonry walls such as windows and doors.

Live Loads Loads that change position and magnitude. They move or are moved. Examples: trucks, people, wind, rain, earthquakes, temperature changes, and so on.

Load Factor A number almost always larger than 1.0 used to increase the estimated loads a structure has to support to account for the uncertainties involved in estimating loads.

Local Buckling The buckling of the part of a larger member that precipitates failure of the whole member.

Long Columns Columns that buckle elastically and whose buckling loads can be predicted accurately with the Euler formula if the axial buckling stress is below the proportional limit.

Mild Steel A ductile low-carbon steel.

Milled Surfaces Those surfaces that have been accurately sawed or finished to a smooth or true plane.

Mill Scale An iron oxide that forms on steel when it is reheated for rolling.

Modulus of Elasticity or Young's Modulus The ratio of stress to strain in a member under load. It is a measure of the stiffness of the material.

Net Area Gross cross-sectional area of a member minus any holes, notches, or other indentations.

Nominal Loads The magnitudes of loads specified by a particular code.

Nominal Strength The theoretical ultimate strength of a member or connection.

Noncompact Section A section that cannot be stressed to a fully plastic situation before buckling occurs. The yield stress can be reached in some but not all of the compression elements before buckling occurs.

Open-Web Joist A small parallel chord truss whose members are often made from bars (hence the common name *bar joist)* or small angles or other rolled shapes. These joists are very commonly used to support floor and roof slabs (see Fig. 19-5).

P-Delta Effect Changes in column moments and deflections due to lateral deflections.

Partially Composite Section A section whose flexural strength is governed by the strength of its shear connectors.

Pitch The longitudinal spacing of bolts measured parallel to the long direction of a member (see Fig. 12-4).

Plane Frame A frame that for purposes of analysis and design is assumed to lie in a single (or two-dimensional) plane.

Plastic Design A method of design that is based on a consideration of failure conditions.

Plastic Modulus The statical moment of the tension and compression areas of a section taken about the plastic neutral axis.

Plastic Moment   The yield stress of a section times its plastic modulus. It's the nominal moment that the section can theoretically resist if it is braced laterally.

Plastic Strain   The strain that occurs in a member after its yield stress is reached with no increase in stress.

Plate Girder   A built-up steel beam (see Fig. 18-4).

Poisson's Ratio   The ratio of lateral strain to axial or longitudinal strain in a member under load.

Ponding   A situation on a flat roof where water accumulates faster than it runs off.

Post-Buckling Strength   The load a member or frame can support after buckling occurs.

Proportional Limit   Largest strain for which Hooke's law applies, or the highest point on the straight-line portion of the stress-strain diagram.

Purlins   Roof beams that span between trusses (see Fig. 4-4).

Quenching   Rapid cooling of steel with water or oil.

Residual Stresses   The stresses that exist in an unloaded member after it's manufactured.

Resistance Factor ϕ A number almost always less than 1.0 multiplied by the ultimate or nominal strength of a member or connection to take into account the uncertainties in material strengths, dimensions, and workmanship. Also called *overcapacity factor*.

Rigid Frame   A structure whose connectors keep substantially the same angles between members before and after loading.

Sag Rods   Steel rods that are used to provide lateral support for roof purlins. They also may be used for the same purpose for girts on the sides of buildings (see Fig. 4-5).

St. Venant Torsion   The part of the torsion in a member that produces only shear stresses in the member.

Scuppers   Large holes or tubes in walls or parapets that enable water above a certain depth to quickly drain from roofs.

Second-Order Analysis   Analysis of a structure for which equilibrium equations are written that include the effect of the deformations of the structure.

Section Modulus   The ratio of the moment of inertia taken about a particular axis of a section divided by the distance to the extreme fiber of the section measured perpendicular to the axis in question.

Seismic   Of or having to do with an earthquake.

Serviceability   The ability of a structure to maintain its appearance, comfort, durability, and function under normal loading conditions.

Service Loads   The loads that are assumed to be applied to a structure when it is in service (also called *working loads*).

Shape Factor   The ratio of the plastic moment of a section to its yield moment.

Shear Center   The point in the cross section of a beam through which the resultant of the transverse loads must pass so that no torsion will occur.

Shear Lag   A nonuniformity of stress in the parts of rolled or built-up sections occurring when a tensile load is not applied uniformly.

Shear Wall   A wall in a structure that is specially designed to resist shears caused by lateral forces such as wind or earthquake in the plane of the wall.

Shims   Thin strips of steel that are used to adjust the fitting of connections. Finger shims are shims that are installed after the bolts already are in place.

Short Columns   Columns whose failure stress will equal the yield stress, and for which no buckling will occur. For a column to fall into this class it would have to be so short as to have no practical application.

Sidesway   The lateral movement of a structure caused by unsymmetrical loads or by an unsymmetrical arrangement of building members.

Skeleton Construction   Building construction in which the loads are transferred for each floor by beams to the columns and thence to the foundations.

Slenderness Ratio   The ratio of the effective length of a column to its radius of gyration, both pertaining to the same axis of bending.

Slender Section   A member that will buckle locally while the stress still is in the elastic range.

Slip-Critical Joint   A bolted joint that is designed to have resistance to slipping.

Space Frame   A three-dimensional structural frame.

Spandrel Beams Beams that support the exterior walls of buildings and perhaps part of the floor and hallway loads (see Fig. 19-3).

Steel An alloy consisting almost entirely of iron (usually over 98 percent). It also contains small quantities of carbon, silicon, manganese, sulfur, phosphorus, and other materials.

Stiffened Element A projecting piece of steel whose two edges parallel to the direction of a compression force are supported (see Fig. 5-6).

Stiffener A plate or an angle usually connected to the web of a beam or girder to prevent failure of the web (see Figs. 18-11, 18-12, and 18-13).

Story Drift The difference in horizontal deflection at the top and bottom of a particular story.

Strain-Hardening Range beyond plastic strain in which additional stress is necessary to produce additional strain.

Stringers The beams in bridge floors that run parallel to the roadway.

Tangent Modulus The ratio of stress to strain for a material that has been stressed into the inelastic range.

Tension Field Action The behavior of a plate girder panel which, after the girder initially buckles, acts much like a truss. Diagonal strips of the web act similarly to the diagonals of a parallel chord truss. The stiffeners keep the flanges from coming together and the flanges keep the stiffeners from coming together (see Fig. 18-9).

Toughness The ability of a material to absorb energy in large amounts. As an illustration, steel members can be subjected to large deformations during fabrication and erection without fracture, thus allowing them to be bent, hammered, sheared, and have holes punched in them without visible damage.

Unbraced Frame A frame whose resistance to lateral forces is provided by its members and their connections.

Unbraced Length The distance in a member between points that are braced.

Unstiffened Element A projecting piece of steel having one free edge parallel to the direction of a compression force, with the other edge in that direction unsupported (see Fig. 5-6).

Upset Rods Rods whose ends are made larger than the regular bodies of the rods. Threads are cut into the upset ends, but the area at the root of the thread in each rod is larger than that of the regular part of the bar (see Fig. 4-3).

Warping Torsion The part of the resistance of a member to torsion that is provided by the warping resistance of the member cross section.

Weathering Steel A high-strength low-alloy steel whose surface when exposed to the atmosphere (not a marine one) oxidizes and forms a tightly adherent film that prevents further oxidation and thus eliminates the need for painting.

Web Buckling The buckling of the web of a member (see Fig. 10-10).

Web Crippling The failure of the web of a member near a concentrated force (see Fig. 10-10).

Working Loads See *Service Loads.*

Wrought Iron An iron with a very high carbon content.

Yield Moment The moment that will just produce the yield stress in the outermost fiber of a section.

Yield Stress The stress at which there is a decided increase in the elongation or strain in a member without a corresponding increase in stress.

Index